MULTI-CARRIER SPREAD SPECTRUM 2007

Lecture Notes Electrical Engineering (Series)

Volume 1

Multi-Carrier Spread Spectrum 2007

Proceedings from the 6th International Workshop on Multi-Carrier Spread Spectrum, May 2007, Herrsching, Germany

Edited by

Simon Plass
German Aerospace Center (DLR)
Germany

Armin Dammann
German Aerospace Center (DLR)
Germany

Stefan Kaiser
DoCoMo Euro-Labs
Germany

Khaled Fazel
Ericsson GmbH
Germany

Springer

A C.I.P. Catalogue record for this book is available from the Library of Congress.

ISBN 978-1-4020-6128-8 (HB)
ISBN 978-1-4020-6129-5 (e-book)

Published by Springer,
P.O. Box 17, 3300 AA Dordrecht, The Netherlands.

www.springer.com

Printed on acid-free paper

Preface

Since the principle of multi-carrier code division multiple access (MC-CDMA) was simultaneously proposed by Khaled Fazel *et al.* and Nathan Yee *et al.* at the IEEE International Symposium on Personal, Indoor and Mobile Radio Communications (PIMRC) in the year 1993, multi-carrier spread spectrum (MC-SS) has rapidly become one of the most wide spread independent research topics on the field of mobile radio communications. Therefore, the International Workshop on Multi-Carrier Spread Spectrum (MC-SS) was initiated in the year 1997. Multi-carrier and spread spectrum systems with their generic air interface and adaptive technologies are considered as potential candidates to fulfill the requirements of next generation mobile communications systems.

The material summarized in this volume was selected for the 6th International Workshop on Multi-Carrier Spread Spectrum (MC-SS 2007) held in Herrsching, Germany from 07-09 May 2007. The workshop was organized by the Institute of Communications and Navigation of the German Aerospace Center (DLR) in Oberpfaffenhofen, Germany.

During the workshop, many hot topics in the field of MC-SS were covered. First, several presentations regarding *General Issues* were given. *Iterative Decoding* strategies were also discussed. Furthermore, a special session was addressed on *Interleave Multiple Division Access* (IDMA). In the areas of *Spectrum and Interference*, *Data Transmission*, *Adaptive Transmission*, *Multiple Antennas*, *Multi-Cell and Multi-Link*, and *Synchronization and Tracking* new and innovative algorithms, strategies, and systems were proposed. Finally, a dominant hot topic during this workshop was the problem of *Channel Estimation*.

We would like to thank all those who have contributed to the success of the conference. First, the authors of paper submissions, this year reaching a total of 54 submissions from 22 countries with 39 accepted papers. Second, the plenary session speakers, the invited papers' authors and session's chairs. Third, the members of the Technical Program Committee and all reviewers, whose expertise and work have contributed to configure a very attractive conference program. All papers were reviewed by at least 2 referees. We would like to thank all sponsors (DLR, DoCoMo Euro-Labs, Nomor Research, and IEEE Communications Society Germany Chapter) and the other organizations for their logistic support.

May 2007

Simon Plass

Armin Dammann

Stefan Kaiser

Khaled Fazel

Program Committee

Y. Bar-Ness (USA)
G. P. Fettweis (Germany)
G. B. Giannakis (USA)
S. Hara (Japan)
H. Imai (Japan)
K. D. Kammeyer (Germany)
A. Klein (Germany)
W. Koch (Germany)
W. A. Krzymien (Canada)
J. Lindner (Germany)
U. Mengali (Italy)
L. B. Milstein (USA)
M. Moeneclaey (Belgium)
W. Mohr (Germany)
M. Nakagawa (Japan)
S. Pasupathy (Canada)
S. Plass (Germany)
R. Prasad (Denmark)
M. Renfors (Finland)
H. Rohling (Germany)
H. Sari (France)
M. Sawahashi (Japan)
M. Sternad (Sweden)
R. E. Ziemer (USA)

Additional Reviewer

Alexander	Arkhipov
Gunther	Auer
Paul Walter	Baier
Marco	Breiling
Armin	Dammann
Markus	Dangl
Robert	Elliott
Harald	Ernst
Khaled	Fazel
Yu	Fu
Atilio	Gameiro
Ingmar	Groh
Rainer	Gruenheid
Stefan	Kaiser
Lutz	Lampe
Christian	Mensing
Michele	Morelli
Patrick	Nickel
Li	Ping
Ronald	Raulefs
Stephan	Sand
Christian	Sgraja
Shreeram	Sigdel
Anja	Sohl
Andreas	Springer
Heidi	Steendam
Tobias	Weber
Stefan	Werner

Table of Contents

General Issues

Iterative Decoding

Interleave Division Multiple Access (IDMA)

* *invited paper*

Spectrum and Interference

Data Transmission

Adaptive Transmission

Multiple Antennas

Multi-Cell and Multi-Link

Synchronization and Tracking

Channel Estimation

FREQUENCY-DOMAIN EQUALIZATION FOR BLOCK CDMA TRANSMISSION

F. Adachi, A. Nakajima, K. Takeda, L. Liu, H. Tomeba, and K. Fukuda
Tohoku University
Electrical and Communication Engineering, Graduate School of Engineering
6-6-05 Aza-Aoba, Aramaki, Aoba-ku, Sendai, 980-8579 Japan
adachi@ecei.tohoku.ac.jp

Abstract: Frequency-domain equalization (FDE) technique may play an important role for broadband packet transmission using MC- and DS-CDMA. The downlink performance is significantly improved with FDE; however, the uplink performance is limited by the multi-access interference (MAI). To remove the MAI while gaining the frequency diversity effect through the use of FDE, frequency-domain block spread CDMA can be used. The performance can be further improved by the use of multi-input/multi-output (MIMO) antenna technique. Recently, particular attention has been paid to MIMO space division multiplexing (SDM) to significantly increase the throughput without expanding the signal bandwidth. In this paper, we present a comprehensive performance comparison of MC- and DS-CDMA using FDE.

Key words: MC-CDMA, DS-CDMA, frequency-domain equalization, MIMO, HARQ

1. INTRODUCTION

The deployment speed of 3rd generation (3G) wireless networks based on direct-sequence code division multiple access (DS-CDMA) technique[1] has accelerated. 3G wireless networks will be continuously evolving for providing packet data services of 50~100Mbps. The evolution of 3G wireless networks will be followed by the development of 4th generation (4G) wireless networks that can support extremely higher-speed packet data services of e.g., 100M~1Gbps[2]. The received signal spectrum is severely

S. Plass et al. (eds.), Multi-Carrier Spread Spectrum 2007, 1–10.

distorted due to frequency-selective fading and thus, equalization techniques are indispensable. Since DS-CDMA using coherent rake combining provides very poor performance in a strong frequency-selective fading channel, multi-carrier CDMA (MC-CDMA)[3] with frequency-domain equalization (FDE) has long time been considered as a broadband multi-access technique. However, it was recently shown[4] that FDE can replace the coherent rake combining to improve the DS-CDMA downlink (base-to-mobile) transmission performance. The uplink (mobile-to-base) performance is limited by the multi-access interference (MAI). To remove the MAI while gaining the frequency diversity effect through the use of FDE, frequency-domain block spread CDMA can be used. Some form of error control technique is indispensable for broadband packet transmission. Frequency-domain hybrid ARQ (HARQ), which is a combination of HARQ and FDE, will be a promising error control technique. Further throughput improvement is possible by the multi-input/multi-output (MIMO) antenna technique. Recently, particular attention has been paid to MIMO space division multiplexing (SDM) to increase the throughput without expanding the signal bandwidth[5]. To exploit the channel frequency-selectivity, signal separation/detection can be combined with FDE. In this paper, first, we discuss similarity of both CDMA techniques and then, show that both are comparable and can remain as a promising wireless access.

2. SIMILARITY OF MC- AND DS-CDMA

FDE based on minimum mean square error (MMSE) criterion can exploit the channel frequency-selectivity to improve the bit error rate (BER) performance. Figure 1 illustrates the transmitter/receiver structure of multicode CDMA using MMSE-FDE. A data-modulated symbol sequence to be transmitted is serial-to-parallel (S/P) converted to U parallel symbol streams, and then, multicode spreading is done using U orthogonal spreading codes $\{c_u(t);\ t=0\sim(SF-1)\}$, $u=0\sim(U-1)$, with spreading factor SF, and further multiplied by a scrambling sequence $c_{scr}(t)$. For MC-CDMA, N_c-point inverse fast Fourier transform (IFFT) is applied to generate the MC-CDMA signal with N_c subcarriers. In DS-CDMA transmission, however, no IFFT is required. Each signal block is transmitted after inserting a cyclic prefix of N_g samples into the guard interval (GI). At the receiver, the received signal block is transformed by N_c-point FFT into N_c subcarrier components $\{R(k);\ k=0\sim(N_c-1)\}$. FDE is carried out as $\hat{R}(k)=w(k)R(k)$ for $k=0\sim(N_c-1)$, where $w(k)$ is the MMSE-FDE weight given by

$$w(k) = H^*(k) \Bigg/ \left(|H(k)|^2 + \left(\frac{U}{SF} \frac{E_s}{N_0} \right)^{-1} \right), \tag{1}$$

where $H(k)$ is the channel gain at the kth subcarrier and E_s/N_0 is the average received signal energy per data symbol-to-AWGN power spectrum density ratio. For DS-CDMA, the time-domain chip sequence is recovered by applying N_c-point IFFT to $\{\hat{R}(k); k = 0 \sim (N_c - 1)\}$, while it is not required for MC-CDMA. Descrambling and multicode despreading are carried out to get a soft-decision symbol sequence for data demodulation. As seen from Fig. 1, the difference between MC- and DS-CDMA is a location of IFFT function; IFFT is required at the transmitter for MC, while it is required at the receiver for DS. This leads to a new transceiver, based on software-defined radio technology, which can flexibly switch between MC-CDMA (or OFDMA) and DS-CDMA.

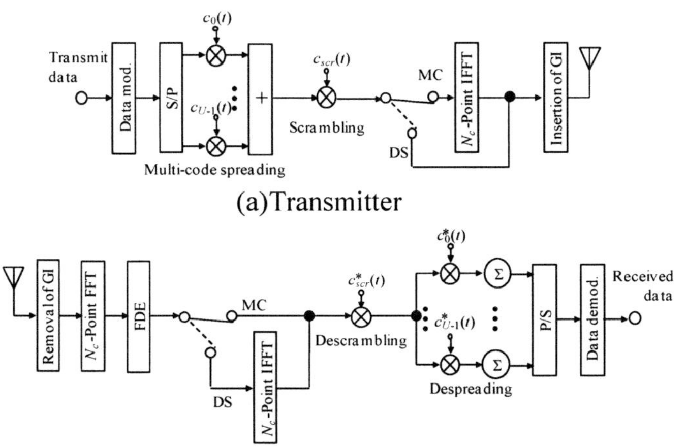

(a)Transmitter

(b) Receiver

Fig. 1 CDMA transmitter/receiver structure.

3. FREQEUNCY-DOMAIN TRANSMIT DIVERSITY

Antenna diversity is a well-known technique to improve the transmission performance. Space-time block coded joint transmit/receive diversity (STBC-JTRD) was proposed[6] that can use an arbitrary number of transmit antennas while limiting the maximum number of receive antennas to 4. Frequency-domain pre-equalization[7] can be introduced to STBC-JTRD for both CDMA[8]. At the transmitter, the multicode CDMA signal is divided into a sequence of G information blocks. For DS-CDMA, N_c-point FFT is applied to decompose the gth chip block, g=0~(G-1), into N_c subcarrier components

$\{S_g(k); k=0\sim(N_c-1)\}$, while it is not required for MC-CDMA. The resulting G consecutive components $\{S_0(k),...,S_g(k),..,S_{G-1}(k)\}$ of the kth subcarrier are encoded into N_t parallel codewords as shown in Fig. 2; the n_tth codeword consisting of a sequence of Q subcarrier components $\{\widetilde{S}_{0,n_t}(k),..., \widetilde{S}_{q,n_t}(k),...,\widetilde{S}_{Q-1,n_t}(k)\}$ is transmitted from the n_tth transmit antenna after performing N_c-point IFFT. Table 1 shows the number G of information chip blocks per codeword, the number Q of coded chip blocks per codeword, and the code rate R for N_t=1~4. A superposition of N_t codewords is received via a frequency-selective fading channel. A simple STBC-JTRD decoding is carried out by using N_r parallel received codewords $\{r_{q,n_r}(t); q=0\sim(Q-1), n_r=0\sim(N_r-1)\}$[8]. For MC-CDMA, after transforming the decoder output $\{\hat{r}_g(t); t=0\sim(N_c-1)\}$ into the N_c subcarrier components by N_c-point FFT, the descrambling and despreading are carried out to get a sequence of the decision variables. For DS-CDMA, FFT is not required.

STBC-JTRD encoding for N_t=2 is represented as

$$\begin{pmatrix}\widetilde{S}_{0,n_t}(k)\\ \widetilde{S}_{1,n_t}(k)\end{pmatrix}=\sqrt{\frac{N_c}{\sum_{n_r=0}^{N_r-1}\sum_{n_t=0}^{N_t-1}\sum_{k=0}^{N_c-1}|w_{n_r,n_t}(k)|^2}}\begin{pmatrix}S_0(k)w_{0,n_t}(k)+S_1(k)w_{1,n_t}(k)\\ S_0^*(k)w_{1,n_t}(k)-S_1^*(k)w_{0,n_t}(k)\end{pmatrix}, \quad (2)$$

where $w_{n_r,n_t}(k)$ is the MMSE pre-equalization weight, given as

$$w_{n_r,n_t}(k)=H^*_{n_r,n_t}(k)\Bigg/\left(\frac{1}{N_r}\sum_{n_r=0}^{N_r-1}\sum_{n_t=0}^{N_t-1}|H_{n_r,n_t}(k)|^2+\left(\frac{U}{SF}\frac{E_s}{N_0}\right)^{-1}\right), \quad (3)$$

where $H_{n_r,n_t}(k)$ is the channel gain between the n_tth transmit antenna and the n_rth receive antenna at the kth subcarrier. The corresponding STBC-JTRD decoding is represented as

$$\begin{pmatrix}\hat{r}_0(t)\\ \hat{r}_1(t)\end{pmatrix}=\begin{pmatrix}r_{0,0}(t)+r^*_{1,1}(N_c-t)\\ r_{0,1}(t)-r^*_{1,0}(N_c-t)\end{pmatrix}, \text{ for } t=0\sim(N_c-1) \quad . \quad (4)$$

The turbo-coded BER performance using frequency-domain STBC-JTRD is plotted in Fig. 3 as a function of the average total transmit energy per information bit-to-AWGN power spectrum density ratio E_b/N_0. For comparison, the BER performance of the space-time transmit diversity (STTD)[9] jointly used with FDE (called frequency-domain STTD) is also plotted. Almost the identical BER performance can be achieved for MC- and DS-CDMA. As N_t increases, frequency-domain STBC-JTRD consistently improves the BER performance while frequency-domain STTD provides the same BER performance irrespective of N_t. The use of frequency-domain STBC-JTRD is advantageous for the downlink applications, where the allowable number of receive antennas at a mobile terminal is limited. Frequency-domain STTD is a good option for the uplink applications.

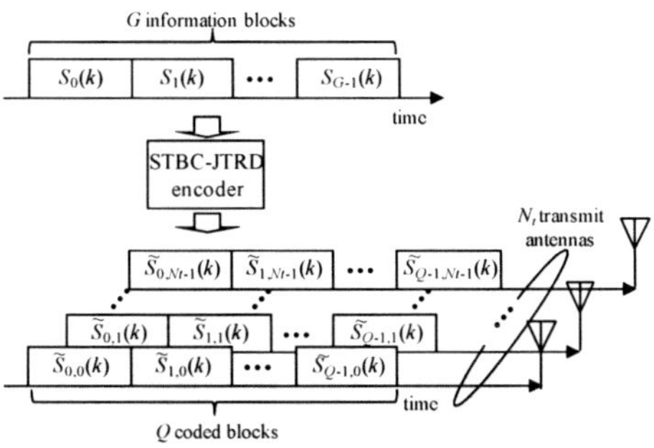

Fig. 2 STBC-JTRD encoding.

Table 1 G, Q and R for N_r=1~4

N_r	G	Q	R
1	1	1	1
2	2	2	1
3	3	4	3/4
4	3	4	3/4

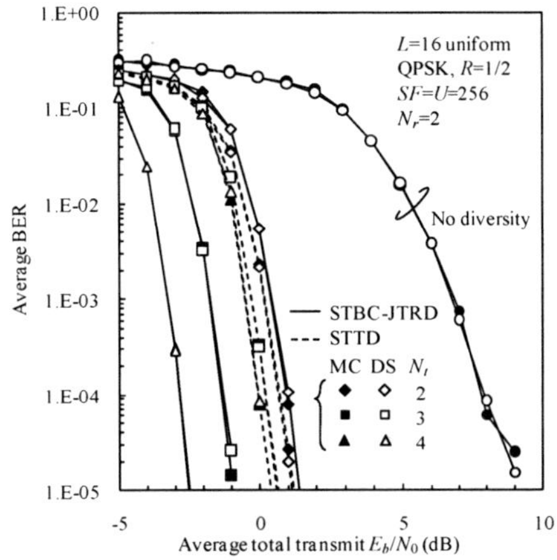

Fig. 3 BER performance of frequency-domain STBC-JTRD with N_r=2.

4. TIME/FREQUENCY-DOMAIN BLOCK SPREAD CDMA

Using MMSE-FDE, users of different data rates can be code-multiplexed without causing significant performance difference in the downlink case. However, as the number of users increases, the BER performance degrades since the inter-code interference (ICI) gets stronger due to orthogonality distortion in a severe frequency-selective fading channel. This can be avoided to certain extent by introducing 2-dimensional (2D) spreading (frequency-time domain spreading) to MC-CDMA[10]. This 2D spreading can also be applied to DS-CDMA, resulting in 2D block spread DS-CDMA[11]. In the uplink case, the orthogonality among different users is always distorted, resulting in MAI. 2D block spreading can be applied to both MC- and DS-CDMA in order to remove the MAI while gaining the frequency diversity effect by MMSE-FDE. As shown in Fig. 4, each data symbol to be transmitted is spread in both frequency- and time-domain using 2D block spreading code. The 2D block spreading code is a product code of two orthogonal spreading codes. It is represented in a matrix form as $\mathbf{C}_u = \mathbf{c}_u^t(\mathbf{c}_u^f)^T$ with $SF_u = SF_u^t \times SF_u^f$ for the uth user, where $\mathbf{c}_u^t$ and $\mathbf{c}_u^f$ are the column vectors representing time-domain and frequency-domain spreading codes with spreading factor SF_u^t and SF_u^f, respectively. The time-domain spreading code is used to remove the MAI; up-to-as many as SF_u^t users can be multiplexed without causing MAI. The frequency-domain spreading code can be used to gain the frequency diversity effect through MMSE-FDE. The optimum choice of (SF_u^t, SF_u^f) for the given spreading factor SF_u is (U, SF_u/U), where U should be a power of 2.

Figure 5 plots the uplink BER performance of rate-1/2 turbo-coded 2D block spread CDMA with SF_u=16. The BER performance of conventional CDMA is also plotted. MC- and DS-CDMA using 2D block spreading can achieve almost the same BER performance. As U increases, the uplink BER performance degrades. This is because the frequency diversity effect decreases due to reduced spreading factor SF_u^f in the frequency-domain. However, the BER performance of 2D block spread CDMA is significantly better than that of the conventional CDMA.

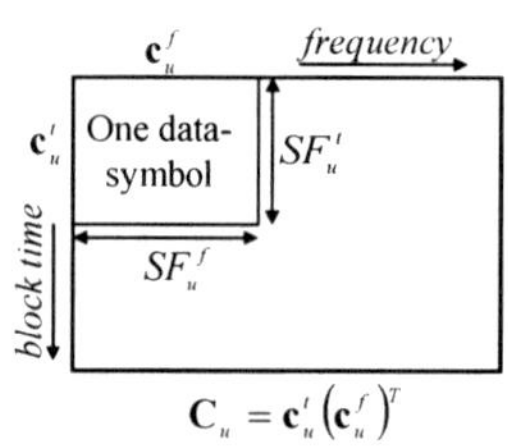

Fig. 4 2D block spreading for MC-CDMA.

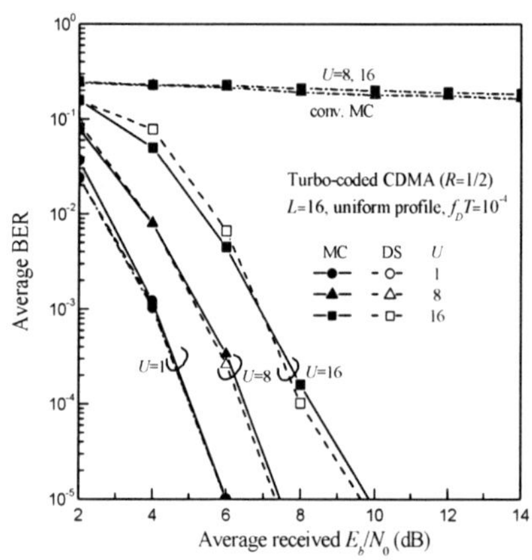

Fig. 5 BER performance.

5. FREQUENCY-DOMAIN HARQ

We consider turbo-coded type II HARQ S-P2[12], as shown in Fig. 6. Rate-1/3 turbo encoder outputs the systematic bit sequence and two parity bit sequences, each has a length of K bits. The 1st transmit packet consists of the systematic bit sequence only and the 2nd and 3rd are taken from two punctured parity bit sequences. Because of the uncoded transmission of the first packet, the throughput performance of full code-multiplexed CDMA is higher than that of OFDM in a higher E_s/N_0 region owing to the frequency diversity effect obtained through MMSE-FDE. However, in a lower E_s/N_0 region, the throughput performance of full code-multiplexed CDMA is worse than that of OFDM owing to the presence of residual ICI after

MMSE-FDE. To reduce the residual ICI, iterative frequency-domain ICI cancellation (FDICIC) technique can be applied (see Fig. 7). MMSE-FDE for the *i*th iteration is performed as $\hat{R}^{(i)}(k) = w^{(i)}(k)R(k)$ for k=0~(N_c−1), where $w^{(i)}(k)$ is the MMSE-FDE weight and can be derived as

$$w^{(i)}(k) = H^*(k) \Bigg/ \left(|H(k)|^2 + \left\{ \frac{1}{SF} \frac{E_s}{N_0} \sum_{u=0}^{U-1} \rho_u^{(i)} \left(\left\lfloor \frac{k}{SF} \right\rfloor \right) \right\}^{-1} \right) \quad , (5)$$

where $\rho_u^{(i)}(n)$ represents the contribution from the residual ICI $I^{(i)}(k)$. $I^{(i)}(k)$ should be removed for improving the throughput performance. FDICIC is carried out as

$$\widetilde{R}^{(i)}(k) = \hat{R}^{(i)}(k) - \widetilde{I}^{(i)}(k) \quad , (6)$$

where $\widetilde{I}^{(i)}(k)$ is the residual ICI replica generated from *a-posteriori* log likelihood ratio (LLR) of turbo decoder output. After performing multicode despreading and turbo decoding, the soft symbol replica is generated using the decoder output (*a-posteriori* LLR). This replica is fed back to update the MMSE-FDE weight and to generate the residual ICI replica for the (i+1)th iteration.

The use of six iterations (i=6) achieves a sufficient performance improvement. The throughput performance of full code-multiplexed (U=SF) CDMA with FDICIC is plotted in Fig. 8 as a function of the average received E_s/N_0. The achievable throughput is higher than that of OFDM.

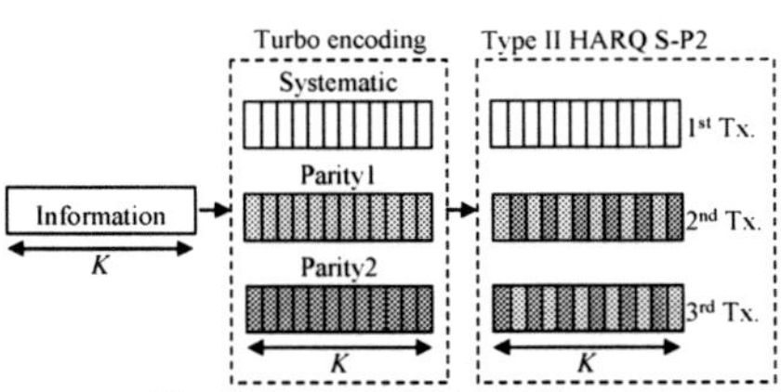

Fig. 6 Type II HARQ S-P2.

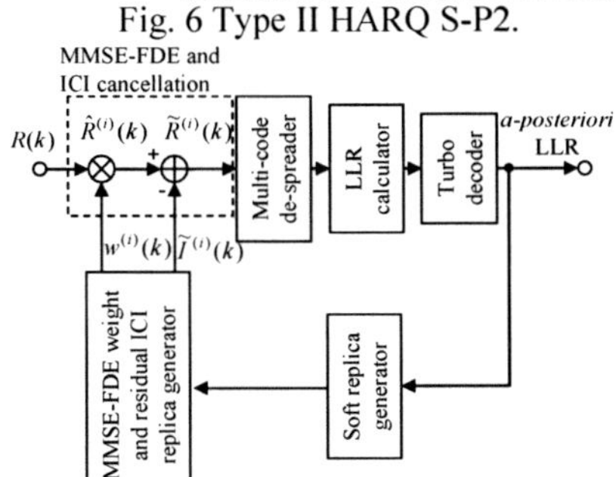

Fig. 7 Iterative FDICIC.

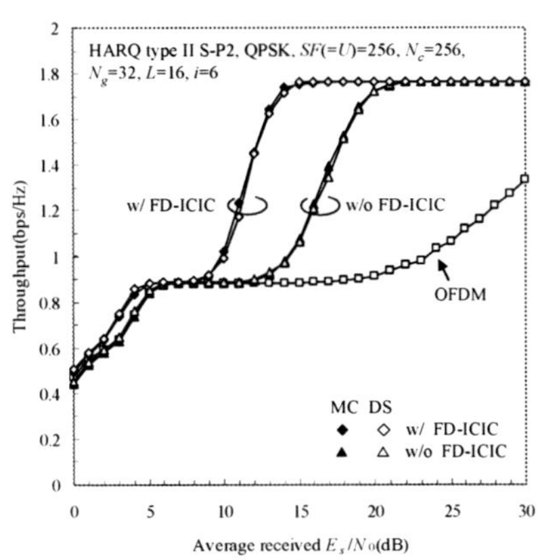

Fig. 8 Throughput performance.

6. FREQUENCY-DOMAIN MIMO SDM

We consider (N_t, N_r) MIMO SDM using the iterative frequency-domain interference cancellation (FDIC) combined with MMSE-FDE[13,14]. A data-modulated symbol sequence is S/P converted into N_t parallel symbol streams $\{d_{n_t,u}(n); n=0\sim(N_c/SF-1), u=0\sim(U-1)\}$, n_t=0~(N_t−1). Then, multicode spreading using U orthogonal spreading codes with spreading factor SF is applied to each symbol stream to obtain the multicode chip sequence. Each resultant sequence is transformed by N_c-point IFFT into the MC-CDMA signal $\{s_{n_t}(t)$; t=0~(N_c−1)}. In the case of DS-CDMA, IFFT is not required. After inserting the GI, N_t CDMA signals are transmitted simultaneously from N_t transmit antennas. After the removal of GI, the received CDMA signal $\{r_{n_r}(t)$; t=0~N_c−1} at the n_rth receive antenna is decomposed by N_c-point FFT into N_c subcarrier components $\{R_{n_r}(k)$; k=0~N_c−1} as

$$R_{n_r}(k)=\sqrt{\frac{2E_s}{SF\cdot T_c}}\sum_{n_t=0}^{N_t-1}H_{n_r,n_t}(k)S_{n_t}(k)+\Pi_{n_r}(k), \qquad (7)$$

where $S_{n_t}(k)$ is the kth frequency component of the multicode CDMA signal transmitted from the n_tth transmit antenna and $\Pi_{n_r}(k)$ is the noise.

In iterative FDIC (see Fig. 9), 2D-MMSE FDE is first carried out as $\hat{R}^{(i)}_{n_t}(k)=\mathbf{w}^{(i)}_{n_t}(k)\mathbf{R}(k)$ to suppress the inter-antenna interference (IAI) and ICI simultaneously, where $\mathbf{R}(k)=[R_0(k),\cdots,R_{N_r-1}(k)]^T$ is N_r-by-1 received signal vector and $\mathbf{w}^{(i)}_{n_t}(k)=[w^{(i)}_0(k),\cdots,w^{(i)}_{N_r-1}(k)]^T$ is 1-by-N_r 2D-MMSE weight vector. $\mathbf{w}^{(i)}_{n_t}(k)$ is given as

$$\mathbf{w}^{(i)}_{n_t}(k)=\mathbf{H}^H_{n_t}(k)\left[\mathbf{H}(k)\mathbf{G}^{(i)}(k)\mathbf{H}^H(k)+\left(\frac{E_s}{SF\cdot N_0}\right)^{-1}\mathbf{I}_{N_r}\right]^{-1}, \qquad (8)$$

where $\mathbf{I}_{N_r}$ is N_r-by-N_r identity matrix, $\mathbf{H}_{n_t}(k)$ and $\mathbf{H}(k)$ are respectively N_r-by-1 channel gain vector and N_r-by-N_t channel gain matrix, and $\mathbf{G}^{(i)}(k)=diag[g^{(i)}_0(k),\cdots,g^{(i)}_{N_t-1}(k)]$ is N_t-by-N_t matrix with $g^{(i)}_{n'_t}(k)$ reflecting the contribution of interference from the n'_tth antenna[14]. Note that $g^{(i)}_{n_t}$ and $g^{(i)}_{n'_t}$ ($n'_t\neq n_t$) correspond to the residual ICI and IAI, respectively. The residual IAI and ICI replicas $\{\widetilde{S}^{(i-1)}_{n'_t}(k)\}$ are generated by feeding back the (i−1)th iteration result and subtracted from $\hat{R}^{(i)}_{n_t}(k)$ as

$$\widetilde{R}^{(i)}_{n_t}(k)=\hat{R}^{(i)}_{n_t}(k)-\sqrt{\frac{2E_s}{SF\cdot T_c}}\sum_{n'_t=0}^{N_t-1}H'^{(i)}_{n'_t}(k)\widetilde{S}^{(i-1)}_{n'_t}(k) \quad , \qquad (9)$$

where $H'^{(i)}_{n'_t \neq n_t}(k) = \mathbf{w}^{(i)}_{n_t}(k)\mathbf{H}^{(i)}_{n'_t}(k)$ is the equivalent channel gain for IAI and $H'^{(i)}_{n'_t = n_t}(k) = \mathbf{w}^{(i)}_{n_t}(k)\mathbf{H}_{n_t}(k) - \tilde{H}_{n_t}(\lfloor k/SF \rfloor)$ for ICI with

$$\tilde{H}_{n_t}(n) = (1/SF)\sum_{k=nSF}^{(n+1)SF-1} \mathbf{w}^{(i)}_{n_t}(k)\mathbf{H}_{n_t}(k) \quad . \tag{10}$$

After descrambling and multicode despreading, the LLR associated with each transmitted bit is computed[15], from which the soft symbol replicas are generated. Then, multicode spreading and scrambling are performed to obtain the CDMA signal replicas $\{\tilde{S}^{(i)}_{n_t}(k); k = 0 \sim (N_c - 1)\}$, n_t=0~(N_t-1), for the next iteration. The above operations are repeated a sufficient number of times to sufficiently suppress the IAI and ICI.

The HARQ throughput performance of full code-multiplexed (U=SF) MC-CDMA (4,4)SDM with iterative FDIC using i=4 is plotted in Fig. 10 as a function of the average received E_s/N_0 per receive antenna. $N_t \times N_r$ channels are independent Rayleigh fading channels having an L=16-path uniform power delay profile. Iterative FDIC significantly improves the throughput performance. The throughput performance of MC-CDMA is almost as same as that of DS-CDMA and is better than OFDM.

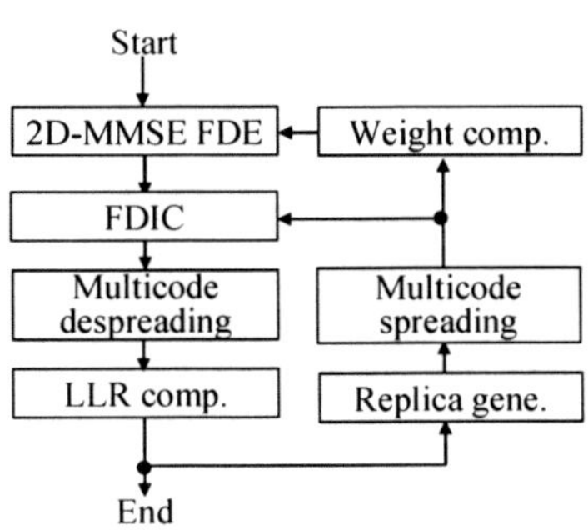

Fig.9 Iterative FDIC.

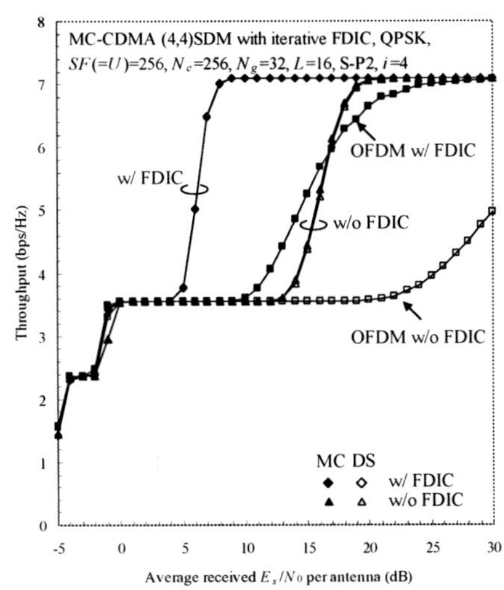

Fig. 10 Throughput performance.

7. CONCLUSION

In this paper, we have presented a comprehensive performance comparison of DS- and MC-CDMA with frequency-domain equalization (FDE). Using FDE, both MC- and DS-CDMA provide almost the identical performance. Since both CDMA transceiver structures are similar, a new CDMA transceiver which can flexibly switch between MC-CDMA (or

OFDMA) and DS-CDMA can be implemented using software defined radio technology. Although OFDMA has recently been attracting attention, CDMA still remains as a promising multi-access technique. Since DS-CDMA signal has less peak-to-average power ratio (PAPR), DS-CDMA is more appropriate for the uplink application than MC-CDMA.

Reference

1. F. Adachi, M. Sawahashi and H. Suda, "Wideband DS-CDMA for next generation mobile communications systems," IEEE Commun. Mag., Vol. 36, No. 9, pp. 56-69, Sept. 1998.
2. Y. Kim, B.J. Jeong, J. Chung, et al., "Beyond 3G: vision, requirements, and enabling technologies," IEEE Commun. Mag., Vol. 41, No. 3, pp.120-124, Mar. 2003.
3. S. Hara and R. Prasad, "Overview of multicarrier CDMA," IEEE Commun. Mag., Vol. 35, pp. 126-133, Dec. 1997.
4. F. Adachi, D. Garg, S. Takaoka, and K. Takeda, "Broadband CDMA techniques," IEEE Wireless Communications Magazine, Vol. 12, No. 2, pp.8-18, Apr. 2005.
5. G. J. Foschini, "Layered space-time architecture for wireless communication in a fading environment when using multi-element antennas," Bell Lab. Tech. Journal, Vol. 1, No. 2, pp.41-59, 1996.
6. H. Tomeba, K. Takeda, and F. Adachi, "Space-time block coded joint transmit/receive diversity in a frequency-nonselective Rayleigh fading channel," IEICE Trans. Commun., Vol. E89-B, No. 8, pp. 2189-2195, Aug. 2006.
7. I. Cosovic, M. Schnell, and A. Springer, "On the performance of different channel pre-compensation techniques for uplink time division duplex MC-CDMA," Proc. IEEE VTC'03 Fall, Vol. 2, pp. 857-861, Oct. 2003.
8. H. Tomeba, K. Takeda, and F. Adachi, "Frequency-domain space-time block coded-joint transmit/receive diversity for the single carrier transmission," Proc. the 10th ICCS, pp. 1-5, Singapore, 30 Oct to 1 Nov 2006.
9. S. Alamouti, "A simple transmit diversity technique for wireless communications", IEEE Journal on Selected Areas in Commun., Vol. 16, No. 8, pp. 1451–1458, Oct. 1998.
10. H. Atarashi, N. Maeda, Y. Kishiyama, and M. Sawahashi, "Broadband wireless access based on VSF-OFCDM and VSCRF-CDMA and its experiments," European Trans. Telecommun., Vol. 15, No. 3, pp.159-172, May-Jun. 2004.
11. L. Liu and F. Adachi, "2-dimensional OVSF spread/chip-interleaved CDMA," IEICE Trans. Commun., Vol. E89-B, No. 12, pp. 3363-3375, Dec. 2006.
12. D. Garg and F. Adachi, "Throughput comparison of turbo-coded HARQ in OFDM, MC-CDMA and DS-CDMA with frequency-domain equalization," IEICE Trans. Commun., Vol.E88-B, No.2, pp.664-677, Feb. 2005.
13. A. Nakajima, D. Garg and F. Adachi, "Frequency-domain iterative parallel interference cancellation for multicode DS-CDMA-MIMO multiplexing," Proc. IEEE VTC'05 Fall, Vol.1, pp. 73-77, Dallas, U.S.A., 26-28 Sept. 2005.
14. A. Nakajima and F. Adachi, "Iterative FDIC using 2D-MMSE FDE for turbo-coded HARQ in SC-MIMO multiplexing," IEICE Trans. Commun. Vol. E90-B, No.3, pp.693-695, Mar. 2007.
15. A. Stefanov and T. Duman, "Turbo coded modulation for wireless communications with antenna diversity," Proc. IEEE VTC'99 Fall, pp.1565-1569, Netherlands, Sept. 1999.

DIFFERENT GUARD INTERVAL TECHNIQUES FOR OFDM: PERFORMANCE COMPARISON

Heidi Steendam
TELIN Dept., Ghent University, Sint-Pietersnieuwstraat 41, 9000 GENT, BELGIUM
Heidi.Steendam@telin.UGent.be

Marc Moeneclaey
TELIN Dept., Ghent University, Sint-Pietersnieuwstraat 41, 9000 GENT, BELGIUM
Marc Moeneclaey@telin.UGent.be

Abstract In this paper, we consider different types of guard intervals for OFDM systems, i.e. cyclic prefix (CP), zero padding (ZP) and known symbol padding (KSP). We compare the different OFDM systems with respect to their SNR performance. We show that CP-OFDM and ZP-OFDM have exactly the same performance, whereas KSP-OFDM has a slightly worse performance. Further, we consider data aided channel estimation for the three OFDM systems; the MSE of the estimators is compared to the corresponding Cramer-Rao bound. It turns out that in practice, channel estimation for CP-OFDM slightly outperforms the one for ZP-OFDM. The practical channel estimation techniques for KSP-OFDM perform worst.

Keywords: Guard intervals, ML channel estimation, Cramer-Rao bounds

1. Introduction

In multicarrier (MC) systems, where the data symbols are transmitted in parallel on N different carriers, the length T of a symbol is extended with a factor N [1]. This extension of the symbol length causes the MC system to be less sensitive to channel dispersion than a single carrier system transmitting data symbols at the same data rate. However, at the edges of a MC symbol, the channel dispersion still causes distortion, and hence introduces interference between successive MC symbols (i.e. intersymbol interference, ISI) and interference between different carriers within the same MC symbol (i.e. intercarrier interference, ICI). To reduce the effect of the ISI, each MC symbol is extended with a guard

S. Plass et al. (eds.), Multi-Carrier Spread Spectrum 2007, 11–24.

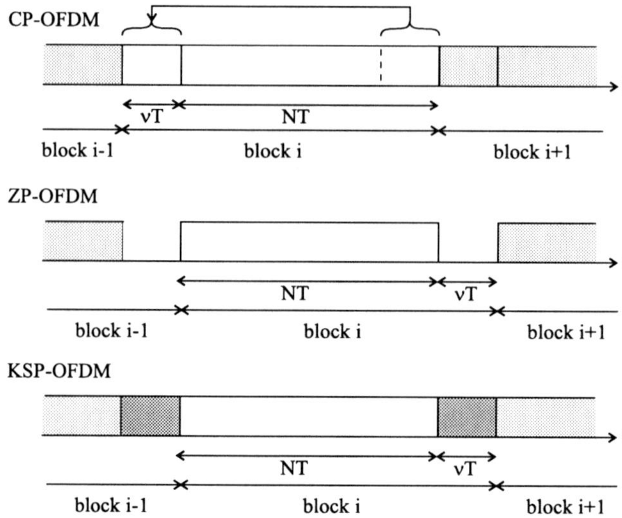

Figure 1. Transmitted signal for CP-OFDM, ZP-OFDM and KSP-OFDM.

interval. When the length of the guard interval is longer than the duration of the channel impulse response, ISI can completely be removed. However, as the transmission efficiency reduces with the insertion of the guard interval (during the guard interval, no new information can be transmitted), the guard interval must be chosen sufficiently small.

The most commonly used guard interval is the cyclic prefix (CP) [1]. In CP-OFDM, the last ν samples of each OFDM symbol of N samples are copied and added in front of the OFDM symbol, as shown in figure 1. At the receiver, the samples in the CP are discarded, as they are affected by interference; the N samples outside the CP are kept for further processing (see figure 2). Because during the guard interval signal is transmitted, the CP-OFDM system suffers from a power efficiency loss with a factor $\frac{N}{N+\nu}$. To avoid this power efficiency loss of CP-OFDM, the zero-padding guard interval was introduced [2]. In ZP-OFDM, a guard interval of ν samples is introduced after each OFDM symbol. During this guard interval, no signal is transmitted, as shown in figure 1. At the receiver, the ν samples of the guard interval are added to the first ν samples of the data part of N samples (see figure 2); the resulting N samples are then further processed by the receiver. Although the power efficiency loss is avoided in ZP-OFDM, we will show in the next section that the noise power will be enhanced with a factor $\frac{N+\nu}{N}$.

Another recently proposed guard interval is the known symbol padding (KSP) [3]-[4]. In KSP-OFDM, a guard interval consisting of ν known samples long is added after each OFDM symbol (corresponding to the

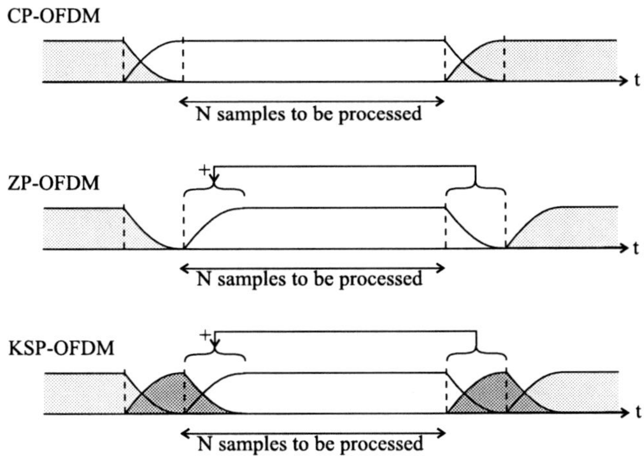

Figure 2. Received signal for CP-OFDM, ZP-OFDM and KSP-OFDM.

dark gray area in figure 1). Assuming the energy per sample is the same for the guard interval and the data part, KSP-OFDM will suffer, like CP-OFDM, from a power efficiency loss of $\frac{N}{N+\nu}$. At the receiver, first the signal corresponding to the known symbols is subtracted from the received signal (i.e. the dark gray areas in figure 2 are removed). Then, like in ZP-OFDM, the samples in the guard interval are added to the first part of the OFDM symbol, and the resulting N samples are further processed. Similarly to ZP-OFDM, KSP-OFDM will suffer from a noise power enhancement with a factor $\frac{N+\nu}{N}$.

The paper is organized as follows. In section 2, we will compare the SNR performance for the three OFDM systems. Then, in section 3, practical data-aided channel estimation techniques will be considered. The MSE performance of the estimators will be compared, and the corresponding Cramer-Rao lower bounds (CRLB) are derived. The conclusions will be drawn in section 4.

2. System Performance

2.1 CP-OFDM

The data symbols to be transmitted during the ith CP-OFDM block are defined as $\mathbf{a}_i = \{a_i(n)|n = 0, \ldots, N-1\}$. The data symbols are assumed to be statistically independent with $E[a_i(n)a^*_{i'}(n')] = E_s\delta_{i,i'}\delta_{n,n'}$. The data symbols are modulated on the carriers using an inverse fast Fourier transform (inverse FFT, IFFT), and the cyclic prefix is inserted.

The resulting samples transmitted during block i are given by

$$\mathbf{s}_{i,CP} = \sqrt{\frac{N}{N+\nu}}\boldsymbol{\Omega}\mathbf{F}^{+}\mathbf{a}_i \tag{1}$$

where $\mathbf{s}_{i,CP} = \{s_{i,CP}(k)|k = -\nu, \ldots, N-1\}$, $\mathbf{F}$ is the $N \times N$ matrix corresponding to the FFT operation, i.e. $\mathbf{F}_{k,\ell} = \frac{1}{\sqrt{N}}e^{-j2\pi\frac{k\ell}{N}}$, and $\boldsymbol{\Omega}$ is the $(N+\nu) \times N$ matrix operator that adds the CP, i.e.

$$\boldsymbol{\Omega} = \begin{pmatrix} \mathbf{0}_{\nu\times(N-\nu)} & \mathbf{I}_{\nu} \\ \multicolumn{2}{c}{\mathbf{I}_N} \end{pmatrix} \tag{2}$$

where $\mathbf{0}_{a\times b}$ is the $a \times b$ all-zero-matrix, $\mathbf{I}_M$ is the $M \times M$ identity matrix and $\mathbf{X}^{+}$ is the Hermitian transpose of $\mathbf{X}$.

The time-domain samples (1) are transmitted over a doubly selective channel [5]. The channel is modeled as a tapped delay line with channel coefficients $h_{ch}(k;\ell)$. We assume that the channel contains a line-of-sight (LOS) component and a zero-mean multipath (MP) fading component, i.e. $h_{ch}(k;\ell) = h_{LOS}(k;\ell) + h_{MP}(k;\ell)$. The LOS component is modeled as $h_{LOS}(k;\ell) = \alpha e^{j\phi(\ell)}\delta(k)$, where the phase $\phi(\ell)$ depends on the time-selectivity of the channel, and the quasi-static amplitude α is assumed to be constant over a number of OFDM symbols. The channel taps of the multipath component are assumed to be WSSUS zero-mean Gaussian distributed [6] with autocorrelation function $R_{MP}(k;\ell)$

$$E[h_{MP}(k_1;\ell_1)h^{*}_{MP}(k_2;\ell_2)] = \delta(k_1-k_2)R_{MP}(k_1;\ell_1-\ell_2). \tag{3}$$

At the receiver, the CP is removed, and the remaining N samples are fed to the FFT. Without loss of generality, we consider the detection of the OFDM block with index $i = 0$. The N outputs of the FFT $y_{CP}(n)$ are given by $\mathbf{y}_{CP} = \sum_{i=-\infty}^{+\infty}\mathbf{F}\boldsymbol{\Delta}\mathbf{H}^{(i)}\mathbf{s}_{i,CP} + \mathbf{F}\boldsymbol{\Delta}\mathbf{w}$, where the $(N+\nu) \times (N+\nu)$ channel matrix $\mathbf{H}^{(i)}_{k,k'} = h_{ch}(k-k'-i(N+\nu);k)$, the operator $\boldsymbol{\Delta} = (\mathbf{0}_{N\times\nu} \;\; \mathbf{I}_N)$ removes the prefix and $\mathbf{w} = \{w(k)|k = -\nu, \ldots, N-1\}$ is the vector of time-domain noise samples. The noise components $w(k)$ are assumed to be statistically independent zero-mean Gaussian distributed with variance N_0. The nth FFT output can be rewritten as

$$y_{CP}(n) = \sqrt{\frac{N}{N+\nu}}\sum_{i=-\infty}^{+\infty}\sum_{n'=0}^{N-1} a_i(n')\gamma_{i,CP}(n,n') + W(n) \tag{4}$$

where $W(n) = (\mathbf{F}\boldsymbol{\Delta}\mathbf{w})_n$ and

$$\gamma_{i,CP}(n,n') = \frac{1}{N}\sum_{k=0}^{N-1}\sum_{k'=-\nu}^{N-1} h(k-k'-i(N+\nu);k)e^{-j2\pi\frac{kn-k'n'}{N}} \tag{5}$$

In [5] it is shown that the signal to interference and noise ratio (SINR) at the output of the FFT is independent of the carrier index n and is given by

$$SINR = \frac{\frac{N}{N+\nu} E_s P_U}{\frac{N}{N+\nu} E_s P_I + P_N} \tag{6}$$

where the contributions of the useful component, the interference and noise are given by

$$P_U = |\alpha|^2 |\Phi(0)|^2 + \frac{1}{N} \sum_{k=-\infty}^{+\infty} \sum_{\ell=-\infty}^{+\infty} \tilde{w}(k;\ell) R_{MP}(k;\ell) \tag{7}$$

$$P_I = |\alpha|^2 + \sum_{k=-\infty}^{+\infty} R_{MP}(k;0) - P_U \tag{8}$$

$$P_N = N_0. \tag{9}$$

where $\Phi(n)$ is the nth output of the N-point FFT of the phase $\phi(\ell)$ and the weight function $\tilde{w}(k;\ell)$ is defined in [7, eq. A2].

2.2 ZP-OFDM

In ZP-OFDM, the data symbols $\mathbf{a}_i$ are applied to the inverse FFT and zero padded, resulting in the time-domain samples

$$\mathbf{s}_{i,ZP} = \mathbf{\Xi} \mathbf{F}^{+} \mathbf{a}_i \tag{10}$$

where the $(N+\nu) \times N$ matrix $\mathbf{\Xi} = (\mathbf{I}_N \quad \mathbf{0}_{N\times\nu})^T$ is the zero-padding operator and $\mathbf{s}_{i,ZP} = \{s_{i,ZP}(k) | k = 0, \ldots, N+\nu-1\}$. At the receiver, the guard interval samples are added to the first ν samples of the data part, and the resulting N samples are applied to the FFT. The N outputs $y_{ZP}(n)$ of the FFT can be written as $\mathbf{y}_{ZP} = \sum_{i=-\infty}^{+\infty} \mathbf{F}\mathbf{\Lambda}\mathbf{H}^{(i)}\mathbf{s}_{i,ZP} + \mathbf{F}\mathbf{\Lambda}\mathbf{w}$ where the $N \times (N+\nu)$ matrix $\mathbf{\Lambda}$ performs the addition of the guard interval to the data part

$$\mathbf{\Lambda} = \begin{pmatrix} \mathbf{I}_N & \begin{matrix} \mathbf{I}_\nu \\ \mathbf{0}_{(N-\nu)\times\nu} \end{matrix} \end{pmatrix} \tag{11}$$

The nth FFT output can be rewritten as

$$y_{ZP}(n) = \sum_{i=-\infty}^{+\infty} \sum_{n'=0}^{N-1} a_i(n') \gamma_{i,ZP}(n,n') + W(n) \tag{12}$$

where

$$\gamma_{i,ZP}(n,n') = \frac{1}{N} \sum_{k=0}^{N+\nu-1} \sum_{k'=0}^{N-1} h(k-k'-i(N+\nu);k) e^{-j2\pi \frac{kn-k'n'}{N}} \tag{13}$$

Using a similar analysis as for CP-OFDM in [5], it can be shown that the SINR at the FFT outputs for ZP-OFDM is independent of the carrier index n and yields

$$SINR = \frac{E_s P_U}{E_s P_I + P_N} \tag{14}$$

Although the summation ranges in (13) and (5) differ, it turns out that for ZP-OFDM P_U and P_I are the same as in CP-OFDM and are given by (7)-(8). The noise component $P_N = \frac{N+\nu}{N} N_0$, i.e. the noise power is enhanced with a factor $\frac{N+\nu}{N}$ as compared to CP-OFDM. Taking into account the effect of the power efficiency loss in CP-OFDM and the noise enhancement in ZP-OFDM, it follows that ZP-OFDM and CP-OFDM yield the same value of SINR.

2.3 KSP-OFDM

In KSP-OFDM, the time-domain samples to be transmitted are given by

$$\mathbf{s}_{i,KSP} = \sqrt{\frac{N}{N+\nu}} \begin{pmatrix} \mathbf{F}^{+}\mathbf{a}_i \\ \mathbf{b}_g \end{pmatrix} \tag{15}$$

where it is assumed that the known symbols $\mathbf{b}_g$ have the same energy per sample as the data samples, i.e. $E[|b_g(n)|^2] = E_s$. At the receiver, the signal corresponding to the known symbols $\mathbf{b}_g$ is first subtracted from the received signal. After adding the ν samples of the guard interval to the data part of the OFDM symbol and applying the resulting N samples to the FFT, the FFT outputs are given by

$$\mathbf{y}_{KSP} = \sqrt{\frac{N}{N+\nu}} \sum_{i=-\infty}^{+\infty} \mathbf{F}\mathbf{\Lambda}\mathbf{H}^{(i)}\mathbf{\Xi}\mathbf{F}^{+}\mathbf{a}_i + \mathbf{F}\mathbf{\Lambda}\mathbf{w} \tag{16}$$

where it is assumed that the channel taps are perfectly known, i.e. the contribution of the known symbols can completely be removed from the signal. Hence the outputs of the FFT for KSP-OFDM are the same as for ZP-OFDM, except for the power efficiency loss factor $\sqrt{\frac{N}{N+\nu}}$. The SINR at the outputs of the FFT is defined as (6), where the contributions P_U and P_I are given by (7)-(8), similarly as for CP-OFDM and ZP-OFDM, and the noise component $P_N = \frac{N+\nu}{N} N_0$ is the same as for ZP-OFDM. Comparing KSP-OFDM with CP-OFDM and ZP-OFDM, it can be observed that KSP-OFDM will have worse performance than the two other systems, as it suffers from both power efficiency loss and noise enhancement, whereas CP-OFDM and ZP-OFDM suffer only from one of these effects. However, if the guard interval length is small as compared to the FFT length, as in most practical cases, the difference in performance will be very small. Further, in practical situations, the channel taps have to be estimated and are thus not perfectly known. In KSP-OFDM, the channel estimation error will cause interference from the known symbols, resulting in extra performance loss.

3. Channel Estimation

Reliable channel estimation is necessary in the above mentioned OFDM systems as data detection algorithms for these systems require the knowledge of the channel. Most common channel estimation techniques are data aided: pilot symbols are inserted in the OFDM signal to enable reliable detection of the channel. In this section, we will consider maximum-likelihood (ML) based data-aided channel estimation techniques for the three different OFDM systems, and compare the MSE of the channel estimation techniques. Further, we compare the performance of the estimators with the corresponding Cramer-Rao lower bounds.

In the following, we will assume that the channel changes slowly as compared to the duration of the length of an OFDM symbol, i.e. during the observed OFDM symbol, $h_{ch}(k;\ell) = h(k)$. We assume that the duration of the channel impulse response is L taps, and define the vector of channel taps $\mathbf{h} = (h(0) \ldots h(L-1))^T$. To avoid intersymbol interference the duration of the guard interval exceeds the duration of the channel impulse response, i.e. $\nu \geq L-1$.

In the following, we will show that for all cases, the observation can be written as $\mathbf{z} = \mathbf{D}\mathbf{h} + \boldsymbol{\omega}$, where the matrix $\mathbf{D}$ depends on the inserted pilots and the noise $\boldsymbol{\omega}$ is zero-mean Gaussian distributed with autocorrelation function $\mathbf{R}_\omega$, i.e. $\boldsymbol{\omega} \sim N(\mathbf{0}, \mathbf{R}_\omega)$. Hence, the observation $\mathbf{z}$ given $\mathbf{h}$ is Gaussian distributed: $\mathbf{z}|\mathbf{h} \sim N(\mathbf{D}\mathbf{h}, \mathbf{R}_\omega)$. The ML estimate of the vector $\mathbf{h}$ is defined as [8]:

$$\hat{\mathbf{h}}_{ML} = \arg\max_{\mathbf{h}} p(\mathbf{z}|\mathbf{h}) \tag{17}$$

If $\mathbf{R}_\omega$ is independent of $\mathbf{h}$ and $\mathbf{D}^+\mathbf{R}_\omega^{-1}\mathbf{D}$ is invertible, the ML estimate of the channel is given by

$$\hat{\mathbf{h}}_{ML} = (\mathbf{D}^+\mathbf{R}_\omega^{-1}\mathbf{D})^{-1}\mathbf{D}^+\mathbf{R}_\omega^{-1}\mathbf{z} \tag{18}$$

and the MSE of the estimation yields

$$MSE = E[||\mathbf{h} - \hat{\mathbf{h}}_{ML}||^2] = \text{trace}\left((\mathbf{D}^+\mathbf{R}_\omega^{-1}\mathbf{D})^{-1}\right) \tag{19}$$

The Cramer-Rao lower bound (CRLB) of the estimation is given by $\mathbf{R}_{\mathbf{h}-\hat{\mathbf{h}}} - \mathbf{J}^{-1} \geq 0$ [8] where $\mathbf{R}_{\mathbf{h}-\hat{\mathbf{h}}}$ is the autocorrelation matrix of the estimation error $\mathbf{h} - \hat{\mathbf{h}}$ and the Fisher information matrix $\mathbf{J}$ is defined as

$$\mathbf{J} = E_{\mathbf{z}} \left[\left(\frac{\partial}{\partial \mathbf{h}} \ln p(\mathbf{z}|\mathbf{h}) \right)^{+} \left(\frac{\partial}{\partial \mathbf{h}} \ln p(\mathbf{z}|\mathbf{h}) \right) \right] \tag{20}$$

Hence, the MSE is lower bounded by $E[||\mathbf{h} - \hat{\mathbf{h}}_{ML}||^2] = \text{trace}(\mathbf{R}_{\mathbf{h}-\hat{\mathbf{h}}}) \geq \text{trace}(\mathbf{J}^{-1})$. When equality occurs, i.e. when the MSE equals the CRLB, the estimate is a minimum variance unbiased (MVU) estimate. In the case that $\mathbf{R}_{\omega}$ is independent of $\mathbf{h}$, it can be shown that $\mathbf{J} = \mathbf{D}^{+}\mathbf{R}_{\omega}^{-1}\mathbf{D}$, i.e. the channel estimate is a MVU estimate.

3.1 CP-OFDM

In CP-OFDM, data-aided channel estimation is performed by replacing some data carriers by pilot carriers. In this paper, without loss of generality, we consider the comb-type pilot arrangement [9], where in every OFDM symbol, $M \geq L$ data carriers are replaced by pilot carriers. The analysis however can easily be extended to other types of pilot arrangements.

Assuming that the pilot symbols $b_c(p)$ are located on carriers n_p, $p = 1, \ldots, M$, it can be shown that the FFT outputs at positions n_p contain sufficient information for the ML estimation of the channel vector $\mathbf{h}$. Defining $z(p) = y_{CP}(n_p)/(\sqrt{\frac{N}{N+\nu}} b_c(p))$, the $M \times 1$ vector $\mathbf{z}$ of observations can be written as $\mathbf{z} = \mathbf{Ah} + \mathbf{W}$, where the $M \times L$ matrix $\mathbf{A}$ has entries $A_{k,\ell} = e^{-j2\pi \frac{n_k \ell}{N}}$, and the noise components $\mathbf{W}$ are zero-mean Gaussian distributed with autocorrelation matrix $\mathbf{R}_{CP} = \frac{N+\nu}{N} \frac{N_0}{E_s} \mathbf{I}_M$. Hence, the observation $\mathbf{z}$ given $\mathbf{h}$ is Gaussian distributed: $\mathbf{z}|\mathbf{h} \sim N(\mathbf{Ah}, \mathbf{R})$. Taking into account that $\mathbf{A}^{+}\mathbf{A}$ is invertible when $M \geq L$, it follows from (18) and (19) that $\hat{\mathbf{h}}_{ML} = (\mathbf{A}^{+}\mathbf{A})^{-1}\mathbf{A}^{+}\mathbf{z}$ and $MSE_{CP} = \frac{N+\nu}{N} \frac{N_0}{E_s} \text{trace}((\mathbf{A}^{+}\mathbf{A})^{-1})$. When M divides N and the pilots are equally spaced over the carriers, i.e. $n_m = n_0 + (m-1)\frac{N}{M}$, it follows that $\text{trace}((\mathbf{A}^{+}\mathbf{A})^{-1}) = \frac{L}{M}$, i.e. the MSE is proportional to the number of channel taps to be estimated, and inversely proportional to the number of pilots. As $\mathbf{R}_{CP}$ is independent of $\mathbf{h}$, it follows that the ML estimate is MVU.

3.2 ZP-OFDM

Similarly as in CP-OFDM, data-aided channel estimation is performed by replacing some data carriers by pilot carriers. We consider the same pilot arrangement as for CP-OFDM. As in CP-OFDM, it can be shown that the FFT outputs at the positions of the M pilots $b_c(p)$ contain sufficient information to perform the ML estimation. Defining $z(p) = y_{ZP}(n_p)/b_c(p)$, the observations can be written as $\mathbf{z} = \mathbf{Ah} + \tilde{\mathbf{W}}$, where $\mathbf{A}$ is the same as for CP-OFDM and the noise contribution $\tilde{\mathbf{W}}$ is zero-mean Gaussian distributed with autocorrelation function $\mathbf{R}_{ZP} = \frac{N_0}{E_s}\mathbf{F}\mathbf{\Lambda}\mathbf{\Lambda}^{+}\mathbf{F}^{+}$, thus $\mathbf{z}|\mathbf{h} \sim N(\mathbf{Ah}, \mathbf{R}_{ZP})$. Hence, in the case that $\mathbf{A}^{+}\mathbf{R}_{ZP}^{-1}\mathbf{A}$ is invertible, the ML estimate of the channel is given by $\hat{\mathbf{h}}_{ML} = (\mathbf{A}^{+}\mathbf{R}_{ZP}^{-1}\mathbf{A})^{-1}\mathbf{A}^{+}\mathbf{R}_{ZP}^{-1}\mathbf{z}$ and the MSE of the estimation yields $MSE_{ZP} = \frac{N_0}{E_s}\text{trace}((\mathbf{A}^{+}\mathbf{R}_{ZP}^{-1}\mathbf{A})^{-1})$. Further, the ML estimate of the channel taps for ZP-OFDM is a MVU estimate, as $\mathbf{R}_{ZP}$ is independent of $\mathbf{h}$.

3.3 KSP-OFDM

In the two previous techniques, we have assumed that there are M pilot symbols to estimate the L channel taps, $M \geq L$. In KSP-OFDM, the known symbols in the guard interval can be used as pilot symbols to estimate the channel. However, the guard interval contains only ν samples, with typically $\nu \approx L$. To increase the number of pilot symbols to M, we can consider two approaches: in the first approach, the guard interval length is kept to ν samples, and $M-\nu$ data carriers are replaced by pilot carriers. In the second approach, we increase the length of the guard interval to M samples, i.e. $\nu = M$.

Approach 1. As in the previous techniques, the comb-type pilot arrangement for the $M-\nu$ pilot carriers in the data part is considered. It can be shown that an observation interval corresponding to the $N+\nu$ time-domain samples of the ith OFDM block (as shown in figure 3) contains sufficient information for the estimation. Defining $\mathbf{z}$ as the vector of $N+\nu$ observed samples, the observation can be written as $\mathbf{z} = \mathbf{Bh} + \epsilon$ where $\mathbf{B} = \mathbf{B}_g + \mathbf{B}_c$ is a $(N+\nu)\times L$ matrix. The matrix $\mathbf{B}_g$ corresponds to the contributions of the pilot symbols $b_g(p)$ in the guard interval with $(\mathbf{B}_g)_{i,j} = b_g(|i-j+\nu|_{N+\nu})$, $|x|_K$ is the modulo-K operation on x and $b_c(i) = 0$ for $i \geq \nu$ and $i < 0$. The matrix $\mathbf{B}_c$ corresponds to the contributions of the pilot symbols $b_c(p)$ transmitted on the carriers with $(\mathbf{B}_c)_{i,j} = s_c(i-j)$; $\mathbf{s}_c = \mathbf{X}\mathbf{b}_c$ where the $N \times (M-\nu)$ matrix $\mathbf{X}$ consists of the subset of columns of the IFFT matrix $\mathbf{F}^{+}$ corresponding to the positions of the pilot carriers, $\mathbf{b}_c$ is the vector of the pilot symbols transmitted on the carriers, and $s_c(i) = 0$ for $i \geq N$ and $i < 0$, i.e. $\mathbf{s}_c$ corresponds to the N-point IFFT of the pilot carriers only. Further, $\epsilon = \mathbf{H}\mathbf{s}_d + \mathbf{w}$, where $(\mathbf{H})_{i,j} = h(i-j)$ is a $(N+\nu)\times N$ matrix, $\mathbf{s}_d = \bar{\mathbf{X}}\mathbf{a}$, a

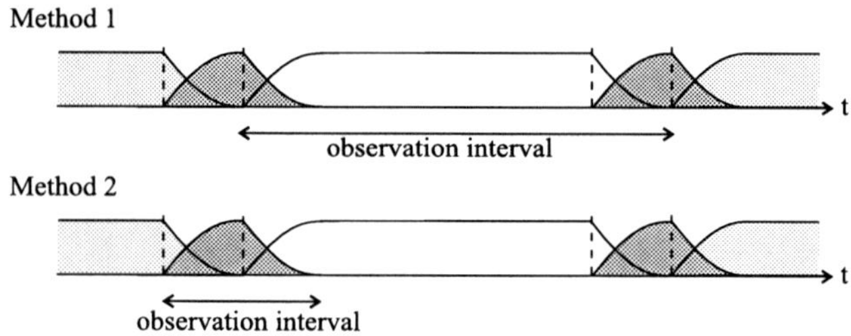

Figure 3. Observation interval for channel estimation in KSP-OFDM.

is the data vector, the $N \times (N - M + \nu)$ matrix $\bar{\mathbf{X}}$ consists of the subset of columns of the IFFT matrix $\mathbf{F}^+$ corresponding to the positions of the data carriers, i.e. $\mathbf{s}_d$ corresponds to the N-point IFFT of the data carriers only, and $\mathbf{w}$ is the additive Gaussian noise component. When $N - M + \nu$ is large, the data contribution $\mathbf{H}\mathbf{s}_d$ to $\boldsymbol{\epsilon}$ can, according to the central limit theorem, be modeled as zero-mean Gaussian distributed. Hence, $\boldsymbol{\epsilon}$ is zero-mean Gaussian distributed with autocorrelation matrix $\mathbf{R}_\epsilon = E_s \frac{N}{N+\nu} \frac{N+\nu-M}{N} \mathbf{H}\mathbf{H}^+ + N_0 \mathbf{I}_{N+\nu}$. As the autocorrelation matrix $\mathbf{R}_\epsilon$ depends on the parameters to be optimized, the ML estimator is very complex. In [4], a suboptimal ML-based solution is suggested: the autocorrelation matrix $\mathbf{R}_\epsilon$ is first estimated from the received signal, and the estimate $\hat{\mathbf{R}}_\epsilon$ is then used to find the estimate for $\mathbf{h}$:

$$\hat{\mathbf{h}}_{KSP,1a} = (\mathbf{B}^+ \hat{\mathbf{R}}_\epsilon^{-1} \mathbf{B})^{-1} \mathbf{B}^+ \hat{\mathbf{R}}_\epsilon^{-1} \mathbf{z} \tag{21}$$

where it is assumed that $\mathbf{B}^+ \hat{\mathbf{R}}_\epsilon^{-1} \mathbf{B}$ is invertible. To evaluate the performance of this estimate, we assume a genie-aided estimator for $\mathbf{R}_\epsilon$, i.e. $\mathbf{R}_\epsilon$ is perfectly known. In this case, the MSE of the estimation of $\mathbf{h}$ yields

$$MSE_{KSP,1a} = \text{trace}\left((\mathbf{B}^+ \mathbf{R}_\epsilon^{-1} \mathbf{B})^{-1} \right) \tag{22}$$

When $\mathbf{R}_\epsilon$ is not perfectly known, the MSE will be increased as compared to (22).

A disadvantage of the estimator (21) is that the autocorrelation matrix $\mathbf{R}_\epsilon$ needs to be known or estimated to be able to estimate the channel. As this autocorrelation matrix depends on the unknown channel, the estimation of this matrix from the received signal is not obvious. Therefore, a simplification of (21) can be made by ignoring the statistics of the interfering data, yielding the estimate

$$\hat{\mathbf{h}}_{KSP,1b} = (\mathbf{B}^+ \mathbf{B})^{-1} \mathbf{B}^+ \mathbf{z} \tag{23}$$

and the corresponding MSE is given by

$$MSE_{KSP,1b} = \text{trace}\left((\mathbf{B}^+ \mathbf{B})^{-1} \mathbf{B}^+ \mathbf{R}_\epsilon \mathbf{B} (\mathbf{B}^+ \mathbf{B})^{-1} \right) \tag{24}$$

As $\mathbf{R}_\epsilon$ depends on $\mathbf{h}$, also the CRLB is very complex. Therefore, we consider the low SNR limit of the CRLB. When E_s/N_0 is low, it can easily be verified that $\mathbf{R}_\epsilon \approx N_0\mathbf{I}_{N+\nu}$. In that case, the CRLB reduces to $E[||\mathbf{h}-\hat{\mathbf{h}}_{ML}||^2] \geq N_0\text{trace}((\mathbf{B}^+\mathbf{B})^{-1})$ i.e. for low SNR, the MSE's (22) and (24) of the estimates (21) and (23) reach the CRLB.

Approach 2. It can easily be verified that the observation interval shown in figure 3 contains sufficient information for the estimation in the second approach, where the pilot symbols are all located in the guard interval. The vector of $\nu+L-1$ observed samples equals $\mathbf{z} = \mathbf{Th} + \boldsymbol{\epsilon}'$ where $(\mathbf{T})_{i,j} = b_g(i-j)$ with $b_g(i)$ denoting the pilot symbols in the guard interval; note that $b_g(i) = 0$ for $i < 0$ or $i \geq \nu$. The noise component can be written as $\boldsymbol{\epsilon}' = \mathbf{H}^{(0)}\mathbf{s}_0 + \mathbf{H}^{(1)}\mathbf{s}_1 + \mathbf{w}$, where $\mathbf{s}_0 = (s_{0,KSP}(N-L+1)\ldots s_{0,KSP}(N-1))^T$ and $\mathbf{s}_1 = (s_{1,KSP}(0)\ldots s_{0,KSP}(L-2))^T$ are the contributions from the data parts of previous and next OFDM symbol, respectively, $(\mathbf{H}^{(0)})_{i,j} = h(L-1-(i-j))$ and $(\mathbf{H}^{(1)})_{i,j} = h(i-j-\nu)$ with $h(i) = 0$ for $i < 0$ and $i > L-1$. When N is large, $\mathbf{s}_0$ and $\mathbf{s}_1$ can be modeled as zero-mean Gaussian distributed. Hence, $\boldsymbol{\epsilon}' \sim N(\mathbf{0}, \mathbf{R}_{\epsilon'})$ with $\mathbf{R}_{\boldsymbol{\epsilon}'} = E_s(\mathbf{H}^{(0)}(\mathbf{H}^{(0)})^+ + \mathbf{H}^{(1)}(\mathbf{H}^{(1)})^+) + N_0\mathbf{I}_{\nu+L-1}$. Similarly as for the first approach, the autocorrelation matrix $\mathbf{R}_{\epsilon'}$ depends on the parameters $\mathbf{h}$ to be estimated. As in the previous method, a suboptimal ML based solution can be proposed [4] by assuming that the autocorrelation is first estimated from the received signal, and then used to estimate $\mathbf{h}$. The estimate $\hat{\mathbf{h}}_{KSP,2a}$ and the $MSE_{KSP,2a}$ (assuming $\mathbf{R}_{\epsilon'}$ is perfectly known) are then given by (21) and (22), where $\mathbf{B}$ and $\mathbf{R}_\epsilon$ are replaced by $\mathbf{T}$ and $\mathbf{R}_{\epsilon'}$, respectively.

As for the previous approach, the knowledge of $\mathbf{R}_{\boldsymbol{\epsilon}'}$ is needed to estimate $\mathbf{h}$. To simplify the estimator, the statistics of the interfering data can be ignored. The resulting estimate $\hat{\mathbf{h}}_{KSP,2b}$ and the $MSE_{KSP,2b}$ are given by (23) and (24) by replacing $\mathbf{B}$ and $\mathbf{R}_\epsilon$ by $\mathbf{T}$ and $\mathbf{R}_{\epsilon'}$.

Further, similarly as for the first method, a low SNR limit for the CRLB can be found: $E[||\mathbf{h}-\hat{\mathbf{h}}_{ML}||^2] \geq N_0\text{trace}((\mathbf{T}^+\mathbf{T})^{-1})$. At low SNR, the MSE of the two estimates reach the CRLB.

3.4 Performance Comparison

In table 1, the energy per information data symbol $E_{s,i}$ and the bandwidth efficiency η_{BW} are given for the different guard interval techniques and pilot positions. When N is large, it is clear that the differences in $E_{s,i}$ and η_{BW} are small for the different techniques.

In figure 4, the normalized MSE, i.e. $NMSE = \frac{E_s}{N_0}MSE$, is shown as function of M for $N = 1024$, $L = 8$ and $\nu = 7$ (except for KSP2, where

	$E_{s,i}$	η_{BW}
CP-OFDM	$\frac{N+\nu}{N-M}E_s$	$\frac{N-M}{N+\nu}\frac{1}{T}$
ZP-OFDM	$\frac{N}{N-M}E_s$	$\frac{N-M}{N+\nu}\frac{1}{T}$
KSP1-OFDM	$\frac{N+\nu}{N+\nu-M}E_s$	$\frac{N+\nu-M}{N+\nu}\frac{1}{T}$
KSP2-OFDM	$\frac{N+M}{N}E_s$	$\frac{N}{N+M}\frac{1}{T}$

Table 1. Energy per information data symbol $E_{s,i}$ and bandwidth efficiency η_{BW} for the different guard interval techniques and pilot positions

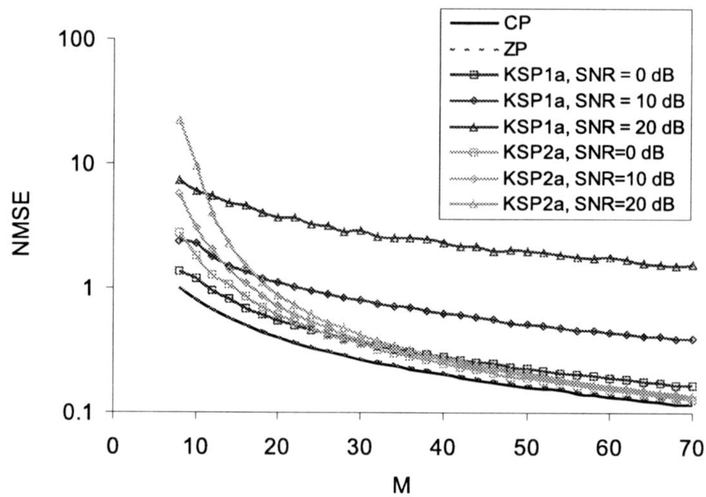

Figure 4. NMSE for $N = 1024$, $L = 8$, $\nu = 7$.

$\nu = M$). As can be observed, the MSE for CP-OFDM outperforms all other techniques. The MSE for ZP-OFDM however is very close to that of CP-OFDM: when $\nu \ll N$, the difference between channel estimation for CP-OFDM and ZP-OFDM will be very small. Further, it can be observed that the first method to estimate the channel in KSP-OFDM (KSP1), by adding pilots in the guard interval and on the carriers, has a much worse MSE performance than CP and ZP-OFDM, especially when the $SNR = \frac{E_s}{N_0}$ is increasing. Indeed, when SNR increases, the interference of the data symbols on the pilot symbols becomes more important, affecting the estimation of the channel. Also, the MSE for the second method for KSP-OFDM (KSP2) is shown. It follows from the figure that for increasing M, the MSE for KSP2 comes close to the ones for CP and ZP-OFDM, and is independent of the SNR. This can be explained as when the guard interval length increases, the relative importance of the interfering data symbols reduces.

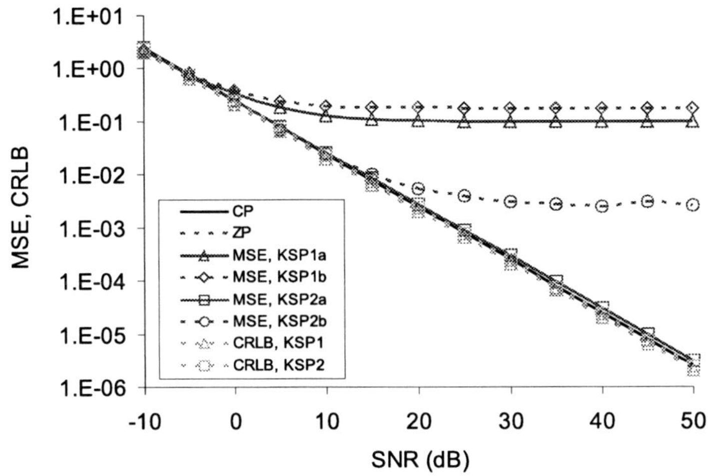

Figure 5. MSE and CRLB for $N = 1024$, $L = 8$, $\nu = 7$, $M = 40$.

In figure 5, the MSE and the CRLB are shown as function of the SNR for $N = 1024$, $L = 8$, $M = 40$ and $\nu = 7$ (except for KSP2, where $\nu = M$). It can be observed that the MSE for CP, OFDM and KSP2 almost coincide. The KSP1a technique performs worse than the three former techniques, especially at high SNR where the interference from the data symbols dominates: the MSE shows an error floor. Also in the KSP1b and KSP2b techniques, an error floor is present at high SNR: the estimates ignore the presence of the data symbols and at high SNR the data symbols are the dominant disturbance. The low SNR limit Cramer-Rao bounds for KSP1 and KSP2 are very close to each other and to the CRLBs of CP-OFDM and ZP-OFDM (which coincide with the MSE for these techniques).

4. Conclusions

In this paper, we have considered three guard interval techniques for OFDM systems, i.e. CP, ZP and KSP. First, we have compared the three techniques with respect to their SNR performance. It is shown that CP-OFDM and ZP-OFDM have the same SNR performance. The KSP-OFDM technique has a slightly worse performance, as it suffers from both power efficiency loss and noise enhancement, whereas the other two techniques suffer from only one of these effects.

Further, we have considered ML based channel estimation techniques for the three systems. We compared the MSE of the estimates with the corresponding Cramer-Rao bounds. It turns out that CP-OFDM channel estimation outperforms the other techniques. However, the difference between the MSE performance of the ZP-OFDM technique and that of CP-OFDM is marginally small when $N \gg \nu$: i.e. ZP-OFDM and CP-OFDM have virtually the same MSE performance. KSP-OFDM has worse MSE performance than the other two techniques; the proposed KSP2 technique, where the guard interval length is extended, outperforms the KSP1 technique, where pilot symbols are placed in the guard interval and on carriers.

References

[1] J. A. C. Bingham, "Multicarrier modulation for data transmission, an idea whose time has come," IEEE Comm. Mag., Vol. 31, pp. 514, May 1990.

[2] B. Muquet, Z. Wang, et. Al., "Cyclic Prefixing or Zero Padding for Wireless Multicarrier Transmissions?," IEEE Trans. on Comm., Vol. 50, no 12, Dec 2002, pp. 2136-2148.

[3] L. Deneire, B. Gyselinckx, M. Engels, "Training Sequence versus Cyclic Prefix – A New Look on Single Carrier Communication," IEEE Comm. Letters, Vol.5, no 7, Jul 2001, pp. 292-294.

[4] O. Rousseaux, G. Leus, M. Moonen, "Estimation and Equalization of Doubly Selective Channels Using Known Symbol Padding," IEEE Trans. on Signal Proc., Vol. 54, no 3, March 2006.

[5] H. Steendam, "Multicarrier Parameter Optimization in Doubly-Selective Fading Channels with Line-of-sight Components," 9th Int. Symp. on Spread Spectrum Techniques and Applications, ISSSTA2006, Manaus, Brazil, August 2006.

[6] R. Steele, Mobile Radio Communications. London, U.K.: Pentech, 1992.

[7] H. Steendam, M. Moeneclaey, 'Analysis and Optimization of the Performance of OFDM on Frequency-Selective Time-Selective Fading Channels', IEEE Trans. on Comm., Vol. 47, no 12, Dec 1999, pp. 1811-1819.

[8] H.L. Van Trees, Detection, Estimation and Modulation Theory, New York, Wiley, 1968.

[9] F. Tufvesson, T. Maseng, "Pilot Assisted Channel Estimation for OFDM in Mobile Cellular Systems," Proc. of IEEE Veh. Tech. Conf., VTC'97, Phoenix, U.S.A, May 4-7, 1997, pp. 1639-1643.

AN ANALYSIS OF OFDMA, PRECODED OFDMA AND SC-FDMA FOR THE UPLINK IN CELLULAR SYSTEMS

Cristina Ciochina[1,2], David Mottier[1] and Hikmet Sari[2]
[1]*Mitsubishi Electric ITE-TCL, 1 Allée de Beaulieu, 35708 Rennes Cedex 7, France*
[2]*Supélec, Plateau de Moulon, 1-3 Rue Joliot Curie, 91192 Gif sur Yvette, France*

Abstract: Orthogonal Frequency Division Multiple Access (OFDMA) appears today as a strong candidate in on-going standardization. Despite its attractive features, it has several drawbacks when employed in the uplink: The high peak-to-average power ratio (PAPR) inherited from Orthogonal Frequency Division Multiplexing (OFDM) and the inherent frequency diversity loss. The loss of frequency diversity can be combated by precoding. Variants of Precoded OFDMA include Spread Spectrum Multi-Carrier Multiple Access (SS-MC-MA) and the frequency-domain implementation of single-carrier (SC) FDMA called DFT-Spread OFDM. The latter transforms OFDMA into a SC transmission system, avoiding the PAPR problem. SC-FDMA with uniformly spaced carriers can be also generated by time-domain processing, a technique known as Interleaved Frequency Division Multiple Access (IFDMA). This paper analyzes OFDMA, several variants of Precoded OFDMA as well as SC-FDMA in its time- and frequency-domain implementations and compares them for uplink transmission. The spectrum and performance analysis confirm the benefits of SC-FDMA in terms of both high-power amplifier (HPA) output back-off and bit error rate (BER) performance on selective channels.

Key words: OFDMA; SC-FDMA; DFT-Spread OFDM; IFDMA; SS-MC-MA; Peak to Average Power Ratio (PAPR); HPA Back-off.

1. INTRODUCTION

Future broadband cellular systems should meet stringent requirements such as high data rates over dispersive channels, coexistence of different services, good coverage, robustness to interference, high flexibility and high

S. Plass et al. (eds.), Multi-Carrier Spread Spectrum 2007, 25–36.

performance. These requirements turn the design of such a system into a real challenge, especially for the uplink, where low-cost and low-complexity mobile terminals are demanded.

Although the principle of multi-carrier (MC) systems is not new, it is only in the past decade that this technique gained recognition and became a key component of many standards. Coded OFDM schemes are today used for terrestrial digital video broadcasting (DVB-T), digital audio broadcasting (DAB), wireless local area networks (IEEE 802.11a, ETSI Hiperlan2) and wireless metropolitan area networks (IEEE 802.16). Almost all current proposals for the air interface of Beyond Third Generation (B3G) and Fourth Generation (4G) cellular systems involve OFDM, OFDMA or one of its derivatives (e.g., MC code division multiple access MC-CDMA, SS-MC-MA). Nevertheless, MC systems suffer from one major problem: The high PAPR. This counterbalances the well-known advantages of MC techniques, particularly on the uplink of cellular systems, since the output power of user terminals is strictly limited and must be efficiently utilized in order to increase coverage. At this point, the debate on the choice between SC and MC systems is not closed. The SC transmission alleviates the PAPR problem. On the other hand, MC transmission opens the way to OFDMA[1], which significantly increases the cell range compared to a SC system or an OFDM system that uses time division multiple access. Indeed, FDMA concentrates the available transmit power in a fraction of the channel bandwidth, which improves the signal-to-noise ratio (SNR).

The considerations above lead to the conclusion that the multiple access technique best suited for the uplink is SC-FDMA, as it combines the low-PAPR characteristics of SC transmission with the advantages of FDMA. This can be precisely achieved by the IFDMA technique[2-4]. Its basic principle is the following: The input data stream is split into symbol blocks, each block is repeated a predetermined number of times and multiplied with a user specific phase ramp. This results in interleaving different users signals in the frequency domain without having to make any transformations between the frequency and the time domains. The 3GPP (Third Generation Partnership Project), which focuses on the Long Term Evolution (LTE) of UMTS (Universal Mobile Terrestrial Systems) radio access, has favored a frequency-domain implementation of SC-FDMA, which is actually a Precoded OFDMA scheme, where precoding is carried out by means of a DFT matrix. The major argument in favor of this implementation, also called DFT-Spread OFDM, is its flexibility in terms of sharing the spectrum between different users.

This paper provides an analysis of these different multiple access schemes. The paper is organized as follows: Section 2 gives the fundamentals of the multiple access techniques reviewed above. Section 3 presents the system model used in our simulations. Performance is evaluated and the results are discussed and compared in section 4. Finally, the conclusions are given in section 5.

2. MULTIPLE ACCESS TECHNIQUES FOR THE UPLINK

Fig. 1 presents the baseband structure of a general MC transmitter, which applies to all types of SC or MC modulation signals transmitted in blocks[5]. Data blocks of size M are precoded with the $[M\times M]$ matrix $\mathbf{P}$. The M-sized output vector is then mapped on M out of N inputs of the inverse DFT according to the subcarrier mapping $[N\times M]$ matrix $\mathbf{Q}$. To combat the effect of the frequency selective channel, a cyclic prefix of length CP is inserted in front of each N-sized block delivered by the inverse DFT. Transmission with different rates among users is available according to each user's requirement, as a different number of subcarriers and a different modulation and coding scheme can be assigned to each user. Let us denote by $s(n)$ the information symbols which are parsed into data blocks of size M. The i-th data block $\mathbf{s}^i$ can thus be written as:

$$\mathbf{s}^i = \left[s(iM),...,s(iM+M-1)\right]^T . \tag{1}$$

The index i will be omitted in the sequel. Let us denote by $\otimes$ the Kronecker product, by $\mathbf{0}_{M\times N}$ the all-zero matrix of size $[M\times N]$ and by $\mathbf{I}_M$ the $[M\times M]$ identity matrix. For clarity, we assume that the size N of the inverse DFT is a multiple of the block size, i.e., $N = MK$.

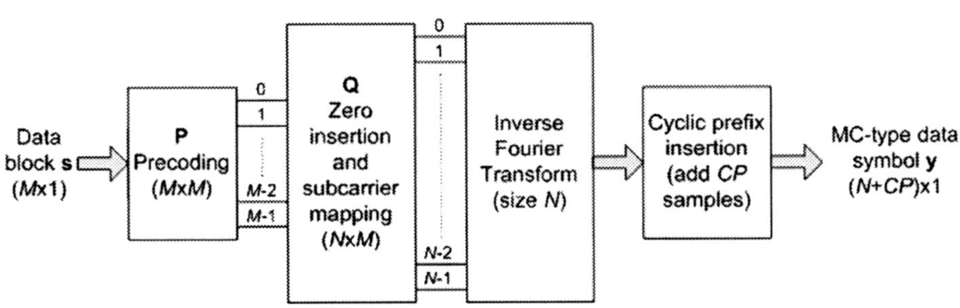

Figure 1. General MC transmitter.

2.1 OFDMA

The trivial case when **P** is the identity matrix, $\mathbf{P} = \mathbf{I}_M$, leads to OFDMA. The user-specific data block **s** is directly mapped onto a subset of M subcarriers, conveniently chosen by the user-specific subcarrier mapping matrix **Q**. The vector $\mathbf{Qs}$ is fed to the entries of the inverse DFT. The form of the matrix **Q** might lead to either a *localized* (Eq. 2) or a *distributed* (Eq. 3) subcarrier mapping:

$$\mathbf{Q}_{N\times M} = \begin{pmatrix} \mathbf{0}_{q\times M} \\ \mathbf{I}_M \\ \mathbf{0}_{(N-q-M)\times M} \end{pmatrix}, \tag{2}$$

$$\mathbf{Q}_{N\times M} = \mathbf{I}_M \otimes \begin{pmatrix} \mathbf{0}_{n\times 1} \\ 1 \\ \mathbf{0}_{(K-n-1)\times 1} \end{pmatrix}. \tag{3}$$

By assigning different groups of subcarriers to different users, each user's transmit power can be concentrated in a restricted part of the channel bandwidth, resulting in significant coverage increase. Different user signals remain orthogonal only if carrier synchronization is maintained and an appropriate cyclic prefix is appended to compensate for timing misalignment at the reception. In order to keep good performance on frequency-selective channels, efficient forward error correction must necessarily be employed.

2.2 Precoded OFDMA

Precoded OFDMA consists of using a precoding matrix **P** that spreads the energy of symbols over the subcarriers allocated to the user. Uniform energy distribution is favored in practice. One of the most well known precoding matrices is the Walsh-Hadamard (WH) matrix:

$$\mathbf{P} = \left[\mathbf{p}_0, \mathbf{p}_1, \ldots, \mathbf{p}_{M-1}\right]^T, \tag{4}$$

where the row vectors $\mathbf{p}_i$, $i = 0 \ldots M-1$, are orthogonal WH sequences of length M. This type of Precoded OFDMA was coined SS-MC-MA[6,7]. The precoding operation $\mathbf{Ps}$ consists in spreading the data symbols by multiplication with orthogonal WH sequences and superimposing them on the same set of subcarriers according to matrix **Q**. Another precoding matrix that spreads the symbol energy uniformly is the DFT matrix. We will discuss this precoding in the following section. Precoded OFDMA conserves the advantages of OFDMA in terms of cell range extension and spectrum spreading, which is expected to provide some robustness against cellular interference. With respect to MC-CDMA, it loses some frequency diversity, but this loss can be compensated by frequency interleaving or frequency hopping techniques[8].

The well-known advantages of MC systems are sometimes counter-balanced by their high PAPR. If we want to avoid nonlinear effects, the input signal must lie in the linear region of the HPA. In order to avoid the use of extremely high back-offs and costly amplifiers, occasional clipping and/or soft thresholding must be allowed. This leads to in-band distortion (which degrades the bit error rate) and to spectral widening that increases adjacent channel interference. Many PAPR reduction algorithms have been developed in order to alleviate this problem. Unfortunately, they do not always yield significant performance gains in practical applications[9].

2.3 SC-FDMA

As an alternative to MC-FDMA, SC-FDMA schemes have been envisioned, since a single-carrier system with an OFDMA-like multiple access would combine the advantages of the two techniques: low PAPR and high coverage. The first SC-FDMA concept[2] was IFDMA, which is based on compression and block repetition in the time domain of the modulated signal. As theoretically proven[4], this manipulation has a direct interpretation in the frequency domain. The spectrum of the compressed and K-times repeated signal has the same shape as the original signal, with the difference that it presents exactly K-1 zeros between two data subcarriers. This feature enables us to easily interleave different users in the frequency domain by simply applying to each user a specific frequency shift, or equivalently, by multiplying the time-domain sequence by a specific phase ramp. Besides, as for OFDMA, robustness to cellular interference can be achieved by coordinating resource allocation between adjacent cells.

The same waveform can be obtained in the frequency domain. Indeed, by using in Fig. 1 the discrete Fourier matrix:

$$\mathbf{P} = \left[p_{k,n} \right], \; p_{k,n} = e^{j2\pi \frac{kn}{M}} \tag{5}$$

as precoding matrix, we obtain a DFT-Precoded OFDMA, which is mathematically identical to IFDMA in a distributed scenario. The precoding operation $\mathbf{Ps}$ is equivalent to an M-point DFT transform. With a mapping matrix $\mathbf{Q}$ as given in Eq. (3), the spectrum of the distributed DFT-Precoded

OFDMA signal is identical to the IFDMA signal spectrum, and thus it corresponds to the same waveform. This is also called DFT-Spread OFDM. The two techniques are just different implementations of SC-FDMA. The advantage of DFT-Precoded OFDMA is its more flexible structure: While IFDMA imposes a distributed signal structure, DFT-Precoded OFDMA allows us to choose the mapping matrix **Q** as desired. Localized versions of implementation or channel-dependent mappings are possible. Also, pulse shape filtering can eventually be performed in the frequency-domain, with a lower complexity than the time-domain filtering. Note that in case of a frequency selective channel, interference may occur within the M elements of each data block. This degradation, which is more important in a distributed subcarrier mapping, impacts WH-Precoded OFDMA as well.

3. SYSTEM MODEL

In what follows, WH-Precoded OFDMA will simply be referred to as Precoded OFDMA. The simulated system model employs OFDMA, Precoded OFDMA and SC-FDMA transmission. We use a signal with $N = 512$ subcarriers, among which 300 are data carriers, split into 25 resource units of 12 subcarriers. 1 DC is added in the case of OFDMA transmission, and the remaining 211 (OFDMA) or 212 (Precoded OFDMA, SC-FDMA) are guard carriers. With these parameters, the sampling frequency corresponding to a 5 MHz channel is 7.68 MHz. The signal constellation is 16QAM with Gray mapping. We employ a $(753,531)_8$ convolutional code with rate 1/2, 3/4 or 5/6. The codes of rate 3/4 and 5/6 are obtained by puncturing the 1/2 code. The data is scrambled before coding and interleaved prior to QAM mapping. Groups of 7 OFDMA-type symbols are encoded together in order to take advantage of the channel diversity. We used soft Viterbi decoding. Frequency-domain equalization is performed when necessary.

The HPA is Rapp's solid state power amplifier model[10]:

$$v_{OUT} = \frac{v_{IN}}{\left(1 + \left(|v_{IN}| / v_{SAT}\right)^{2p}\right)^{\frac{1}{2p}}}, \tag{6}$$

where v_{IN}, v_{OUT} are respectively the complex input and complex output signals (baseband equivalent, normalized) and v_{SAT} corresponds to the output saturation level normalized to unity, $P_{SAT} = |v_{SAT}|^2$. We consider a Rapp model HPA with knee factor $p = 2$, since it is reported[11] to be a good representation of typical HPAs in the sub-10GHz frequency range. We also define the input back-off (IBO) and output back-off (OBO) with respect to the saturation values:

$$IBO\big|_{dB} = -10\log_{10}\frac{E\left\{ \left|v_{IN}(t)\right|^2 \right\}}{P_{SAT,IN}}, \tag{7}$$

$$OBO\big|_{dB} = -10\log_{10}\frac{E\left\{ \left|v_{OUT}(t)\right|^2 \right\}}{P_{SAT,OUT}}. \tag{8}$$

4. IMPACT OF NONLINEARITIES AND PERFORMANCE CONSIDERATIONS

In this section, we compare the performance of OFDMA, Precoded OFDMA and SC-FDMA. We separately evaluate the impact of nonlinearities on these multiple access techniques and study their behavior on fading channels.

4.1 Signal Envelope Variations, Output back-off (OBO) and Total Degradation

In order to illustrate the PAPR performance, we evaluate the Complementary Cumulative Distribution Function (CCDF) of the instantaneous normalized power (INP), which is defined as:

$$\text{CCDF}(INP(v_{IN})) = \Pr\left(\frac{\left|y_i\right|^2}{P_{avg,IN}} > \gamma^2\right), \tag{9}$$

where v_{IN} is the signal present at the input of the HPA, y_i denotes its samples and $P_{avg,IN} = E\left\{ \left|v_{IN}(t)\right|^2 \right\}$ is its average power. The CCDF of INP[9] is a more relevant performance criterion than the widely used CCDF of PAPR: It takes into account all of the signal samples that are susceptible of causing degradation when passing through the HPA, and not only the highest peak of each OFDM symbol. The signal is oversampled by a factor of 4. Fig. 2 presents the CCDF of INP for the three transmission schemes with $M = 24$ distributed subcarriers, 16QAM signal mapping and 3/4 coding rate. The SC properties of SC-FDMA results in better PAPR performance: 2 dB better than OFDMA at a clipping probability per sample of 10^{-4}. OFDMA and Precoded OFDMA have similar PAPR. It has to be noted that the performance gain predicted by the CCDF of INP can be directly interpreted in terms of (input) back-off difference only in the case of an ideal (clipper-type) HPA[9]. With a realistic HPA, a back-off difference given by the CCDF curves should only be perceived as an upper bound of performance gain.

For practical system design, the main relevant evaluation criterion is the necessary amount of OBO that is needed to reach some performance, e.g., BER=10^{-4}, while complying with the spectrum mask requirements and the out-of-band (OOB) radiation limits. In order to evaluate OOB radiation, the adjacent channel leakage ratio (ACLR) is defined as the ratio between the in-band signal power and the power radiated in the adjacent band. Fig. 3 presents the spectrum of a SC-FDMA signal complying with the UMTS uplink 5 MHz spectrum mask[12] and an ACLR restriction of at least 33 dB. In order to comply with the spectrum mask, current regulations require the verification of the case when all resource units are allocated to the same user which is a rather unlikely scenario in practice. A verification of typical allocation scenarios would be preferable, but no such scenarios were defined

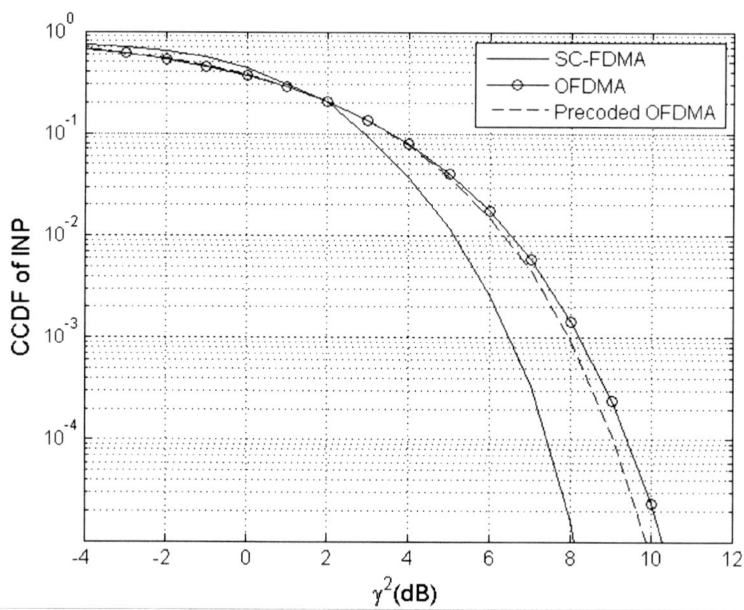

Figure 2. CCDF of INP for SC-FDMA, OFDMA and Precoded OFDMA with 16QAM 3/4, 24 distributed subcarriers, oversampling by a factor of 4.

so far. Here, actual specifications were treated (300 occupied subcarriers); an OBO of 5.76 dB is necessary. From Fig. 2, the back-off advantage of SC-FDMA over OFDMA is at most 1.6 dB (in order to maintain the clipping probability of $7 \cdot 10^{-3}$ corresponding to 5.3 dB of IBO in the SC-FDMA case). Indeed, by reproducing the results in Fig. 3 for OFDMA, we find that 7.17 dB of OBO (6.75 dB of IBO) are necessary in order to maintain the same constraints. This leads to a relative gain of 1.4 dB. The results obtained for Precoded OFDMA are similar to those of OFDMA.

Nonlinear effects also cause BER degradation. In order to approximate the real performance of a practical system, an estimation of the total degradation ($OBO + E_b / N_0$ loss) must be performed. Fig. 4 shows the total degradation at the BER target of 10^{-4}. BER simulations take into account an AWGN channel in order to evaluate the impact of nonlinearities separately from the channel fading effects. We can notice that OFDMA and Precoded OFDMA have similar behavior. SC-FDMA can achieve a minimum total degradation that is 1.2 dB lower than that of OFDMA. However, this result cannot be directly interpreted, as the operating point of the HPA is imposed by three conditions: Spectral requirements (mask, ACLR), BER degradation and amplifier type. For SC-FDMA, the minimum total degradation is attained when working with an OBO of 4 dB, which corresponds to an E_b/N_0

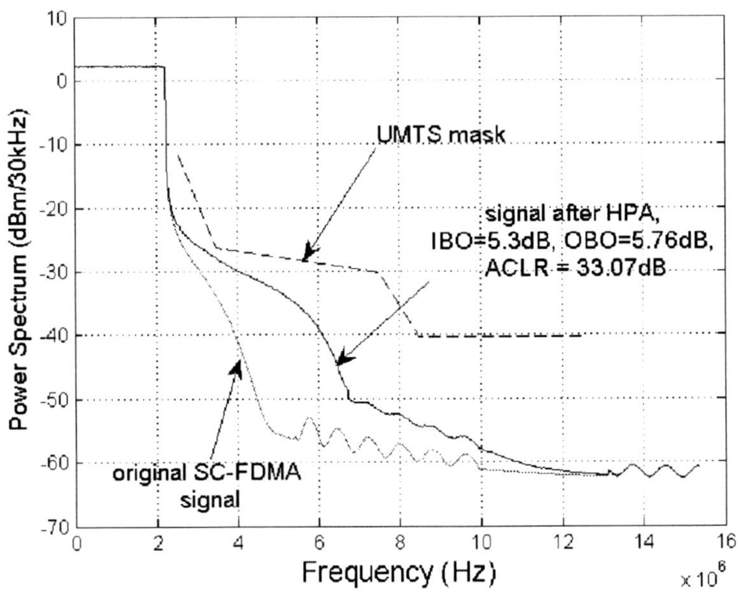

Figure 3. SC-FDMA spectrum, 300 occupied subcarriers, complying with UMTS spectrum mask and ACLR constraints[13]

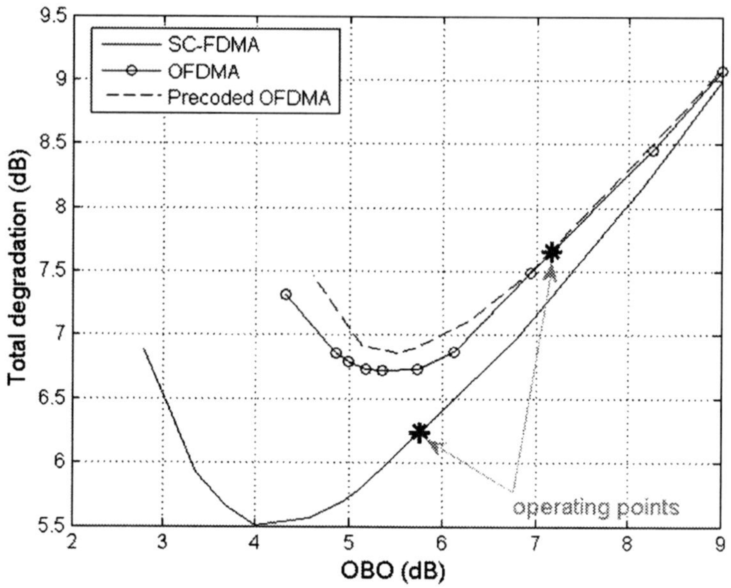

Figure 4. Total degradation vs. OBO for SC-FDMA, OFDMA and Precoded OFDMA, 16QAM 3/4, 24 distributed subcarriers, AWGN channel, target BER = 10^{-4}.

loss of 1.6 dB. But in practice (see Fig. 3,) an OBO of 5.76 dB is required in order to comply with the spectral constraints. Therefore, the performances of the three schemes have to be compared by considering their respective operating points (marked on Fig. 4,) given by the necessary OBO values to fit with the spectrum constraints. This corresponds to a total degradation of 6.25 dB for SC-FDMA and of 7.65 dB for its two counterparts, thus a gain of 1.4 dB in the favor of SC-FDMA.

4.2 BER Performance on Frequency Selective Channels

Fig. 5 presents the BER performance in the case of a transmission over a frequency selective COST 259 channel[13] in the absence of HPA. SC-FDMA and Precoded OFDMA have similar BER performance because they both recover the diversity of the system thanks to the spreading component. OFDMA performance is very dependent on the coding rate. When low coding rate (5/6) is employed, OFDMA performs poorly because coding does not manage to compensate the influence of carriers with a low SNR. When stronger coding is present (1/2), OFDMA benefits from the coding diversity and thus it recovers the difference and even outperforms SC-FDMA / Precoded OFDMA transmissions, whose performances are limited by inter-symbol interference.

5. CONCLUSIONS

In this paper, we have reviewed the fundamentals of three multiple access techniques which are potential candidates for the uplink of future cellular communications systems: OFDMA, WH-Precoded OFDMA and SC-FDMA. Precoded OFDMA combines the advantages of spectrum spreading with OFDMA-type multiple access. SC-FDMA is based on SC transmission and FDMA-type multiple access. The second technique can be implemented in the time domain (IFDMA) or in the frequency domain (DFT-Precoded OFDMA, also called DFT-Spread OFDM). The two implementations are completely equivalent in the distributed case. DFT-Precoded OFDMA has the advantage of the flexible subcarrier allocation.

A pertinent analysis of realistic HPA's impact on performance was conducted taking into account the constraints of practical systems (spectrum mask, ACLR). BER performance on frequency selective channels was also

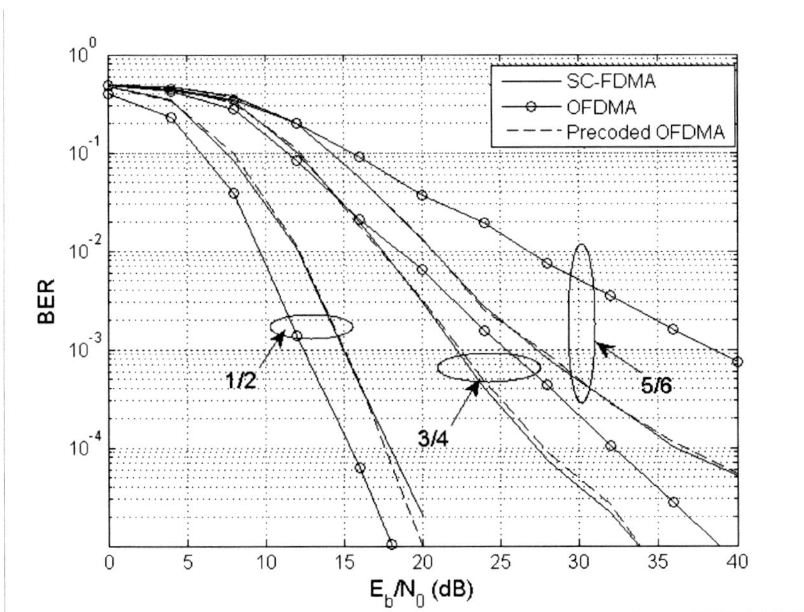

Figure 5. BER performance of OFDMA, Precoded OFDMA and SC-FDMA with 16QAM and 24 distributed subcarriers.

investigated. With 16QAM over distributed sub-carriers, OFDMA was found to have good BER performance only in the presence of robust channel coding. It is outperformed by both Precoded OFDMA and SC-FDMA when less powerful coding is employed. Precoded OFDMA has similar performance to SC-FDMA on frequency selective channels, since their spreading properties recover frequency diversity. On the other hand, it inherits the high PAPR disadvantages of OFDMA. Thanks to its single-carrier signal structure, SC-FDMA has the advantage of lower PAPR and thus lower back-off requirements than OFDMA and Precoded OFDMA.

ACKNOWLEDGEMENT

The present work was carried out in the framework of the CELTIC project WISQUAS[14] (CP2-035).

REFERENCES

[1] H. Sari and G. Karam, "Orthogonal frequency-division multiple access and its application to CATV networks," *European Transactions on Telecommunications (ETT)*, vol. 9, no. 6, pp. 507-516, Nov. – Dec. 1998.

[2] U. Sorger, I. De Broeck and M. Schnell, "Interleaved FDMA – A new spread-spectrum multiple-access scheme," *ICC'98*, Atlanta, Georgia, USA, June 7-11 1998.

[3] M. Schnell and I. De Broeck, "Application of IFDMA to Mobile Radio Transmission," *ICUPC'98*, Florence, Italy, Oct. 1998.

[4] T. Frank, A. Klein, E. Costa and E. Schultz, "Robustness of IDFMA as Air Interface Candidate for Future High Rate Mobile Radio Systems," *Advances in Radio Science*, vol. 3, pp. 265-270, 2005.

[5] Wireless World Research Forum, "*Technologies for the Wireless Future*," vol. 2, John Wiley & Sons, Ltd, 2006.

[6] S. Kaiser and K. Fazel, "A flexible spread-spectrum multi-carrier multiple-access system for multi-media applications," *PIMRC'97*, Helsinki, Finland, Sept. 1997.

[7] S. Kaiser and W.A. Krzymien, "An asynchronous spread spectrum multi-carrier multiple access system," *GLOBECOM'99*, vol. 1, pp. 314-319, Dec. 1999.

[8] N. Chapalain, D. Mottier and D. Castelain, "Performance of Uplink SS-MC-MA Systems with Frequency Hopping and Channel Estimation Based on Spread Pilots," *PIMRC'05*, Berlin, Germany, Sept. 2005.

[9] C. Ciochina, F. Buda and H. Sari, "An Analysis of OFDM Peak Power Reduction Techniques for WiMAX Systems," *ICC'06*, Istanbul, Turkey, June 2006.

[10] C. Rapp, "Effects of the HPA-nonlinearity on a 4-DPSK/OFDM signal for a digital sound broadcasting system," *Tech. Conf. ECSC'91*, Luettich, Oct. 1991.

[11] T. Kaitz, "Channel and interference model for 802.16b Physical Layer," contribution to the IEEE 802.16b standard, 2001.

[12] 3GPP TS 25.101 V6.11.0: "User Equipment (UE) radio transmission and reception (FDD)," Mar. 2006.

[13] Luis Correia, "Wireless Flexible Personalized Communications: Cost 259: European Co-Operation in Mobile Radio Research," John Wiley & Sons Jan. 2001.

[14] CELTIC project WISQUAS website: www.celtic-initiative.org/projects/wisquas.

INFLUENCE OF A CYCLIC PREFIX ON THE SPECTRAL POWER DENSITY OF CYCLO-STATIONARY RANDOM SEQUENCES

Michel T. Ivrlač and Josef A. Nossek
Munich University of Technology
Institute for Circuit Theory and Signal Processing
Arcisstr. 21, D-80333 Munich, Germany
ivrlac@tum.de, josef.a.nossek@tum.de

Abstract The influence of a cyclic prefix on the spectral power density of discrete-time, zero-mean, cyclo-stationary, random sequences is investigated. Analytical expressions for the time-varying autocorrelation function, the average autocorrelation function, and the spectral power density of the cyclic-prefixed sequence are provided. Application to discrete Fourier transform based orthogonal frequency division multiplexing is presented, and shown that the cyclic prefix introduces a strong ripple in the spectral power density within the main frequency band. Besides the loss of spectral efficiency caused by the cyclic prefix, this ripple may require reduction of transmit power in order to obey regulatory spectrum masks.

1. Introduction

Following the introduction of multi-carrier modulation (MCM) in the late fifties of the twentieth century [1], its famous variant, the so-called orthogonal frequency division multiplexing (OFDM) [2], is now widely recognized for high data rate, high bandwidth communications. It has been selected as the air-interface for a large number of communication applications, including Wireless LAN, WiMAX, and DVB-T [3], and considered a prominent candidate for the downlink of future fourth-generation cellular communication systems [4].

Current OFDM communication systems are implemented in discrete-time and based on the discrete Fourier transform (DFT). In addition to the payload data, a special guard signal, the so-called cyclic prefix is transmitted [5]. If the cyclic prefix is at least as long as the discrete-time channel impulse response, a time-invariant multi-path channel is effectively transformed into a periodically time-varying additive white Gaussian noise (AWGN) channel [6]. In this way, inter-symbol interference (ISI) due to the multi-path propagation is completely

S. Plass et al. (eds.), Multi-Carrier Spread Spectrum 2007, 37–46.

avoided, without the need of channel equalization. However, this elegant way of avoiding ISI comes, of course, at the price of reduced bandwidth efficiency, since the cyclic prefix adds redundancy to the signal.

In this paper, we look at another loss in performance, besides the loss in bandwidth efficiency, which the cyclic prefix is also responsible for. The cyclic prefix affects the spectral power density of the transmitted signal, such that a ripple is introduced inside the main frequency band. This ripple may require a reduction of the transmit power in order to obey regulatory spectrum masks, that is, the transmit power cannot be made as large as it potentially could, because of the "overshoot" in the power density. Because of these reasons, multi-carrier systems which do not need a cyclic prefix, may be advantages to DFT-based OFDM [7].

In the following, we will focus on the mentioned influence of a cyclic prefix on the spectral power density of a discrete-time, zero-mean, cyclo-stationary, random sequence. Hereby, we develop closed-form expressions for the time-varying autocorrelation function, the average autocorrelation function, and the spectral power density of the cyclic-prefixed sequence and compare to the sequence without cyclic prefix.

2. Signal Model

Let us consider a random sequence $(\cdots,\ s_{-1},\ s_0,\ s_1,\ \cdots)$ of complex numbers $s_i \in \mathbb{C}$ and partition it into consecutive blocks of size M:

$$\boldsymbol{s}[k] = \begin{bmatrix} s_{kM} & s_{kM+1} & \cdots & s_{kM+M-1} \end{bmatrix}^{\mathrm{T}} \in \mathbb{C}^{M\times 1}\,, \tag{1}$$

where $k \in \mathbb{Z}$ is the block index, and $(.)^{\mathrm{T}}$ denotes transposition. We assume:

$$\mathrm{E}\left[\boldsymbol{s}[k]\right] = \boldsymbol{0}\,, \tag{2}$$

$$\mathrm{E}\left[\boldsymbol{s}[k]\boldsymbol{s}^{\mathrm{H}}[k']\right] = \boldsymbol{R_s}\cdot\delta_{k,k'}\,, \tag{3}$$

that is, the sequence has zero mean, different blocks are mutually uncorrelated, and the covariance matrix $\boldsymbol{R_s} \in \mathbb{C}^{M\times M}$ is the same for all blocks. Here, $(.)^{\mathrm{H}}$ denotes complex conjugate transpose, $\mathrm{E}[.]$ is the expectation operation, while $\delta_{k,k'}$ equals unity for $k = k'$ and is zero else. This sequence is now processed by a linear block transformation such that a new sequence $\boldsymbol{x}[k]$ of block-size N is generated:

$$\boldsymbol{x}[k] = \boldsymbol{F}\boldsymbol{s}[k] \in \mathbb{C}^{N\times 1}\,, \tag{4}$$

where $\boldsymbol{F} \in \mathbb{C}^{N\times M}$, with $N \geq M > 0$. In OFDM, the columns of $\boldsymbol{F}$ are discrete Fourier transformation vectors, and N can be interpreted as the number of subcarriers, while M is the number of used subcarriers. That is, there are $(N - M) \geq 0$ zero-carriers.

The blocks $\boldsymbol{x}[k] \in \mathbb{C}^{N\times 1}$ are now prepended with a cyclic prefix of size C, such that a sequence

$$\boldsymbol{y}[k] = \left[\boldsymbol{x}^{\mathrm{T}}_{N-C+1:N}[k] \;\; \boldsymbol{x}^{\mathrm{T}}[k]\right]^{\mathrm{T}} \in \mathbb{C}^{(N+C)\times 1}\,, \tag{5}$$

results. Herein

$$\boldsymbol{x}^{\mathrm{T}}_{N-C+1:N}[k] = \begin{bmatrix} \mathbf{e}^{\mathrm{T}}_{N-C+1}\boldsymbol{x}[k] \\ \mathbf{e}^{\mathrm{T}}_{N-C+2}\boldsymbol{x}[k] \\ \vdots \\ \mathbf{e}^{\mathrm{T}}_{N}\boldsymbol{x}[k] \end{bmatrix}^{\mathrm{T}} \in \mathbb{C}^{1\times C} \tag{6}$$

contains the last C elements of $\boldsymbol{x}[k]$, while $\mathbf{e}_i$ is the i-th unit vector. By reading out all elements of all vectors $\boldsymbol{y}[k]$, the discrete-time sequence

$$y[n] = \boldsymbol{e}^{\mathrm{T}}_{n+1-k\cdot(N+C)}\boldsymbol{y}[k] \tag{7}$$

is obtained, where

$$k = \mathrm{floor}\left[\frac{n}{N+C}\right]\,, \tag{8}$$

and $n \in \mathbb{Z}$ is the time-index. The function floor(.) returns the largest integer which is smaller than or equal to its argument. The discrete-time sequence $y[n]$ is now mapped onto a continuous-time, finite-energy, pulse shape $g(t)$, such that a continuous-time signal

$$y(t) = T\sum_{n=-\infty}^{\infty} y[n]\cdot g(t-nT)\,, \tag{9}$$

is generated. Herein, $t \in \mathbb{R}\cdot\mathrm{sec}$ denotes time, while $T \in \mathbb{R}\cdot\mathrm{sec}$ is the time between which two pulses are generated, i.e. $1/T$ is the symbol rate. The expected energy in the pulse associated with $y[n]$ is given by

$$\mathcal{E}_y[n] = \mathrm{E}\left[|T\cdot y[n]|^2\right]\cdot\int_{-\infty}^{\infty}|g(t)|^2\,\mathrm{d}t\,. \tag{10}$$

With the normalization:

$$\int_{-\infty}^{\infty}|g(t)|^2\,\mathrm{d}t \;=\; \frac{1}{T}\,, \tag{11}$$

and the fact that $y[n]$ is cyclo-stationary with a period of $N+C$, we can write the average transmit power of the continuous-time sequence $y(t)$ as the average pulse energy divided by the symbol time T, where averaging is performed over one period of cyclo-stationarity, i.e. over $N + C$ consecutive symbols. Hence, we obtain the average transmit power given below:

$$\overline{P_{\mathrm{T}}} = \frac{1}{N+C} \sum_{n=0}^{N+C-1} \frac{\mathcal{E}_y[n]}{T} \tag{12}$$

$$= \frac{1}{N+C} \sum_{n=0}^{N+C-1} \mathrm{E}\left[|y[n]|^2\right] . \tag{13}$$

3. Autocorrelation Function Without Cyclic Prefix

The autocorrelation function of the sequence

$$x[n] = \boldsymbol{e}_{n+1-k'N}^{\mathrm{T}} \boldsymbol{x}\left[k'\right] , \quad \text{with } k' = \text{floor}\left[n/N\right] , \tag{14}$$

that is, the sequence *without* the cyclic prefix, is defined as

$$\tilde{\rho}[n,m] \stackrel{\text{def}}{=} \mathrm{E}[x[n] \cdot x^*[n+m]] , \tag{15}$$

where $n \in \mathbb{Z}$ is the time-index, while $m \in \mathbb{Z}$ denotes the temporal displacement. Herein, $(.)^*$ denotes complex conjugation. Because the matrix $\boldsymbol{F}$ in (4) is constant and the covariance matrix $\boldsymbol{R_s}$ is the same for all blocks (see (3)), the sequence $x[n]$ is cyclo-stationary with period N. Without loss of generality, we can therefore restrict the focus to the zero-th block , for which we find:

$$\tilde{\rho}[n,m] = \mathbf{e}_{n+1}^{\mathrm{T}} \boldsymbol{F} \boldsymbol{R_s} \boldsymbol{F}^{\mathrm{H}} \mathbf{e}_{n+m+1} , \tag{16}$$

where we need to restrict the range of n and m such that both n and $(n+m)$ remain inside the zero-th block:

$$\begin{aligned} n &\in \{0, 1, \ldots, N-1\} , \\ m &\in \{-n, -n+1, \ldots, -n+N-1\} . \end{aligned} \tag{17}$$

Finally, the average autocorrelation function is given by:

$$\tilde{\rho}[m] = \frac{1}{N} \sum_{n=0}^{N-1} \tilde{\rho}[n,m] . \tag{18}$$

4. Autocorrelation Function With Cyclic Prefix

The autocorrelation function of the sequence $y[n]$ from (7) is defined as

$$\rho[n,m] \stackrel{\text{def}}{=} \mathrm{E}[y[n] \cdot y^*[n+m]] . \tag{19}$$

Because $y[n]$ is cyclo-stationary with period $(N + C)$, the average autocorrelation function is given by

$$\rho[m] = \frac{1}{N+C} \sum_{n=0}^{N+C-1} \rho[n, m] \; . \tag{20}$$

In order to compute $\rho[n, m]$ from $\tilde{\rho}[n, m]$, it is helpful to look at the relationship between $x[n]$ and $y[n]$:

n	0	1	$\cdots$	$C-1$	C	$C+1$	$\cdots$	$N+C-1$
$y[n]$	$x[N-C]$	$x[N-C+1]$	$\cdots$	$x[N-1]$	$x[0]$	$x[1]$	$\cdots$	$x[N-1]$

which we can also write as:

$$y[n] = \begin{cases} x[N-C+n] & \text{for} \quad 0 \le n \le C-1 \, , \\ x[n-C] & \text{for} \quad C \le n \le N+C-1 \, . \end{cases} \tag{21}$$

We can distinguish the following four cases:

❶ Time $0 \le n \le C - 1$ and displacement $-n \le m \le C - 1 - n$

As $y[n] = x[n + N - C]$ and $y[n + m] = x[n + m + N - C]$, we have

$$\rho[n, m] = \tilde{\rho}[n + N - C \, , \, m] \; . \tag{22}$$

❷ Time $0 \le n \le C - 1$ and displacement $C - n \le m \le C - n + N - 1$

As $y[n] = x[n + N - C]$ and $y[n + m] = x[n + m - C]$, we have

$$\rho[n, m] = \tilde{\rho}[n + N - C \, , \, m - N] \; . \tag{23}$$

❸ Time $C \le n \le N + C - 1$ and displacement $-n \le m \le C - 1 - n$

As $y[n] = x[n - C]$ and $y[n + m] = x[n + m + N - C]$, we have

$$\rho[n, m] = \tilde{\rho}[n - C \, , \, m + N] \; . \tag{24}$$

❹ Time $C \le n \le N + C - 1$ and displacement $C - n \le m \le N + C - 1 - n$

As $y[n] = x[n - C]$ and $y[n + m] = x[n + m - C]$, we have

$$\rho[n, m] = \tilde{\rho}[N - C \, , \, m] \; . \tag{25}$$

Note that, for other values of the displacement m, which are not covered by the four cases above, the autocorrelation is zero, since the blocks are mutually uncorrelated (see (3)).

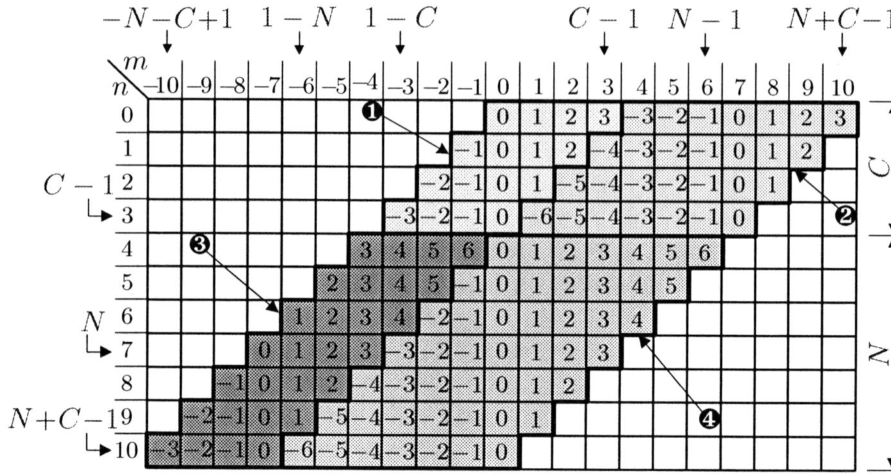

Figure 1. The entry $\beta_{n,m}$ in the table means that $\rho[n,m] = \tilde{\rho}[n_*, \beta_{n,m}]$, where n_* is defined in (26). Table entries without a number mean $\rho[n,m] = 0$.

In order to derive a closed form expression for the average autocorrelation function from (20), let us first look at an example case. Assuming $N = 7$ and $C = 4$, the resulting autocorrelation $\rho[n,m]$ is displayed in Figure 1. The numerical value $\beta_{n,m}$ in the n-th row and the m-th column of the table indicates that $\rho[n,m] = \tilde{\rho}[n_*, \beta_{n,m}]$, where

$$n_* = \begin{cases} n + N - C\,, & \text{for the cases ❶ and ❷}\,, \\ n - C\,, & \text{for the cases ❸ and ❹}\,. \end{cases} \tag{26}$$

For example, $\rho[3,2] = \tilde{\rho}[6,-5]$, while $\rho[7,-5] = \tilde{\rho}[3,2]$. The encircled numbers indicate the four ranges of (n,m) that are introduced above. By close observation of Figure 1, it is not difficult to compute the sums $\sum_n \rho[n,m]$ inside these ranges. The average autocorrelation function then turns out to be:

① For $m = 0$: (average transmit power)

$$\rho[0] = \frac{1}{N+C}\left(\sum_{n'=0}^{C-1} \tilde{\rho}[n' + N - C, 0] + \sum_{n'=C}^{N+C-1} \tilde{\rho}[n' - C, 0]\right), \tag{27}$$

② For $1 \le m \le C-1$:

$$\rho[m] = \frac{\sum\limits_{n'=C}^{N+C-1-m} \tilde{\rho}[n' - C, m]}{N+C} + \frac{\sum\limits_{n'=C-m}^{C-1} \tilde{\rho}[n' + N - C, m - N]}{N+C} + \frac{\sum\limits_{n'=0}^{C-1-m} \tilde{\rho}[n' + N - C, m]}{N+C}, \tag{28}$$

③ For $C \le m \le N-1$:

$$\rho[m] = \frac{1}{N+C}\left(\sum_{n'=C}^{N+C-1-m} \tilde{\rho}[n'-C,m] + \sum_{n'=0}^{C-1} \tilde{\rho}[n'+N-C,m-N]\right), \tag{29}$$

④ For $N \le m \le N+C-1$:

$$\rho[m] = \frac{1}{N+C}\sum_{n'=0}^{N+C-1-m} \tilde{\rho}[n'+N-C,m-N], \tag{30}$$

⑤ In general:

$$\rho[-m] = (\rho[m])^* . \tag{31}$$

For other values of the displacement m, which are not covered above, the average autocorrelation function is zero.

5. Spectral Power Density

The spectral power density can be computed from the average autocorrelation functions $\rho[m]$ and $\tilde{\rho}[m]$, for the signal with or without cyclic prefix, respectively as [8]:

① Spectral power density WITHOUT cyclic prefix

$$\widetilde{\Phi}(f) = T \cdot |G(f)|^2 \cdot \left(\tilde{\rho}[0] + 2\sum_{m=1}^{N-1} \mathrm{Re}\left\{\tilde{\rho}[m]\cdot \mathrm{e}^{\mathrm{j}2\pi f m T}\right\}\right). \tag{32}$$

② Spectral power density WITH cyclic prefix

$$\Phi(f) = T \cdot |G(f)|^2 \cdot \left(\rho[0] + 2\sum_{m=1}^{N+C-1} \mathrm{Re}\left\{\rho[m]\cdot \mathrm{e}^{\mathrm{j}2\pi f m T}\right\}\right), \tag{33}$$

$$= \widetilde{\Phi}(f)\cdot \frac{\rho[0] + 2\sum_{m=1}^{N+C-1} \mathrm{Re}\left\{\rho[m]\cdot \mathrm{e}^{\mathrm{j}2\pi f m T}\right\}}{\tilde{\rho}[0] + 2\sum_{m=1}^{N-1} \mathrm{Re}\left\{\tilde{\rho}[m]\cdot \mathrm{e}^{\mathrm{j}2\pi f m T}\right\}}. \tag{34}$$

Herein, $f \in \mathbb{R}\cdot\mathrm{Hz}$ denotes frequency, while $G(f)$ is the continuous Fourier transform of the pulse shape $g(t)$. Note that $G(f)$ is dimensionless, hence, the spectral power densities $\Phi(f)$ and $\widetilde{\Phi}(f)$ have dimension of energy, or equivalently, power per frequency. By substituting (18), (27), (28), (29), and (30) into (34), we have arrived at the closed-form solution for the problem of how the spectral power density is influenced by the inclusion of a cyclic prefix.

Because of $\mathrm{E}[|y[n]|^2] = \rho[n,0]$, we can observe from (27) and (13) that, in general, the average transmit power $\overline{P_\mathrm{T}} = \rho[0]$, is influenced by the cyclic prefix. However, if $\tilde{\rho}[n,m] = \tilde{\rho}[m]$ holds true inside a block, i.e. for block-wise (wide-sense) stationary $x[n]$, it follows from (27) that $\rho[0] = \tilde{\rho}[0]$. In this case, the average transmit power is not affected by the cyclic prefix. We will discuss this special case in more detail in the next Section.

It is also interesting to note that, in general, for given $|G(f)|$ and T, it is not possible to compute $\Phi(f)$ from $\widetilde{\Phi}(f)$. This comes about, since for given $|G(f)|$ and T, the spectral power density $\widetilde{\Phi}(f)$ only determines the *average* autocorrelation function $\tilde{\rho}[m]$, however, to compute $\rho[m]$, we need $\tilde{\rho}[n,m]$, instead. The only exception occurs when the sequence $x[n]$, that is, the sequence without the cyclic prefix, is stationary inside each block. Because in this case, $\tilde{\rho}[n,m] = \tilde{\rho}[m]$ holds true inside a block, we can indeed find $\Phi(f)$ from only knowing $\widetilde{\Phi}(f)$, $|G(f)|$ and T. This exceptional case will be discussed in the next Section.

6. OFDM with Stationary and Uncorrelated Data

Let as assume thatboth

① the matrix $\boldsymbol{F}$ from (4) is a (partial) DFT matrix, and

② $\boldsymbol{R_s} = \mathrm{const} \cdot \mathbf{I}_M$, i.e. the original data sequence $s[n]$, that is, the sequence before application of the (partial) DFT, is both *uncorrelated* and (wide-sense) *stationary*.

This corresponds to an OFDM system (possibly with zero-carriers) driven by an uncorrelated, stationary, data sequence $s[n]$. Let

$$\tilde{\boldsymbol{F}} = \frac{1}{\sqrt{N}} \cdot \begin{bmatrix} 1 & 1 & 1 & \dots & 1 \\ 1 & w & w^2 & \dots & w^{N-1} \\ 1 & w^2 & w^4 & \dots & w^{2(N-1)} \\ \vdots & \vdots & \vdots & \vdots & \\ 1 & w^{N-1} & w^{2(N-1)} & \dots & w^{(N-1)(N-1)} \end{bmatrix}, \tag{35}$$

with $w = \mathrm{e}^{\mathrm{j}\frac{2\pi}{N}}$, be a unitary DFT matrix, and

$$\boldsymbol{F} = \begin{bmatrix} \tilde{\boldsymbol{f}}_{p(1)} & \tilde{\boldsymbol{f}}_{p(2)} & \cdots & \tilde{\boldsymbol{f}}_{p(M)} \end{bmatrix} \in \mathbb{C}^{N \times M}, \tag{36}$$

where $\tilde{\boldsymbol{f}}_i$ is the i-th column vector of $\tilde{\boldsymbol{F}}$, and $p(.)$ is an index function with which we select M pair-wise different columns of $\tilde{\boldsymbol{F}}$ to obtain the partial DFT matrix $\boldsymbol{F}$. Hence, the remaining $(N - M)$ columns of $\tilde{\boldsymbol{F}}$ correspond to zero-carriers. Note that, $\mathrm{rank}(\boldsymbol{F}) = M$. By setting $\boldsymbol{R_s} = (N/M) \cdot \mathbf{I}_M$, we obtain from evaluating (16) the following autocorrelation function:

$$\tilde{\rho}[n,m] = \frac{1}{M}\sum_{i=1}^{M} w^{-m\cdot(p(i)-1)}, \quad \text{for} \quad |m| \le N-1. \tag{37}$$

As we can observe from (37), the autocorrelation function $\tilde{\rho}[n,m]$ does *not* depend on the time-index n, hence $\tilde{\rho}[m] = \tilde{\rho}[n,m]$, inside a block (of size N). Furthermore, $\tilde{\rho}[0] = 1$, independent of the particular value of $M > 0$. Because of $\tilde{\rho}[n,m] = \tilde{\rho}[m]$, the expression for $\rho[m]$ from Section 4 simplifies considerably. It turns out that:

$$\rho[m] = \frac{1}{N+C}\cdot\begin{cases} \tilde{\rho}[0]\cdot(N+C), & m=0 \\ \tilde{\rho}[m-N]\cdot m + \tilde{\rho}[m]\cdot(N+C-2m), & 1\le m\le C-1 \\ \tilde{\rho}[m-N]\cdot C+\tilde{\rho}[m]\cdot(N-m), & C\le m\le N \\ \tilde{\rho}[m-N]\cdot(N+C-m), & N+1\le m\le N+C-1 \\ 0 & |m|\ge N+C \\ (\rho[-m])^* & \forall m \end{cases} \tag{38}$$

As an example, let us have a look at a DFT-based OFDM system with $N = 64$ subcarriers and duration of one OFDM symbol of 3.2×10^{-6} seconds, which translates to $T = 5 \times 10^{-8}$ seconds. A root-raised-cosine pulse shape [8] with Nyquist bandwidth of $1/T = 20$ MHz, and a roll-off factor $\alpha = 0.19$ is used. The transmit power is set to $P = 1$W. In order to improve the sharpness of transition between the used band and out-of-band, not all subcarriers, but only a number $M = 52$ are used to transmit data. The remaining 12 subcarriers are constantly set to zero, namely the DC-carrier and the 11 subcarriers with highest frequencies. The $M = 52$ signals are zero-mean, mutually uncorrelated random variables with variance $64/52$. Due to the zero-carriers, the signal after DFT, i.e. $x[n]$, is correlated, but it remains block-wise stationary (see (37)).

The resulting spectral power density with and without a cyclic prefix of size $C = 16$, is shown in logarithmic scaling in Figure 2. Notice that the cyclic prefix introduces a strong ripple of about 3.7dB in the spectral power density $\Phi(f)$ inside the main frequency band. Notice also the "overshoot" in spectral power density of about 1.5dB compared to the case without cyclic prefix. In this way, the transmit power of the signal with cyclic prefix has to be reduced by 1.5dB in order to have the same *peak* spectral power density as the signal without cyclic prefix. Therefore, the inclusion of the cyclic prefix not only lowers the bandwidth efficiency, but may also lead to a loss in the signal to noise ratio at the receiver, since the transmit power may have to be reduced to stay inside regulatory spectrum masks.

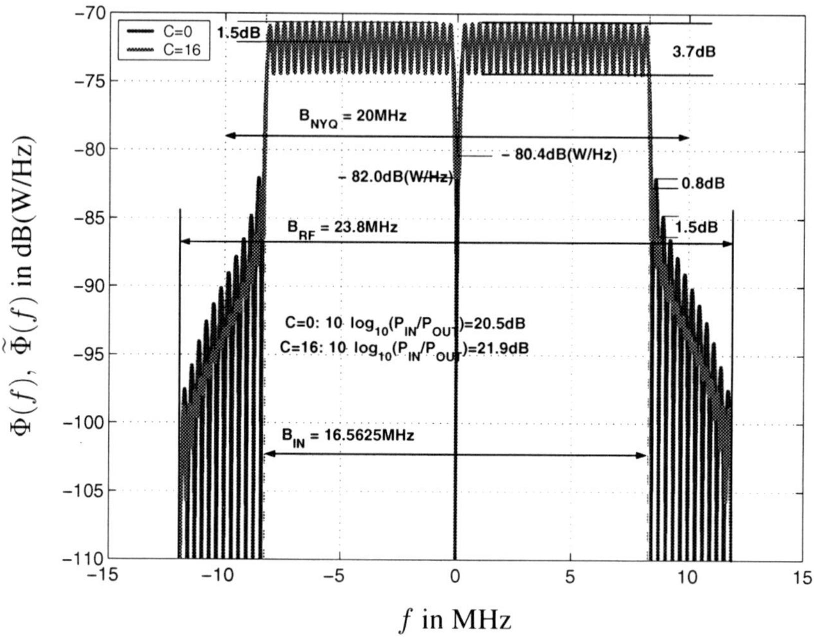

Figure 2. Example spectral power density for zero-carrier OFDM driven by an uncorrelated, stationary sequence. Both the result with and without cyclic prefix is shown.

References

[1] R. R. Mosier and R. G. Clabaugh, "Kineplex a Bandwidth-Efficient Binary Transmission System", *AIEE Transactions, Part I: Communications and Electronics*, Vol. 76, pp. 723-728, January, 1958.

[2] R. W. Chang, "High Speed Multi-channel Data Transmission with Band-limited Orthogonal Signals", *Bell Systems Technical Journal*, Vol. 45, pp. 1775-1796, December, 1966.

[3] A. N. Akansu, P. Duhamel, X. Lin, and M. de Courville, "Orthogonal Transmultiplexers in Communication: A Review", *IEEE Transactions on Signal Processing*, Vol. 46, no. 4, pp. 979-995, April 1998.

[4] A. Bria, F. Gessler, O. Queseth, R. Stridh, M. Unbehaun, W. Jiang, J. Zander, and M. Flament, "4th-generation Wireless Infrastructures: Scenarios and Research Challenges", *IEEE Personal Communications*, Vol. 8, no. 6, pp. 25-31, December, 2001.

[5] S. B. Weinstein, and P. M. Ebert, "Data Transmission by Frequency Division Multiplexing Using the Discrete Fourier Transform", *IEEE Transactions on Communications*, Vol 19, no. 5, pp. 628-34, October, 1971.

[6] M. T. Ivrlač, *Wireless MIMO Systems – Models, Performance, Optimization*, Shaker Verlag, 2005.

[7] D. S. Waldhauser, L. G. Baltar, and J. A. Nossek. "Comparison of Filter Bank based Multicarrier Systems with OFDM", *Proc. IEEE Asia Pacific Conference on Circuits and Systems*, APCCAS 2006, pages 1978-1981, 04-07 December, 2006.

[8] J. G. Proakis, *Digital Communications*, 3rd Edition, McGraw-Hill, 1995.

DOWNLINK SCHEDULING FOR MULTIPLE ANTENNA MULTI-CARRIER SYSTEMS WITH DIRTY PAPER CODING VIA GENETIC ALGORITHMS

Robert C. Elliott and Witold A. Krzymień
University of Alberta / TRLabs
Edmonton, Alberta, Canada
{rce,wak}@ece.ualberta.ca

Abstract: MIMO systems are of interest to meet the expected demand for higher data rates and lower delays in future wireless systems. In addition, multi-carrier systems are of interest to combat frequency-selective fading experienced over the larger bandwidth these future broadband systems will use. The introduction of these spatial and frequency resources adds extra dimensions and complexity for any scheduling algorithm in a multiuser system. In this paper, we investigate scheduling through utility functions implemented via genetic algorithms in order to reduce complexity. This paper extends our prior work for a single-carrier MIMO system using dirty paper coding to the multi-carrier case. Here, the users that are scheduled, the order they are encoded in, and the subcarrier frequencies they are assigned to will all affect the performance of the scheduling algorithm. We demonstrate that the genetic algorithm is still able to achieve a similar near-optimal performance relative to an exhaustive search with the same relative reduction in complexity. Additionally, with the use of OFDM, an increase in capacity is seen relative to the single-carrier case.

Key words: Scheduling; genetic algorithms; dirty paper coding; multiuser MIMO; multi-carrier systems; OFDM.

1. INTRODUCTION

In order to meet the expected demands of future wireless systems for higher data rates and lower latency, multiple-input multiple-output (MIMO) systems are increasingly of interest in the design of future wireless systems. It is well-known

S. Plass et al. (eds.), Multi-Carrier Spread Spectrum 2007, 47–56.

that the capacity of a single-user MIMO system scales linearly with the minimum of the number of transmit and receive antennas in a sufficiently rich scattering environment and at a sufficiently high signal-to-noise ratio (SNR) [1,2]. In a multiuser system, the concept of multiuser diversity can be exploited via a suitable scheduling algorithm in order to obtain further gains in capacity.

In a MIMO system, the multiuser broadcast channel (BC) capacity is achieved through a process known as dirty-paper coding (DPC) [3], wherein the transmitter performs a sort of successive interference pre-cancellation. The signal of user k is encoded such that the signals of the k–1 users previously encoded do not interfere. While the actual algorithm of DPC is currently infeasible to implement, it can be closely approximated by methods such as vector quantization or trellis preshaping [4]. From the point of view of scheduling, it is important to note that capacity is achieved by transmitting to several users simultaneously, as opposed to transmission to a single user in a single-antenna system [5]. In fact, to obtain most of the DPC BC capacity, it is sufficient to transmit to at most the same number of users as there are transmit antennas; transmitting to additional users will not significantly increase the throughput [6].

Broadband MIMO systems are expected to use a much larger bandwidth than current systems. As a result, frequency-selective fading becomes a problem. To combat this, multi-carrier solutions such as orthogonal frequency division multiplexing (OFDM) [7,8] are of interest. Such systems split the available bandwidth into many subcarriers, each of which can be considered to undergo approximately flat fading.

Considering scheduling, the use of DPC in a MIMO system means that the scheduler must not only select which users to transmit to, but also their encoding order, as both factors will affect the rates the individual users receive. Introducing a multi-carrier component to the system adds a third dimension of scheduling complexity, as users are allocated resources across time, space, and frequency. This scheduling often takes the form of optimizing the value of some sort of utility function that incorporates the relevant parameters (e.g. throughput, delay, queue length, etc.) and any constraints thereupon. Two well-known scheduling criteria are maximum throughput, which maximizes the sum of the users' throughputs, and proportional fairness (PF) [9], which maximizes the sum of the logarithms of the users' average throughputs.

With the added complexity caused by the spatial and frequency allocations, it becomes extremely difficult to perform this optimization within a scheduling interval (on the order of milliseconds in current systems, e.g. [10]). One possible suboptimal solution to the complexity issue is the use of a genetic algorithm (GA) [11]. GAs are known for finding a very good solution to an optimization problem in a short amount of time.

This paper extends the work on MIMO GA scheduling we present in [12] to a multi-carrier scenario. We aim to demonstrate that a genetic algorithm can provide near-optimal results as in the single-carrier scenario with a similar reduction in computational complexity.

2. SYSTEM MODEL AND SIMULATION SETUP

2.1 System Model

Our methodology is largely the same as that described in [12] and [13], but extended to a simplified multi-carrier system model with four non-interfering subcarriers. We assume there is a base station with a total transmit power limitation of P_T, which divides that power equally among the subcarriers. The base station has N_T transmit antennas, and schedules transmissions to at most N_T users per subcarrier from a pool of K active users, with $K>N_T$. Power is divided among the scheduled users such that $\sum_k p_{jk} \leq P_T/4$, where j is the subcarrier index. Each user in the pool has N_R receive antennas. All the users experience the same path loss and are statistically identical in terms of noise, shadowing and fading conditions. The channel gains for all transmit-receive antenna pairs, and at all frequencies and scheduling instances, are modeled as independent zero-mean circularly symmetric complex Gaussian processes with unit variance (Rayleigh fading). When analyzing the PF algorithm, we also add a log-normal shadowing component [14] with a standard deviation of 8 dB so as to create a variation in the average rates of the users. It is assumed the base station knows the channel matrices between it and the users perfectly at all frequencies.

Because determining the supportable rates under DPC is quite complex, we instead perform our calculations on the dual multiple access channel (MAC) [15]. The dual MAC considers the users as transmitters and the base station as a receiver that performs successive interference cancellation in the reverse order of the BC DPC encoding. The channel matrices between the users and the base station are the Hermitian transposes of those in the BC. The individual rates and sum-capacity region for the dual MAC are identical to those of the original BC when the sum transmit power constraint of the MAC users equals the base station power constraint for the BC. However, the optimal power covariance matrices (and hence the user rates and system capacity) can be found more easily on the dual MAC as the optimization problem is convex (versus non-convex for the BC). We calculate the MAC covariance matrices using the iterative power waterfilling method described in [16], and thereby find the supportable rates for the users for the BC.

2.2 Details of Genetic Algorithm

Scheduling is implemented via a genetic algorithm, much the same as in [12]. A GA represents several possible solutions to an optimization problem through data structures referred to as chromosomes, and with a set of chromosomes referred to as a population. The form of the chromosome can vary, but is commonly represented by a vector of bits that somehow encodes the parameters being optimized. The chromosomes are then crossbred and combined with each other in such a way that the population "evolves" towards the optimal solution with each "generation" of the population (i.e. each iteration of the algorithm). Chromosomes that represent better solutions to the optimization problem (as defined by the fitness or cost function of the optimization) are favored in the evolution, and are more likely to interbreed and hence pass on their characteristics (i.e. parameters for the optimal solution) to the next generation. The specifics of the implementation of the algorithm depend on the problem to be solved, but most GAs follow the following general procedure.

1. Initialization: A population of N_p chromosomes is initialized, usually randomly.
2. Selection: Chromosomes are selected for breeding based on their fitness, with more fit chromosomes having a higher probability of being selected.
3. Breeding: A pair of the selected chromosomes (the "parents") exchange information and recombine into two new chromosomes (the "offspring" or "children") through crossover and mutation operations. In this paper, crossover is performed by randomly selecting a crossover point in the chromosome. The two parents are split at that point, and the sections after the split are interchanged between the parents to form the children as shown in Figure 1. This operation occurs with probability p_c, which we set to 1. For the mutation operation, each bit in the newly-formed children has a p_m chance of being toggled, where p_m is the mutation rate.
4. Iteration: Once a new population of N_p chromosomes has been formed through selection and breeding, the new population replaces the old one. This process continues until the number of generations reaches N_g.

We extend the chromosomes we use in [12] to the multi-carrier case by replacing the bit vector with a 4-row bit matrix as seen in Figure 1. The jth row represents the scheduling on the jth subcarrier. The first K bits in each row (i.e. the "head" of the chromosome) represent scheduling decisions for the K active users, with a '1' indicating a scheduled user and a '0' indicating an unscheduled user. The maximum weight of the head per row (i.e. the number of '1's) is N_T, while the minimum weight is 1. The remaining $N_T \times \lceil \log_2(N_T) \rceil$ bits per row (i.e. the "tail") denotes the encoding order of the scheduled users. The binary number represented by each group of $\lceil \log_2(N_T) \rceil$ bits in the tail indicates the relative order of encoding, with the first group (or "order number") corresponding to the first '1' in the head

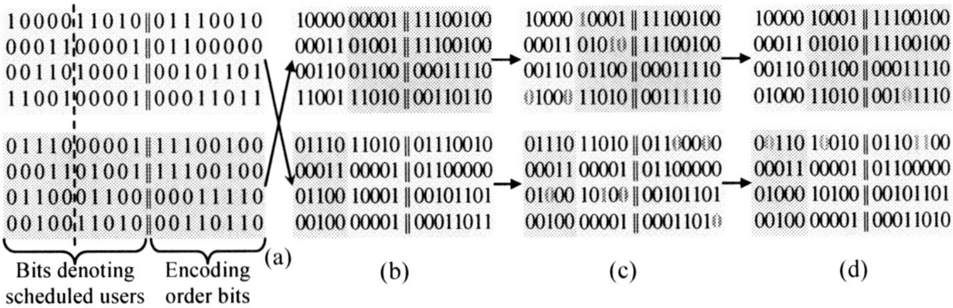

Figure 1. Example of genetic algorithm chromosomes for the multi-carrier system with 4 transmit antennas and 10 users, and typical algorithm operation during one generation. (a) Two typical chromosomes. (b) Crossover. (c) Mutation. (d) Repair of invalid genes.

of the chromosome, the second group to the second '1', and so on. For example, with N_T=4 with 10 active users, the row [0101001010|10110001] indicates users 2, 4, 7, and 9 are to be scheduled on that subcarrier, with user 2 ordered 3rd, user 4 4th, user 7 1st, and user 9 2nd. The first w order numbers in the tail of any row must be unique, where w is the weight of the head of that row, but their values do not necessarily have to be less than w. For example, the row [0001000100|11011111] is valid, as two users (4 and 8) are scheduled, and the first two order numbers ('11' and '01') are unique. As '01' is smaller than '11', user 8 is encoded before user 4. The remaining bits in the tail of that row are ignored. Invalid weights per row (i.e. w=0 or w>N_T) are corrected after the crossbreeding and mutation process by toggling '1's in the head of the chromosome to '0's at random until the weight of that row equals N_T, or by setting one bit at random to a '1' in the case of a weight of zero. Similarly, invalid orderings are corrected by randomly changing duplicate order numbers to a non-duplicate value until the first w order numbers in the row are unique.

A chromosome i is selected for breeding from the population with probability $G_i/\sum_i G_i$, where G_i is the value of the utility function for chromosome i. During the breeding process, to create a new generation of N_p chromosomes, N_p–2 chromosomes are created via crossbreeding and mutation as described above. An additional chromosome is inserted into the new generation as a copy of the previous best solution (a practice known as elitism, which prevents the previous best solution from being destroyed in the breeding process). The remaining chromosome is added as a copy of the previous best solution, but with the order of two users swapped at random in each row. For the mutation operation, we use the same adaptive mutation rate as in [12,13], where $p_m = 1/(\beta_1 + \beta_2\sigma_G/\mu_G)$, μ_G and σ_G are the mean and standard deviation of the fitness of the current generation's population before breeding, and β_1=1.2 and β_2=10 are constants. This adaptive rate increases when the variance in the population's fitness becomes small, and hence forces more diversity into the population to prevent the solution from converging on a local maximum. At the end of each generation, we find the best rows out of all the chromosomes and combine them into a single chromosome. Then, if that chromosome does not yet exist in the population, we replace the chromosome with the smallest total utility function with the new chromosome.

For the maximum throughput algorithm, the scheduling metric is $G_{MT} = \sum_{\forall j,k} r_{jk}$, where r_{jk} is the rate of the kth user on the jth subcarrier. Under DPC, scheduling two groups of the same users with different orders would result in the same sum-rate, and hence the same value of the utility function. We break the tie in these cases based on a max-min criterion: the ordering that maximizes the smallest individual rate(s) to the scheduled users is preferred, as this provides more fairness in throughput across those users.

For the PF algorithm, the utility function metric is $U_{PF} = \sum_k \log_2\left(\overline{R}_k\right)$, where $\overline{R}_k$ is the average throughput of user k. It is shown in [17] that maximizing $G_{PF} = \sum_{\forall j,k} r_{jk} / \overline{R}_k$ also maximizes U_{PF}. We use this form of the metric in the GA. It can be seen that maximizing G_{PF} is a weighted sum-rate maximization, which requires different power allocations to be maximized than an equal-weighted sum. We use the algorithm in [18] to determine the power allocations in this case. $\overline{R}_k$ is approximated as in [9] by a moving average given by $\overline{R}_k(t+1) = (1-1/t_c)\overline{R}_k(t) + (1/t_c) r_k(t)$, where $r_k = \sum_j r_{jk}$ and t_c is the time constant of the averaging window; we use t_c=100 slots.

3. SIMULATION RESULTS

We analyze the performance of the genetic algorithm in terms of the value of the utility function obtained. It is assumed that the total usable bandwidth W_T of the multi-carrier system is the same as for the single-carrier system in [12].

Figure 2 shows the overall spectral efficiency of the system relative to a single-carrier system that performs the same number of iterations of the GA when using the maximum throughput algorithm. For reference, we also include the performance of an exhaustive search for the single-carrier system [12]. It can be seen in Figure 2(a) that an ordinary FDM system achieves virtually the same capacity as the single-carrier system. However, in the case of a 4-channel OFDM system in Figure 2(b), there is an improvement in spectral efficiency compared to the single-carrier system, thanks to the overlap in adjacent subcarriers in an OFDM system. Furthermore, it should be noted that the GA obtains a significant portion of the overall capacity, still at a much lower computational complexity than an optimal exhaustive search. Given that the GA obtains about 94-99% of the maximum achievable throughput in the single-carrier case, and that the FDM

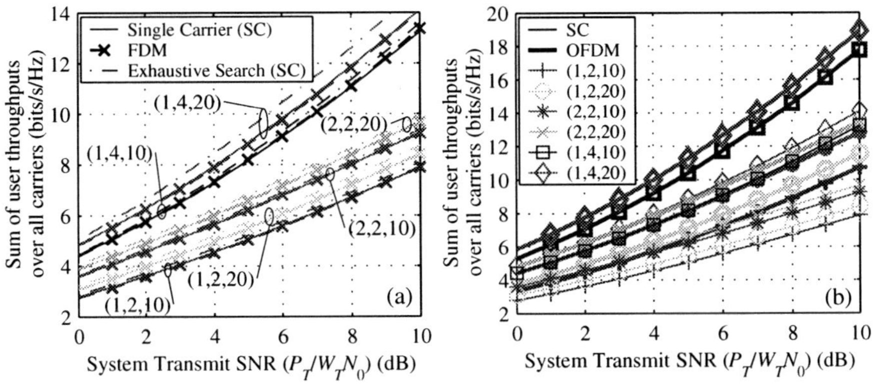

Figure 2. Performance of maximum throughput scheduling algorithm vs. SNR for a (N_R,N_T,K) MIMO multi-carrier system implemented via genetic algorithm. (a) Results for FDM compared to the single-carrier case and an exhaustive search in a single-carrier system. (b) Results for OFDM compared with the single-carrier case.

results are virtually identical to the single-carrier results, it is very likely that the multi-carrier GA results are also about 94-99% of the maximum value. Gains in throughput as N_R, N_T, or K increase can also be observed, as expected due to the respective increases in spatial and multiuser diversity.

Figure 3 shows the performance of the GA vs. SNR when using the PF algorithm in terms of the value of the utility function U_{PF} achieved. Note that the values are negative since the average rate per user is generally less than 1 bit/s/Hz, which yields a negative value for the logarithm. As with the maximum throughput algorithm, the value of the utility function obtained in the single-carrier and FDM cases are very similar, indicating that the average rates achieved by users (in bits/s/Hz) in both scenarios are almost identical. In the case of OFDM, there is again an increase in spectral efficiency and hence the value of the utility function relative to the single-carrier case. More notably, comparing the GA results to an exhaustive search demonstrates that the GA performance is again approximately 0.5 dB away from the best possible.

In theory, the multi-carrier system could schedule up to $4\times N_T$ different users simultaneously, and the likelihood of doing so increases with the number of active users due to multiuser diversity. Our simulations showed that the system does tend to schedule relatively large numbers of users (e.g. usually 5-7 users for the (N_R,N_T,K) = (1,2,10) case and 8-10 users for the (1,4,10) case). This has two main implications; first, each user is most often assigned to just one or occasionally two carriers per scheduling interval. Hence, their instantaneous rates would be lower than the single-carrier case, as their assigned bandwidth is lower. However, it also means their delays are much lower, as the users are scheduled more often. Our simulations indicate that the distributions of user delays unsurprisingly are 4 times lower than those of the single-carrier case. Since any given user is scheduled

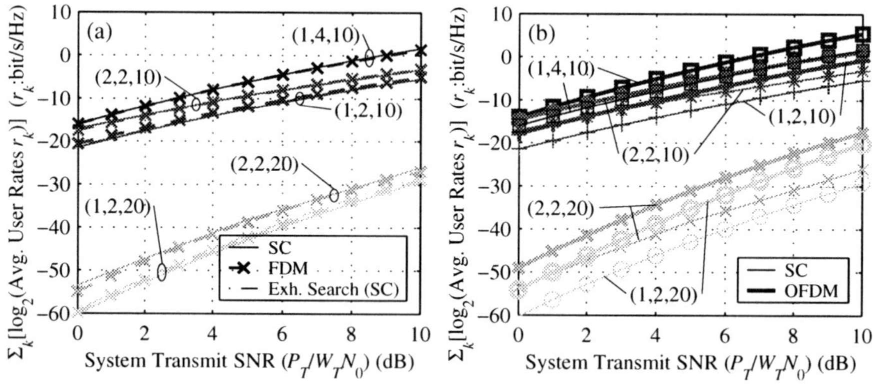

Figure 3. Performance of proportionally fair scheduling algorithm vs. SNR for a (N_R,N_T,K) MIMO multi-carrier system implemented via genetic algorithm. (a) Results for FDM compared to the single-carrier case and an exhaustive search in a single-carrier system. (b) Results for OFDM compared with the single-carrier case.

more often, their average rate increases, which compensates for the drop in instantaneous throughput.

Table 1 compares the computational complexity of the genetic algorithm to that of an optimal exhaustive search in terms of the number of utility function evaluations. Note that the number of evaluations is independent of the number of receive antennas at the mobiles. However, as the number of receive antennas increases, each evaluation will require more computations. Note also that the complexity of the GA is significantly less dependent on the number of active users and/or transmit antennas than that of an exhaustive search; within a certain range of users, the GA population size (i.e. the number of function evaluations) can be kept constant without significantly affecting the GA performance. It must be noted that the multi-carrier case performs 4 times as many calculations for an overall function evaluation as the single-carrier case. (Essentially, the function evaluation for the single-carrier case is carried out 4 times, once for each carrier.)

The proportionally fair algorithm, being a weighted sum-rate maximization, is a special case, as the optimal encoding order is known beforehand; users should be encoded in the reverse order of their relative weights [18]. That is, the user with the largest weight should be encoded last such that he experiences no interference from any other user (thereby maximizing his rate), the user with the second-highest weight is encoded second-to-last, and so on. Thus in this case, an exhaustive search does not have to search through all possible encoding orders, which reduces the complexity. The complexity of the maximum throughput exhaustive search could also be reduced, as the value of the utility function (i.e. the sum-rate) does not depend on the encoding order. Hence, the exhaustive search could simply first find the set of users that maximizes the sum-rate, then search through the ordering of that set to maximize the individual rates.

Table 1. Complexity comparison of genetic and optimal algorithms in terms of utility function evaluations

(K,N_T)	GA ($N_p \times N_g$)	Optimal (General)	Optimal (PF)	Optimal (Max. Throughput)
(10,2)	10×5=50	100	55	57
(10,4)	10×10=100	5860	385	409
(20,2)	10×10=100	400	210	212
(20,4)	20×10=200	123520	6195	6219

There is a large complexity reduction for the GA algorithm compared to the optimal algorithm. With (K,N_T) = (10,4) and (20,4), there is a complexity reduction by a factor of about 4 and 31 respectively for both the maximum throughput and PF algorithms. For a general exhaustive search, the complexity is reduced by a factor of 58.6 and 617.6, respectively.

4. CONCLUSION

This paper has investigated the use of a genetic algorithm to perform scheduling in a multi-carrier MIMO system, extending on previous work for a single-carrier system. We have considered a base station operating on four subcarriers and with N_T antennas transmitting to a pool of K active users each with N_R antennas. It has been observed that the performance of the GA is very close to that of an optimal exhaustive search, obtaining on average about 94-99% of the optimal utility function value in the case of the maximum throughput algorithm. In terms of SNR, the GA is about 0.5 dB away from optimal. The performance of an FDM system is approximately the same as for the single-carrier system, but an OFDM system displays a noted improvement in performance due to its more efficient use of spectrum through overlapping orthogonal subcarriers. The results for the PF algorithm are much the same, again being about 0.5 dB away from the optimum. These results come at a greatly reduced complexity relative to an exhaustive search, with a reduction by a factor of up to 31 for the examined algorithms and up to 617.6 for a general exhaustive search.

Future work in this area will involve the addition of factors such as Doppler shift and packet delay, and the consideration of more quality of service requirements such as a minimum throughput or maximum delay.

ACKNOWLEDGMENT

The authors gratefully acknowledge funding for this research provided by the Natural Sciences and Engineering Research Council (NSERC) of Canada, the Alberta Informatics Circle of Research Excellence (iCORE), the Alberta Ingenuity Fund, TRLabs, and the Rohit Sharma Professorship.

REFERENCES

1. G. J. Foschini and M. J. Gans, "On limits of wireless communications in a fading environment when using multiple antennas," *Wireless Personal Communications*, vol. 6, no. 3, pp. 311-335, Mar. 1998.
2. İ. E. Teletar, "Capacity of multi-antenna Gaussian channels," *European Trans. Telecommun.*, vol. 10, no. 6, pp. 585-595, Nov./Dec. 1999.
3. M. H. M. Costa, "Writing on dirty paper," *IEEE Trans. Inform. Theory*, vol. 29, no. 3, pp. 439-441, May 1983.
4. W. Yu and J. M. Cioffi, "Trellis and convolutional precoding for transmitter-based interference presubtraction," *IEEE Trans. Commun.*, vol. 53, no. 7, pp. 1220-1230, Jul. 2005.
5. R. Knopp and P. A. Humblet, "Information capacity and power control in single-cell multiuser communications," in *Proc. IEEE Int. Conf. Commun.* (*ICC '95*), Seattle, WA, Jun. 1995, vol. 1, pp. 331–335.
6. D. J. Mazzarese and W. A. Krzymień, "Linear space-time transmitter and receiver processing and scheduling for the MIMO broadcast channel," in *Proc. IEEE Veh. Technol. Conf.* (*VTC 2004-Spring*), Milan, Italy, May 2004, vol. 2, pp. 752-756.
7. Y. (G.) Li and G. L. Stüber, eds., *Orthogonal Frequency Division Multiplexing for Wireless Communications*, New York, NY: Springer, 2006.
8. K. Fazel and S. Kaiser, *Multi-Carrier and Spread Spectrum Systems*, Chichester, West Sussex, England: John Wiley & Sons Ltd., 2003.
9. A. Jalali, R. Padovani, and R. Pankaj, "Data throughput of CDMA-HDR a high efficiency-high data rate personal communication wireless system," in *Proc. IEEE Veh. Technol. Conf.* (*VTC 2000-Spring*), Tokyo, Japan, May 2000, vol. 3, pp. 1854-1858.
10. 3GPP2 C.S0024-A, *cdma2000 High Rate Packet Data Air Interface Specification*, Version 3.0, 3rd Generation Partnership Project 2 (3GPP2), Sept. 2006.
11. J. H. Holland, *Adaptation in Natural and Artificial Systems*, 1st ed. Ann Arbor, MI: Univ. of Michigan Press, 1975.
12. R. C. Elliott and W. A. Krzymień, "Downlink scheduling for multiple antenna systems with dirty paper coding via genetic algorithms," in *Proc. IEEE Veh. Technol. Conf.* (*VTC2007-Spring*), Dublin, Ireland, Apr. 2007, 5 pages.
13. V. K. N. Lau, "Optimal downlink space-time scheduling design with convex utility functions – multiple-antenna systems with orthogonal spatial multiplexing," *IEEE Trans. Veh. Technol.*, vol. 54, no. 4, pp. 1322-1333, Jul. 2005.
14. G. L. Stüber, *Principles of Mobile Communication*, 2nd ed. Boston/Dordrecht: Kluwer Academic Publishers, 2001.
15. S. Vishwanath, N. Jindal, and A. Goldsmith, "Duality, achievable rates, and sum-rate capacity of Gaussian MIMO broadcast channels", *IEEE Trans. Inform. Theory*, vol. 49, no. 10, pp. 2658-2668, Oct. 2003.
16. N. Jindal, W. Rhee, S. Vishwanath, S. A. Jafar, and A. Goldsmith, "Sum power iterative water-filling for multi-antenna Gaussian broadcast channels," *IEEE Trans. Inform. Theory*, vol. 51, no. 4, pp. 1570-1580, Apr. 2005.
17. F. P. Kelly, A. K. Maulloo, and D. K. H. Tan, "Rate control for communication networks: shadow prices, proportional fairness, and stability," *J. of the Operational Research Soc.*, vol. 49, no. 3, pp. 237-252, Mar. 1998.
18. H. Viswanathan, S. Venkatesan, and H. Huang, "Downlink capacity evaluation of cellular networks with known-interference cancellation," *IEEE J. Select. Areas Commun.*, vol. 21, no. 5, pp. 802-811, Jun. 2003.

EXIT-CHART ANALYSIS OF ITERATIVE SPACE-TIME-FREQUENCY CODED MULTI-CARRIER RECEIVERS

Stephan Sand and Armin Dammann
German Aerospace Center (DLR)
Institute of Communications and Navigation
Oberpfaffenhofen, 82234 Wessling, Germany
Email: {stephan.sand,armin.dammann}@dlr.de

Abstract In this paper, we analyze a multi-carrier (MC) system employing combined space-time-frequency (STF) codes and an iterative receiver with extrinsic information transfer (EXIT) charts and bit-error rate plots. As STF codes, we examine orthogonal space-time block codes, cyclic delay diversity and discontinuous Doppler diversity as well as combinations of these codes. Moreover, we show that EXIT charts are not always a suitable tool to analyze the performance of iterative receivers for STF coded MC systems even with differential code doping or spreading.

1. Introduction

Multi-carrier (MC) transmission in the form of orthogonal frequency-division multiplexing (OFDM) in combination with bit-interleaved coded modulation (BICM) has turned out a robust yet implementation efficient technique for reliable communication over fading channels without channel state information (CSI) at the transmitter [1]. It is also beneficial to spread the symbols in one or more dimensions of time, frequency, and space to gain diversity through spreading [2] and space-time-frequency (STF) coding [3]. Different STF techniques can exploit spatial diversity or can generate transmit diversity. For example, the well known orthogonal space-time block codes (OSTBCs), such as the Alamouti scheme, take advantage of the spatial diversity. Other techniques, such as cyclic delay diversity (CDD) [3] and discontinuous Doppler diversity (DDoD) [4] increase the frequency and time diversity.

At the receiver, the system performance can be further improved by iteratively exchanging extrinsic information between the demodulator (DMOD) and decoder [5]. The critical design parameter for a BICM

S. Plass et al. (eds.), Multi-Carrier Spread Spectrum 2007, 57–66.

receiver with iterative demodulation and iterative decoding (IDEM) is the choice of the symbol alphabet mapping [6]. To predict and analyze the performance of IDEM, extrinsic information transfer (EXIT) charts are a well established tool [2, 5–7].

In this paper, we study a STF coded BICM MC system with IDEM at the receiver and optional differential code doping or spreading at the transmitter. We briefly explain the system model and the computation of EXIT charts in Sections 2 and 3. Then, we present the considered STF codes, namely OSTBCs, CDD, and DDoD in Section 4. The next section analyzes the influence of these STF codes on the inner decoder. As we will demonstrate, it is not possible in any case for the classical EXIT chart to predict the performance of IDEM correctly. We then investigate the performance of IDEM for the STF coded BICM MC system both with EXIT charts and bit-error rate (BER) plots in Section 6, which confirms our analysis. Further results show that a combination of the Alamouti scheme with CDD and DDoD can gain significant compared to pure Alamouti while keeping the receiver complexity the same.

2. System Model

Figure 1 represents the block diagram of the STF coded MC transmitter. At the transmitter, a binary signal is encoded by a channel coder and interleaved by a code-bit interleaver. After the interleaver, the bits c_μ are mapped to the symbol alphabet, e.g., 16-QAM and set partitioning (SP) mapping [6]. Before the modulator, the bits c_μ can be optionally doped with a differential code, i.e., a recursive systematic convolutional code (RSCC) punctured to rate 1 [6]. The modulated signal is serial-to-parallel converted to form the single-input single-output (SISO) OFDM frame $S_{n,k}$. Here, n denotes the subcarrier index and k the OFDM symbol number. Additionally, each data symbol can be spread with a Walsh-Hadamard code and then the spread chips of the different data symbols can be summed [2]. The data symbols are space-time-frequency (STF) coded to obtain $S^m_{n,k}$ $(m = 1, \ldots, N_{\mathrm{TX}})$ in Figure 1. Here, m denotes the transmit antenna index. As possible STF codes we consider OSTBCs, CDD, or DDoD. In our system, we will apply the OSTBC code in frequency direction over consecutive subcarriers.The resulting N_{TX} OFDM frames with N_{c} active subcarriers and N_{s} OFDM symbols per frame are transformed by an inverse fast Fourier transform (IFFT) of size N_{FFT} in the time domain and cyclically extended by the guard interval (GI) of length N_{GI} before they are transmitted over a time-variant MIMO multipath channel.

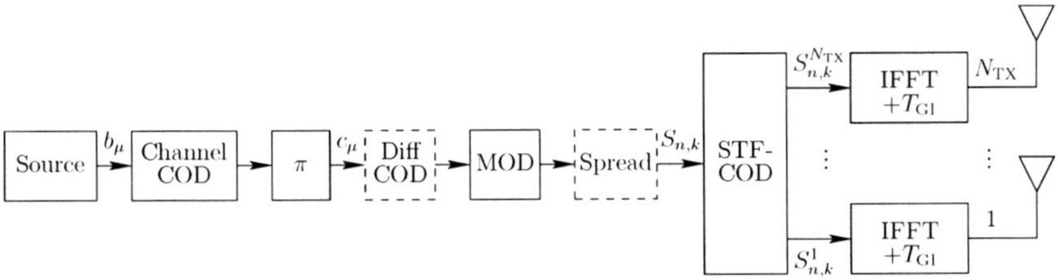

Figure 1. MIMO multi-carrier transmitter

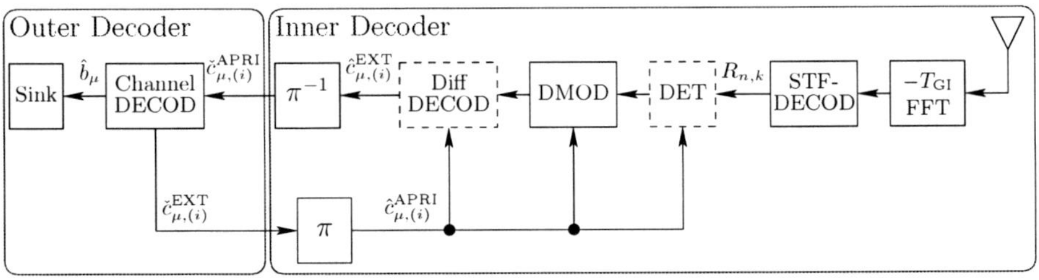

Figure 2. Iterative MIMO multi-carrier receiver

At the receiver, the GI is removed from the received symbols and the frequency domain signal is computed by an FFT (Figure 2). After STF decoding and parallel-to-serial conversion of the received data symbols $R_{n,k}$, the DMOD computes from the received symbols soft-demodulated extrinsic log-likelihood ratio values $\hat{c}_{\mu,(i)}^{\mathrm{EXT}}$ (L-values [5]). To obtain the L-values, the DMOD exploits the knowledge of a-priori L-values $\check{c}_{\mu,(i)}^{\mathrm{APRI}}$ coming from the decoder. In the initial iteration ($i = 0$), the DMOD assumes that the L-values $\check{c}_{\mu,(i)}^{\mathrm{APRI}}$ are zero. If the transmitter uses spreading, a detector (DET) that benefits from a-priori L-values returns the despread signal [2]. Similar, the differential decoder computes the extrinsic L-values from the DMOD output and the a-priori L-values [6] if the transmitter uses a differntial encoder. After deinterleaving, the extrinsic L-values $\hat{c}_{\mu,(i)}^{\mathrm{EXT}}$ become the a-priori L-values to the channel decoder. The channel decoder computes for all code bits the L-values $\check{c}_{\mu,(i)}^{\mathrm{EXT}}$ using the MAP algorithm. The extrinsic L-values are then interleaved to become the a-priori L-values $\hat{c}_{\mu,(i)}^{\mathrm{APRI}}$ used in the next iteration in the DMOD as well as in the optional DET and differential decoder (Figure 2). The above described IDEM can be repeated several times. In the final iteration, the decoder returns hard decision estimates $\hat{b}_\mu$ of the originally transmitted bits using the MAP algorithm.

3. EXIT Charts

EXIT charts have been introduced by [5] for an iterative receiver with BICM assuming perfect CSI. A fundamental assumption to compute EXIT charts is that the a-priori L-values c_μ^{APRI} can be modeled as independent identically distributed Gaussian random variables, i.e.,

$$c_\mu^{\mathrm{APRI}} = \mathrm{E}\left\{c_\mu^{\mathrm{APRI}}|c_\mu\right\} \cdot c_\mu + n_{c_\mu^{\mathrm{APRI}}}. \tag{1}$$

Here c_μ is the transmitted code bit (see Figure 1) and $n_{c_\mu^{\mathrm{APRI}}}$ additive white Gaussian noise (AWGN) with mean zero and variance σ^2_{APRI}. The mean of c_μ^{APRI} given c_μ needs to fulfill $\mathrm{E}\{c_\mu^{\mathrm{APRI}}|c_\mu\} = \frac{\sigma^2_{\mathrm{APRI}}}{2}$.

To generate an EXIT characteristic for a component such as the decoder or DMOD, we first compute the variance σ^2_{APRI} from the mutual information I^{APRI}. Then, we generate the random variable c_μ^{APRI} according to (1) and use c_μ^{APRI} as a-priori L-value in the component. At the output of the component, we estimate the extrinsic mutual information I^{EXT} from the extrinsic L-values c_μ^{EXT} according to [7]

$$I^{\mathrm{EXT}} \approx 1 - \frac{1}{N}\sum_{\mu=1}^{N} H_{\mathrm{b}}\left(\frac{1}{1+e^{-|c_\mu^{\mathrm{EXT}}|}}\right). \tag{2}$$

The underlying assumptions in (2) are that the random processes such as the noise and channel variations are ergodic and a large number of samples N have been observed. Having obtained an estimate for I^{EXT}, we can plot the point $(I^{\mathrm{APRI}}, I^{\mathrm{EXT}})$ in the EXIT chart.

4. Space-Time-Frequency Coding

In the following, we study three different STF codes, i.e., OSTBCs, CDD, and DDoD. Without loss of generality, we examine the Alamouti scheme [3] as an example for OSTBCs in the sequel. In this case, the received signal after STF decoding of two adjacent subcarriers is

$$\begin{aligned}\begin{pmatrix} R_{2n',k} \\ R_{2n'+1,k}\end{pmatrix} &= \frac{1}{\sqrt{2}}\left(|H^1_{2n',k}|^2 + |H^2_{2n',k}|^2\right)\begin{pmatrix} S_{2n',k} \\ S_{2n'+1,k}\end{pmatrix} + \\ &\begin{pmatrix} H^1_{2n',k} & H^2_{2n',k} \\ \left(H^2_{2n',k}\right)^* & -\left(H^1_{2n',k}\right)^*\end{pmatrix}^H \cdot \begin{pmatrix} Z_{2n',k} \\ Z_{2n'+1,k}\end{pmatrix},\end{aligned} \tag{3}$$

where the index n' maps to the subcarrier index n by $n = 2n'$. Here, we assume that the channel is constant over two consecutive subcarriers. The AWGN $Z_{2n',k}$ and $Z_{2n'+1,k}$ have zero mean and variance σ^2. Defining an effective channel coefficient

$$\tilde{H}^{\mathrm{OST}}_{n,k} = \frac{1}{\sqrt{2}}\left(|H^1_{2n',k}|^2 + |H^2_{2n',k}|^2\right) \tag{4}$$

and an effective AWGN $\tilde{Z}_{n,k}$ with zero mean and variance $\sigma^2_{Z_{n,k}} = (|H^1_{2n',k}|^2 + |H^2_{2n',k}|^2)\sigma^2$, we can rewrite (3) as

$$R_{n,k} = \tilde{H}^{\mathrm{OST}}_{n,k} \cdot S_{n,k} + \tilde{Z}_{n,k}. \tag{5}$$

In the following, we assume w.r.t transmit antennas uncorrelated Gaussian distributed channel coefficients with zero mean and unit variance. To compute the mean and power of $\tilde{H}^{\mathrm{OST}}_{n,k}$, we note that the distribution of $\tilde{H}^{\mathrm{OST}}_{n,k}$ is chi-square with four degrees of freedom [8]. Thus, the mean and the power of $\tilde{H}^{\mathrm{OST}}_{n,k}$ are $\mathrm{E}\{\tilde{H}^{\mathrm{OST}}_{n,k}\} = \sqrt{2}$ and $\mathrm{E}\{|\tilde{H}^{\mathrm{OST}}_{n,k}|^2\} = 3$.

Alternative to OSTBCs, we can apply CDD to the MC system as described in detail in [3]. After the IFFT the time domain signal $s_{u,k}$ is

$$s_{u,k} = \frac{1}{\sqrt{N_{\mathrm{FFT}}}} \sum_{n=1}^{N_{\mathrm{FFT}}} S_{n,k} e^{j \frac{2\pi}{N_{\mathrm{FFT}}} nu}, \tag{6}$$

where u denotes the sample time index of the kth OFDM symbol. The CDD transmit signal for antenna $m = 1, \ldots, N_{\mathrm{TX}}$ is then equal to

$$s^m_{u,k} = \frac{1}{\sqrt{N_{\mathrm{TX}}}} s_{((u-\delta^{\mathrm{cyc}}_m) \bmod N_{\mathrm{FFT}}),k}, \tag{7}$$

where $x \bmod y$ is the modulo operator returning the remainder of x divided by y. Transforming $s^m_{u,k}$ into the frequency domain, we obtain

$$S^m_{n,k} = \frac{1}{\sqrt{N_{\mathrm{TX}}}} S_{n,k} e^{-j \frac{2\pi}{N_{\mathrm{FFT}}} \delta^{\mathrm{cyc}}_m n}. \tag{8}$$

Consequently, the received signal $R_{n,k}$ is given by

$$R_{n,k} = \frac{1}{\sqrt{N_{\mathrm{TX}}}} S_{n,k} \sum_{m=1}^{N_{\mathrm{TX}}} H^m_{n,k} e^{-j \frac{2\pi}{N_{\mathrm{FFT}}} \delta^{\mathrm{cyc}}_m n} + Z_{n,k} = S_{n,k} \tilde{H}^{\mathrm{CDD}}_{n,k} + Z_{n,k}, \tag{9}$$

where $H^m_{n,k}$ is the CSI between transmit antenna m and the receive antenna, $\tilde{H}^{\mathrm{CDD}}_{n,k}$ the equivalent CSI experienced by the receiver, and $Z_{n,k}$ AWGN. The mean and variance of $\tilde{H}^{\mathrm{CDD}}_{n,k}$ are $\mathrm{E}\{\tilde{H}^{\mathrm{CDD}}_{n,k}\} = 0$ and $\mathrm{E}\{|\tilde{H}^{\mathrm{CDD}}_{n,k}|^2\} = 1$.

From (9), we infer that CDD causes no ISI even if the cyclic delays δ^{cyc}_m are larger than the guard interval. Further, CDD does not require any additional signal processing at the receiver and hence, is a standard conformable antenna diversity technique [3]. The spatial diversity of the transmit antennas is transformed into the equivalent CSI $\tilde{H}^{\mathrm{CDD}}_{n,k}$ with increased frequency diversity. For a large number of uncorrelated transmit antennas, the channel fading coefficients $\tilde{H}_{n,k}^{\mathrm{CDD}}$ between neighboring subcarriers become uncorrelated.

Besides increasing the frequency diversity with CDD, DDoD can increase the time diversity without causing inter-carrier interference (ICI) [4]. In that case, the time domain signal for antenna m becomes

$$s_{u,k}^{m} = \frac{1}{\sqrt{N_{\mathrm{TX}}}} s_{u,k} \cdot e^{j\frac{2\pi(N_{\mathrm{FFT}}+N_{\mathrm{GI}})}{N_{\mathrm{FFT}}} T_{\mathrm{s}} \Delta f_m k}, \tag{10}$$

where $T_{\mathrm{s}} = 1/F_{\mathrm{s}}$ denotes the OFDM symbol duration and Δf_m the antenna specific spectrum shift. Consequently, the frequency domain signal $S_{n,k}^{m}$ is given by

$$S_{n,k}^{m} = \frac{1}{\sqrt{N_{\mathrm{TX}}}} S_{n,k} e^{j\frac{2\pi(N_{\mathrm{FFT}}+N_{\mathrm{GI}})}{N_{\mathrm{FFT}}} T_{\mathrm{s}} \Delta f_m k}. \tag{11}$$

Similar to (9), we obtain the received signal

$$R_{n,k} = \frac{1}{\sqrt{N_{\mathrm{TX}}}} S_{n,k} \sum_{m=1}^{N_{\mathrm{TX}}} H_{n,k}^{m} e^{j2\pi\left(1+\frac{N_{\mathrm{GI}}}{N_{\mathrm{FFT}}}\right) T_{\mathrm{s}} \Delta f_m k} + Z_{n,k} = S_{n,k} \tilde{H}_{n,k}^{\mathrm{DDoD}} + Z_{n,k}, \tag{12}$$

where $\tilde{H}_{n,k}^{\mathrm{DDoD}}$ is the effective CSI experienced by the receiver. Thus, the mean and variance of $\tilde{H}_{n,k}^{\mathrm{DDoD}}$ become $\mathrm{E}\{\tilde{H}_{n,k}^{\mathrm{DDoD}}\} = 0$ and $\mathrm{E}\{|\tilde{H}_{n,k}^{\mathrm{DDoD}}|^2\} = 1$. Compared to OSTBCs, CDD and DDoD can support an arbitrary number of transmit antennas and do not require any additional processing at the receiver. Moreover, CDD and DDoD have no rate loss for more than two transmit antennas as it is the case for OSTBCs [3]. Note, we can combine OSTBCs with CDD and DDoD to increase diversity.

5. Inner decoder and MAP demodulator

In Figure 2, the inner decoder includes the STF decoder and the DMOD. When employing OSTBCs, CDD, or DDoD at the transmitter, the STF decoder cannot benefit from a-priori L-values from the decoder. However, it influences the performance of the DMOD. In the most general case the MAP DMOD computes the extrinsic L-value c_{μ}^{EXT} by

$$c_{\mu}^{\mathrm{EXT}} = \ln \frac{\sum\limits_{\underline{S}(\underline{c}):c_{\mu}=1} p(\underline{R}|\underline{S})p(\underline{S})}{\sum\limits_{\underline{S}(\underline{c}):c_{\mu}=0} p(\underline{R}|\underline{S})p(\underline{S})} - c_{\mu}^{\mathrm{APRI}}. \tag{13}$$

Here, $\underline{c}$ denotes a code word, $\underline{S}(\underline{c})$ and $\underline{S}$ the symbol sequence obtained from mapping the code word $\underline{c}$ to the symbol alphabet, and $\underline{R}$ the received symbols after STF decoding corresponding to $\underline{c}$. Assuming that the received signal is only perturbed by AWGN, the conditional channel probability in (13) can be factored into $p(\underline{R}|\underline{S}) = \prod_{l=1}^{N} p(R_l|S_l)$, where R_l and S_l are the lth received and transmitted symbol. The index l is related to the subcarrier index n and OFDM symbol index k by $l = n+k\cdot N_c$. When considering memoryless or redundancy free modulation, the a-priori probability in (13) can be factored into $p(\underline{S}) = \prod_{l=1}^{N} p(S_l)$. With the above results, (13) simplifies to [2]

$$c_\mu^{\mathrm{EXT}} = \ln \frac{\sum\limits_{S_l(\underline{c}_l):c_\mu=1} p(R_l|S_l)p(S_l)}{\sum\limits_{S_l(\underline{c}_l):c_\mu=0} p(R_l|S_l)p(S_l)} - c_\mu^{\mathrm{ARPI}}, \tag{14}$$

where $p(R_l|S_l) = \frac{1}{\sqrt{2\pi\sigma^2}} e^{-\frac{|R_l - \tilde{H}_l S_l|^2}{2\sigma^2}}$. $\tilde{H}_l$ denotes the effective channel coefficient the DMOD experiences due to STF coding and decoding as well as OFDM modulation and demodulation or spreading and DET.

In case of the Alamouti scheme, the effective channel coefficient $\tilde{H}_l^{\mathrm{OST}}$ is given by (4) and its mean and variance are $\mathrm{E}\{\tilde{H}_l^{\mathrm{OST}}\} = \sqrt{2}$ and $\mathrm{E}\{|\tilde{H}_l^{\mathrm{OST}}|^2\} = 3$. When employing CDD or DDoD, the effective channel coefficients $\tilde{H}_l^{\mathrm{CDD}}$ and $\tilde{H}_l^{\mathrm{DDoD}}$ are obtained from (9) and (12) with mean and variance $\mathrm{E}\{\tilde{H}_l^{\mathrm{CDD}}\} = \mathrm{E}\{\tilde{H}_l^{\mathrm{DoD}}\} = 0$ and $\mathrm{E}\{|\tilde{H}_l^{\mathrm{CDD}}|^2\} = \mathrm{E}\{|\tilde{H}_l^{\mathrm{DDoD}}|^2\} = 1$. Thus, the mean and variance of the effective channel are unaltered by CDD or DDoD.

Consequently, we expect that the mutual information $I_{\mathrm{DMOD}}^{\mathrm{EXT}}$ in (2) and thus the DMOD characteristic in the EXIT chart will not change for CDD compared to SISO, but for the Alamouti scheme. Furthermore, $I_{\mathrm{DMOD}}^{\mathrm{EXT}}$ will be the same whether neighboring subcarriers or OFDM symbols are correlated or not. Hence, EXIT charts cannot always accurately predict the performance of MIMO communication systems with iterative receivers.

6. Simulation Results

To obtain the results in Figure 3, we investigate a small office environment according to the IEEE 802.11n C channel model [9] with non-line-of-sight propagation, maximum delay 200ns, and bell shaped Doppler spectrum (maximum Doppler frequency 29 Hz at a carrier frequency of

5.25 GHz). A BICM-MC scheme is assumed employing a rate 1/2 RSCC with generators $(7,5)_8$ with 16-QAM and SP mapping [6], $N_c = 900$ active subcarriers with $F_s = 20$ kHz occupying a bandwidth of 18 MHz. The resulting OFDM symbol duration is $T_s = 50\mu s$ and the sampling time $T_{samp} = T_s/N_{FFT} = 48.828$ns, where $N_{FFT} = 1024$. We choose a guard interval $T_{GI} = 21T_{samp} \approx 1.03\mu s$. The system transmits $N_s = 101$ OFDM symbols per frame resulting in a frame duration of 5.15ms and a data rate of 38.8Mbps.

In Figures 3(a), 3(b), and 3(c), we plot the decoder and the DMOD characteristics of different STF codes and the IEEE 802.11n C channel model as well as a channel where each subcarrier and OFDM symbol is subject to independent Rayleigh (IR) fading. As expected from the analysis in Section 5, the EXIT chart in Figure 3(a) predicts that the 1-transmit-antenna (TX) scheme over the IEEE 802.11n channel yields the same performance as over the IR channel or a combined CDD and DDoD scheme (4 TX). In contrast, the Alamouti scheme performs as anticipated better than the SISO system. However, different channels or a combination of the STF codes (8 TX, rate 1, Alamouti, CDD, and DDoD) apparently results in the same performance. When spreading the transmit signal (Figure 3(b)), we consider ideal spreading and despreading over the complete OFDM frame. In this case, the different schemes show slightly different performances. Only for the IR channel, the 1 TX and the Alamouti system perform exactly the same. Regarding differential code doping only every 60th redundancy bit is transmitted. In Figure 3(c), the EXIT chart predicts similar behavior as in Figure 3(a) except that the (1, 1) point in the EXIT chart is reached and thus all schemes should transmit error free.

Comparing the predicted performance from the EXIT chart for IDEM with the simulated BERs at 6 dB in Figure 3(d), we see that the EXIT chart is not very accurate. For instance, at 6 dB E_b/N_0, the 1 TX system in combination with the IEEE 802.11n channel is worst whereas the 1 TX scheme with the IR channel achieves a two orders of magnitude lower BER. Similar, the combined STF codes over the IEEE 802.11n channel perform about three orders of magnitude worse than the Alamouti scheme over the IR channel. Note, in all cases we used $i = 10$ iterations in the IDEM. Comparing the combined STF code with the pure Alamouti code, we see that at a BER of 10^{-5} the combined scheme gains about 4 dB SNR while having the same receiver complexity. As expected, all schemes show a BER floor after the turbo cliff [6]. Employing spreading, the error floors are further lowered, but still exist (Figure 3(e)). Through differential code doping before the modulator and a corresponding decoder in the IDEM receiver, the error floors due to the SP

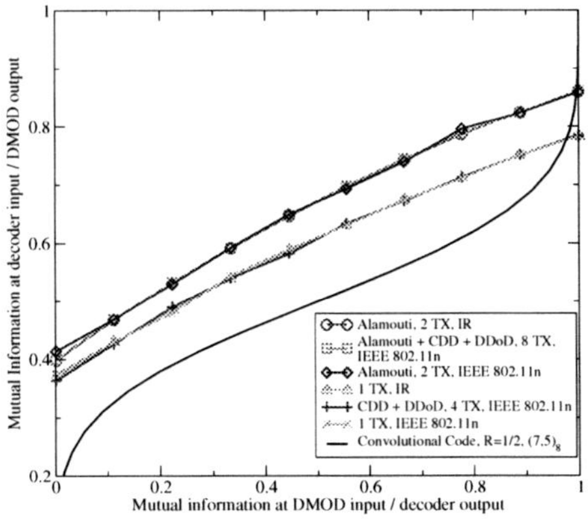

(a) no differential code doping or spreading

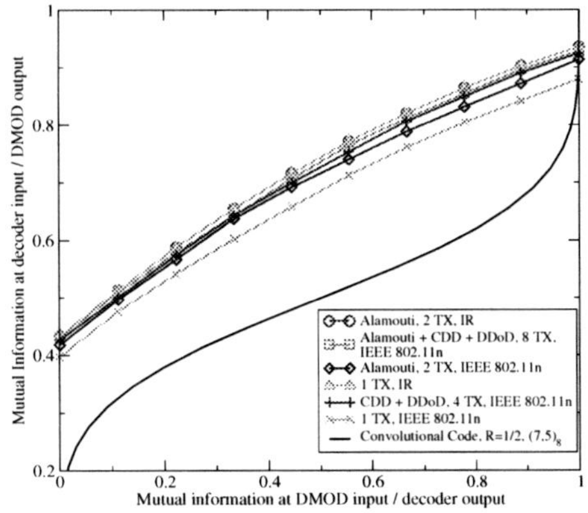

(b) spreading

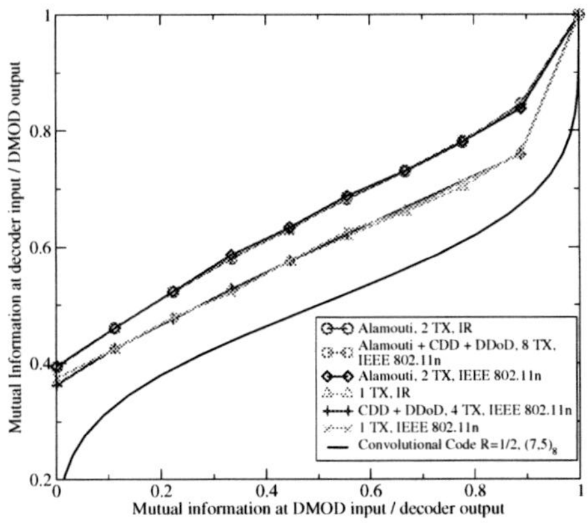

(c) differential code doping

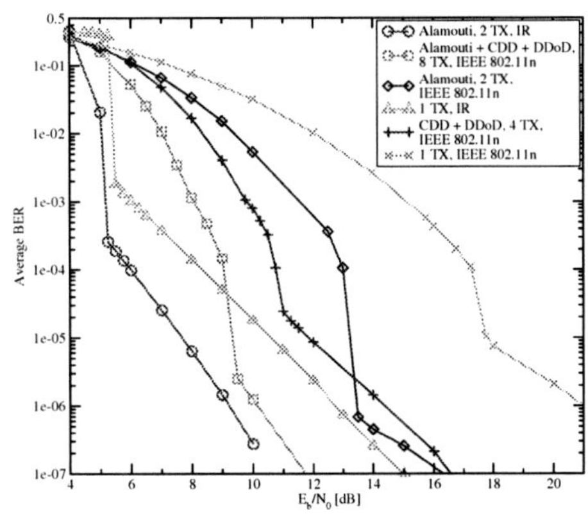

(d) no differential code doping or spreading

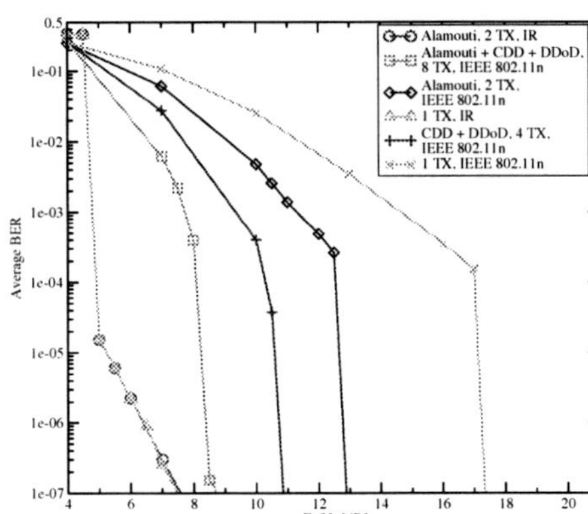

(e) spreading

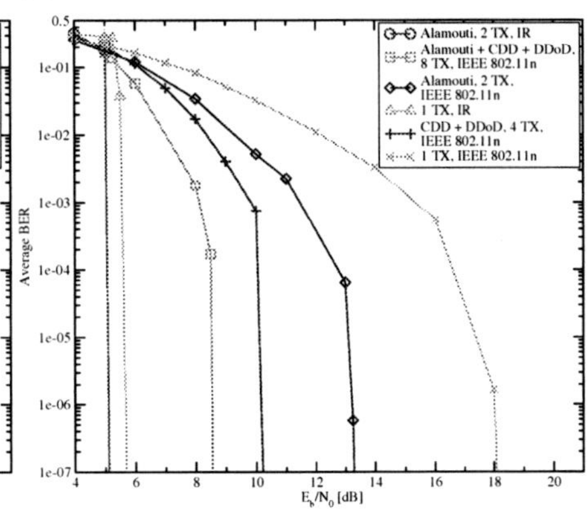

(f) differential code doping

Figure 3. EXIT charts ((a), (b), and (c)) at 6 dB E_b/N_0 and BER plots ((d), (e), and (f)) of iterative receivers for BICM MC systems.

mapping can be completely removed. Comparing Figures 3(b) and 3(c) with 3(e) and 3(f) at 6 dB E_b/N_0, we infer that the EXIT chart predictions are only correct for the IR channel and the 1 TX or Alamouti scheme. All other schemes do not achieve a low BER.

7. Conclusions

This paper analyzes an MC system employing combined STF codes with optional differential code doping or spreading and an iterative receiver over correlated and uncorrelated channels. As STF codes, we examine OSTBCs, CDD, and DDoD. We present a combined STF code for eight transmit antennas with rate one. The analysis of the MAP demodulator and STF codes reveals that EXIT charts are not always a suitable tool to analyze iterative receivers for STF coded MC systems. The theoretical findings are confirmed by simulation results. The combined STF code gains about 4 dB SNR at a BER of 10^{-5} compared to the Alamouti code while having the same receiver complexity.

References

[1] G. Caire, G. Taricco, and E. Biglieri. Bit-interleaved coded modulation. *IEEE Trans. Inform. Theory*, May 1998.

[2] Armin Dammann et al. On iterative detection, demodulation, and decoding for OFDM-CDM. In *Proc. ISTC 2006*, Munich, Germany, April 2006.

[3] Armin Dammann et al. Comparison of space-time block coding and cyclic delay diversity for a broadband mobile radio air interface. In *Proc. WPMC 2003, Yokosuka, Japan*, October 2003.

[4] Armin Dammann and Ronald Raulefs. Increasing time domain diversity in OFDM systems. In *Proc. IEEE GLOBECOM 2004, Dallas, TX, USA*, November–December 2004.

[5] Stephan ten Brink. Designing iterative decoding schemes with the extrinsic information transfer chart. *AEÜ Int. J. of Electr. Commun.*, December 2000.

[6] Stephan Pfletschinger and Frieder Sanzi. Error floor removal for bit-interleaved coded modulation with iterative detection. *IEEE Trans. Wireless Commun.*, November 2006.

[7] Joachim Hagenauer. The EXIT chart. In *Proc. EUSIPCO 2004, Vienna, Austria*, September 2004.

[8] John G. Proakis. *Digital Communications*. McGraw-Hill, 3rd edition, 1995.

[9] 802.11n Task Group. TGn channel models. Technical Report 802.11-03 940r2, IEEE P802.11, January 2004.

INTERLEAVER FOR HIGH PARALLELIZABLE TURBO DECODER

Laurent Boher, Jean-Baptiste Doré, Maryline Hélard, and Christian Gallard
France Telecom R&D Division
4 rue du Clos Courtel, 35512 Cesson-Sévigné, France
{laurent.boher;jeanbaptiste.dore;maryline.helard;christian.gallard}@orange-ftgroup.com

Abstract In this paper a new interleaver is proposed as turbo codes intern interleaver. This interleaver designed from shifted identity matrices has been developed to facilitate implementation of parallelized turbo decoding. The structure is first illustrated and then, implementation aspects are discussed. We demonstrate that the proposed interleaver is contention free and enables low complex implementation without performance loss.

1. Introduction

Since the introduction of Turbo-codes [1], high data rate and low complexity decoders have been required by communication systems such as 3GPP Long Term Evolution (LTE) [2]. In order to improve decoding throughput without memories duplication, a parallel decoding is done on a single frame. The frame is divided into segments which are decoded in parallel using a sliding window algorithm [3]. An architecture with memory banks is used to be associated with a parallel decoding. Memory conflicts, which occur when two processors try to access the same memory bank, can appear with such an architecture. This problem has been addressed in the literature (for instance, a state of art is done in [4]), and many solutions have been proposed. A first solution is a particular memory mapping [5] preventing from the emergence of any memory conflict. Another way is to jointly design the interleaver and the architecture in order to take into account possible memory conflicts. These interleavers are labelled contention free interleavers.

In this paper we propose a new family of contention free interleavers, easily characterizable with a high parallelization level and enabling a fast and simple address generation when decoders are parallelized. The

S. Plass et al. (eds.), Multi-Carrier Spread Spectrum 2007, 67–75.

paper is divided into five sections. In section 2 we first introduce the new interleaver and describe main parameters. Section 3 demonstrates how the interleaver allows an efficient implementation of a parallelized decoder. Performance of the structure are illustrated in section 4 and we conclude in section 5.

2. Interleaver Structure

2.1 Interleaver definition

The interleaver can be described by four successive permutation laws. First the frame of $K = mz$ elements is interleaved by a z-row m-column permutation. This permutation can be expressed by:

$$\Pi_1(i) = (i \bmod m)\, z + \left\lfloor \frac{i}{m} \right\rfloor \tag{1}$$

In a second time, each group k, $k \in [0, m-1]$ of z elements is right circularly shifted by $\delta(k)$ positions, corresponding to the following permutation law:

$$\Pi_2(i) = \left(\delta\left(\left\lfloor \frac{i}{z} \right\rfloor\right) + i\right) \bmod z + \left\lfloor \frac{i}{z} \right\rfloor z \tag{2}$$

Then, the groups of z elements are interleaved: the group in position k becomes the group in position $P(k)$. The last step of the interleaving process consists in a m-row z-column permutation, dual to the first one. The resulting permutation law is then described by:

$$\Pi_{tot}(i) = P\,(i \bmod m) + m\left(\delta\,(i \bmod m) + \left\lfloor \frac{i}{m} \right\rfloor\right) \bmod z \tag{3}$$

The whole interleaver is characterized thanks to $2m + 2$ parameters, m, z, $P(k)$ and $\delta(k)$ for $k \in [0, m-1]$.

The interleaver can also be described using a matrix of size $K \times K$:

$$\mathbf{\Pi} = \tilde{\mathbf{Q}}\mathbf{S}\mathbf{Q} \tag{4}$$

where the $K \times K$ matrix $\mathbf{Q}$ represents the z-row m-column permutation and $\tilde{\mathbf{Q}}$ its dual. The matrix $\mathbf{S}$ of size $K \times K$ is composed of m circularly right shifted $z \times z$ identity matrices. Considering a $m \times m$ matrix, identity matrix of line k is permuted by $\delta(k)$ positions and located in the $P(k)$ column. For instance, if we consider the following parameters: $m = 3$,

$z = 4$, $P = [1;0;2]$ and $\delta = [1;3;0]$, matrix $\mathbf{S}$ is :

$$\mathbf{S} = \left[\begin{array}{cccc|cccc|cccc} 0&0&0&0&0&1&0&0&0&0&0&0 \\ 0&0&0&0&0&0&1&0&0&0&0&0 \\ 0&0&0&0&0&0&0&1&0&0&0&0 \\ 0&0&0&0&1&0&0&0&0&0&0&0 \\ \hline 0&0&0&1&0&0&0&0&0&0&0&0 \\ 1&0&0&0&0&0&0&0&0&0&0&0 \\ 0&1&0&0&0&0&0&0&0&0&0&0 \\ 0&0&1&0&0&0&0&0&0&0&0&0 \\ \hline 0&0&0&0&0&0&0&0&1&0&0&0 \\ 0&0&0&0&0&0&0&0&0&1&0&0 \\ 0&0&0&0&0&0&0&0&0&0&1&0 \\ 0&0&0&0&0&0&0&0&0&0&0&1 \end{array}\right]$$

As illustrated in Figure 1, the interleaving function obtained with this matrix transforms the sequence in natural order $\{0, 1, 2, 3, 4, 5, 6, 7, 8, 9, 10, 11\}$ into the interleaved sequence:

$$\{4, 9, 2, 7, 0, 5, 10, 3, 8, 1, 6, 11\}$$

2.2 Interleaver properties

Performance of turbo-codes are highly correlated by the quality of the interleaver. First the correlation between the extrinsic information under an iterative decoding can be minimized if short length cycles are avoided. Moreover low weight codewords can be generated when two bits which are relatively closed in the natural order are also closed after the interleaving process.

In order to characterize the quality of an interleaver, the notion of spread was introduced to design S-random interleavers [6]. The spread between two positions i and j is the sum of the distance between both i and j and the projection of these two positions after interleaving. Using tail biting termination, the minimum spread $S_{\min}$ between two bits i and j is defined by:

$$S_{\min} = \min_{i,j} \left(\Delta(i,j) + \Delta(\Pi(i), \Pi(j))\right) \tag{5}$$

where $\Delta(i,j) = \min\left(|i-j|, K-|i-j|\right)$. In the case of the proposed interleaver, the spread is at least $2m$ if

$$(\delta(i) - \delta(j)) \mod z \geq 3 \quad \forall i \neq j \tag{6}$$

This constraint must be taken into account for the design of interleaver. From equation (6), parameter z has to be greater than $3m$ to guarantee a spread of at least $2m$ resulting in the following constraint:

$$m < \sqrt{\frac{K}{3}} \tag{7}$$

Otherwise, when puncturing is used to increase the rate of the turbo codes, it is preferable to equally spread the punctured parity bit over the frame of information bits. For instance, if we consider a half-rate binary turbo code, the most common puncturing pattern is to alternate puncturing of the parity bits from each of the component codes. If the interleaver is designed in such a way that odd and even systematic bits are respectively kept in odd and even position after interleaving, one of the two parity bits associated with each information bit will be punctured. In the proposed interleaver, this condition can be satisfied if m is even by imposing that $P(i)$ and i are of same parity. It is possible to extend this property to other puncturing pattern by imposing other rules on the coefficient $P(i)$.

Definition and properties of our proposed interleaver have been discussed in this section. However the major interest of such a structure is the implementation simplicity. This aspect is investigated in the next section.

3. Implementation aspects

3.1 Avoiding memory collision

We consider implementation of a turbo decoder composed of two convolutional decoders. One way to avoid simultaneous memory access consists in dividing the memory in multiple banks. If we suppose that each decoder is able to decode p sections of the trellis in parallel, the interleaving process must not access more than one value of a given memory bank among the p decoded values. With this specific structure, a first condition to get a parallelism of p sets that K must be a multiple of p.

In particular, when z is a multiple of p, memory allocation is made easier. The p elements decoded simultaneously by decoder one are in position $v + uK/p$, where $u \in [0, p-1]$ is the index of the subframe and $v \in [0, K/p-1]$ is the position of the bit into this subframe. Considering $v = i + lm$, we notice that after the first row-column permutation these p elements are read/written from the same group i of z elements in positions:

$$(l + u\,\frac{z}{p}) \bmod z, \quad u \in [0, p-1]$$

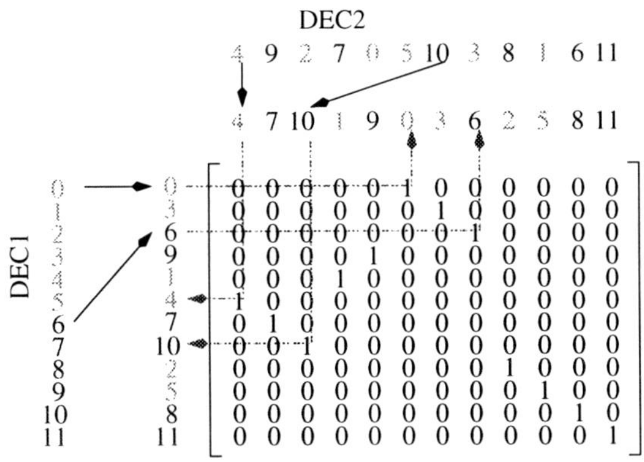

Figure 1. Collision free access in a two-bank memory. At each cycle one data is read/written in each memory bank for both decoders. The two-memory-banks data are respectively represented in light green and dark blue colors.

Thus, the decoder two reads/writes these p elements in same group $P(i)$ at positions:

$$\left(\delta(i) + l + u\ \frac{z}{p}\right) \quad \text{mod } z, \quad u \in [0, p-1]$$

Then if memory is divided into p banks, each bank x containing elements from xK/p to $((x+1)K/p) - 1$, memory collisions are avoided when simultaneous read/write access is needed as shown in Figure 1. In this example, the frame of 12 bits is decoded with a parallelism factor of 2. Thus, data in position 0 and 6 are accessed simultaneously by the first decoder. These two data are loaded from two different memory banks. At the decoder 2 side, data loaded simultaneously are always in two memory banks, as the data in position 4 and 10.

3.2 On simple address generation

In the particular case where z is a multiple of p, the number of memory addresses simultaneously required is reduced. Addresses are easily computed from i, l and u for decoder one and $P(i)$, $(\delta(i) + l)$ and u for decoder two. Thanks to properties of circularly shifted matrices, parameters i and l (resp. $P(i)$ and $\delta(i) + l$) are constant for the p accesses. These p addresses are so obtained by considering $u = 0$ to $\mathrm{p} - 1$: in each memory bank the address is determined from i and l (resp. $P(i)$ and $\delta(i) + l$), and the memory bank number is determined by parameter u. In the example of Figure 1, data in position 0 and 6 loaded by decoder 1 are both in position 0 in their own memory bank and the results will be

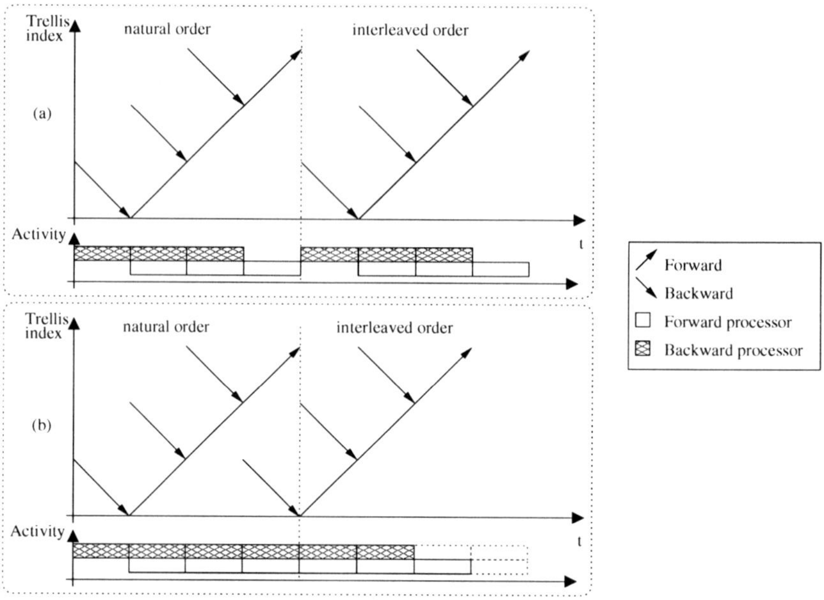

Figure 2. Scheduling of the classical sliding windows decoding algorithm and activities of the two sub processors (a). Part (b) illustrates how to increase activities of each processor. Forward and backward processors are activated continuously during the decoding of a frame.

rewritten into the same position. Identically, the first values read and written by decoder 2 are both in position 4 of their memory bank.

3.3 Reducing the number of interleaver parameters

As previously explain, the whole interleaver is characterized thanks to $2m+2$ parameters, m, z, $P(k)$ and $\delta(k)$ for $k \in [0, m-1]$. In many applications such as 3GPP, the number of block size and coding rate configurations required is large. The system has to store the set of parameters for each configuration. Under this assumption, interleavers with only a few set of configuration parameters are preferable. In order to reduce the number of parameters characterizing the interleaver, $P(k)$ and $\delta(k)$ can be defined through a simple permutation law. First $P(k)$ is described by a Relative Prime (RP) law characterized by only two integers p_0 and p_1:

$$P(k) = (p_0 + p_1 k) \bmod m \tag{8}$$

In a similar manner, the set of m parameters $\delta(k)$ can be computed through a RP law with two integers d_0 and d_1:

$$\delta(k) = (d_0 + d_1 k) \bmod z \tag{9}$$

With this proposed characterization, the whole interleaver is defined thanks to 6 parameters.

3.4 Interleaver design rules for a particular scheduling

In an efficient implementation of a turbo decoder, the same decoding processor is used for both constituent sub-decoders. This decoding processor is divided into backward and forward metric calculators. To optimize the latency of decoding, it is preferable to use in a continuous way these two calculators without waiting between the decoding of each constituent code (see Fig. 2). As studied in [4], this scheduling imposes constraints on the interleaver to enable the last forward processing of a sub-decoder in parallel with the first backward processing of the other sub-decoder. The interleaver must be designed in such a way that the extrinsic information needed for a decoding has been already computed by the other decoding. This constraint can easily be respected by our interleaver by fixing forbidden values on the parameters $\delta(i)$.

We assume that the length of the decoding window is equal to the parameter m of the interleaver. In order to impose that the inputs of the first window of a constituent sub-decoder are not present in the last window processed by the other sub-decoder, it is sufficient to forbid the value 1 and $z-1$ for the parameters $\delta(i)$.

More generally, for a parallelism factor of p, the forbidden values for parameter $\delta(i)$ are:

$$\left\{q\frac{z}{p} + 1; (q+1)\frac{z}{p} + 1 \mid q = 0 \cdots p-1\right\} \tag{10}$$

Other design rules can be derived very simply for various scheduling of turbo decoding.

4. Simulation Results

In this section we evaluate performance of turbo decoding with the proposed interleaver. We compare on Figure 3 performance of 3GPP interleaver [7] and the proposed structure for a rate 1/3 turbo code decoded using max-log-map algorithm with 8 iterations. On the first subfigure, the size of the interleaver is $K = 640$ with parameters $m = 10$ and $z = 64$. The parameters $\delta(i)$ and $P(i)$ are chosen for maximizing the

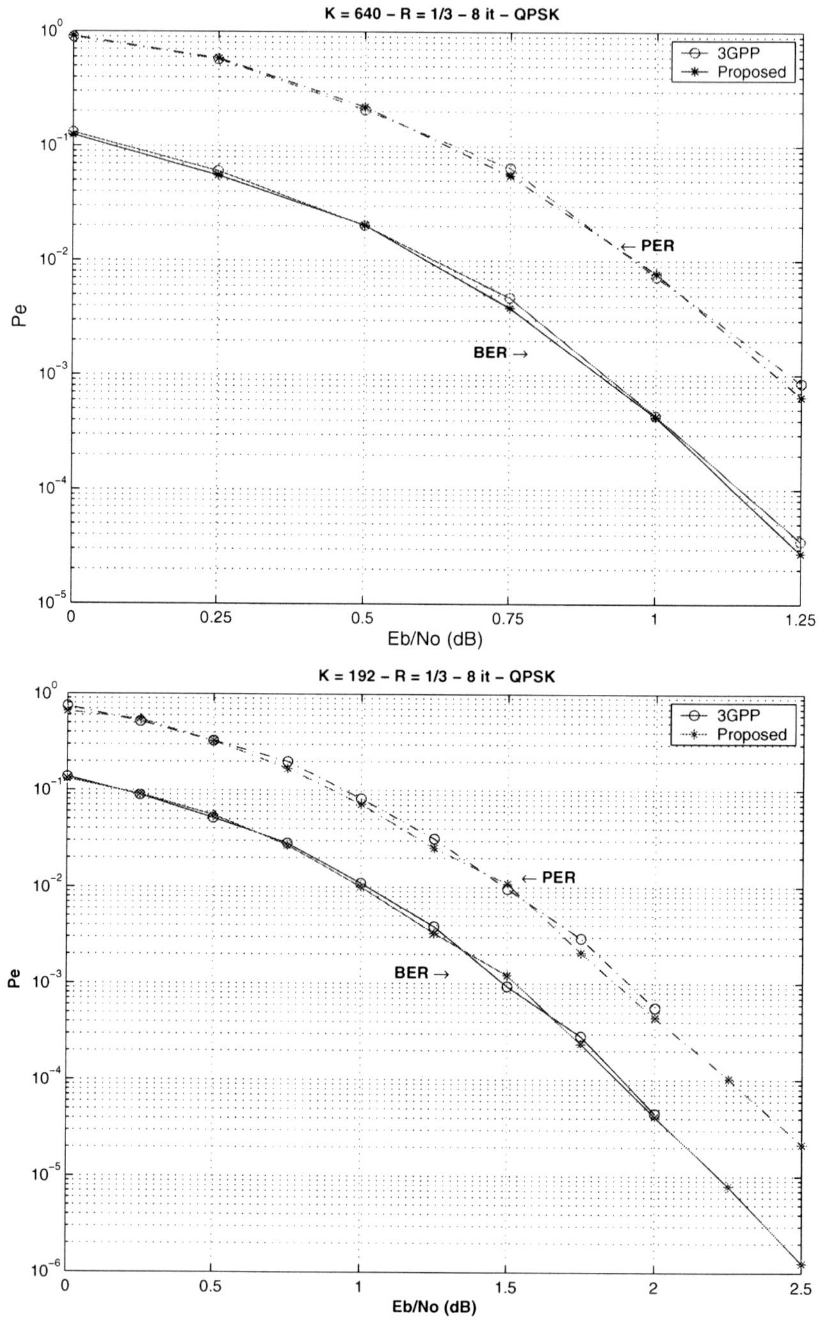

Figure 3. Performance of the proposed interleaver for AWGN channel. $R = 1/3$, circular 8-states binary turbo-code, 8 iterations of Max Log-Map

spread without other constraints. On the second subfigure, the interleaver of size $K = 192$ has for parameters $m = 8$, $z = 24$, $P(i) = i$ and $\delta(i) = (7i + 12) \mod 24$.

In both cases the proposed interleavers provide no degradation compare to the corresponding 3GPP interleavers. Moreover they enable a parallelization level of any divisor of both 640 and 192 without any memory conflict.

5. Conclusion

In this paper, we have proposed a new interleaver for turbo-codes. This interleaver is designed from shifted identity matrices, enabling a simple characterization. In particular case, the interleaver can be described through 6 parameters enabling an inline computation of the addresses. We also proposed design rules for interleaver to guarantee a minimum spread and to adapt to a puncturing pattern. We demonstrate that a high parallelized turbo decoding is possible without memory contention. The interleaving function can be easily adapted to an optimized decoding scheduling. Simulation results illustrate that the proposed structure enable both good performance and a low-complex implementation.

References

[1] C. Berrou, A. Glavieux, and P. Thitimajshima. Near shannon limit error-correcting coding and decoding:turbo-codes. *IEEE International Conference on Communications*, 2, may 1993.

[2] http://www.3gpp.org/highlights/lte/lte.htm.

[3] S. Benedetto, D. Divsalar, G. Montorsi, and A. F. Pollara. Soft output decoding algorithms for continuous decoding of parallel concatenated convolutional codes. *IEEE International Conference on Communications*, June 1996.

[4] D. Gnaedig, E. Boutillon, J. Tousch, and M. Jezequel. Towards an optimal parallel decoding of turbo codes. *4th International Symposium on Turbo-Codes and Related Topics*, April 2006.

[5] A. Tarable, S. Benedetto, and G. Montorsi. Mapping interleaving laws to parallel turbo and LDPC decoder architectures. *IEEE Transactions on Information Theory*, Sept 2004.

[6] S. Dolinar and D. Divsalar. Weight distributions for turbo codes using random and nonrandom permutations. *TDA Progress Report 42-122, JPL*, Aug 1995.

[7] 3GPP TS 25.212 v6.4.0. *Multiplexing and Channel Coding (FDD) (Release 6)*, (2005-03).

MULTI-USER DETECTION TECHNIQUES FOR POTENTIAL 3GPP LONG TERM EVOLUTION (LTE) SCHEMES

Qinghua Guo, Xiaojun Yuan and Li Ping
Department of Electronic Engineering, City University of Hong Kong, Hong Kong SAR
eeliping@cityu.edu.hk

Abstract: This paper is concerned with multi-user detection techniques for orthogonal frequency division multiplexing – interleave division multiple access (OFDM-IDMA) and IDMA with cyclic prefix (IDMA-CP). These two schemes have attractive features such as low-cost multi-user detection/equalization and flexible rate adaptation. We will show that significant performance improvement (i.e., the so-called multi-user gain) can be achieved by allowing concurrent multiple user transmission on common carries.

Key words: Multi-carrier, Single-carrier, multi-user detection, IDMA, OFDM, FDE.

1. INTRODUCTION

Multiple access interference (MAI) and inter-symbol interference (ISI) are major hazards in wireless communication systems. Conventional multi-user detection and time domain equalization techniques are very costly. It has been shown that ISI can be efficiently resolved by orthogonal frequency division multiplexing (OFDM) [1] and single carrier transmission with frequency domain equalization (SC-FDE) [8] techniques. OFDM has the advantage of allowing water-filling among frequency sub-carriers and SC-FDE has the advantage of low peak-to-average-power-ratio. These two

This work was supported by a grant from the Research Grant Council of the Hong Kong Special Administrative Region, China, under Project CityU 116706.

S. Plass et al. (eds.), Multi-Carrier Spread Spectrum 2007, 77–86.

techniques are currently considered for the 3rd Generation Partnership Project Long Term Evolution (3GPP LTE).

In this paper, we examine multi-user transmission options for OFDM and SC-FDE. The interleave division multiple access (IDMA) [4] principle is combined with OFDM and SC-FDE respectively, resulting in the so-called OFDM-IDMA [2][3] and IDMA with cyclic prefix (IDMA-CP) schemes. In both schemes, we assume that the synchronization errors between users are within the range of CP extension so that it can be treated together with the multi-path delay spread of individual users. MAI and ISI in both schemes can be resolved efficiently using iterative frequency domain detection techniques. The related detection costs are very low. For both OFDM-IDMA and IDMA-CP receivers, the normalized cost (per user) is independent of simultaneous user number and path number. We will show that significant performance improvement (i.e., the so-called multi-user gain (MUG) [9][10]) can be achieved by allowing concurrent multi-user transmission on common carries, which is justified by both information theory study and simulation results.

Notations: ENC, SP, DEC and DES denote encoder, spreader, decoder and de-spreader, respectively. Π_k denotes interleaver for user k and Π_k^{-1} the corresponding de-interleaver. DFT and IDFT indicate the discrete Fourier transform and its inverse respectively. ACP denotes the operation of adding cyclic prefix and RCP removing cyclic prefix.

2. MOTIVATION

Fig. 1 shows the required average transmitted sum power versus system sum rate for different multiple access schemes over quasi-static flat fading channels. From this figure, we can see that there are significant gaps between the theoretical limit and the required transmitted sum power of orthogonal frequency division multiple access (OFDMA) and conventional CDMA with single user detection. Comparing Fig. 1(a) with Fig. 1(b), we can also see that the gap increases with the number of simultaneous users. The theoretical limit shown in Fig. 1 can only be achieved by schemes with multi-user transmission and detection [9][10]. This advantage of multi-user systems is referred to as multi-user gain (MUG). Although only flat fading channels are considered in this figure, the trend is the same for frequency selective channels, which is demonstrated by the simulations in section 5.

Note that Fig. 1 is for a single-cell environment. Since transmission power saving can lead to interference reduction, we expect that MUG can bring about significant improvement in multi-cell performance. Also, other considerations arise for a multi-cell environment. For example, the spreading gain of a CDMA type system can mitigate the worst case cross-cell interference, which can be beneficial to curb fluctuation in carrier to interference (C/I) ratio. This is partially the reason why CDMA is used in 3G systems. However, a detailed discussion on multi-cell scenario is beyond the scope of this paper.

The above provides the motivation of work, i.e., how to realize MUG, which is the focus of the discussion below.

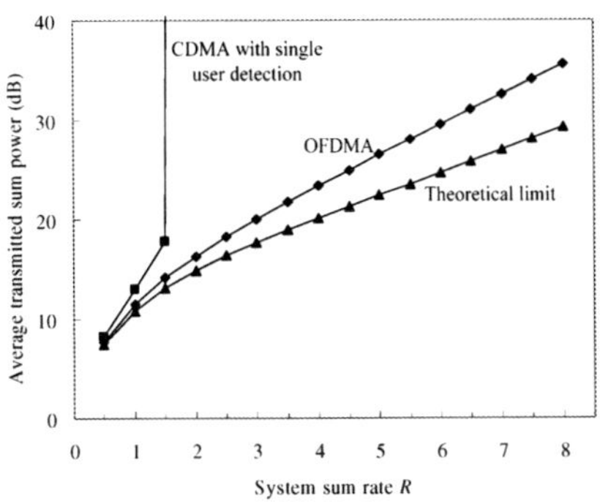

(a) 2 simultaneous users

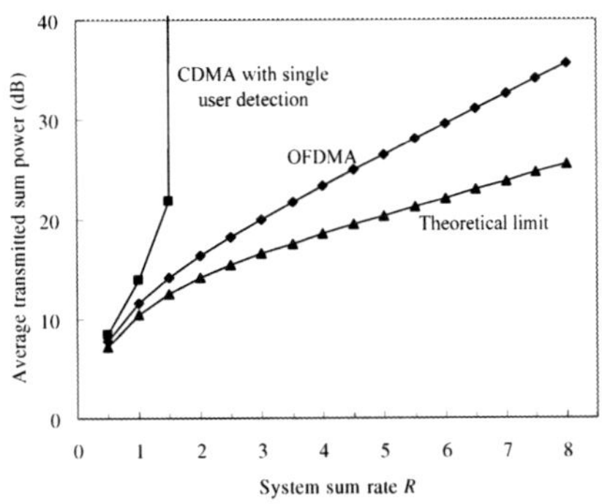

(b) 4 simultaneous users

Figure 1. The average transmitted sum power versus system sum rate for different multiple access schemes in a single-cell environment. The users are uniformly distributed in a single hexagon cell. All the users have the same rate R/K (K denotes the user number). The channel effect includes path loss (fourth law), Rayleigh fading and lognormal fading (standard deviation = 8dB). The distance from the farthest corner of the cell to the base station is normalized to 1. Transmitters have the knowledge of channel information and an outage probability of 0.01 is allowed.

3. OFDM-IDMA

The OFDM-IDMA system shown in Fig. 2 follows the principles proposed by Mahafeno, Langlais and Jego. Supposition coded modulation [5] is employed when high single user rate is required. The input data sequence $\boldsymbol{d}_k$ for user k is encoded into a coded bit sequence $\boldsymbol{c}_k$[1]. Then, the coded bits are interleaved by a user-specific interleaver Π_k, and mapped to a complex sequence $\{X_k(n)\}$ by using quadrature-phase-shift-keying (QPSK). The sequence is modulated onto sub-carriers via IDFT before transmission. At receiver side, the received signals after OFDM demodulation are given by

[1] In order to achieve high frequency diversity in an OFDM-IDMA system, a low-rate code is needed. A simple way to realize a low-rate code is to concatenate a common FEC code (e.g., rate-1/2 convolutional code) with a repetition code.

$$Y(n) = \sum_{k=1}^{K} H_k(n) X_k(n) + Z(n), \quad n = 0,1,\cdots,N-1 \tag{1}$$

where N is the number of sub-carriers and $\{Z(n)\}$ are samples of a complex additive white Gaussian noise (AWGN) with zero mean and variance $N_0/2$ per dimension. The coefficients $\{H_k(n)\}$ are the DFT of the L channel taps $\{h_k(l),\ l = 0, 1,\ldots L\text{-}1\}$ seen by user k. We focus on a particular bit carried by $X_k^{\mathrm{Re}}(n)$. Let $H_k(n) = |H_k(n)| e^{j\theta}$. Then (1) can be rewritten as

$$\mathrm{Re}\left(e^{-j\theta} Y(n)\right) = |H_k(n)| X_k^{\mathrm{Re}}(n) + \xi_k^{\mathrm{Re}}(n), \tag{2}$$

where $\xi_k^{\mathrm{Re}}(n) = \mathrm{Re}(e^{-j\theta} \sum_{m\neq k} H_m(n) X_m(n) + e^{-j\theta} Z(n))$ represents the interference and noise components in Re($e^{-j\theta}Y(n)$) with respect to $X_k^{\mathrm{Re}}(n)$. From the central limit theorem, $\xi_k^{\mathrm{Re}}(n)$ can be approximated by a Gaussian random variable.

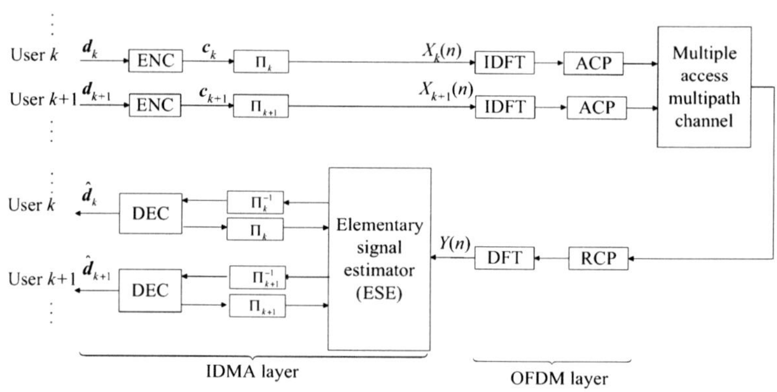

Figure 2. The transmitter/receiver structure for OFDM-IDMA.

The core of the OFDM-IDMA receiver shown in Fig. 2 consists of the following iteration process carried out by the elementary signal estimator (ESE) and *a posteriori* probability decoders (APP DECs). The discussion below is brief. The interesting readers may refer to [2][4] for details.

The ESE Operations: The ESE computes the extrinsic log-likelihood ratio (LLR) about $X_k^{\mathrm{Re}}(n)$, $\forall n, k$, given by

$$e_{ESE}\left(X_k^{\mathrm{Re}}(n)\right) = \ln \frac{\Pr(\mathrm{Re}(e^{-j\theta}Y(n)) \mid X_k^{\mathrm{Re}}(n) = +1)}{\Pr(\mathrm{Re}(e^{-j\theta}Y(n)) \mid X_k^{\mathrm{Re}}(n) = -1)} = \frac{2|H_k(n)|}{\mathrm{Var}\left(\xi_k^{\mathrm{Re}}(n)\right)} \left(\mathrm{Re}\left(e^{-j\theta}Y(n)\right) - \mathrm{E}\left(\xi_k^{\mathrm{Re}}(n)\right)\right). \tag{3}$$

The method for calculating $E\left(\xi_k^{\mathrm{Re}}(n)\right)$ and $Var\left(\xi_k^{\mathrm{Re}}(n)\right)$ can be found in [4]. The treatment for $X_k^{\mathrm{Im}}(n)$ is similar.

The DEC Operations: The extrinsic LLRs calculated by ESE are delivered to the DEC for APP soft output decoding. The outputs of DECs are then used to refine $\mathrm{E}\left(\xi_k^{\mathrm{Re}}(n)\right)$ and $\mathrm{Var}(\xi_k^{\mathrm{Re}}(n))$ in the next iteration.

OFDM-IDMA supports flexible allocation of resources among users. Suppose that a repetition code is employed for frequency diversity. High-rate transmission can be supported by the superposition coding techniques [5]. Superposition coding provides a convenient means for rate adaptation by assigning an appropriate number of code streams to each user. The adjustment step size can be made very fine by using low-rate repetition codes. This is useful for resource allocation as well as water-filling. For example, each user may superimpose a different number of code streams on different sub-carriers according to channel conditions. As a comparison, rate adaptation using conventional coded modulation schemes can be a difficult issue: a large collection of different encoders/decoders may be needed if we want a large adaptation range and/or a fine adjustment step size.

4. IDMA-CP

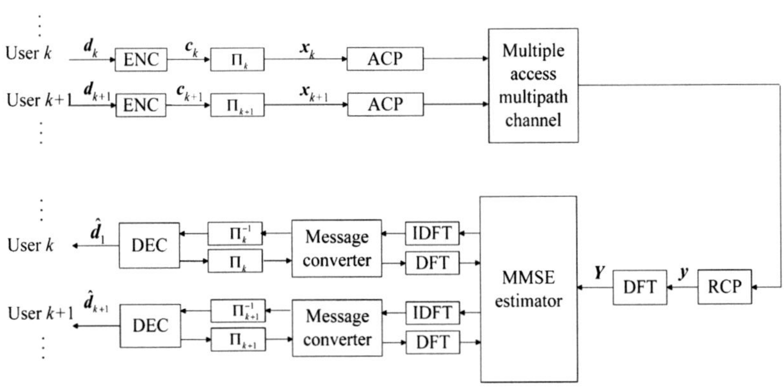

Figure 3. Transmitter/receiver structure for IDMA-CP.

SC-FDE is an alternative technique to combat severe ISI in wireless communications [8]. The IDMA-CP scheme shown in Fig.3 is a multiple access technique based on iterative SC-FDE techniques [11]. The encoded bit stream $\boldsymbol{c}_k$ is interleaved by a user-specific interleaver Π_k, QPSK modulated and padded with cyclic prefix before transmission. Let N be the length of $\boldsymbol{x}_k = \{x_k(n)\}$ and $\boldsymbol{X}_k = \{X_k(n)\}$ be DFT of $\boldsymbol{x}_k$, i.e.,

$$X_k(n) = \frac{1}{\sqrt{N}} \sum_{m=0}^{N-1} x_k(m) e^{-j2\pi mn/N}, \quad n = 0, 1, \ldots, N\text{-}1 \tag{4}$$

where $j = \sqrt{-1}$. At the receiver side, CP is first removed. DFT is then applied to the received signal $\boldsymbol{y}$, producing $\boldsymbol{Y} = \{Y(n)\}$:

$$Y(n) = \sum_{k=1}^{K} H_k(n) X_k(n) + Z(n), \qquad n = 0, 1, \ldots, N\text{-}1. \tag{5}$$

Similar to (2), we rewrite (5) in a signal plus distortion form as

$$Y(n) = H_k(n) X_k(n) + \xi_k(n), \quad n = 0, 1, \ldots N\text{-}1 \tag{6}$$

where $\xi_k(n) = \sum_{m \neq k} H_m(n) X_m(n) + Z(n)$ represents the signal plus noise component. We again approximate $\xi_k(n)$ by a Gaussian random variable.

Before going into the details in the detection algorithm, we first explain the computation of statistical parameters (i.e., means and variances) for the signals in Fig. 3. Let us treat the entries in $\boldsymbol{x}_k$ as random variables and assume that their means $\mathrm{E}(\boldsymbol{x}_k)$ are given. Let $\boldsymbol{F}$ be the normalized DFT matrix with the (m, n)-th entry given by $N^{-1/2}\exp(-j2\pi mn/N)$. Then the means of $\boldsymbol{X}_k = \{X_k(n)\}$ and $\boldsymbol{Y} = \{Y(n)\}$ can be computed as

$$\mathrm{E}(\boldsymbol{X}_k) = \boldsymbol{F}\mathrm{E}(\boldsymbol{x}_k) \tag{7a}$$

$$\mathrm{E}(\boldsymbol{Y}) = \sum_{k=1}^{K} \boldsymbol{H}_k \mathrm{E}(\boldsymbol{X}_k), \tag{7b}$$

where $\boldsymbol{H}_k$ is a diagonal matrix, i.e., $\boldsymbol{H}_k = \mathrm{diag}[H_k(0), \ldots, H_k(N\text{-}1)]$. We assume that all $\{x_k(n)\}$ have symmetric Gaussian distributions, i.e., their real and imaginary parts have the same variance. Furthermore, we assume these variances are the same and denoted by

$$\mathrm{Var}(x_k(n)) = v_{x_k}, \quad \forall n \text{ and for a fixed } k. \tag{8}$$

(We will come back to this assumption later when we discuss message converters.) With this assumption, the variances of $\{X_k(n)\}$ are given by

$$\mathrm{Var}(X_k(n)) = v_{x_k}, \quad \forall n \text{ and for a fixed } k. \tag{9a}$$

The variance of $Y(n)$ can be computed as

$$\mathrm{Var}(Y(n)) = \sum_{k=1}^{K} |H_k(n)|^2 \, \mathrm{Var}(X_k(n)) + N_0. \tag{9b}$$

We are now ready to present the detection algorithm. Note that $X_k(n)$ is not a binary variable. Instead, it can be approximated by a Gaussian random variable by applying the central limit theorem to (4). Therefore, MMSE estimation is used below to estimate $X_k(n)$. This is the key difference between the detection algorithm in Section 3 and the one presented here.

The MMSE Operations: Generate the MMSE estimate (characterized by the *a posterior* mean and variance) for $X_k(n)$ [6]

$$\mathrm{E}^{post}\left(X_k(n)\right) = \mathrm{E}\left(X_k(n)\right) + \frac{\mathrm{Var}\left(X_k(n)\right) H_k^*(n)\left(Y(n) - \mathrm{E}\left(Y(n)\right)\right)}{\mathrm{Var}(Y(n))} \tag{10a}$$

$$\mathrm{Var}^{post}\left(X_k(n)\right) = \mathrm{Var}\left(X_k(n)\right) - \frac{\left(\mathrm{Var}\left(X_k(n)\right)\right)^2 |H_k(n)|^2}{\mathrm{Var}\left(Y(n)\right)}. \tag{10b}$$

DFT is applied to transform the estimates of $\{X_k(n)\}$ to those of $\{x_k(n)\}$:

$$\mathrm{E}^{post}\left(\boldsymbol{x}_k\right) = \boldsymbol{F}^H \mathrm{E}^{post}\left(\boldsymbol{X}_k\right), \tag{11a}$$

$$\mathrm{Var}^{post}\left(x_k(n)\right) = v_{x_k}^{post}, \quad \forall n \text{ and for a fixed } k \tag{11b}$$

where $v_{x_k}^{post} = (1/N)\sum_{n=0}^{N-1} \mathrm{Var}^{post}\left(X_k(n)\right)$. Due to the use of interleavers, we will assume that the correlation among $\{\mathrm{E}^{post}(x_k(n))\}$ can be ignored.

The Message Conversion Operations: The kth message converter in Fig. 3 performs the following functions.

- In the direction from the DEC to the MMSE estimator, it converts the DEC outputs in terms of LLR into the corresponding means $\{\mathrm{E}(x_k(n))\}$ and variances $\{\mathrm{Var}(x_k(n))\}$ [4]. As mentioned for (8), we assume that $\mathrm{Var}(x_k(n))$ are the same for a fixed k and different n. In practice, this is realized by approximating $\{\mathrm{Var}(x_k(n)), \forall n\}$ using their average v_{x_k}.
- In the direction from the MMSE estimator to the DEC, it converts the *a posteriori* means and variances of $\{x_k(n)\}$ to the extrinsic LLRs of $\{x_k^{\mathrm{Re}}(n)\}$ and $\{x_k^{\mathrm{Im}}(n)\}$ (assuming the *a posteriori* variances of $x_k^{\mathrm{Re}}(n)$ and $x_k^{\mathrm{Im}}(n)$ are the same) as [6]

$$e\left(x_k^{\mathrm{Re}}(n)\right) = 4\left(\frac{\mathrm{Re}\left(\mathrm{E}^{post}\left(x_k(n)\right)\right)}{v_{x_k}^{post}} - \frac{\mathrm{Re}\left(\mathrm{E}\left(x_k(n)\right)\right)}{v_{x_k}}\right). \tag{12}$$

Similarly, $e(x_k^{\mathrm{Im}}(n))$ can be obtained.

<u>The DEC Operations</u>: The kth DEC performs soft output decoding based on inputs $\{ e(x_k^{\mathrm{Re}}(n)) \}$ and $\{ e(x_k^{\mathrm{Im}}(n)) \}$. Its outputs are converted into means and variances by the kth message converter and used by the MMSE estimator as *a priori* information in the next iteration.

Similar to OFDM-IDMA, IDMA-CP can also be implemented based on superposition coding techniques for flexible resource allocation among users.

5. SIMULATION RESULTS AND DISCUSSIONS

We first show that OFDM-IDMA and IDMA-CP have similar complexity as well as performance.

For complexity analysis, the computational costs in all the operations (i. e., ESE, DEC, MMSE and message conversion) are independent of user number and path number. The total cost of DFT (using FFT) is $O(N\mathrm{Log}_2N)$, or equivalently, $O(\mathrm{Log}_2N)$ per bit. Also, the receiver cost increases linearly with iteration number.

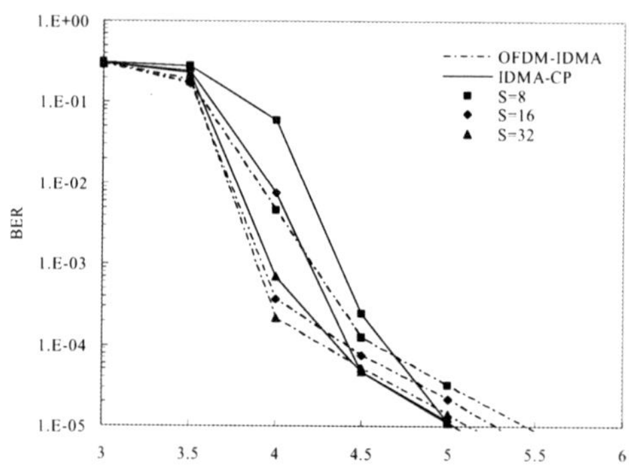

Figure 4. Performance comparison between OFDM-IDMA and IDMA-CP with different length of repetition code *S*. The information block length for both systems is 512. The system sum rate is 2. The number of iterations for both systems is 15.

Fig.4 shows the performance of OFDM-IDMA and IDMA-CP. The coding scheme for both systems is a rate-1/2 convolutional code (with generator $(23, 35)_8$) followed by a length-*S* repetition code and QPSK modulation. The uncorrelated fading channel model [1] is used. The system sum rate is fixed to 2, so the number of simultaneous users is 2*S*.

Overall, OFDM-IDMA and IDMA-CP can offer similar performance. There are some subtle differences. OFDM-IDMA performs slightly worse in the low BER range due to the frequency selectiveness among different

OFDM sub-carriers, but the difference reduces when spreading length S increases. On the other hand, IDMA-CP performs slightly worse in the waterfall range, which may be caused by the approximation in (8).

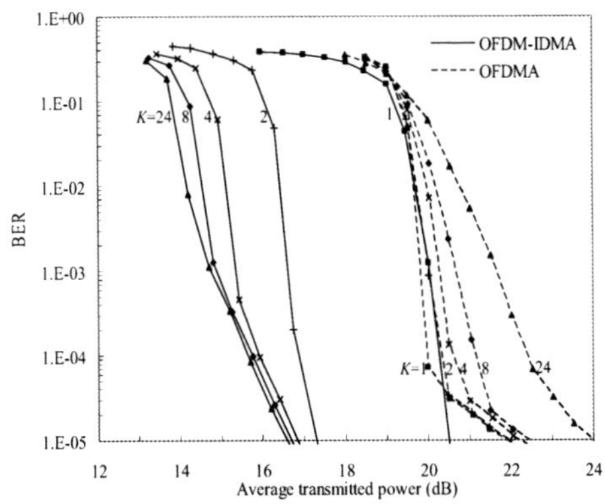

Figure 5. Performance comparison between OFDM-IDMA and OFDMA with throughput 3. The number of iterations for both systems is 10. In OFDM-IDMA, the information block length for each code stream is 512 (i. e., 1024 coded bits after a rate-1/2 FEC coding). The total information block length for each user is $(512\times24)/K$ for both OFDM-IDMA (with $24/K$ code streams per user) and OFDMA.

Next, we examine MUG as discussed in Section 2. Fig. 5 shows the performance of the worst user (with the lowest power) in an OFDM-IDMA system (with superposition techniques [5]) using a rate-1/2 convolutional code (with generator $(23, 35)_8$) followed by a length-8 repetition code and QPSK modulation. We fix the system sum rate R to 3, so 24 code streams are involved. We assume that all users have the same number of streams. Thus each user is assigned one stream when the number of users $K = 24$, and two streams when $K = 12$, etc. The channel conditions considered here are the same as those in Fig. 1 except that the effect of frequency selectiveness is added (The uncorrelated fading channel model [1] is used). The unequal power allocation approach follows from [4], and the channel matching strategy [4] are used based on the received power profile shown in table 1. For comparison, we also show the OFDMA performance (with the same throughput 3) based on a BICM scheme using 16-QAM modified set partitioning mapping [7].

It can be observed from Fig. 5 that the performance of OFDMA degrades with increasing K. (This performance loss is caused by the decrease of block length when K increases.) In contrast, the power efficiency of OFDM-IDMA improves as K increases due to MUG. We observe that, when $K = 4$, OFDM-IDMA can achieve about 4 dB gain compared with OFDMA. This agrees with the gap shown in Fig. 1(b). Clearly, the impact of MUG is significant. Similar gain can also be shown for IDMA-CP.

Table 1 THE RECEIVED POWER PROFILE FOR AN OFDM-IDMA SYSTEM WITH THROUGHPUT 3

Code stream number	13	2	2	7
Power level (dB)	0	3.1677	3.9596	5.5430

6. CONCLUSIONS

In this paper, we have outlined the basic principles of OFDM-IDMA and IDMA-CP for wireless communications and showed their merits including low-complexity MUD detection, flexible resource allocation among users. We also show their significant power efficiency improvement (MUG) compared with OFDMA. In conclusion, OFDM-IDMA and IDMA-CP offer two attractive options for future wireless communications.

REFERENCES

[1] K. Fazel and S. Kaiser, *Multi-Carrier and Spread Spectrum Systems*. Chichester: John Wiley & Sons, 2003.

[2] I. Mahafeno, C. Langlais, and C. Jego, "OFDM-IDMA versus IDMA with ISI cancellation for quasi-static Rayleigh fading multipath channels," in *Proc. 4th Int. Symp. on Turbo Codes & Related Topics*, Munich, Germany, Apr. 3-7, 2006.

[3] S. Zhou, Y. Li, M. Zhao, X. Xu, J. Wang, and Y. Yao, "Novel techniques to improve downlink multiple access capacity for beyond 3G, " *IEEE Commun. Mag.*, vol. 43, pp. 61-69, Jan. 2005.

[4] L. Liu, J. Tong, and Li Ping, "Analysis and optimization of CDMA systems with chip-level interleavers," IEEE *J. Select. Areas Commun.*, vol. 24, pp. 141-150, Jan. 2006.

[5] J. Tong, Li Ping, and X. Ma, "Superposition coding with peak-power limitation," in *Proc. IEEE Int. Conf. on Commun.* (ICC'06), Istanbul, Turkey, 11-15 June 2006.

[6] Q. Guo, Li Ping, and H. Loeliger, "Turbo equalization based on factor graphs," in *Proc. IEEE Int. Symp. Inform. Theory* (ISIT), pp.2021-2025, Australia, 4-9 Sept. 2005.

[7] J. Tan and G. L. Stüber, "Analysis and design of symbol mappers for iteratively decoded BICM," *IEEE Trans. Wireless Commun.*, vol. 4, pp. 662-672, Mar. 2005.

[8] D. Falconer, S. L. Ariyavisitakul, A. Benyamin-Seeyar and B. Eidson, "Frequency domain equalization for single-carrier broadband wireless systems," *IEEE Commun. Mag.*, pp. 58-66, April 2002.

[9] P. Wang, J. Xiao, and Li Ping, "Comparison of orthogonal and non-orthogonal approaches to future wireless cellular systems," *IEEE Vehicular Technology Magazine*, vol. 1, no. 3, pp. 4-11, Sept. 2006.

[10] D. Tse and P. Viswanath, *Fundamentals of Wireless Communication*, Cambridge: Cambridge University Press, 2005.

[11] M. Tüchler, and J. Hagenauer, "Turbo equalization using frequency domain equalizers," in *Proc. Of Allerton Conference*, Monticello, IL, USA, Oct. 2000.

ANALYSIS AND PERFORMANCE OF AN EFFICIENT ITERATIVE DETECTION STRATEGY FOR IDMA SYSTEMS

Petra Weitkemper
and Karl-Dirk Kammeyer
University of Bremen
Dept. of Communications Engineering
Otto-Hahn-Allee NW 1, D-28359 Bremen
{weitkemper,kammeyer}@ant.uni-bremen.de

Abstract IDMA is a multiple access scheme closely related to CDMA and of great interest for future mobile networks. In contrast to orthogonal schemes like TDMA or FDMA collisions are not avoided but the interference is treated by iterative detection exploiting channel coding. By tolerating a certain amount of interference a higher spectral efficiency is achieved. The number of supportable users can be further increased by an optimized power profile. In order to decode all users with reasonable complexity, efficient detection strategies should be considered. This paper investigates a complexity reduction for parallel interference cancellation (PIC) by means of decoding only promising users in the first iterations, which means a user- and iteration-wise switching strategy. The optimal strategy is found by variance transfer charts (VTC) and confirmed by bit error rate simulations. Furthermore, a scheduling strategy for successive interference cancellation (SIC) is considered and shown to be most efficient in terms of channel decoder evaluations.

1. Introduction

Iterative multiuser detection is widely used in CDMA systems and has been improved in the last years. IDMA can be regarded as a special case of CDMA where the spreading is done by an arbitrary low rate channel code equal for all users and the separation is done by user-specific interleavers as in Ping et al., 2006. Depending on the channel code additional coding gain can be introduced. IDMA has many advantages known from CDMA but some interesting differences. Interleaving is done on chip-level and therefore, the interleaver length is the spreading factor times higher than for CDMA which is advantageous for iterative detection. The channel decoder and the multi-layer detector

S. Plass et al. (eds.), Multi-Carrier Spread Spectrum 2007, 87–96.

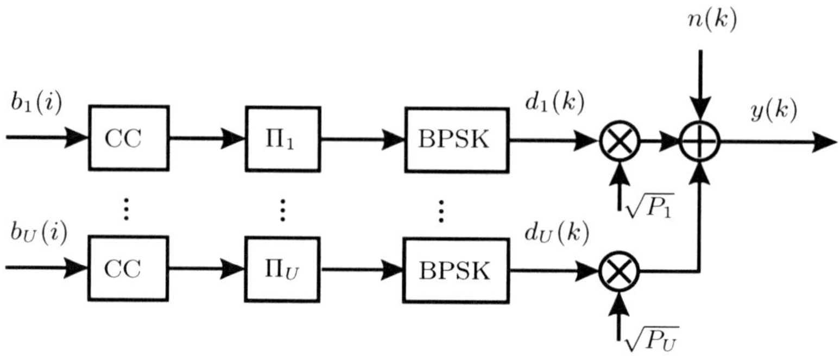

Figure 1. Transmitter of IDMA uplink

can be regarded as a serially concatenated system, so turbo-like detection is applied. Variance analysis of the parallel interference cancellation (PIC) in CDMA systems has been done e.g. in Schlegel and Shi, 2004 and can be easily applied to IDMA.

Although iterative detection has a complexity only linearly increasing e.g. with the number of users and iterations, it is desirable to further decrease detection complexity. In Kusume and Bauch, 2006, some simple complexity reduction techniques are investigated to be used in ad hoc networks. In this case not all users have to decode all layers of the other users. The channel code consists of a convolutional code followed by a repetition code. It is suggested to split up this concatenated channel code and only decode the repetition code for some users or some iterations to save complexity.

In this paper an uplink scenario is considered where the base station needs to decode all layers completely. For this case a user- and iteration-wise strategy is proposed and the choice of parameters is motivated by VTC analysis. This scheme does not effect the overall performance as long as the total number of iterations is sufficiently large. It will be shown that despite an increased number of iterations the overall complexity in terms of channel decoder evaluations will be significantly decreased. Furthermore, successive interference cancellation (SIC) will be considered as it is known to require less iterations than PIC. For this case a scheduling strategy for further complexity reduction will be derived. The paper is organized as follows: Section 2 introduces the system model of the considered IDMA uplink. In Section 3 the basics of analysis with variance transfer charts (VTC) are given and are used for complexity reduction of PIC and SIC in Section 4. Finally, Section 5 concludes this paper.

2. System Model

In this section the structure of a multiuser IDMA system is introduced. For simplicity a synchronous IDMA system over an AWGN channel with real val-

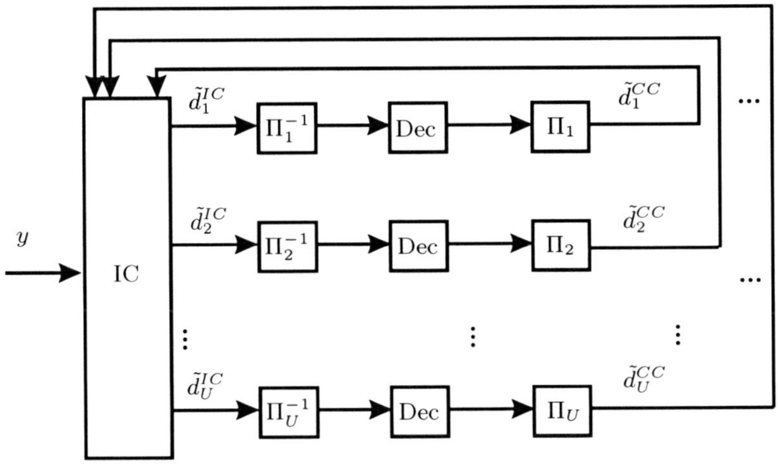

Figure 2. Iterative receiver for IDMA uplink

ued noise of variance σ_n^2 is assumed where each user transmits only one layer. The channel code consists of a serial concatenation of a $[5,7]_o$ outer convolutional code and an inner repetition code of rate $1/8$ and has the overall rate $R_c = 1/16$. The number of active users is denoted by U. The information bits $b_u(i)$ of user u are encoded, interleaved by a user-specific random interleaver Π_u and BPSK-modulated. The BPSK symbol of user u at a particular time instant k is denoted by $d_u(k)$. So the received signal can be written as

$$y(k) = \sum_{u=1}^{U} \sqrt{P_u} \cdot d_u(k) + n(k) \tag{1}$$

with P_u denoting the received power of user u and the total power $P_{tot} = \sum_u P_u$. Without loss of generality the users are assumed to be sorted according to decreasing powers P_u. Each user is interfered by noise and the signals of all other users. This multi-user interference (MUI) degrades performance significantly even for low system loads $\beta = U \cdot R_c$. At the receiver iterative soft interference cancellation is applied as in Kai et al., 2005, estimating the multi-user interference and canceling it before detection. The soft bit estimates after interference cancellation are denoted as $\tilde{d}_u^{IC}$. After soft output decoding the extrinsic soft bit estimates $\tilde{d}_u^{CC}$ are used as a-priori-information for the multi-user detector (MUD) in the next iteration. In this way, the reliability of the estimates can be improved in each iteration.

3. Analysis of MUD

Iterative detection schemes for serially concatenated systems can be analyzed by variance transfer charts (VTC) as done in Schlegel and Shi, 2004 for CDMA. The relation between input and output error variance of each component is called transfer function. For the channel decoder Monte Carlo simulations may be necessary, but the transfer function of the MUD can be easily calculated. The effective variance σ_{eff}^2 of the u-th user after interference cancellation is given by

$$\sigma_{\text{eff},u}^2 = \mathrm{E}\{|\sqrt{P_u} d_u - \tilde{d}_u^{IC}|^2\} = \sigma_n^2 + \sum_{v \neq u} P_v \, \sigma_{d,v}^2 \tag{2}$$

and the error variance at the output of the channel decoder is defined as

$$\sigma_{d,u}^2 = \mathrm{E}\{|d_u - \tilde{d}_u^{CC}|^2\} = f\left(\frac{\sigma_{\text{eff},u}^2}{P_u}\right) \tag{3}$$

which is a function of the SINR at the channel decoders input denoted as $f(\cdot)$. If all users have equal power $P_u = P$ the variances are the same for all users and (2) becomes

$$\sigma_{\text{eff}}^2 = \sigma_n^2 + (U-1)\, P \, \sigma_d^2 \quad . \tag{4}$$

In this paper we won't restrict to equal powers for the users and therefore use average variances to describe a system with unequal powers in a similar way. The average error variance at the output of the interference canceler equals

$$\begin{aligned} \bar{\sigma}_{\text{eff}}^2 &= \frac{1}{U} \sum_u \sigma_{\text{eff},u}^2 = \sigma_n^2 + \frac{U-1}{U} \sum_u P_u \, \sigma_{d,u}^2 \\ &= \sigma_n^2 + (U-1)\, \bar{P} \, \bar{\sigma}_d^2 \end{aligned} \tag{5}$$

with $\bar{P} = P_{\text{tot}}/U$ and

$$\bar{\sigma}_d^2 = \frac{1}{P_{\text{tot}}} \sum_u P_u \, \sigma_{d,u}^2 \quad . \tag{6}$$

This is analog to the definition for equal powers in (4). The error variance σ_d^2 at the decoder output depends only on the SINR at the input as stated in (3)

$$\begin{aligned} \sigma_{d,u}^2 &= f\left(\frac{1}{\text{SINR}_u}\right) = f\left(\frac{\sigma_{\text{eff},u}^2}{P_u}\right) \\ &= f\left(\frac{\bar{\sigma}_{\text{eff}}^2 + \bar{P}\,\bar{\sigma}_d^2 - P_u \, \sigma_{d,u}^2}{P_u}\right) \approx f\left(\frac{\bar{\sigma}_{\text{eff}}^2}{P_u}\right) \quad . \end{aligned} \tag{7}$$

Assuming that $\bar{\sigma}_{\text{eff}}^2 \gg \bar{P}\ \bar{\sigma}_d^2 - P_u\ \sigma_{d,u}^2$ for a large number of users, the approximation in (7) is tolerable. Taking (6) and (7), $\bar{\sigma}_d^2$ can be approximately expressed in terms of $\bar{\sigma}_{\text{eff}}^2$ by

$$\bar{\sigma}_d^2 = \frac{1}{P_{\text{tot}}} \sum_u P_u\ \sigma_{d,u}^2 \approx \frac{1}{P_{\text{tot}}} \sum_u P_u\ f\left(\frac{\bar{\sigma}_{\text{eff}}^2}{P_u}\right) \quad . \tag{8}$$

The VTC of a PIC reduces to a two-dimensional problem analog to the equal power case as shown in Figure 3.

4. Complexity Reduction

Complexity Reduction for PIC

Parallel interference cancellation in general means decoding all users at the same time in a parallel fashion.

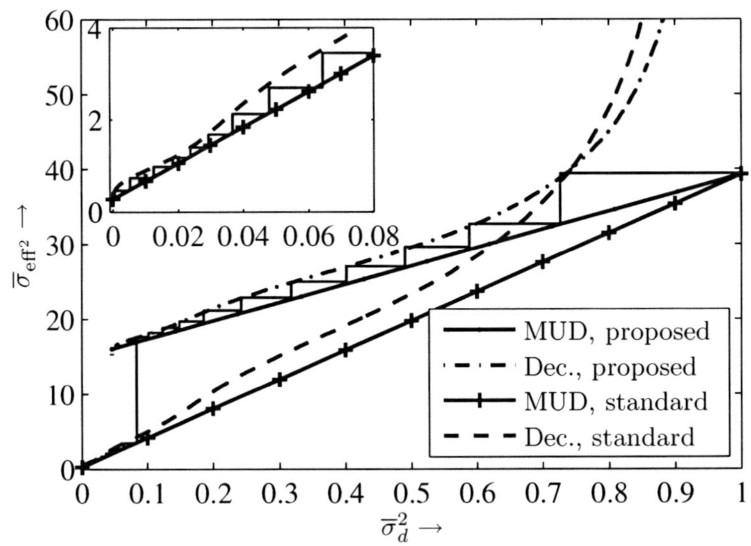

Figure 3. Transfer function and simulated trajectory for PIC of an IDMA system with 40 users

When considering unequal receive powers of the users, some may benefit more from decoding in a particular iteration and some user's benefit may be only marginal. In this case it is questionable, if it is worth computational effort to decode these users although their contribution to the overall convergence in a particular iteration is marginal. In Kusume and Bauch, 2006, a simple complexity reduction technique is considered suitable for at-hoc networks. It is suggested to split up the concatenated channel code and only decode the

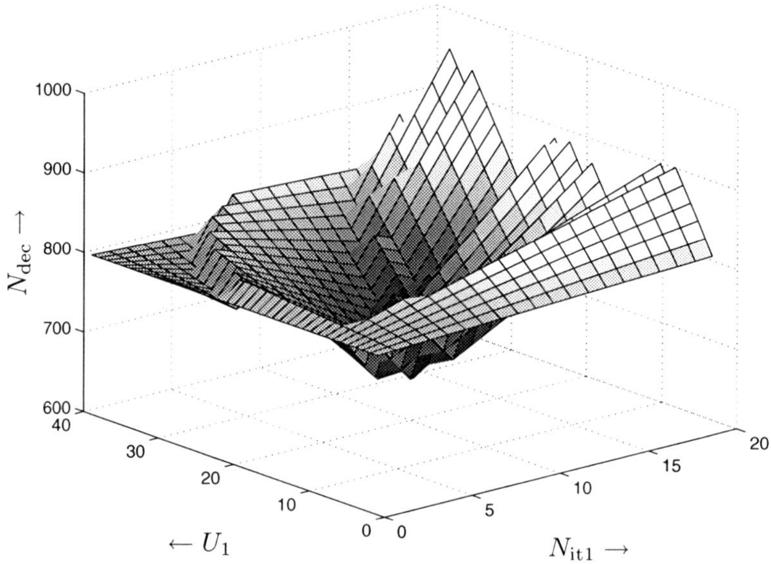

Figure 4. Number of channel decoder evaluations

repetition code for some users or some iterations to save complexity. Iteration-wise switching is analyzed by EXIT-Charts, but user-wise switching is only shown by simulations. A combination of both is not considered and the choice of parameters like number of layers not decoded completely is not motivated. In this paper a user- and iteration-wise switching strategy is proposed and the choice of parameters is motivated by VTC analysis. In the first N_{it1} iterations only the U_1 strongest users are considered. After a certain number of iterations the detector switches to the standard detection considering all layers.

In Figure 3 the resulting transfer functions for a system with 40 users, resulting in a load of $\beta = 2.5$ are depicted. In addition to the transfer functions of standard detection the resulting functions for the case of considering only U_1 users is shown.

Although the weaker users are not decoded during the first iterations they may be considered for interference cancellation. For theoretical considerations it makes no difference if the estimated interference is subtracted from the weaker users already in the first iterations or not until they are also decoded. If the weaker users are considered as noise during the first N_{it1} iterations, the calculations and transfer functions are much easier. Furthermore, a simulated trajectory is plotted to show the tightness of the prediction used.

The simulated trajectory matches the new detection in the first 11 iterations. Then the detection is switched and the trajectory follows the transfer function of the standard detection. Due to the tightness of the VTC analysis the number

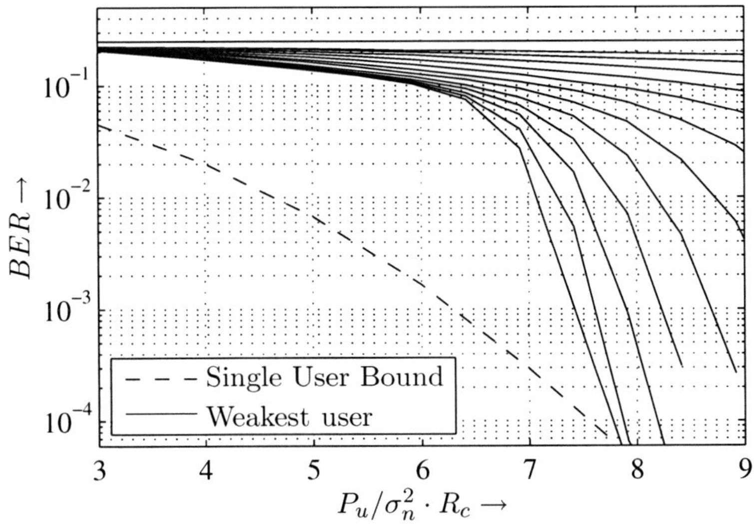

Figure 5. Bit error rate vs. E_b/N_0 for the weakest user

of channel decoding steps can be calculated by prediction of the trajectories without Monte Carlo simulations for every constellation of N_{it1} and U_1. Finally the case of minimum complexity can be predicted as shown in Figure 4. The minimum number of channel decoding steps is obtained at $N_{it1} = 11$ and $U_1 = 25$ and is equal to $N_{dec} = 675$. For this case BER simulation results for the weakest user are shown in Figure 5. Although the proposed strategy works for arbitrary power distributions in this paper only one special case is considered due to limited space. The power profile used in this paper was optimized as in Weitkemper et al., 2006 for a BER of 10^{-4} and 20 iterations considering standard detection.

Complexity Reduction for SIC

The analysis of successive interference cancellation schemes was investigated for the first time in Weitkemper et al., 2005, for CDMA and can also be applied to IDMA. Averaging over all users at a particular iteration is no longer justified for SIC. Already the second user in the first iteration sees an improved interference situation. The variance of the remaining MUI of the u-th user in iteration m can be calculated by

$$\sigma_{\text{eff},u}^{2\,(m)} = \sigma_n^2 + \sum_{v=1}^{u-1} P_v\, \sigma_{d,v}^{2\,(m)} + \sum_{w=u+1}^{U} P_w\, \sigma_{d,w}^{2\,(m-1)} \tag{9}$$

where the superscript (m) denotes the iteration index. For the u-th user the improved estimation of the previous users of the current iteration is already used. Therefore, each user has different variances in each iteration and the behavior can not be predicted with the help of transfer functions displayed in a two dimension diagram as for PIC. Asymptotically SIC and PIC show the same performance if $m \rightarrow \infty$. In other words, the behavior of SIC only differs from PIC with respect to required number of iterations. It is well known that successive interference cancellation needs less iterations than PIC to reach the same performance. The reason for still considering PIC is the possibility of parallel processing, but if the constraint is not the overall time delay but the overall number of operations, SIC is superior to PIC. For that reason in this paper also the complexity of SIC is considered. Mostly the detection order is the natural ordering $1, 2, ..., U, 1, 2, .., U$. In this case the gain in convergence speed is only due to improved estimates within one iteration. Intuitively one would choose a descending, but fixed ordering. In this paper an optimized scheduling is derived. For simplicity, one channel decoding operation is called sub-iteration. In each sub-iteration the detector looks for and decodes that user which would gain most in terms of average error variance. Considering $\overline{\sigma_d^2}$ the improvement depending on the user w to be decoded next can be expressed as

$$\left.\overline{\sigma_d^2}\right|_w = \frac{1}{P_{\text{tot}}} \left(\sum_{u \neq w} P_u \sigma_{d,u}^2 + P_w f\left(\frac{\sigma_{\text{eff},w}}{P_w}\right) \right) . \tag{10}$$

So the performance difference depending on the user to be decoded next is simply the weighted improvement of σ_d^2. Finding the user delivering the best improvement concerning average error variance is given by

$$\max_u P_u \left(\sigma_{d,u}^2 - f\left(\frac{\sigma_{\text{eff},u}^2}{P_u} \right) \right) . \tag{11}$$

It can be calculated online because decisions are made for each iteration in time. The savings in terms of decoder steps of this scheme can be seen in Figure 6. The average error variance $\overline{\sigma_{\text{eff}}^2}$ is plotted over the number of channel decoder runs. For PIC the function looks like a staircase because the decoding step causes one improvement of the error variance after U simultaneous decoding steps. The scheme proposed in this paper improves slower in terms of iterations but faster in terms of channel decoder steps. For SIC with natural scheduling the improvement within one iteration (decoding user 1 up to user U) is obvious. Also the decreasing gain for the weaker users can be seen. The improved scheduling for SIC has a more heterogeneous slope and reaches the required SINR first. Even standard SIC with natural ordering requires much lower complexity than PIC to reach the same required SINR. For

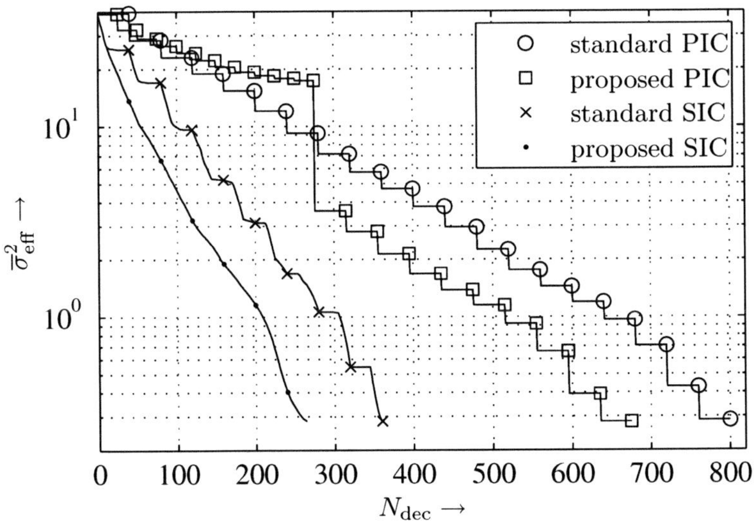

Figure 6. Comparison of complexity for SIC and PIC

the same system parameters standard SIC requires only 360 channel decoding steps whereas the two PIC schemes require 800 and 675, respectively. If the proposed scheduling is used for the SIC scheme, only 264 channel decoding steps are necessary which corresponds to a reduction of $\approx 27\%$ compared to standard SIC and even 67% compared to standard PIC.

5. Conclusion

In this paper, complexity reduction strategies for IDMA uplink detection are presented which are based on semi-analytical analysis with Variance Transfer Charts. A user- and iteration-wise detection strategy is proposed for parallel interference cancellation. The choice of parameters is based on VTC analysis. The overall complexity is reduced up to 16% without any performance degradation. Monte Carlo simulations verify the results by means of bit error rates. Furthermore, SIC is considered known to require less iterations than PIC achieving the same performance. In addition to the natural scheduling an improved scheduling is investigated for SIC. This improved SIC scheduling enables a complexity reduction of 67% compared to standard PIC without any performance degradation. Furthermore, the proposed scheduling for SIC can be calculated online during the detection process.

References

Kai, L., Xiaodong, W., and Ping, L. (2005). Analysis and optimization of interleave-division multiple-access communication systems. In *ICASSP 2005 - IEEE International Conference on Acoustics, Speech, and Signal Processing*.

Kusume, K. and Bauch, G. (2006). Some Aspects of Interleave Division Multiple Access in Ad Hoc Networks. In *4th International Symposium on Turbo Codes & Related Topics*.

Ping, L., Liu, L., Wu, K. Y., and Leung, W. K. (2006). Interleave Division Multiple Access. *IEEE Transactions on Wireless Communications*, 5:938–947.

Schlegel, C. and Shi, Z. (2004). Optimal power allocation and code selection in iterative detection of random CDMA. In *Zurich Seminar on Digital Communications*.

Weitkemper, P., Kuehn, V., and Kammeyer, K.-D. (2005). Analysis of Iterative Successive Interference Cancellation in SC-CDMA Systems. In *Fifth International Workshop on Multi-Carrier Spread-Spectrum*, Oberpfaffenhofen, Germany.

Weitkemper, P., Zielinski, K., Kammeyer, K.-D., and Laur, R. (2006). Optimization of Interference Cancellation in Coded CDMA Systems by Means of Differential Evolution. In *4th International Symposium on Turbo Codes & Related Topics*.

JOINT NAVIGATION & COMMUNICATION BASED ON INTERLEAVE-DIVISION MULTIPLE ACCESS

Peter Adam Hoeher and Kathrin Schmeink
Information and Coding Theory Lab
Faculty of Engineering, University of Kiel
Kaiserstr. 2, D-24143 Kiel, Germany
{ph,kas}@tf.uni-kiel.de

Abstract In this paper, a full-duplex joint navigation and communication system with a unified signal structure based on interleave-division multiple access (IDMA) is proposed. At the transmitter side, the navigation data and the communication data are linearly superimposed and transmitted on the same carrier frequency at the same time. Both data rates are adjustable in a wide range, including high data rates. At the receiver side, the navigation data can be separated near perfectly from the communication data. The position is estimated by means of a channel estimator rather than a delay-locked loop. No oversampling is necessary. Navigation and communication is feasible in the presence of low signal-to-noise ratios, interference, and multipath – situations where conventional navigation systems may fail. Possible applications include, but are not restricted to, indoor scenarios. Monte-Carlo simulations illustrate the performance of the navigation and the communication unit.

Keywords: Navigation, Positioning, Localization, Geolocation, Communication, IDMA, CDMA

1. Introduction

Precise localization and navigation has been of importance for mankind since thousands of years. Demand, expansion, and significance of wireless navigation systems is growing steadily. In many situations, it is desirable to have a pocket-type wireless navigation system including a *return link* for the purpose of communication. Examples include security applications, logistics, transportation/avionics, and leisure. Since pure navigation systems do not offer a return link for the purpose of communication, navigation systems (such as GPS, GLONASS, Galileo, etc.) are sometimes complemented by a cellular radio system (such as GSM, UMTS, etc.). This adds additional cost, however. The other extreme is an extension of cellular radio systems or wireless LANs

S. Plass et al. (eds.), Multi-Carrier Spread Spectrum 2007, 97–106.

for the purpose of navigation. Typically, the precision is poor, especially if existing systems are extended.

As opposed to these approaches, in this paper a full-duplex joint navigation and communication system with a unified signal structure is proposed, offering the following features:

- Flexible configuration
- High data rates of the communication unit
- Robustness with respect to noise and interference
- Compatibility with respect to DS-CDMA systems
- Possibility of encryption of all or selected layers
- Low-cost transmitter and receiver design (no oversampling)
- High precision of the navigation unit

The flexible configuration facilitates a trade-off between navigation (NAV) and communication (COM) unit, and adaptivity with respect to (w.r.t.) changing channel conditions and user requirements (e.g., data rate, QoS). Channel coding is an integral part of the signal structure. Robustness w.r.t. noise and interference is particularly important in factory buildings and warehouses. The compatibility w.r.t. DS-CDMA systems offers the possibility of a terrestrial supplementation of GPS and Galileo, or the possibility of an enhancement of IS-95, cdma-2000, UMTS, and TD-SCDMA. Only authorized users having access to the interleaver patterns are able to perform navigation and/or communication.

The system proposal is based on a generalization of IDMA [1], called *multi-layer IDMA* [2]. A related joint navigation and communication system proposal based on DS-CDMA has been published in [3].

The remainder of this paper is organized as follows: In Section 2, the system concept under investigation is introduced. Pulse shaping, physical channel, and receive filtering can be represented by an equivalent discrete-time channel model. As shown in Section 3, the channel coefficients of this discrete-time channel model are a function of the propagation delay. Hence, positioning can be performed by means of channel estimation. The proposed receiver design is introduced in Section 4. Numerical results are shown in Section 5, and conclusions are drawn in Section 6.

2. System Concept

Throughout this paper, the complex baseband notation is used. The transmitter structure is illustrated in *Figure 1*. The basic idea is to superimpose NAV and COM data streams, subsequently referred to as *layers*, according to the principle of multi-layer IDMA [2]. Without loss of generality, we may assume that the first NAV layer covers the navigation signal, which is known at the receiver side. This layer may also be used for *pilot-aided channel estimation* (PLACE) [4]. All other NAV layers may carry supporting data for navigation purposes, such as the exact positions of the transmitters. The COM layers belong to the return link and are unknown to the receiver. In multi-layer IDMA, each uncoded data stream is encoded by a low-rate encoder (FEC). The encoded symbols are referred to as chips. Each encoded data stream is fed into a different chip-by-chip interleaver (INT). A pseudo-random or a deterministic interleaver may be applied. Then, scrambling (SCR) is optionally performed in order to obtain a zero-mean data stream, independent of the uncoded data. After scrambling, the chips are mapped onto the complex plane (MAP). After proper power allocation, the layers are linearly superimposed. Since all data streams are generated in the same unit, synchronous multiplexing is feasible (and mandatory). Finally, pulse shaping is performed. It is important to note that all interleavers must be mutually different. All other units (FEC, SCR, MAP) may be identical without performance loss. It is also important to note that after mapping binary sequences are sufficient from an information theory point of view [2]. From a practical point of view, binary sequences are superior w.r.t. complexity and performance.

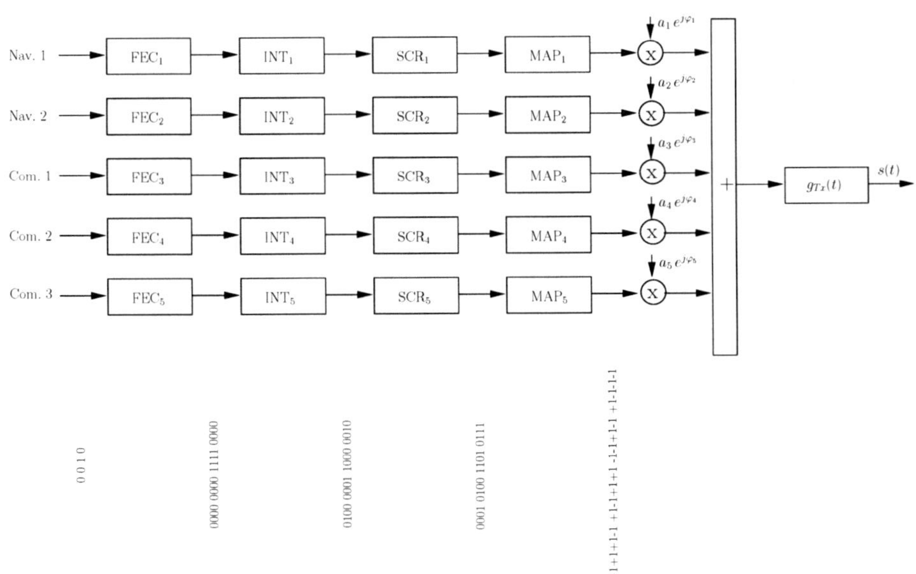

Figure 1. Transmitter structure for the example of two navigation and three communication layers. (FEC$_m$: channel encoder, INT$_m$: interleaver, SCR$_m$: scrambler, MAP$_m$: mapper, a_m: amplitude, φ_m: phase of the mth data stream.)

In the illustrative example in *Figure 1*, an uncoded (NAV or COM) data stream consists of 4 bits. In the simple example, a binary repetition code of length 4 is applied. Hence, 16 coded bits are obtained. The ordering of these 16 coded bits is permuted by means of an interleaver of length 16. A deterministic scrambler inverts every second bit. Finally, binary antipodal mapping, multiplexing, and pulse shaping are performed.

For the purpose of 3D positioning, at least four transmitters having a known position must be seen by a receiver. Subsequently, these transmitters are referred to as base stations. All transmitters simultaneously operate on the same carrier frequency. In order to separate uplink and downlink, either frequency-division duplexing (FDD) or time-division duplexing (TDD) may be applied.

The system setup is quite flexible. The number of layers can be adapted to the channel conditions. A rule-of-thumb for the supportable number of binary layers in a single-antenna scenario is about $4 \cdot SF$, where SF is the reciprocal code rate of the FEC encoder, corresponding to a bandwidth efficiency of about 4/(1+r) bps/Hz, where r is the rolloff factor. The bandwidth efficiency is particularly important for the COM unit, but also for the NAV unit a larger data rate compared to state-of-the-art NAV systems is feasible (e.g., 50 bps in GPS), since the layers are linearly superimposed rather than modulo two.

3. Equivalent Discrete-Time Channel Model

In this section, the equivalent discrete-time channel model for the joint navigation and communication system is derived. As several base stations are needed for positioning, channel coefficients for every link between a transmitter and the receiver have to be determined. The discrete-time channel model is determined by the pulse shaping filter of the transmitter, the physical channel, the receive filter, and the sampling phase and frequency, cf. *Figure 2*. For simplicity, the physical channel is assumed to be linear and non-dispersive, and hence is completely defined by the impulse response

$$h_b(\tau, t) = \alpha_b(t) \cdot \delta(\tau - \tau_b(t)) \cdot e^{j\phi_b(t)}, \tag{1}$$

where $\tau_b(t)$ is the propagation delay between the bth transmitter and the receiver, $\alpha_b(t)$ is an attenuation factor, and $\phi_b(t)$ is a random phase. The overall channel impulse response is given by

$$\begin{aligned} h_b^T(\tau, t) &= g_{Tx}(\tau) * \left(\alpha_b(t) \cdot \delta(\tau - \tau_b(t)) \cdot e^{j\phi_b(t)}\right) * g_{Rx}(\tau) \\ &= \alpha_b(t) \cdot g_{TxRx}(\tau - \tau_b(t)) \cdot e^{j\phi_b(t)}, \end{aligned} \tag{2}$$

where $g_{TxRx}(\tau)$ represents the convolution of the transmitter filter $g_{Tx}(\tau)$ and the receiver filter $g_{Rx}(\tau)$. For illustrative purposes, an ideal rectangular filter is assumed for $g_{Tx}(\tau)$ and for $g_{Rx}(\tau)$. In the simplest case of no oversampling, the overall impulse response is sampled at multiples of the chip duration T_c.

Given the assumptions that the sampling phase is zero and the maximum propagation delay, $\tau^{\max}$, is less than T_c, the discrete-time channel is completely described by two channel coefficients (see *Figure 2*). For every propagation delay between zero and T_c there exists a unique pair of channel coefficients $(h_{b,0}, h_{b,1})$, i.e., the propagation delay is an invertible function of these channel coefficients:

$$\tau_b = f\left(h_{b,0}, h_{b,1}\right). \tag{3}$$

Hence, it is possible to calculate the propagation delay by means of channel estimation. If $\tau^{\max} > T_c$ and/or the sampling phase is non-zero, τ_b depends on more than two channel coefficients.

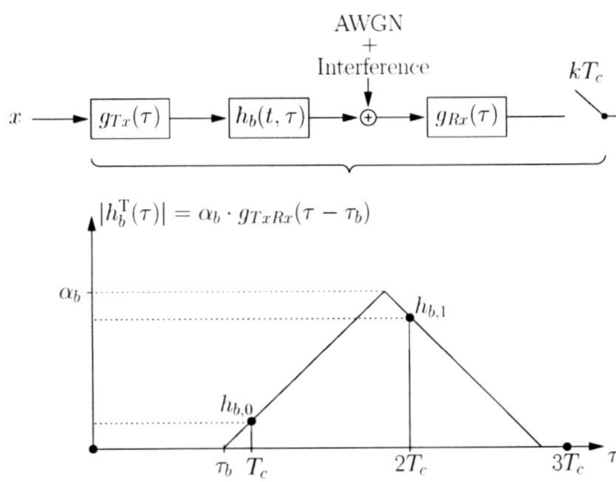

Figure 2. Overall channel impulse response and discrete-time channel coefficients.

So far, we assumed that the received signal is not synchronized. The position of the receiver can be determined by the *time of arrival* (TOA) concept (see [5] and [6]). A drawback of this *asynchronous mode* is that due to the lack of synchronization a power loss occurs.

An alternative is the so-called *synchronous mode*, where the ith transmitter is synchronized by means of an adaptive receive filter. Correspondingly, the overall impulse response $h_b^T(\tau, t)$ of every base station b has to be shifted by $-\tau_i$. The resulting propagation delays $\tau_b' = \tau_b - \tau_i$ are no longer absolute but relative w.r.t. the ith transmitter. In this case, three channel coefficients are necessary to completely describe the channel. Every possible triple $(h_{b,0}', h_{b,1}', h_{b,1}')$ is unique. The relative propagation delays are again an invertible function of the channel coefficients:

$$\tau_b' = f\left(h_{b,0}',\, h_{b,1}',\, h_{b,2}'\right). \tag{4}$$

In the synchronous mode, the *time difference of arrival* (TDOA) concept (see [5]) has to be used for positioning.

4. Receiver Design

There are two basic receiver designs according to the two discrete-time channel models considered above. The first design corresponds to the asynchronous mode: The receiver consists of only one branch with one non-adaptive receive filter. In the synchronous mode, the receiver is designed in parallel: Each branch contains an adaptive filter, which synchronizes the sampling to a certain transmitter. The higher complexity of the second receiver is compensated by a better performance.

The receiver performs two tasks: data detection and channel estimation. In order to separate the superimposed layers, an *iterative chip-by-chip multiuser detector* as proposed in [1] may be applied. Since the first NAV layer is assumed to be known at the receiver side, this layer does not have to be detected. The first NAV layer can be taken into account as a priori information. It can be shown that the interference caused by this layer can be canceled out perfectly, if perfect channel knowledge is assumed, otherwise near perfectly. (The proof is omitted here due to space limitations.) As a consequence, the detection of the COM layer(s) improves.

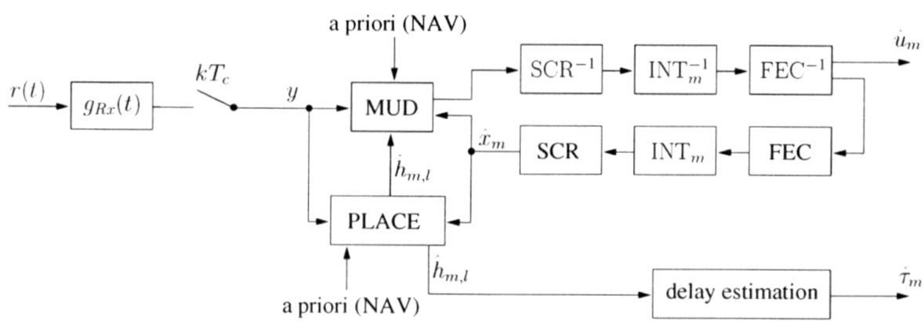

Figure 3. Receiver structure for the mth layer.

As pointed out in the preceding section, channel estimation is of great importance. The better the channel estimates, the better the receiver position can be determined. We propose to apply *pilot-layer aided channel estimation* (PLACE) [4]. In *Figure 3*, an iterative receiver structure with a PLACE unit is illustrated. The pilot layer corresponds to the known NAV layer. Soft-chips obtained by the multiuser detector can be integrated into the channel estimation process. We investigated three different concepts: *training-based channel estimation without interference cancellation* (IC), *training-based channel estimation with IC*, and *semi-blind channel estimation*. These concepts differ in computational complexity and performance, see [4]. A trade-off between simple (cheap) and sophisticated (expensive) handhelds can easily be adjusted.

5. Numerical Results

The numerical results presented in this section are based on the following 2D setup. A quadratic room with $B = 4$ active base stations, one in every corner, is assumed as shown in *Figure 4 a)*. Each base station transmits either $N = 2$ layers (1 NAV and 1 COM layer) or $N = 3$ layers (1 NAV and 2 COM layers). Each layer contains 320 data bits. All $M = B \cdot N$ data streams are generated randomly and are mutually independent. Since the positions of the base stations are assumed to be known at the receiver side, no further NAV data has to be transmitted. All M data streams are encoded by a repetition encoder of length 8, interleaved, scrambled, and afterwards mapped onto the complex plane by means of binary antipodal signaling in conjunction with a uniform phase distribution: $\varphi_n = n \cdot \pi/N$, where $0 \leq n \leq N-1$. The set of M interleavers of length 2560 ($= 320 \cdot 8$) is generated pseudo-randomly. The transmit power $P_T = 0.02$ W per base station is non-uniformly distributed among the N layers: The power of the pilot layer is chosen to be $\rho \cdot P_T$, the remaining power $(1-\rho) \cdot P_T$ is uniformly distributed among the $N-1$ COM layers. Numerical results for $\rho = 0.5$ and $\rho = 0.1$ are reported below.

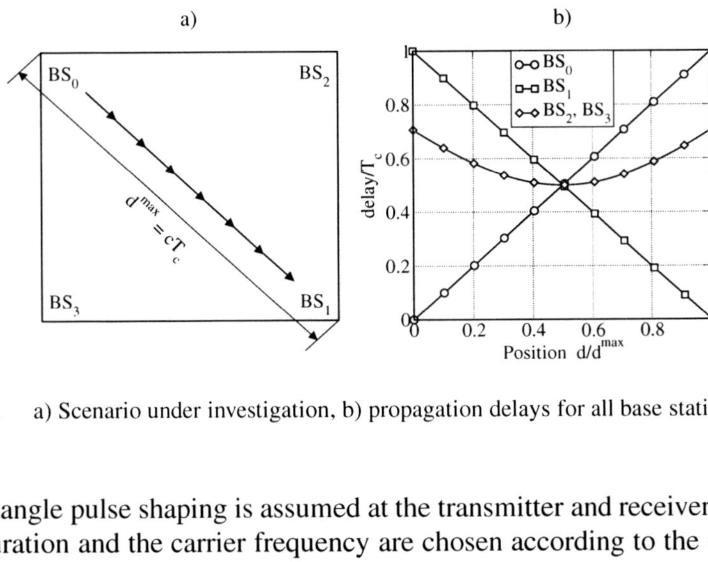

Figure 4. a) Scenario under investigation, b) propagation delays for all base stations.

Rectangle pulse shaping is assumed at the transmitter and receiver side. The chip duration and the carrier frequency are chosen according to the *GPS standard positioning service* (SPS): $T_C = 1$ μs and $f_c = 1575.42$ MHz (L1). The receiver is moving along the diagonal of the room, cf. *Figure 4 a)*. The length of the diagonal, $d^{\max}$, corresponds to the maximum possible propagation delay $\tau^{\max} = T_c$ (here: $d^{\max} = 300$ m). In *Figure 4 b)* the propagation delays for all base stations are shown, where d is the current distance from the starting point. The attenuation is assumed to be proportional to τ_b^4, hence $\alpha_b \sim 1/\tau_b^2$.

The receiver is modeled as described in Section 4. PLACE in conjunction with *joint least squares channel estimation* (JLSCE) is applied [4]. 20 iterations are performed before making hard decisions. In the asynchronous mode, the bit error rate (BER) is determined by all $B \cdot (N-1)$ COM layers. In the synchronous mode, the BER is determined for each receiver branch separately: only the corresponding $N-1$ COM layers are considered.

Subsequently, only results for the synchronous mode (w.r.t. BS_0) are reported. In the following figures, the SNR (left y-axis, dashed lines) and the BER of the receiver branch corresponding to BS_0 or the standard deviation (SD) w.r.t. BS_0 in meters (right y-axis, solid lines) are plotted versus the position. The position is described by the ratio $d/d^{\max}$. The SD is obtained by a one-shot measurement.

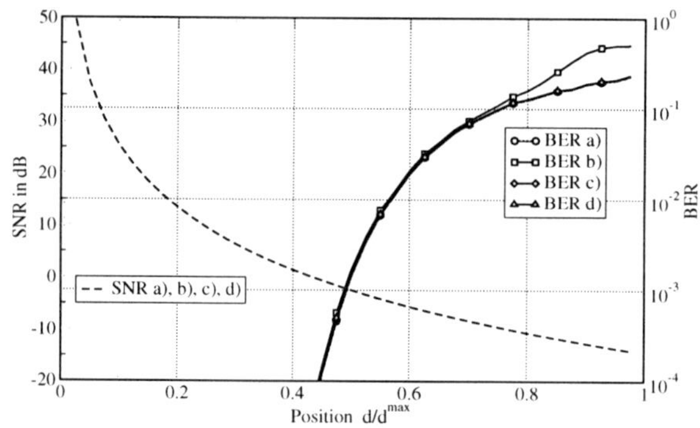

Figure 5. BER for synchronous mode (BS_0), $N = 2$, and $\rho = 0.5$; a) perfect channel knowledge, b) training-based JLSCE, c) training-based JLSCE with IC, d) semi-blind JLSCE.

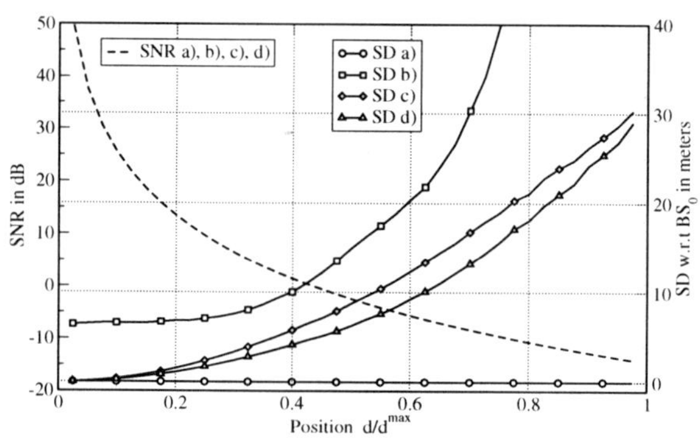

Figure 6. MSE w.r.t. BS_0 for synchronous mode (BS_0), $N = 2$, and $\rho = 0.5$; a) perfect channel knowledge, b) training-based JLSCE, c) training-based JLSCE with IC, d) semi-blind JLSCE.

In *Figure 5* and *Figure 6* the influence of channel estimation is shown for $N = 2$ and $\rho = 0.5$. In both plots, perfect channel knowledge is also depicted to serve as a lower bound. It is interesting to note that the BER is scarcely influenced by the choice of the channel estimator – a near-optimum performance can be achieved except for the case of pure training without IC. The MSE (and hence the ranging precision), however, strongly depends on the channel estimator. Training-based channel estimation without IC shows the worst performance. Because of its low computational complexity (see [4]), it is suitable for cheap and simple receivers that only have to provide rough position estimates. If the requirement for positioning accuracy is high, semi-blind channel estimation should be applied. The superior navigation performance is paid by a high computational complexity. Training-based channel estimation with IC offers a good trade-off between complexity and performance.

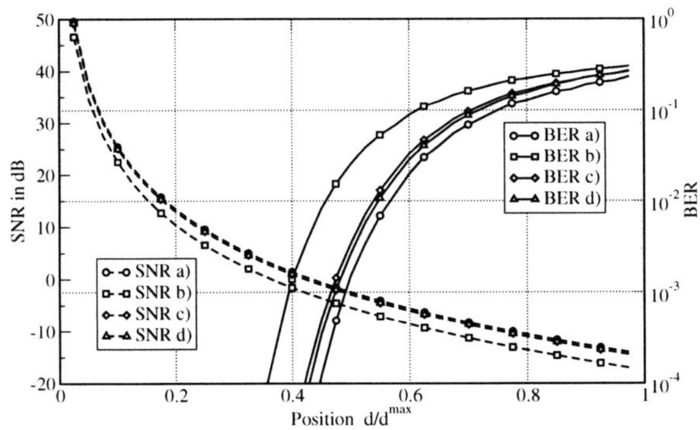

Figure 7. BER for synchronous mode (BS_O); a) $N = 2$, $\rho = 0.5$, training-based JLSCE with IC, b) $N = 3$, $\rho = 0.5$, training-based JLSCE with IC, c) $N = 3$, $\rho = 0.1$, training-based JLSCE with IC, d) $N = 3$, $\rho = 0.1$, semi-blind JLSCE.

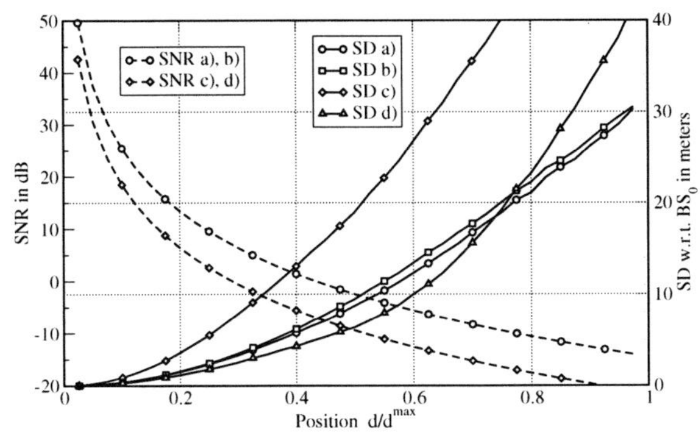

Figure 8. MSE w.r.t. BS_O for synchronous mode (BS_O); a) $N = 2$, $\rho = 0.5$, training-based JLSCE with IC, b) $N = 3$, $\rho = 0.5$, training-based JLSCE with IC, c) $N = 3$, $\rho = 0.1$, training-based JLSCE with IC, d) $N = 3$, $\rho = 0.1$, semi-blind JLSCE.

In *Figure 7* and *Figure 8* it is shown that semi-blind channel estimation outperforms training-based channel estimation with IC in a wide range. Let us consider training-based channel estimation with IC first. If the power of the pilot layer is fixed ($\rho = 0.5$) and the number of COM layers is increased from one to two, the BER degrades because the SNR per COM layer is decreased. The navigation performance degrades a little, because the soft-bits are less reliable. When decreasing the power of the pilot layer to $\rho = 0.1$, the SNR per COM layer is increased and the BER improves, but the MSE degrades a lot.

Next, semi-blind channel estimation is applied for the same setting. Both performances (BER and SD) improve, especially for the NAV unit, because the whole transmit power (instead of the power of the pilot layer only) is used for channel estimation. The advantage of semi-blind channel estimation is that only a small fraction ρ is needed to achieve a good navigation performance. The lower ρ, the higher is the remaining power for the COM layers. Thus, either the BER is improved or a higher data rate (i.e., more COM layers) is feasible.

6. Conclusions

A full-duplex joint navigation and communication system with a unified signal structure based on interleave-division multiple access (IDMA) is proposed. Channel coding is an integral part of the signal structure. Excellent performance can be achieved for iterative chip-by-chip multiuser detection and channel decoding in conjunction with semi-blind pilot-layer aided channel estimation. Navigation and communication units can be separated near perfectly.

References

[1] Li Ping, L. Liu, K. Wu, and W. K. Leung, "Interleave Division Multiple-Access," *IEEE Trans. Wireless Commun.*, vol. 5, no. 4, pp. 938-947, Apr. 2006.

[2] P. A. Hoeher and H. Schoeneich, "Interleave-Division Multiple Access from a Multiuser Point of View," in *Proc. Int. Symp. on Turbo Codes & Related Topics in conjunction with Int. ITG Conf. on Source and Channel Coding*, Munich, Germany, Apr. 2006.

[3] M. Takeda, T. Tereda, and R. Kohno, "Spread Spectrum Joint Communication and Ranging System Using Interference Cancellation Between a Roadside and a Vehicle," in *Proc. IEEE Vehicular Technology Conference (VTC)*, Ottawa, Canada, pp. 1935-1939, May 1998.

[4] H. Schoeneich, and P. A. Hoeher, "Iterative Pilot-Layer Aided Channel Estimation with Emphasis on Interleave-Division Multiple Access Systems," *EURASIP J. Applied Signal Processing*, vol. 2006, article ID 81729, 2006.

[5] J. J. Caffrey and G. L. Stueber, "Overview of Radiolocation in CDMA Cellular Systems," *IEEE Communications Magazine*, vol. 36, no. 4, pp. 38-45, Apr. 2005.

[6] K. Pahlavan, X. Li, and J.-P. Makela, "Indoor Geolocation Science and Technology," *IEEE Communications Magazine*, vol. 40, no. 2, pp. 112-118, Feb. 2002.

OUT-OF-BAND RADIATION IN MULTICARRIER SYSTEMS: A COMPARISON

Leonardo G. Baltar, Dirk S. Waldhauser and Josef A. Nossek
Munich University of Technology
Institute for Circuit Theory and Signal Processing
Arcisstrasse 21, 80290 Munich, Germany
{baltar,waldhauser,nossek}@nws.ei.tum.de

Abstract OFDM systems suffer from high out-of-band radiation. Consequently, they require methods reducing those spectral out-of-band components. Because of adjustable frequency confinement, filter bank based multicarrier systems allow for a lower out-of-band radiation. This paper compares an OFDM system employing one out-of-band reduction method with a filter bank based multicarrier system (FBMC).

Keywords: Multicarrier, OFDM, filter banks, out-of-band radiation, spectrum analysis

1. Introduction

6 Multicarrier systems have been considered as one of the most promising modulation solutions for future wireless communication systems due to their robustness against multipath propagation and the efficient use of bandwidth of the transmission channel. Orthogonal frequency-division multiplexing (OFDM) systems provide this efficiency but suffer from high out-of-band radiation originated from the sidelobes of the modulated subcarriers. Therefore, either a spectral guard band between adjacent services, zero input subcarriers or some kind of out-of-band reduction method need to be employed.

The use of spectrum guard bands or zero input subcarriers result in an undesired loss in the scarce spectrum resource.

Recently, new methods for out-of-band energy reduction in OFDM multicarrier systems have been proposed for the application of overlay systems. They are either based on the use of certain subcarriers as so-called cancellation carriers (CCs) [1] at the ends of the OFDM signal

S. Plass et al. (eds.), Multi-Carrier Spread Spectrum 2007, 107–116.

spectrum, on the employment of weighting coefficients at each subcarrier input [2] or on multiple choice sequences [3]. However, those methods result in an increase in bit error ratio (BER), a loss in bandwidth efficiency, an increased peak-to-average power ratio (PAPR), additional signaling overhead and/or an increased processing complexity at the transmitter and receiver. Multicarrier systems anyway suffer from increased PAPR values compared to single-carrier systems [4].

One alternative to the conventional OFDM system are filter bank based multicarrier systems (FBMC) or transmultiplexer (TMUX) systems. A TMUX based on exponentially modulated filter banks has the advantage of reduced implementation complexity by the use of polyphase decompositions and the Fast Fourier Transform (FFT) [5]. The stopband attenuation of the TMUX prototype filter determines the out-of-band energy of the modulated signal and can, therefore, be adjusted very flexibly in accordance with the requirements. The inevitable prototype filter lengths L with $L > M$, where M is the number of subcarriers, only allow for orthogonality in so-called orthogonally multiplexed QAM (OQAM) systems [6] or Modified DFT (MDFT) filter banks [7].

In this paper we compare the performance of an OFDM system employing the aforementioned cancellation carriers technique for the reduction of out-of-band radiation with a multicarrier system based on a Modified Discrete Fourier Transform transmultiplexer (MDFT-TMUX) [7] in the context of current and future 3GPP specifications [8, 9]. We show how the data throughput can be increased by the employment of an FBMC without substantially increasing complexity and latency while still fitting into the specified spectrum mask. That increase has two origins: there is no need of a prefix and the number of occupied subcarriers can be greater than the recommended.

In Section 2 we describe the filter bank based multicarrier system and present the spectrum modeling for both FBMC and conventional OFDM with and without cyclic prefix. Some spectrum examples are shown in Section 3 along with complexity and latency analysis. We summarize the results and draw some conclusions in Section 4.

2. System model

First, we briefly present the basics of multicarrier systems based on digital filter banks and in the sequel we describe the spectrum models adopted in the simulations.

2.1 Filter bank based multicarrier systems

There are three basic differences between the FBMC and the conventional OFDM system: no inclusion of a (cyclic) prefix; the complex input symbols have their real and imaginary parts interleaved, resulting in what is called OQAM; and there is a filtering step after the complex modulation of each sub-channel, also called polyphase network.

As mentioned before the best choice for a FBMC is the one where a prototype filter is modulated by complex exponentials. The prototype is designed in a way that adjacent subcarriers overlap, but remain orthogonal, and in non-neighboring subcarriers the attenuation guarantees negligible interference. The prototype can be, for example, a truncated root raised cosine filter (RRC) with length L and roll-off ρ. With this kind of prototype intersymbol interference (ISI) is also eliminated, provided that an OQAM stage is included [6, 7].

The modulations can be implemented via a DFT. With this modification, the polyphase components of the protoype filter are placed after each output of the DFT, instead of filtering each subchannel. In this way, an efficient implementation is obtained.

Figure 1 depicts an efficient structure of an MDFT synthesis filter bank. More efficient structures for the MDFT filter bank exist [10], but this topic is out of the scope of this work.

It is worth mentioning that, if the protoype has length M and all coefficients are equal to one, the OQAM stage can be eliminated and the conventional OFDM modulator is obtained.

2.2 Power spectrum density (PSD)

The total instantaneous spectral density of the signal at the output of a general multicarrier modulator results from the sum of the spectral densities of each ℓ-th subcarrier and it is described in the normalized angular frequency domain ω, for $0 \leq \omega < 2\pi$, by

$$S_k(\omega) = \sum_{\ell=0}^{M-1} |H_\ell(\omega)|^2 S_{x_{\ell,k}}(\omega), \quad -\infty \leq k \leq \infty \tag{1}$$

where $x_{\ell,k}$ is the complex QAM symbol modulating the ℓ-th subcarrier at the k-th time instant and $S_{x_{l,k}}(\omega)$ its corresponding spectrum density. This holds true because of the reasonable assumption of uncorrelated input symbols $x_{\ell,k}$, which will be justified because of coding and interleaving in practical systems. We define two types of shaping filters $H_\ell(\omega)$ for each subcarrier:

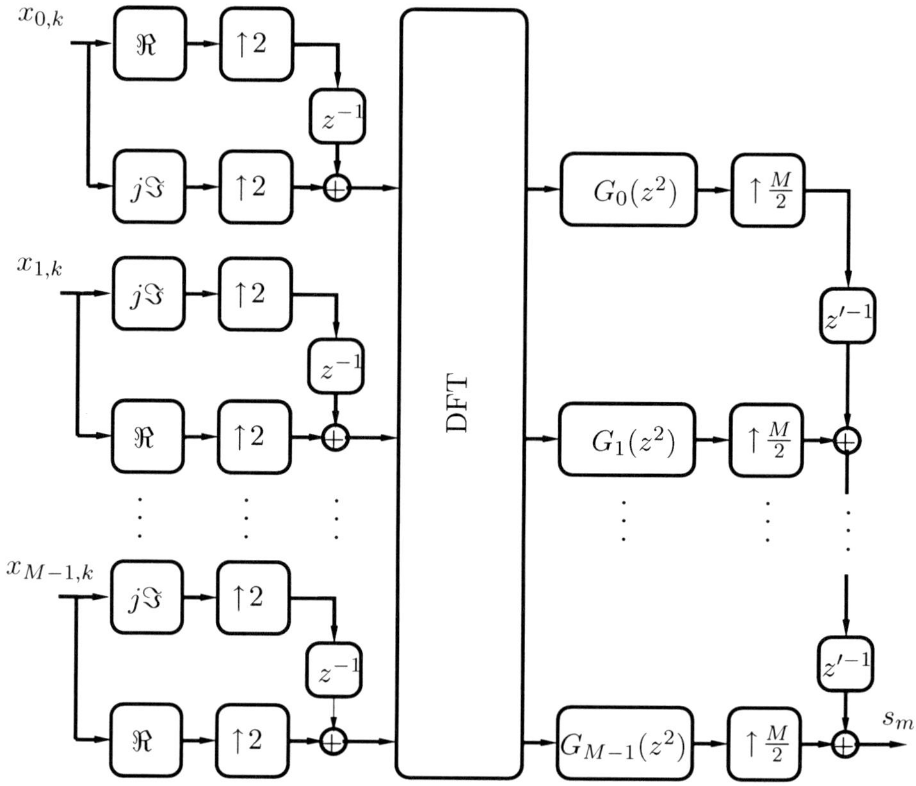

Figure 1. Synthesis Part of an MDFT Filter Bank, where z'^{-1} represents a delay in the output symbol rate

$$H_\ell(\omega) = \begin{cases} E_\ell(\omega) & : \text{ OFDM system,} \\ F_\ell(\omega) & : \text{ FBMC system.} \end{cases} \tag{2}$$

It can be demonstrated that the amplitude of the Fourier transform of the ℓ-th subchannel in the conventional OFDM system is given by the Dirichlet kernel

$$E_\ell(\omega) = \frac{\sin\left(M\left(\frac{\omega}{2} - \frac{\pi\ell}{M}\right)\right)}{M\sin\left(\frac{\omega}{2} - \frac{\pi\ell}{M}\right)}, \quad \ell = 0, ..., M-1. \tag{3}$$

The insertion of the cyclic prefix (CP) modifies (3) according to

$$\tilde{E}_\ell(\omega) = \frac{\sin\left((M + L_{\text{CP}})\left(\frac{\omega}{2} - \frac{\pi\ell}{M}\right)\right)}{M\sin\left(\frac{\omega}{2} - \frac{\pi\ell}{M}\right)}. \tag{4}$$

We can conclude from (4) that the cyclic prefix will change the bandwidth of each subchannel resulting in ripples in the final spectrum [11].

In an FBMC system, we define the prototype coefficients as $h[n]$, with $n = 0, .., L-1$, where $L = KM + 1$ and K is the length of each polyphase components. We define the variable $\mathcal{L} = \frac{KM}{2}$ and assume that the prototype has even symmetry around the $\mathcal{L}$-th coefficient, this means that $h[n] = h[KM - n]$. The individual amplitude of the ℓ-th subcarrier is then given by

$$F_\ell(\omega) = \left[h\left[\mathcal{L}\right] + 2 \sum_{n=1}^{\mathcal{L}-1} h\left[\mathcal{L} - n\right] \cos\left(n \left(\omega - \frac{2\pi\ell}{M} \right) \right) \right]. \tag{5}$$

Each $F_\ell(\omega)$ in (5) is equivalent to a frequency shifted version of the amplitude of the prototype filter.

3. Simulations

The Technical Specification Group for Radio Access Network of the 3GPP decided to focus the Long-Term Evolution feasibility study on multicarrier based downlink. Therefore, we will use the parameters recommended in the document [8], as it defines and describes the potential physical layer for evolved Universal Terrestrial Radio Access (E-UTRA).

The radio access has a hierarchical frame structure. Each radio frame has 10 ms and is composed by 20 subframes. The number of OFDM or FBMC symbols on each subframe depends on other parameters. The document specifies six different possible bandwidths (1.25, 2.5, 5, 10, 15, 20 MHz) and the subcarrier spacing $\Delta f = 15$ kHz is fixed regardless of the system bandwidth. The size of the FFT is chosen according to the desired bandwidth.

We will apply in our example the transmission bandwidth of 5 MHz, which corresponds to an FFT size of $M = 512$ for both OFDM and FBMC, where in both cases only 300 subcarriers are used, with the others having zero input symbols. The sampling frequency for that bandwidth is $f_s = 7.68$ MHz.

For the conventional OFDM using cyclic prefix, a short or a long prefix is possible. When a short prefix is used, each subframe should have seven OFDM symbols, six of which have length $L_{CP} = 36$ and one has $L_{CP} = 40$. With these values, 93.33% of the subframe are used for data transmission.

The recommendation also foresees a longer prefix targeting multi-cell broadcast and very-large-cell scenarios. In this case, six OFDM symbols with $L_{CP} = 128$ are filled into each subframe. With this value 80% of the subframe are used for data transmission.

For the FBMC case, seven and a half blocks of complex symbols (or fifteen blocks of pure real/complex symbols) compose each subframe.

Assuming an input with constant unitary spectral density ($S_{x_{l,k}}(\omega) = 1$) or, equivalently, uncorrelated symbols with unit energy, Figure 2 exhibits the power spectral densities of the FBMC with $K = 4$ and roll-off $\rho = 1$ and of the conventional OFDM without cyclic prefix. Besides that, it depicts the quadratic spectrum of one OFDM block employing the cancellation carriers method with $x_\ell = 1, \forall \ell$. The spectrum mask of the Universal Mobile Telecomunication System (UMTS) for a system with bandwidth of 5 MHz is drawn as a reference, where the multicarrier based E-UTRA physical layer with the same bandwidth has to fit into.

We used 2 CCs at each end of the spectrum for the optimization and didn't consider any power limitation, which means that the quadratic inequality constraint was not applied to the least squares problem [1]. We used 10 sidelobes at each side of the spectrum, which means that 10 samples in the optimization range were used [1].

If we loot at Figure 2, it is clear that the conventional OFDM without cyclic prefix or any method for reducing the out-of-band radiation does not fit into the specified mask.

We can also see that, when the FBMC or the OFDM with CCs is employed, they do not only fit into the mask, but also provide some room for further spectrum utilization. As a consequence, more subcarriers can be occupied, resulting in higher throughput and spectral efficiency.

In Figure 3 $N = 330$ subcarriers were used for the FBMC and $N = 328$ for the conventional OFDM with and without CCs, instead of $N = 300$.

With those new numbers of occupied subcarriers, an increase of 10% in the total throughput can be achieved for the FBMC and of 9.33% for the OFDM employing CCs.

We can notice in both spectrum examples that the OFDM system employing CCs presents strong ripples near the spectrum borders and around the DC subcarrier.

3.1 Complexity

In this section we consider the complexity of the signal generation only at the transmitter side. If we incorporate the modification proposed in [10] into the structure of Figure 1, and assume that M is a power of 2, the number of "flops" (floating-point operations) per output sample for the FBMC is given by

$$\mathsf{flops}_{\mathsf{FBMC}} = \mathsf{flops}_{\mathsf{FFT}} + 2M(4K + 1)$$

where $\mathsf{flops}_{\mathsf{FFT}}$ is the number of flops of the FFT and its current value is exhibited in [12]. If we use the values from the examples presented before

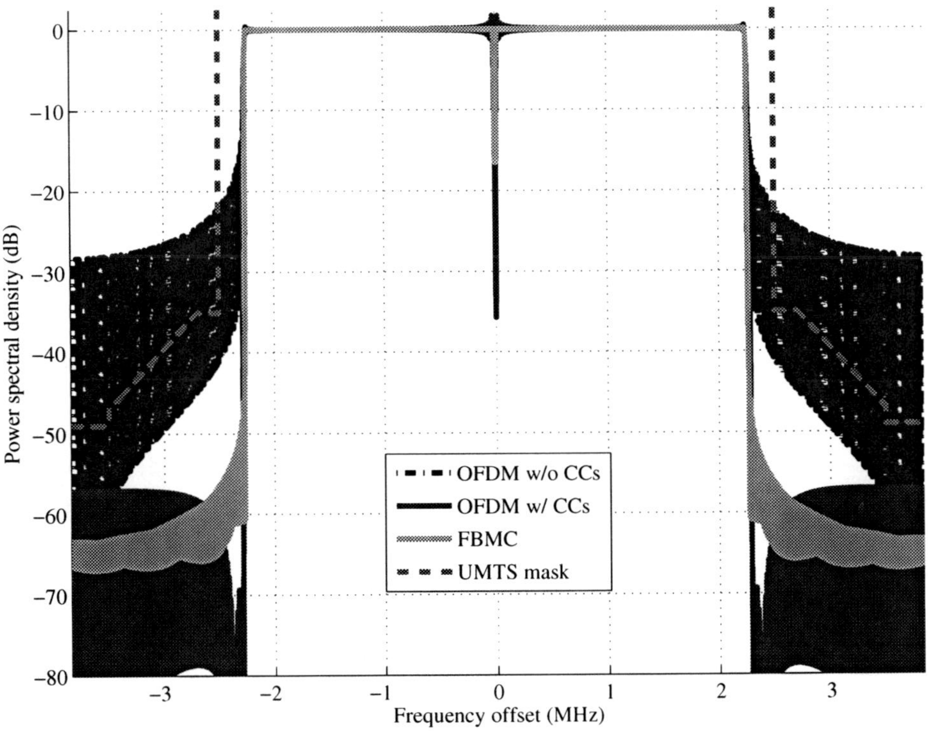

Figure 2. PSDs of FBMC and OFDM without CCs and squared frequency response of OFDM with CCs and input data $x_{\ell,k} = 1$. In all cases $N = 300$ subcarriers are active.

($K = 4$ and $M = 512$), the FBMC transmitter will need around twice the number of flops needed by conventional OFDM. Furthermore, if the prototype is designed to provide perfect reconstruction, the polyphase component pairs ℓ and $\ell + \frac{M}{2}$ can be efficiently realized using lattice structures [10].

When the CCs technique is incorporated into conventional OFDM, there is also an increase in complexity. Only to calculate the spectrum samples in the optimization range [1] for each input block, the same complexity as in the FBMC case is reached. But it still remains the computational burden for calculating the CCs coefficients. The latter will depend on the adoption or not of the quadratic constraint, and on the used method for solving the least squares problem.

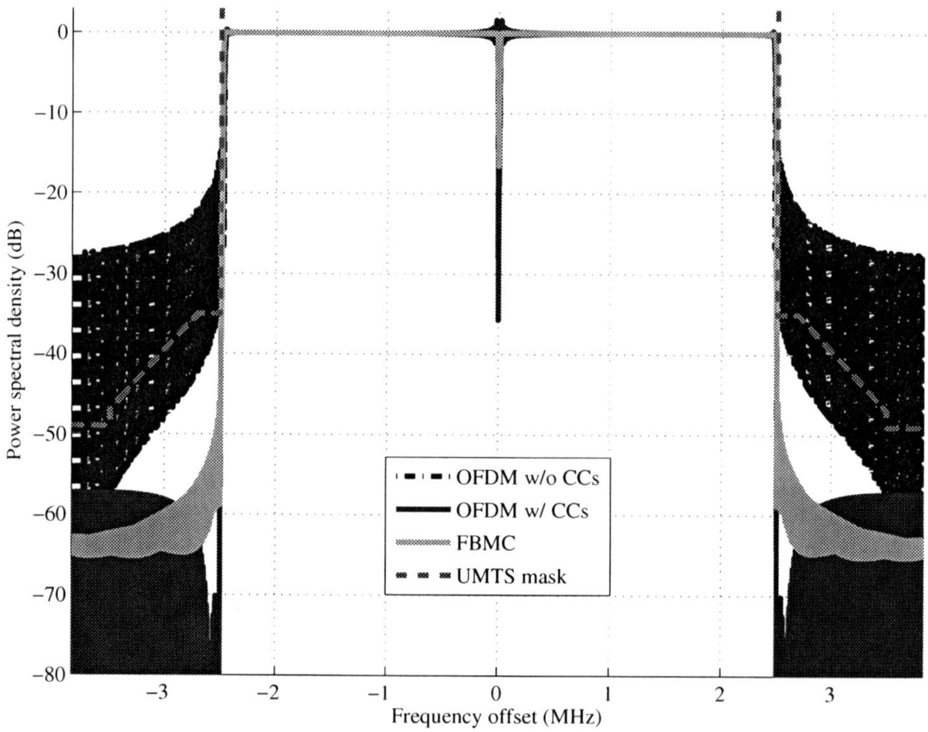

Figure 3. PSD of FBMC ($N = 330$) and OFDM without CCs ($N = 328$) and squared frequency response of OFDM with CCs and input data $x_{\ell,k} = 1$ ($N = 328$)

It is worth mentioning that in the FBMC, the same complexity exists at the receiver side, while for the OFDM with CCs, the receiver keeps the complexity of one FFT.

3.2 Latency

One of the drawbacks of employing an FBMC instead of a conventional OFDM, besides the increased complexity, is the increased latency. This latency is mainly caused by the filtering performed in the polyphase network. It can be demonstrated that the delay of the FBMC is given by $d_{\text{FBMC}} = \frac{(K+1)M}{f_s}$. The resulting latency for $K = 4$ and $M = 512$ is $d_{\text{FBMC}} = 0.33$ ms. This delay added to the average delay of 4.0 ms generated by the adopted protocol architecture of LTE [8] still keeps the total user-plane delay below the recommended 5.0 ms.

When CCs are incorporated into conventional OFDM, some delay will also be inserted. But in this case, the delay will depend on the capabilities of the hardware employed to calculate the CCs.

4. Conclusions

We first briefly described the multicarrier system based on filter banks, then showed how the power spectral density for both FBMC and conventional OFDM can be modeled and explained the effect of the cyclic prefix on the modeling of the latter.

The simulations were performed under the framework of the Long-Term Evolution recommendation from the 3GPP standardization group. We showed two examples of spectral densities: The first adopted the number of subcarriers found in the recommendations and a second used an increased number of active subcarriers for a more efficient spectrum occupation. We proved that if FBMC or OFDM with the cancellation carriers method is employed, more subcarriers can be occupied as defined in the 3GPP recommendations without exceeding the defined mask. If we combine this increase with the lack of prefix, the FBMC will achieve a gain of 17.15% or 35% in data throughput compared to the conventional OFDM system with the short or the long cyclic prefix, respectively.

We showed that the complexity is increased by a factor of two for both FBMC and OFDM with cancellation carriers when compared to conventional OFDM and that the increased latency in the FBMC resulting from polyphase filtering is acceptable and remains below the recommended latency.

Both FBMC and OFDM have the drawback of a high peak-to-average power ratio, but when the cancellation carriers method is included in OFDM, it becomes even higher.

As the FBMC system presents more degrees of freedom, the length of the polyphase components and the roll-off factor can be adjusted to keep the spectrum of the output signal at the transmitter into the specified mask for each regulated frequency band under consideration of allowed complexity and maximum latency.

References

[1] S. Brandes, I. Cosovic, and M. Schnell. Reduction of out-of-band radiation in OFDM systems by insertion of cancellation carriers. *Communications Letters, IEEE*, 10(6):420–422, June 2006.

[2] I. Cosovic, S. Brandes, and M. Schnell. Subcarrier weighting: a method for sidelobe suppression in OFDM systems. *Communications Letters, IEEE*, 10(6):444–446, June 2006.

[3] Ivan Cosovic and Tiziano Mazzoni. Special Issue on MC-SS Suppression of sidelobes in OFDM systems by multiple-choice sequences. *European Transactions on Telecommunications*, 17(6):623–630, 2006.

[4] D. S. Waldhauser, L. G. Baltar, and J. A. Nossek. Comparison of filter bank based multicarrier systems with OFDM. In *Proc. IEEE Asia Pacific Conference on Circuits and Systems, APCCAS 2006*, pages 1978–1981, 04-07 Dec. 2006.

[5] P. P. Vaidyanathan. *Multirate Systems and Filter Banks.* Prentice-Hall, Englewood Cliffs, NJ, 1993.

[6] B. Hirosaki. An orthogonally multiplexed QAM system using the discrete fourier transform. *Communications, IEEE Transactions on*, 29(7):982–989, July 1981.

[7] T. Karp and N.J. Fliege. Modified DFT filter banks with perfect reconstruction. *Circuits and Systems II: Analog and Digital Signal Processing, IEEE Transactions on*, 46(11):1404–1414, Nov. 1999.

[8] 3rd Generation Partnership Project; Technical Specification Group Radio Access Network. Physical layer aspects for evolved Universal Terrestrial Radio Access (UTRA); 3GPP TR 25.814 V7.1.0. Technical report, 2006–09.

[9] 3rd Generation Partnership Project; Technical Specification Group Radio Access Network. Base Station (BS) radio transmission and reception (FDD); 3GPP TS 25.104 V7.5.0. Technical report, 2006–12.

[10] T. Karp and N. J. Fliege. Computationally efficient realization of MDFT filter banks. In *Proc. of the 8th European Signal Processing (EUSIPCO '96)*, volume 2, pages 1183–1186, September 1996.

[11] M. Ivrlac and A. J. Nossek. Influence of a cyclic prefix on the spectral power density of cyclo-stationary random sequences. In *Proc. of the 6th International Workshop on Multi-Carrier Spread Spectrum*, May 2007.

[12] S. G. Johnson and M. Frigo. A modified split-radix FFT with fewer arithmetic operations. *Signal Processing, IEEE Transactions on*, 55(1):111–119, January 2007.

MITIGATION OF DYNAMICALLY CHANGING NBI IN OFDM BASED OVERLAY SYSTEMS

Sinja Brandes and Michael Schnell
German Aerospace Center (DLR)
Institute of Communications and Navigation
Oberpfaffenhofen, 82234 Wessling, Germany
{sinja.brandes,michael.schnell}@dlr.de

Abstract In overlay systems based on orthogonal frequency-division multiplexing (OFDM) narrow-band interference (NBI) from licensed systems transmitting in the same frequency band degrades system performance significantly. In this paper, the impact of NBI is either reduced by means of an optimized compensation matrix or by simply subtracting an estimated NBI signal from the received signal. The required estimation of the NBI signal is performed by means of a few observation subcarriers. Variations in the center frequency of the interferer are tracked by exploiting information from pilot OFDM symbols for channel estimation. Simulation results show that with both techniques system performance is improved significantly even in presence of strong and dynamically changing NBI.

1. Introduction

Spectral scarcity and the simultaneous demand for higher data rates require a more efficient use of the available bandwidth. One approach is the coexistence of an overlay system with an already existing licensed system in the same frequency band [1]. In order not to interfere with the licensed system the overlay system transmits in the spectral gaps that are currently not used by the licensed system. For modulation OFDM is applied as the OFDM transmit spectrum can easily be adapted to the changing spectrum allocation by turning on and off individual subcarriers in order to guarantee that the licensed systems are not disturbed by the overlay system. However, the overlay system is exposed to interference originating from licensed systems. Due to the spectral selectivity of OFDM NBI has a strong impact on the performance of the overlay system and hence has to be mitigated. Existing techniques for NBI

S. Plass et al. (eds.), Multi-Carrier Spread Spectrum 2007, 117–126.

mitigation developed for wireline applications [2, 3] with almost static interference are suitable for overlay systems only to a limited extent. As the interference situation varies significantly due to the changing spectrum allocation of the licensed system NBI mitigation has to be adapted dynamically. In Fig. 1, the interference situation in a 1 MHz band in the very high frequency (VHF) band is depicted. In the following an overlay scenario in this band is considered as example.

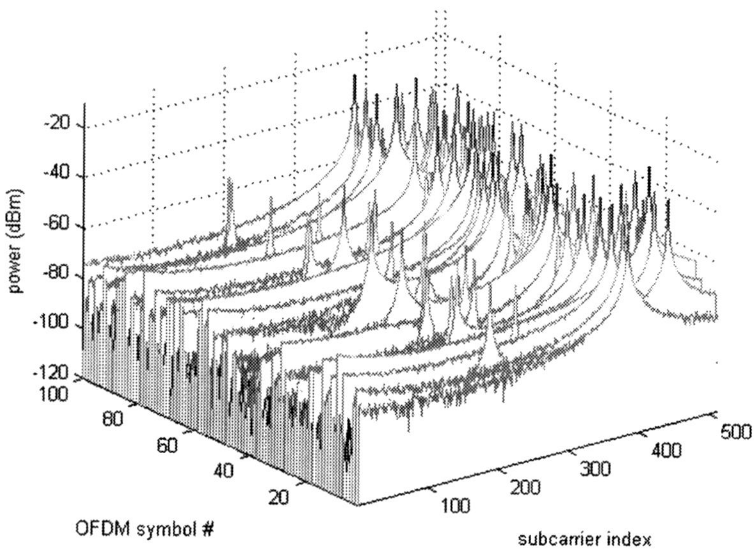

Figure 1. Interference spectrum at overlay system Rx after 512-point FFT, 1 MHz within VHF band.

The remainder of this paper is organized as follows. In Section 2, a typical interference situation in the VHF band as well as the leakage effect that causes the significant influence of NBI are described. In Sections 3 and 4, algorithms for the estimation and mitigation of the NBI signal are presented. The impact of the remaining interference is evaluated in terms of bit error rate (BER) performance of the complete overlay system in Section 5. In Section 6 conclusions are drawn.

2. Description of Interference

2.1 Interference Situation in VHF Band

As a realistic basis for the investigations an overlay system in the VHF band assigned to aeronautical communications is considered, e.g. [4]. The VHF band is subdivided into 760 25 kHz channels mainly used for analog voice communications based on DSB-AM. For the overlay system a bandwidth of 1 MHz spanning 40 25 kHz channels is selected. Each VHF channel is used by 12 OFDM subcarriers resulting in a subcarrier spacing of $\Delta f = 2.083$ kHz. In the chosen 1 MHz band as illustrated in Fig. 1 four interferers are present in four dedicated VHF channels, representing a typical interference situation retrieved from measurement flights at an en-route flight level [5]. Due to variations in the power as well as in the activity of the interferers the interference situation changes significantly from OFDM symbol to OFDM symbol. According to their average power and duty cycle two strong and two weak interferers are distinguished. The strong and weak interferers have a duty cycle of 25.28% and 7.47% and an average power of -78.36 dBm and -83.57 dBm, respectively. The total noise power within the 1 MHz band is -110 dBm. Apart from that the carrier frequency of the interferers may vary within one subcarrier spacing to due Doppler effects and inaccuracies in DSB-AM devices.

2.2 Leakage Effect

The interference in the VHF band is mainly caused by DSB-AM signals that are band-limited to 5.4 kHz, i.e. they only span 3-4 OFDM subcarriers. However, as can be seen in Fig. 1 the spectrum of the DSB-AM signal is spread over the complete bandwidth due to the leakage effect. As the carrier mainly contributes to the leakage effect the influence of the voice part in the sidebands is neglected and only the carrier signal is considered in the following.

Within a frame of M OFDM symbols the interference signal on the mth OFDM symbol after an N-point FFT yields

$$I_m[k] = \frac{A_m}{N} \cdot \exp\left(-j\pi(k-k_c)\frac{N-1}{N} + j\varphi_m\right) \cdot \frac{\sin(\pi(k-k_c))}{N\sin\left(\pi/N(k-k_c)\right)} \quad (1)$$

with $k = 0, \ldots, N-1$ and $\mathbf{I}_m = [I_m[0], \ldots, I_m[N-1]]^T$. The discrete carrier frequency is denoted by k_c. A_m is the amplitude of the carrier signal which is assumed to be constant within one OFDM symbol as the bandwidth of the DSB-AM signal is small compared to the OFDM signal. The phase φ_m is different on consecutive OFDM symbols and equals $\varphi_m = \varphi_0 + m\frac{2\pi N_{GI} k_c}{N}$. This phase shift is caused by cutting out parts of the continuous waveform when extracting OFDM symbols from the Rx data stream. When the guard interval (GI) is removed an additional constant phase shift between the interference signals of consecutive OFDM symbols is induced. The interference signal from (1) can be interpreted as the convolution of the Fourier transform of the sinusoidal carrier signal, i.e. a complex exponential sequence, and the

Fourier transform of the rectangular receive window which is a sinc-function. The leakage effect originating from sampling a shifted sinc-function is illustrated in Fig. 2 for $N=16$. When the carrier frequency k_c of the interferer equals the frequency of one OFDM subcarrier, the sinc-function is sampled in the maximum and all zero-crossings. However, due to inaccuracies of the carrier frequency and the Doppler effect k_c varies within one subcarrier spacing, i.e. $k - k_c$ is not an integer, and the sinc-function is sampled at its non-zero values and the leakage effect occurs. As illustrated in Fig. 2, the maximum leakage occurs when the carrier frequency is in the middle between two adjacent subcarriers, e.g. $k_c=7.5$.

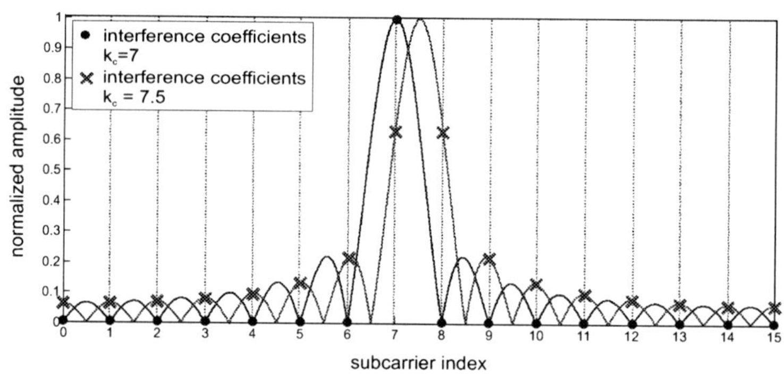

Figure 2. Illustration of leakage effect: interference coefficients for $k_c = 7$, i.e. no leakage effect, and $k_c = 7.5$, i.e. strongest leakage effect, N=16.

3. Estimation of NBI signal

The interference signal $\mathbf{I}_m$ on each OFDM symbol is estimated based on measurements on L observation subcarriers not used for data transmission. Alternatively or additionally, pilot OFDM symbols are exploited as the NBI signal can be observed over the complete bandwidth. As the leakage effect is a property of the FFT the NBI signal can be reconstructed according to (1). Consequently, only the carrier frequency k_c of the interferer, the amplitude A_m, and phase φ_m have to be estimated.

3.1 Estimation of Carrier Frequency

As it can be assumed that the carrier frequency only varies within one subcarrier spacing k_c can be estimated based on one observation subcarrier on the left and right hand side of the carrier of the NBI signal. In case k_c is in the middle between two subcarriers the received (Rx) values $R_m[k_1]$ and $R_m[k_2]$ on the two directly adjacent subcarriers have the same amplitude (c.f. Fig. 2), where k_1 is the subcarrier left and k_2 the subcarrier right to the carrier frequency k_c. When k_c is shifted to the left or to the right, the two subcarriers obtain different amplitudes. Hence, the carrier frequency can be derived from the ratio of the amplitudes defined as

$$C := \frac{|\bar{R}[k_1]|}{|\bar{R}[k_2]|} \tag{2}$$

where $\bar{R}[k]$ are averaged samples from the Rx signal $\mathbf{R}_m = [R_m[0], \ldots, R_m[N-1]]^T$. The samples are averaged over all OFDM symbols within one frame for the case when observation subcarriers for each VHF channel are available in all OFDM symbols. Otherwise the Rx signal is averaged over the pilot OFDM symbols only. Averaging is advantageous as the influence of the channel and noise is diminished hence making the estimation more accurate.

After inserting (1) into (2) and some simplifications an estimation of k_c is obtained

$$\hat{k}_c = \frac{C \cdot k_1 + k_2}{C + 1}. \tag{3}$$

3.2 Estimation of Phase and Amplitude

Once $\hat{k}_c$ is estimated according to (3) a preliminary interference estimation $\tilde{\mathbf{I}}_m$ for the whole OFDM frame is generated from (1) assuming $A_m = 1$ and $\varphi_m = 0$, i.e.

$$\hat{\mathbf{I}}_m = \hat{A}_m \cdot \tilde{\mathbf{I}}_m \cdot \exp(j\hat{\varphi}_m). \tag{4}$$

The amplitude and phase on each OFDM symbol are obtained by comparing the Rx values on the L observation subcarriers and the corresponding values on the preliminary interference signal

$$\hat{A}_m = \frac{1}{L}\sum_{l=1}^{L} \frac{|R_m[l]|}{|\tilde{I}_m[l]|}, \quad \hat{\varphi}_m = \frac{1}{L}\sum_{l=1}^{L} \arg(R_m[l]) - \arg(\tilde{I}_m[l]). \tag{5}$$

This estimation is performed for each OFDM symbol as the amplitude of the NBI signal changes significantly and the phase varies due to the removal of the GI as explained above.

Theoretically, with this algorithm an NBI signal can be estimated based on an arbitrary number and position of observation subcarriers. However, the best estimation accuracy is obtained when the observation subcarriers are positioned close to the carrier frequency of the interferer. A reason for that is that the influence of noise as well as of other NBI signals is relatively small in the maximum of the NBI signal. When more than one interferer is active in the band, all interferers can be estimated separately without a significant loss in estimation accuracy. Finally, the estimated NBI signals of all interferers are summed up and mitigated jointly.

3.3 Detection of Active Interferers

Prior to estimation it is necessary to detect active interferers. This is realized by means of observations on pilot OFDM symbols. After subtracting the known pilot symbols from the Rx values the power of the interference signal on each subcarrier is determined. Active interferers can easily be detected by deciding whether the interference power is below or above a certain threshold. To reduce the probability of false detections the threshold is slightly above the noise level. In order to detect interferers occurring during one OFDM frame, this is repeated for all pilot OFDM symbols within one frame. Note, interferers that disappear within the OFDM frame are tracked as well since the amplitude is estimated for each OFDM symbol. In case an interferer stops transmitting an amplitude close to 0 is estimated.

4. NBI Mitigation

The impact of the NBI signal is either mitigated by subtracting the estimated NBI signal from the Rx signal or by designing an optimized compensation matrix and multiplying the Rx signal with this matrix. In addition, Rx windowing in time domain can be applied in order to further reduce NBI. In that case the NBI signal is not reconstructed according to (1) but to a convolution of the Fourier transform of the windowing function and (1). For more details please refer to [7].

4.1 Subtraction of Estimated NBI Signal

A straightforward approach for NBI mitigation is the subtraction of the estimated NBI signal from the Rx signal $\mathbf{R} = [\mathbf{R}_0, \ldots, \mathbf{R}_{M-1}]$ of one OFDM frame according to

$$\mathbf{R}' = \mathbf{R} - \hat{\mathbf{I}}, \tag{6}$$

where $\hat{\mathbf{I}} = [\hat{\mathbf{I}}_0, \ldots, \hat{\mathbf{I}}_{M-1}]$ is the estimated NBI signal on all M OFDM symbols within one frame.

The performance of the subtraction is shown in Fig. 3 for one DSB-AM interferer with $k_c = 263.5$. The spectrum of the DSB-AM signal before subtraction averaged over 10,000 OFDM symbols is given as reference. When the NBI signal is estimated based on $L = 2$ subcarriers adjacent to the carrier and subtracted the power of the NBI signal is reduced by 14 dB. A further reduction is possible with more observation subcarriers as the estimation becomes more accurate then. Note, the achieved results are always suboptimal as only the carrier of the DSB-AM signal is reconstructed and subtracted. However, the good results and the simple estimation justify this assumption.

When k_c is varied by up to 1 kHz, i.e. $k_c = \pm 0.48$ and the variation is not tracked nearly no reduction of the NBI signal is possible. In some cases the influence of the NBI signal even increases. When the carrier frequency is estimated according to (3) nearly the same reduction is achieved as in the case when k_c is perfectly known, hence indicating that the proposed estimation works properly.

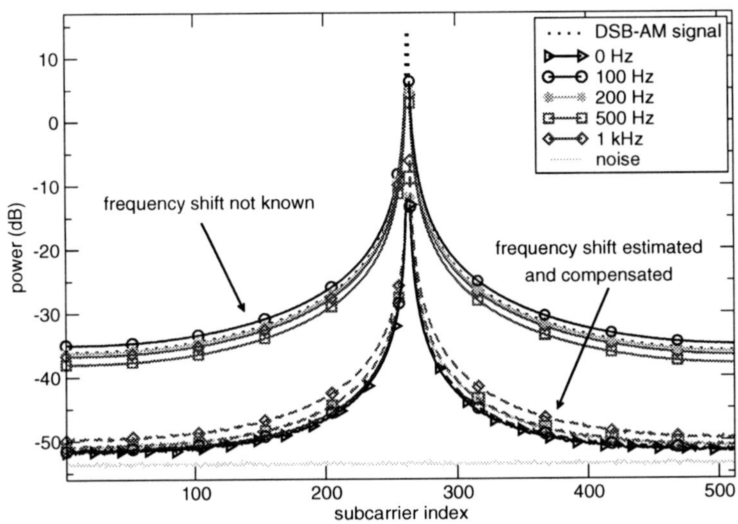

Figure 3. Spectrum of one interferer before and after subtraction of estimated NBI signal, $L = 2$ observation subcarriers, $k_c = 263.5 \pm 0.48$.

4.2 MMSE Based Leakage Compensation

Another approach for compensating the leakage effect of the NBI signals is multiplying the Rx signal $\mathbf{R}$ with a compensation matrix $\mathbf{C}_{\text{NBI}}$

$$\mathbf{R}' = \mathbf{C}_{\text{NBI}} \cdot \mathbf{R}. \tag{7}$$

The optimal compensation matrix yields $\mathbf{C}_{\text{NBI,opt}} \cdot \mathbf{I} = \mathbf{0}$ with $\mathbf{I} = [\mathbf{I}_0, \ldots, \mathbf{I}_{M-1}]$ denoting the interference signal for M OFDM symbols within one frame. However, due to noise and channel influences and an imperfect NBI estimation $\mathbf{C}_{\text{NBI}}$ is optimized such as to minimize the mean square error (MSE) on each subcarrier. Due to the structure of the compensation matrix interference values from the observation subcarriers weighted by compensation coefficients are subtracted from each data symbol. Despite several matrix multiplications the complexity of the algorithm is moderate. For a more detailed description of the algorithm please refer to [6].

The performance of the compensation algorithm is shown in Fig. 4 for the spectrum of one DSB-AM interferer with $k_c = 263.5$ averaged over 10,000 OFDM symbols. When estimating the NBI signal based on $L = 2$ observation subcarriers adjacent to the carrier the NBI signal is reduced by 22 dB. Compared to subtraction an additional reduction of 8 dB is achieved although based on the same estimation. With $L = 4$ observation subcarriers the NBI signal is reduced by 35 dB nearly to the noise level.

A variation of k_c by up to 1 kHz hardly has an impact on the performance of the compensation method indicating its robustness towards estimation errors. As illustrated in Fig. 4 this performance loss is mitigated nearly completely when k_c is estimated according to (3).

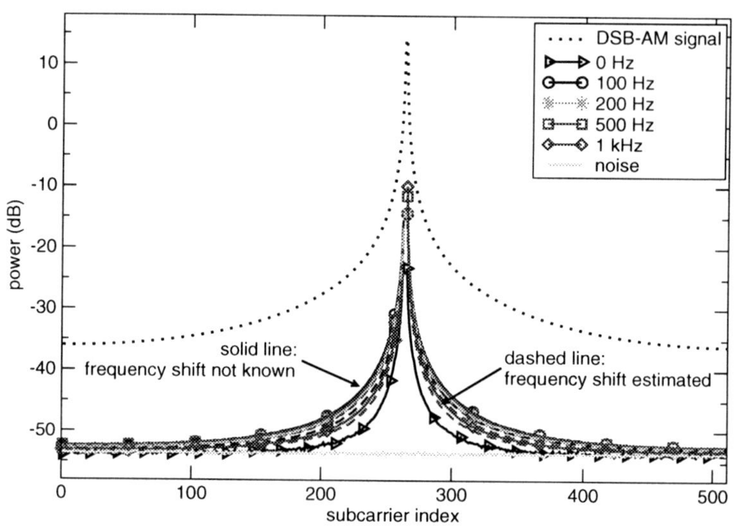

Figure 4. Spectrum of one interferer with and w/o matrix based leakage compensation, $L = 2$ observation subcarriers, $k_c = 263.5 \pm 0.48$.

5. Simulation Results

For the simulations an overlay system in the VHF band is considered. DSB-AM interference is modeled as depicted in Fig. 1 and described in

Section 2. The channels where strong interferers are present are excluded for the overlay system. The channels with weak interferers may be used by the overlay system as the overlay system does not interfere with DSB-AM systems in these channels. In these two channels $L = 2$ observation subcarriers are required that can not be used for data transmission. Each OFDM frame contains three equally distributed pilot OFDM symbols and eight OFDM symbols for data transmission, i.e. the distance of pilot symbols in time direction is equal to 5. For multiple access multi-carrier code-division multiple access (MC-CDMA) with spreading length 4 is applied and a fully loaded system is simulated. Uncoded quadrature phase-shift keying (QPSK) symbols are transmitted. The channel model considers a Doppler spectrum with a line-of-sight component and multi-path propagation. The GI contains 16 samples which is sufficient for the occurring multi-path delays. For all simulations, perfect channel estimation and perfect synchronization are assumed.

In Fig. 5, the BER vs E_b/N_0 for different NBI mitigation techniques is given. The comparison of the performance with and without NBI shows the strong impact of NBI. As the compensation of only strong interferers does not lead to a significant improvement, weak interferers have to be mitigated as well. When subtracting the NBI signal the SNR loss is reduced by 20 dB at BER $= 10^{-3}$. An additional NBI mitigation of 6.1 dB is achieved with the MMSE based leakage compensation. Both results are improved when a triangular window with a roll-off factor according to 52 samples is applied in addition. With subtraction and windowing the SNR loss due to NBI is reduced by 27.7 dB which is approximately the same as for MMSE based leakage compensation without windowing. The combination of MMSE based compensation and windowing mitigates the impact of NBI by 29.6 dB such that the performance without NBI is nearly reached.

In addition, carrier frequencies randomly varying within one subcarrier spacing are considered. A comparison of the results for perfect estimation of k_c and the results for real estimation only shows small differences. As expected an inaccurate estimation of k_c has a stronger impact on subtraction. At BER $= 10^{-3}$ the SNR loss due to remaining NBI is 3.3 dB larger than for perfect estimation of k_c. For MMSE based compensation this loss is negligible.

6. Conclusions

In this paper, a simple method for estimating the carrier of an NBI signal with sufficient accuracy has been presented. Interference is either mitigated by subtracting the estimated NBI signal or by optimizing a

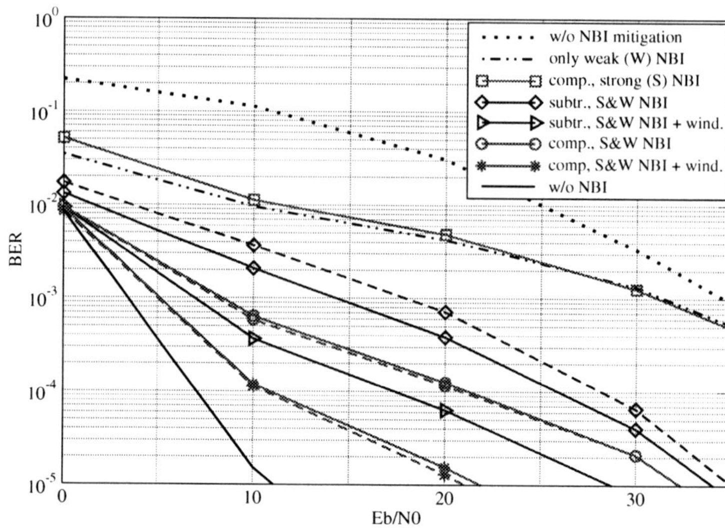

Figure 5. BER vs E_b/N_0 for different NBI mitigation techniques, en-route scenario, QPSK, uncoded.

matrix that compensates the leakage effect. Simulations in a realistic interference scenario have shown that the MMSE based approach performs better at the cost of additional computational complexity. When the MMSE based compensation is combined with Rx windowing the SNR loss due to NBI is diminished by approximately 30 dB such that the impact of NBI is reduced nearly completely.

References

[1] T. Weiss and F. Jondral, "Spectrum pooling - an innovatiove strategy for the enhancement of spectrum efficiency," *IEEE Communications Magazine*, vol. 42, no. 3, pp. S8–S14, March 2004.

[2] J. A. Bingham, *ADSL, VDSL, and Multicarrier Modulation.* John Wiley & Sons, Inc., 2000.

[3] L. de Clercq, M. Peeters, S. Schelstraete, and T. Pollet, "Mitigation of Radio Interference in xDSL Transmission," *IEEE Communications Magazine*, vol. 38, no. 3, pp. 168–173, 2000.

[4] B-VHF project, *www.b-vhf.org.*

[5] B-VHF project, "VHF Channel Occupancy Measurements," available at *www.b-vhf.org*, Tech. Rep. D-12, February 2005.

[6] J. Schwarz, "Enhancing the Spectral Selectivity of Discrete Multi-Tone Modulation," Ph.D. dissertation, University of Mannheim, 2006.

[7] S. Brandes, L. Falconetti, and M. Schnell, "Time and Frequency Domain NBI Mitigation in OFDM Based Overlay Systems," *11th International OFDM Workshop (InOWo '06)*, Hamburg, Germany, Aug. 2006, pp. 51–55.

CANCELLATION OF DIGITAL NARROWBAND INTERFERENCE FOR MULTI-CARRIER SYSTEMS

Mohamed Marey and Heidi Steendam
DIGCOM research group, TELIN Dept., Ghent University
Sint-Pietersnieuwstraat 41, 9000 Gent, BELGIUM
{mohamed, hs} @telin.ugent.be

Abstract Spectrum-overlay scenarios for wideband multi-carrier (MC) systems bring new technical challenges that must be considered during the system design. In such a scenario, the performance of the multi-carrier system is affected by the presence of possibly strong in-band interference. To improve the performance of the MC communication link in an interference corrupted environment (without increasing transmit bandwidth), the interference must be estimated and removed. In this paper, we propose a new low complexity algorithm to estimate and suppress digitally modulated interferers located in the MC spectrum. The performance of the interference suppression is evaluated in an analytical way. Further, simulations have been carried out to verify the validity of approximations in the analysis.

Keywords: Interference cancellation, narrowband interference

1. Introduction

The scarcity of available bandwidth typically necessitates spectrum sharing between legacy and new multi-carrier (MC) systems. The broadband very high frequency (B-VHF) project [1], which aims to develop a new integrated broadband VHF system for aeronautical voice and data link communications based on multi-carrier technology, is a good example of an overlay system. In this project, the MC system is intended to share parts of the VHF spectrum which are currently used by narrowband (NB) systems. Due to the spectral leakage of the discrete Fourier Transform (DFT) demodulation, many MC subcarriers near the interference frequency will suffer from serious interference, limiting the effectiveness of the multi-carrier system. In order to improve the performance of the MC communication link, some means of interference removal should be used.

S. Plass et al. (eds.), Multi-Carrier Spread Spectrum 2007, 127–136.

The techniques to cope with NBI in wireless MC systems can be divided into two categories. The first category is based on NBI suppression. Receiver windowing is a well known NBI suppression technique [2] using samples from the cyclic prefix to construct a window that reduces the NBI component of the received signal without affecting the data component. The result is that the NBI is convolved in frequency domain with a window that has smaller side-lobes than the sinc function, limiting the leakage to other subchannels. However, this technique requires a cyclic prefix length that is sufficiently long, reducing the efficiency of the MC system. Other techniques for NBI suppression are based on spreading the data over the whole MC bandwidth by either using orthogonal carrier interferometry spreading codes as in[3] or using orthogonal Hadamard sequences as in [4]. However, the complexity of both techniques is rather high.

The second category of NBI suppression techniques is based on interference cancellation: the interference is estimated and then subtracted from the received signal. In [5], the linear minimum mean square estimator (LMMSE) is adopted to estimate the NBI in the frequency domain; however this technique requires prior information about the power spectral density (PSD) of the NBI signals. In [6], unmodulated subcarriers close to the NBI central frequency are used to demodulate the NBI signal. Then, the NBI signal is reconstructed by passing the demodulated NBI data through the NBI transmit pulse. The complexity of this technique is rather high. It is worth to mention that former NBI suppression and cancellation techniques are done in the frequency domain, i.e. after FFT. This approach will lead to several practical issues like the design of the D/A converter (especially when the NBI signal is stronger the MC signal) and synchronization of the MC system [7–9]. This motivates us to propose a new low complexity algorithm to estimate NBI signals in time domain, before synchronization. The rest of the paper is organized as follows. In section 2, a MC system model and a narrow-band interference model are described. The proposed narrowband interference cancellation technique is illustrated in section 3. Simulation and analytical results are shown in section 4. Finally, the conclusions are given in section 5.

2. System Description

Fig. 1 shows the simplified Orthogonal Frequency Division Multiplexing (OFDM) system including the digital NBI. The sequence of input data symbols is segmented into blocks of length N_u. Define the input data as $a_l(k)$, $k = 0, ..., N_u - 1$, where the subscript l is used to indicate

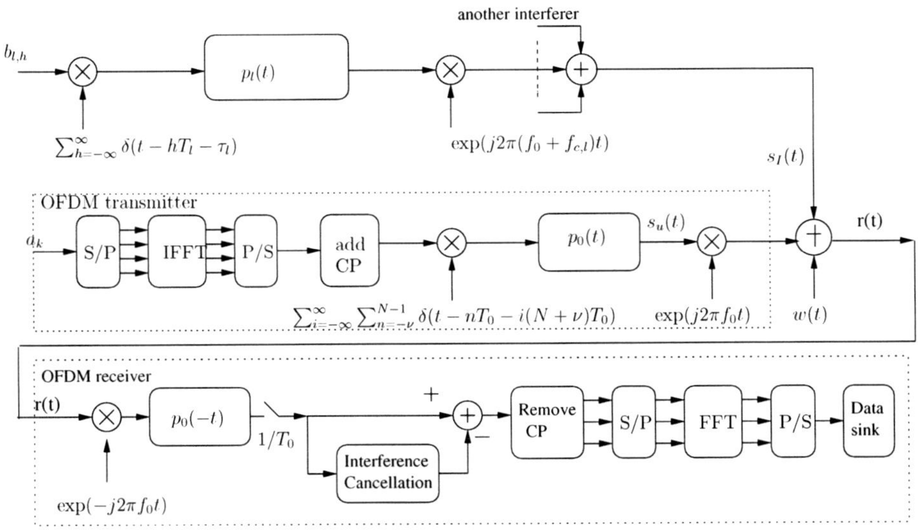

Figure 1. Schematic model of the OFDM system with NBI signals

the lth block of data, and k refers to the kth subchannel. The OFDM transmitter takes an N point IFFT of the lth block, and copies the last ν samples of the result as a cyclic prefix to form $x_l(n)$:

$$x_l(n) = \sqrt{\frac{1}{N+\nu}} \sum_{k \in I_u} a_i(k)\, e^{\frac{j2\pi kn}{N}} \quad -\nu \leq n \leq N-1 \tag{1}$$

where I_u is a set of N_u carrier indices. The data symbols are assumed i.i.d.[1] random values with zero mean and variance $E[|a_l(k)|^2] = E_s$. The time domain baseband OFDM signal $s_u(t)$ consists of the concatenation of all time domain blocks $x_l(n)$:

$$s_u(t) = \sum_{i=-\infty}^{\infty} \sum_{n=-\nu}^{N-1} x_l(n)\, p_0(t - nT_0 - i(N+\nu)T_0) \tag{2}$$

where $p_0(t)$ is the unit-energy transmit pulse of the OFDM system and $1/T_0$ is the sample rate. The baseband signal (2) is up-converted to the radio frequency f_0. At the receiver, the signal is first down-converted. Next, the OFDM receiver discards the first ν samples of the received block, and takes an N-point FFT of the result. The OFDM signal is disturbed by additive white Gaussian noise with uncorrelated real and imaginary parts, each having variance σ_n^2. The signal to noise ratio at the output of the matched filter is defined as $SNR = \sigma_s^2/\sigma_n^2$ where σ_s^2 is the variance of the time domain OFDM signal per real dimension. Further, the signal is disturbed by narrowband interference residing within the same frequency band as the wideband OFDM signal. In this paper, we assume the NBI signal consists of N_I digitally modulated signals. Following the interference model from [7, 8], the interfering signal $s_I(t)$ may be written as

$$s_I(t) = \sum_{l=1}^{N_I} s_l(t) e^{j2\pi(f_0+f_{c,l})t} \tag{3}$$

where $s_l(t)$ is a baseband narrowband signal and $f_{c,l}$ is the carrier frequency deviation for the lth interferer from the MC carrier frequency f_0. The baseband interference $s_l(t)$ is modeled as a digitally modulated signal

$$\begin{aligned} s_l(t) &= \sum_{h=-\infty}^{\infty} b_{h,l} p_l(t - hT_l - \tau_l) \\ &\simeq \sum_{h=0}^{q} b_{h,l} p_l(t - hT_l - \tau_l) \end{aligned} \tag{4}$$

where $p_l(t)$ is the time domain impulse response of the transmit filter of the lth interferer, $b_{h,l}$ is the hth interfering data symbol, τ_l is its delay, and $1/T_l$ its sample rate. Let B_l be the bandwidth of $p_l(t)$ and B_0 the bandwidth of the OFDM signal. In an MC symbol duration T_{FFT}, there are q symbols of $s_l(t)$, where q is an integer equal to or less $\frac{B_l}{B_0}$. Because $p_l(t)$ degrades rapidly in time, symbols $\{b_{h,l}\ \forall h < 0\ or\ h > q\}$ have a negligible effect on the signal $s_l(t)$ $(0 \leq t \leq T_{FFT})$. Therefore, the approximation in (4) is valid. The total NBI signal at the output of the matched filter of the MC receiver yields

$$r_I(t) \simeq \sum_{l=1}^{N_I} \sum_{h=0}^{q} b_{h,l}\, e^{j2\pi f_{c,l} hT_l}\, g_l(t - hT_l) \tag{5}$$

where $g_l(t)$ is the convolution of $p_0(-t)$ and $p_l(t-\tau_l)\exp(j2\pi f_{c,l}t)$. The normalized location of the interferer within the MC spectrum may be defined as $f'_{c,l} = f_{c,l}/B_0$. It is assumed that the interfering symbols are uncorrelated with each other, i.e. $E[b_{h,l} b^*_{h',l'}] = E'_l \delta_{ll'} \delta_{hh'}$, where E'_l is the energy per symbol of the lth interferer. Further, the interfering data symbols $b_{h,l}$ are statistically independent of the OFDM data symbols $a_i(n)$. The signal to interference ratio (SIR) at the input of the receiver is defined as [8]

$$SIR = \frac{2\sigma_s^2/T_0}{\sum_{l=1}^{N_I} \frac{E_l'}{T_l}}. \tag{6}$$

3. Narrowband Interference Cancellation

In the proposed algorithm, each NBI signal $s_l(t)$ is estimated separately and subtracted from the received signal. Let us assume that we use the available samples at the MC receiver to estimate the interference. The sample $r(mT_0)$ at the output of the matched filter of the OFDM receiver consists of a useful signal $r_u(mT_0)$, an interfering $r_I(mT_0)$ component, and noise $w(mT_0)$:

$$r(mT_0) = r_u(mT_0) + r_I(mT_0) + w(mT_0). \tag{7}$$

Further, in our analysis we assume that the frequency $f_{c,l}$ is perfectly known. In practice, a simple estimate of $f_{c,l}$ can be obtained by using the squared magnitude of the FFT outputs, as in a periodogram, searching for the subcarriers with the strongest interference; the estimate of $f_{c,l}$ can then be found by interpolation [5]. Although a perfect knowledge of the frequency $f_{c,l}$ is assumed, it turns out that the proposed algorithm is insensitive to small estimation errors in $f_{c,l}$. The samples (7) are multiplied with $\exp(-j2\pi f_{c,l} mT_0)$ to down-convert the lth NBI to baseband. Next, the samples are averaged over a $(2K+1)$ size sliding window.

$$\hat{s}_l(nT_0) = \frac{1}{2K+1} \sum_{k=-K}^{K} r((n+k)T_0) \cdot e^{-j2\pi(n+k)f_{c,l}T_0} \tag{8}$$

This averaging acts as a low-pass filter with bandwidth $1/(2K+1)T_0$, which reduces the effects of the noise, the OFDM signal, and the contributions of other NBI signals on the estimation of the wanted NBI signal. Increasing K will reduce the bandwidth of the equivalent low-pass filter, and results in a reduction of the effects of noise, OFDM, and other disturbing NBI signals. However, if K is selected too large, the sliding window will not be able to track the small variations of the wanted NBI signal, i.e. the bandwidth of the equivalent low-pass filter must be larger than the bandwidth of the NBI signals. Let $\sigma_l^2(nT_0)$ be the mean squared error of lth interferer at instant nT_0, $\sigma_l^2(nT_0) = E[|s_l(nT_o) - \hat{s}_l(nT_0)|^2]$. After tedious computations, it follows that $\sigma_l^2(nT_0)$ can be written as

$$\sigma_l^2(nT_0) = \sigma_{SI_l}^2(nT_0) + \sigma_{MI_l}^2(nT_0) + \sigma_{AWGN_l}^2(nT_0) \tag{9}$$

where the self noise $\sigma_{SI_l}^2(nT_0)$ is the variance resulting from the unablity to track the small variations of the wanted NBI signal, $\sigma_{MI_l}^2(nT_0)$ is the variance resulting from other NBI signals, and $\sigma^2_{AWGN_l}(nT_0)$ is the variance resulting from AWGN and the OFDM signal[2]. These variances can be written as

$$\begin{aligned}\sigma^2_{SI_l}(nT_0) &= E_l \sum_{h=-\infty}^{\infty} |\xi_{l,h}(n,0)|^2 \\ &+ \left(\frac{1}{2K+1}\right)^2 \sum_{k,k'=-K}^{K} E_l \sum_{h=-\infty}^{\infty} \xi_{l,h}(n,k)\xi^*_{l,h}(n,k') \\ &- Re\left\{\frac{2}{2K+1}\sum_{k=-K}^{K} E_l \sum_{h=-\infty}^{\infty} \xi_{l,h}(n,0)\xi^*_{l,h}(n,k)\ e^{J2\pi f_{c,l}(n+k)T_o}\right\}\end{aligned} \tag{10}$$

where $\xi_{l,h}(n,k) = g_l\left((n+k)T_0 - hT_l\right)$.

$$\sigma^2_{MI_l}(nT_0) = \left(\frac{1}{2K+1}\right)^2 \sum_{k,k'=-K}^{K} \sum_{l'=1,l'\neq l}^{N_I} E_{l'} \sum_{h=-\infty}^{\infty} \xi_{l',h}(n,k)\xi^*_{l',h}(n,k') \tag{11}$$

$$\sigma^2_{AWGN_l}(nT_0) = \frac{2(\sigma_s^2+\sigma_n^2)}{2K+1}. \tag{12}$$

Note that $\sigma_l^2(nT_0)$ is a periodic function with period T_l ,i.e. $\sigma_l^2(nT_0) = \sigma_l^2\left(nT_0 + \frac{T_l}{T_0}T_0\right)$. Therefore, the average variance σ_l^2 can be written

$$\sigma_l^2 = \frac{1}{[[T_l/T_0]]} \sum_{n=0}^{[[T_l/T_0]]-1} \sigma_l^2(nT_0) \tag{13}$$

where $[[x]]$ rounds x to the nearest integer. Assuming that the NBI signals are independent, the total variance σ_I^2 of the estimator equals $\sum_{l=1}^{N_I}\sigma_l^2$.

4. Numerical Results

For the numerical and simulation results, we assume that the number of sub-carriers $N = 256$ and the number of active sub-carriers is $N_u = 256$ i.e. all carriers are modulated. The guard interval equals $\nu = 20$ and the bandwidth of OFDM signal is $B_0 = 1024$ kHz. The bandwidth of NBI signal equals B_l= 25 kHz and $\tau_l = 0$. We use 8-PSK and QPSK modulation for the data symbols of the OFDM and the interferer signals respectively. Transmit filters are square-root raised-cosine filters with roll off factors $\alpha_0 = 0.25$ and $\alpha_I = 0.5$ for OFDM and interfering signals, respectively.

Fig. 2 shows the simulated and analytical variance of the estimation error (σ_I^2) as a function of the window length $(2K + 1)$ assuming that the OFDM system transmits data on all carriers. As can be observed, at high signal to interference SIR values, σ_I^2 decreases with window size. At low SIR, σ_I^2 shows a minimum at intermediate window size. This can be explained with the aid of Fig. 3. In the estimator, there are two types of noise. The first type comes from the OFDM signal and AWGN noise ($N1$) while the second type is the noise ($N2$) that results from the unability of the estimator to track the variations of the NBI signal. At high SIR values, the former dominates. Since increasing the window size reduces the effect of this type of noise, we notice that σ_I^2 decreases with the window size. At low SIR, i.e. high interference power, the first type of noise diminishes but the second type increases with increasing window size. Therefore, σ_I^2 increases again with the window size.

Fig. 4 shows the variance of the estimation error (σ_I^2) as a function of the signal to interference ratio (SIR) at a window size equal to 33 assuming that the OFDM system does not transmit data on M carriers around the location of the NBI. Since at low SIR, the second type of noise dominates, the OFDM signal has only a small effect on the estimator. Therefore, σ_I^2 is approximately independent of M as can be observed in the figure. For large SIR (≥ 0 dB), the effect of the second type of noise diminishes and the first type of noise dominates. Increasing M will reduce the effect of this type of noise, as the the spectral leakage from the OFDM signal to the NBI signal reduces. Therefore, increasing M leads to a reduction of σ_I^2.

Fig. 5 shows the power spectrum of the original interference signal and residual interference signal (after cancellation) in two cases. In the first case, we have assumed that the received signal only consists of NBI signal, i.e. the OFDM signal and noise are not present. This result gives us an indication of the maximum possible interference reduction. In the second case, we also consider the presence of the OFDM signal and noise. We observe a strong reduction in power for frequencies close to the frequency of NBI. The reduction of larger NBI frequency components will be smaller.

Fig. 6 shows the bit error rate (BER) performance of the OFDM system with and without NBI cancellation for different values for M. As can be observed, NBI cancellation achieves a great improvement in

BER performance for small SIR values i.e. when the interference signal is strong as compared to the OFDM signal. At high SIR values, the proposed cancellation algorithm has worse performance than the case of no NBI cancellation. This is explained as at high SIR, the NBI signal is very small as compared to the OFDM signal, and therefore, the NBI is very difficult to extract from the received signal. Further, at high SIR the presence of the estimator will cause a noise enhancement to the OFDM system with variance equal to $\left(\sigma_s^2 + \sigma_n^2\right) / (2K + 1)$ per real dimension (see (12)). However, at high SIR, the effect of the NBI on the OFDM system (without cancellation) becomes negligible and the BER reaches an asymptote. This asymptote depends on the SNR, as AWGN is the dominating disturbance on the OFDM system. Note however that the difference between in BER no cancellation for the proposed algorithm becomes small when the gap M increases. Therefore, we can conclude that the proposed algorithm works well when M is chosen sufficiently large.

Fig 7 shows the total variance σ_I^2 of the estimator as function of the number N_I of NBI signals in two cases. In case 'A', we consider that the SIR is fixed per interferer, so the total SIR decreases inversely proportional to N_I. In case 'B', we consider a fixed total SIR, i.e. SIR per interferer decreases linearly as N_I increases. As can be observed, the variance of the estimator increases with N_I. This is because the noise caused by other NBI signal increases with N_I. The corresponding BER performance of the MC system is shown in Fig. 8. We note that the BER is essentially independent of N_I at M =16: the spectrum leakage from the NBI signals on the MC signal becomes very small when M increases.

5. Conclusions

In this paper, a new NBI cancellation scheme for MC systems has been proposed. The estimator is based on averaging the received baseband samples over a sliding window. It can be used to suppress the spectral leakage that occurs when many digitally modulated NBI signals reside in the same frequency band of the MC signal. Further, we have derived the mean squared error (MSE) of the estimator in an analytical way. Simulation results show that the theoretical expressions for the MSE agree well with the simulation. Moreover, bit error rate performance shows that the proposed estimator performs well especially if the MC system avoids the use of a number of subcarriers around the NBI frequencies.

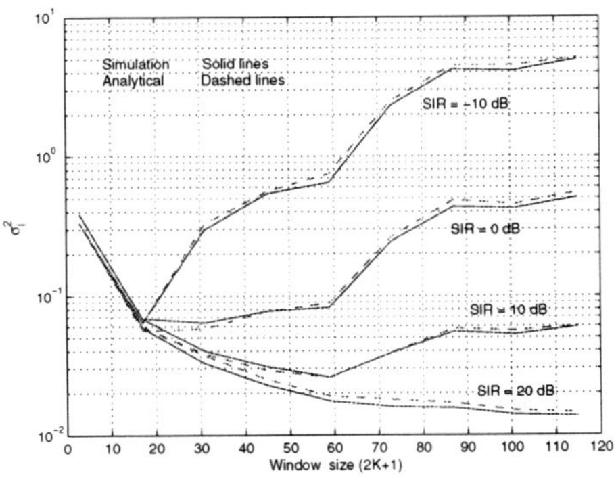

Figure 2. Averaging variance, σ_I^2 at $SNR = 8$ dB, $N_I = 1$.

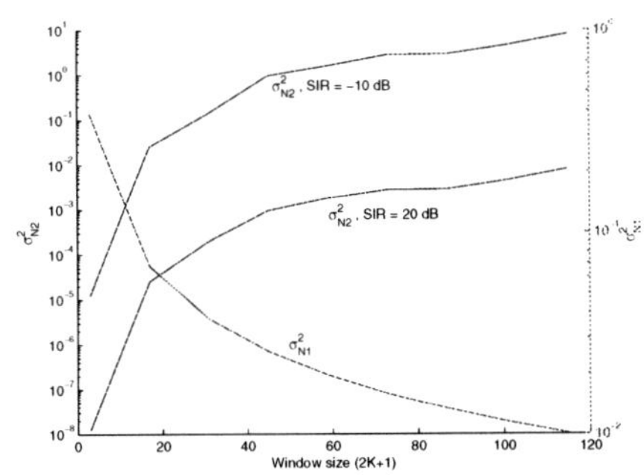

Figure 3. Averaging variance, σ_{N1}^2, σ_{N2}^2 at $SNR = 8$ dB, $N_I = 1$.

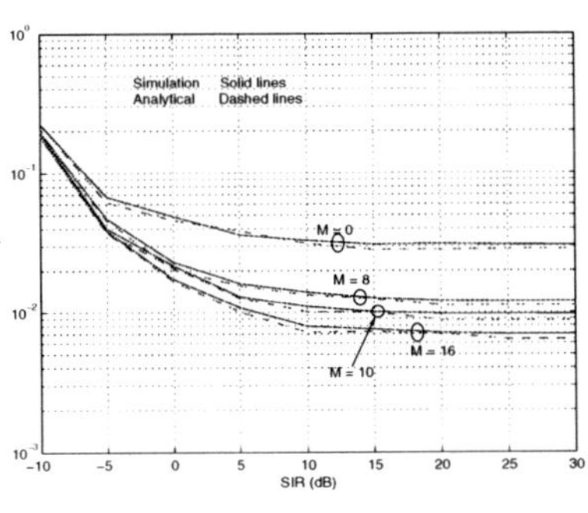

Figure 4. Averaging variance, σ_I^2 at $SNR = 8$ dB, and $N_I = 1$

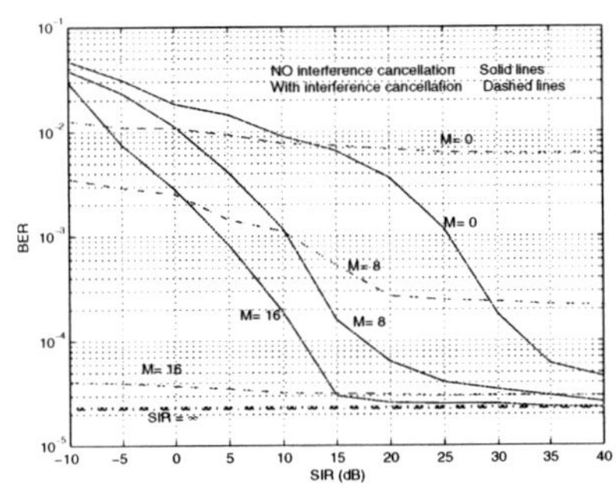

Figure 6. BER at SNR =17dB and $N_I = 1$

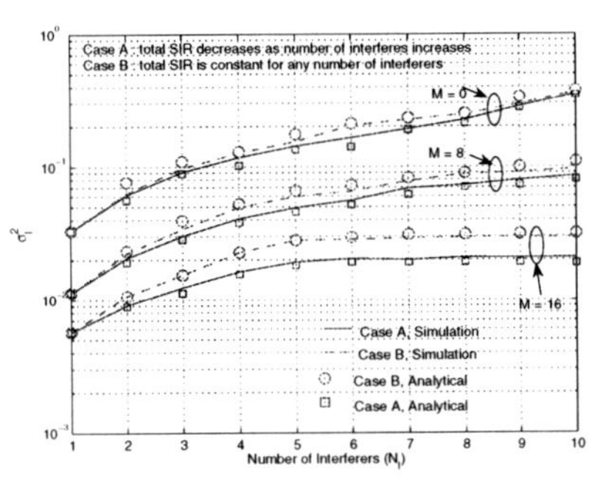

Figure 7. Averaging variance, σ_I^2 at $SIR = 10$ dB and SNR=17dB

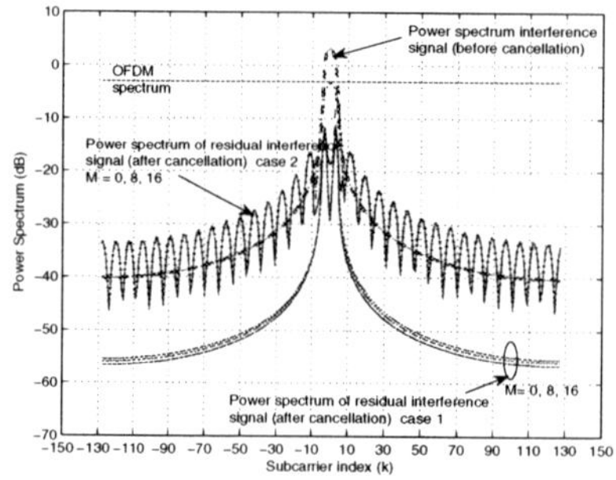

Figure 5. Power spectrum at $SNR = 17$ dB, $SIR = 10$ dB, and $N_I = 1$

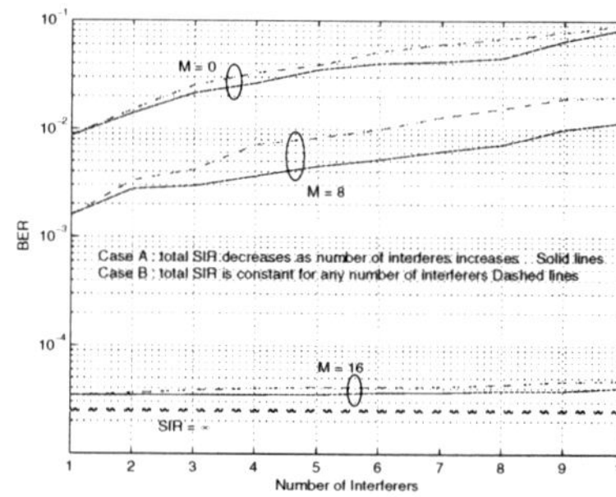

Figure 8. BER at SNR=17 dB and $SIR = 10$ dB

Notes

1. i.i.d. = independently and identically distributed

2. From the estimator viewpoint, the OFDM signal can be modeled (according to central limit theorem) as zero-mean Gaussian distributed.

References

[1] http://www.b-vhf.org.

[2] A. J. Redern. Receiver Window Design for Multicarrier Communication Systems *IEEE Journal on Selected Areas in Comm.* , 20(5):1029–1036, June 2002.

[3] Z. Wu and C. Nassar. Narrowband Interference Rejection in OFDM via Carrier Interferometry Spreading Codes *IEEE Trans. on Wireless Comm.*, 4(4):1491–1505, July 2005.

[4] D. Gerakoulis and P. Salmi. An Interference Suppressing OFDM System for Wireless Communications *Proc. of The IEEE International Conference on Communications, ICC* . USA, Apri 2002.

[5] R. Nilsson, F. Sjoberg, and J. P. LeBlanc. A Rank-Reduced LMMSE Canceller for Narrowband Interference Suppression in OFDM-Based Systems *IEEE Trans. on Comm.*, 51(12): 2126–2140, December 2003.

[6] D. Zhang, P. Fan, and Z. Cao. Interference Cancellation for OFDM Systems in Presence of Overlapped Narrow Band Transmission System *IEEE Trans. on Consumer Electronics.*, 50(1): 108 – 114, Feb. 2004.

[7] M. Marey and H. Steendam. The Effect of Narrowband Interference on the Timing Synchronization for OFDM Systems. *Proc. of the 12th IEEE Benelux Symposium on Communications and Vehicular Technology* . Nov 2005.

[8] M. Marey and H. Steendam. The Effect of Narrowband Interference on Frequency Ambiguity Resolution for OFDM *Proc. of Vehicular Technology Conference Fall 2006.* Sept. 2006.

[9] M. Marey and H. Steendam. The Effect of Narrowband Interference on ML Fractional Frequency Offset Estimator for OFDM *Proc. of International Mobile Multimedia Communications* . Alghero, Italy Sept. 2006.

RAKE RECEPTION FOR SIGNATURE-INTERLEAVED DS CDMA IN RAYLEIGH MULTIPATH CHANNEL

Alexey Dudkov
Turku Centre for Computer Science (TUCS), University of Turku
Joukahaisenkatu 3-5 B, 6th floor, FIN-20520 Turku, Finland
alexey.dudkov@utu.fi

Abstract Performance of RAKE receiver in a multipath channel for direct sequence code division multiple access system (DS CDMA) can be severely impaired due to high level of multiple access interference (MAI), whose value is controlled by correlation functions of user signatures. Due to data modulation MAI becomes dependent on aperiodic user signature correlations of big shifts, which are hard to control and become even less so as multipath delay grows, leading to significant MAI increase. In this article we analyze performance of RAKE receiver for Signature-Interleaved (SI) DS CDMA in synchronous multipath Rayleigh channel in comparison with that of conventional DS CDMA. It is shown that SI DS CDMA provides better performance in terms of bit error rate (BER), especially when using specifically tailored user signatures.

Keywords: Code division multiaccess, multiple access interference, RAKE, equal gain combining, multipath channels.

1. INTRODUCTION

In direct sequence code division multiple access (DS CDMA) communication every user is assigned its specific code sequence called *signature*. Correlation properties of the chosen signatures materially affect system performance, thus making adequate signature construction procedure a critically important point in the system design. When precise synchronism between users' signals can be maintained and no multipath propagation is present in the channel, the signature selection does not look particularly challenging, provided number of users K is no greater than the spreading factor N. Indeed, in such a scenario orthogonal signatures secure total removal of the multiple access interference (MAI) in a single-user (matched-filter) receiver, guaranteeing the latter opti-

S. Plass et al. (eds.), Multi-Carrier Spread Spectrum 2007, 137–145.

mal AWGN performance regardless of the presence of any foreign users' signals [1].

Under the multipath propagation every signature is received accompanied by multiplicity of its time-shifted replicas, and when overall delay spread spans a sufficient number of chips, mutual orthogonality of all the signature replicas cannot be preserved. Thereby MAI appears to be inevitable, and more than this, due to modulation of signatures by the data stream MAI components are caused by both even and odd correlations of signatures. For this reason even if the channel delay spread allows resorting to zero correlation zone ensemble (ZCZ) [2], the latter does not secure a complete MAI elimination. At the same time designing practicable signature ensembles with low level of both even and odd correlations has remained problematic so far.

The performance degradation due to multipath fading may be overcome by employing diversity. A popular way to do so is implementation of a RAKE receiver, which performs combining of the multipath components [3]. However, in this case system performance may be limited by increased MAI, which is combined as well. In this article we analyze efficiency of an equal gain combining (EGC) RAKE receiver for signature-interleaved (SI) DS CDMA [4] in synchronous multipath Rayleigh channel in comparison with that of conventional DS CDMA. It is shown that a RAKE receiver for SI DS CDMA provides better results in terms of bit error rate.

The remainder of the text is organized as follows. In Section 2, a model of the conventional DS CDMA system is considered and an expression for MAI at the RAKE receiver output is obtained. In Section 3, SI DS CDMA concept is reviewed and analyzed, and in Section 4 the systems are compared in terms of bit error rate. Final conclusion and comments are given in Section 5.

2. CONVENTIONAL DS CDMA

Consider a synchronous DS CDMA with K users, each employing a signature of length N. Denoting as $b_k(i)$ the i-th data bit of the k-th user, and as $\mathbf{s}_k = [s_k(0), s_k(1), ..., s_k(N-1)]$ the k-th user signature of unity norm, $i = ..., -1, 0, 1, ...$, $1 \leq k \leq K$, the overall (group) signal of all users on the time interval of bits $b_k(i)$ can be written as

$$s(t) = \sum_{k=1}^{K} b_k(i) \sum_{j=0}^{N-1} s_k(j) c_0(t - j\Delta), \tag{1}$$

where $c_0(t)$ is the waveform of chip of duration Δ (Figure 1a). We focus here on CDMA with periodic signatures having period N equal to the number of chips per data bit and BPSK data modulation combined with real (e.g. binary) signatures.

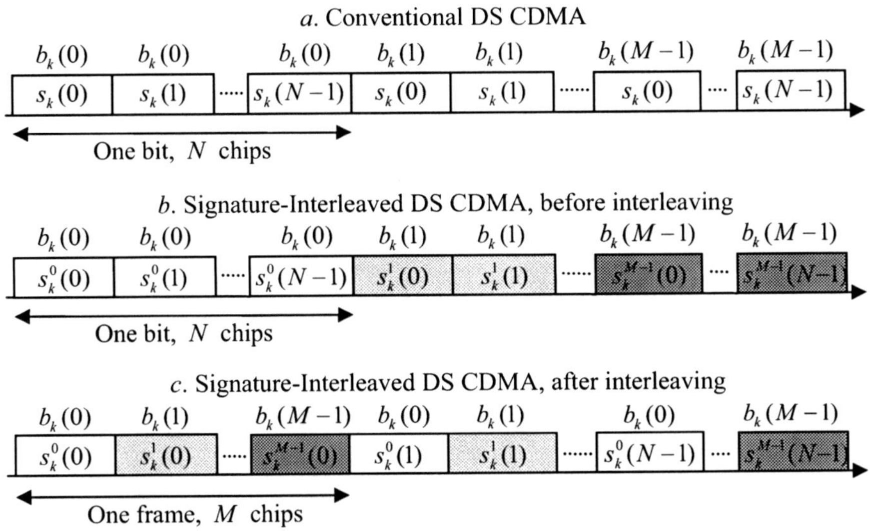

Figure 1. Illustration of the SI DS CDMA

Consider Rayleigh channel with R paths, r-th one having amplitude A_r, phase φ_r and delay τ_r, $1 \le r \le R$. Assume delays of the paths to be nonnegative multiples m_r of the chip duration, $\tau_r = m_r\Delta$, with m_r bounded from above by the constant L: $0 \le m_1 \le ... \le m_r \le ... \le m_R \le L$, where rays are sorted in non-decreasing order of their delays. We will also assume that fading pattern remains stable during the bit interval, that is, A_r and φ_r are treated as constants. Then complex envelope of the received signal is of the form

$$y(t) = \sum_{r=1}^{R} A_r s(t - \Delta m_r) \exp(j\varphi_r) + n(t), \tag{2}$$

where $n(t)$ is the complex white Gaussian noise. We assume that at the receiving end an equal gain combining (EGC) RAKE is implemented with perfect knowledge of the paths phases, which for the k-th user can be described as calculating correlation of (2) and the modified signature

$$s_k'(t) = \sum_{r=1}^{R} s_k(t - \Delta m_r) \exp(-j\varphi_r), \tag{3}$$

providing the following result at the output of the receiver:

$$
\begin{aligned}
z_k(i) &= b_k(i) \sum_{r=1}^{R} A_r \\
&+ \sum_{p=1}^{R} \left[A_p \sum_{\substack{l=1 \\ l \neq k}}^{K} b_l(i) R_{kl}(0) \right. && (4) \\
&+ A_p \sum_{r=1}^{p-1} \exp\left(j(\varphi_p - \varphi_r)\right) \sum_{l=1}^{K} \Big\{ b_l(i) R_{kl}(m_r - m_p) \\
&\qquad + b_l(i+1) R_{kl}(N + m_r - m_p) \Big\} && (5) \\
&+ A_p \sum_{r=p+1}^{R} \exp\left(j(\varphi_p - \varphi_r)\right) \sum_{l=1}^{K} \Big\{ b_l(i-1) R_{kl}(m_r - m_p - N) \\
&\qquad \left. + b_l(i) R_{kl}(m_r - m_p) \Big\} \right] && (6) \\
&+ n_k,
\end{aligned}
$$

where $b_l(i-1)$ and $b_l(i+1)$ are, respectively, information bits of the l-th user preceding and following $b_l(i)$, n_k is the noise sample and $R_{kl}(m)$ is aperiodic cross-correlation function between signatures $\mathbf{s}_k$ and $\mathbf{s}_l$. Finally, a decision on the bit $b_k(i)$ is formed as

$$\hat{b}_k(i) = \operatorname{sgn}\left(\operatorname{Re}\left[\mathrm{z_k(i)}\right]\right). \tag{7}$$

Terms (4), (5) and (6) represent MAI, from foreign users' signals of the same multipath ray p, $1 \leq p \leq R$; from user signals of the multipath rays with delays no greater than the current one $m_r \leq m_p$, $1 \leq r \leq p-1$; and from user signals of the multipath rays with delays no smaller than the current one, $m_r \geq m_p$, $p+1 \leq r \leq R$, respectively. These equations indicate that MAI level is governed by values of aperiodic signatures' cross-correlations of shifts up to $\max(m_r) = L$. For usually implemented signature ensembles (Gold, Kasami, cyclic shifts of m-sequence [1]) values of those are uncontrollable and occur almost at random [5], which may result in high level of MAI and degrade system performance. It is possible, however, to change system design so that only cross-correlations of zero and first shift affect MAI level *independently* of the maximal delay value.

3. SIGNATURE-INTERLEAVED (SI) DS CDMA

A distinguishing feature of SI DS CDMA is that every user utilizes M signatures $\mathbf{s}_k^i = [s_k^i(0), s_k^i(1), ..., s_k^i(N-1)]$ of length N and unity norm instead of a single one [4], as it is the case for conventional DS CDMA, $i = 0, 1, \ldots, M-1$ being number of the signature assigned to the k-th user, $1 \le k \le K$. Supposing again that a data bit covers N signature chips, let the data bit $b_k(0)$ of the k-th user modulate $\mathbf{s}_k^0$, the next one $b_k(1)$ modulate $\mathbf{s}_k^1$, and so on up to the bit $b_k(M-1)$ modulating $\mathbf{s}_k^{M-1}$; bit $b_k(M)$ modulating again $\mathbf{s}_k^0$, etc. This way the data-manipulated chip stream is split into blocks of M bits (further M-blocks), each block spanning MN chips. Then the synchronous group signal of all users is (Figure 1*b*)

$$s(t) = \sum_{k=1}^{K} \sum_{i=0}^{M-1} b_k(i) \sum_{j=0}^{N-1} s_k^i(j) c_0(t - j\Delta - iN\Delta). \tag{8}$$

Every such M-block is then fed into the chip interleaver before the transmission. The output chip pattern for the k-th user will consist of the first elements of all k-th user signatures $s_k^0(0)$, $s_k^1(0)$, $\ldots$ $s_k^{M-1}(0)$ successively modulated by the information bits $b_k(0)$, $b_k(1)$, $\ldots$, $b_k(M-1)$, respectively, followed by the second elements of the signatures modulated by the same bits and so on up to the N-th elements. After these rearrangements chips of every user signature will be distanced from each other by M positions, and the transmitted group signal takes the form (Figure 1*c*)

$$s(t) = \sum_{k=1}^{K} \sum_{i=0}^{M-1} b_k(i) \sum_{j=0}^{N-1} s_k^i(j) c_0(t - i\Delta - jM\Delta). \tag{9}$$

Output of RAKE receiver of the k-th user processing bit $b_k(i)$, $i = 0, 1, ..., M-1$ may then be written as

$$\begin{aligned} z_k^i = b_k(i) \sum_{r=1}^{R} A_r + \sum_{p=1}^{R} \Bigg[& A_p \sum_{\substack{l=1 \\ l \ne k}}^{K} b_l(i) R_{(k,\, i),(l,\, i)}(0) \\ & + A_p \sum_{r=1}^{p-1} \exp\left(j(\varphi_p - \varphi_r)\right) \sum_{l=1}^{K} \mu_{kl}(i, r) \\ & + A_p \sum_{r=p+1}^{R} \exp\left(j(\varphi_p - \varphi_r)\right) \sum_{l=1}^{K} \varsigma_{kl}(i, r) \Bigg] + n_k, \end{aligned} \tag{10}$$

where

$$\mu_{kl}(i,r) = \begin{cases} b_l(i'')R_{(k,i),(l,i'')}(0), & i < M - m_p + m_r, \\ b_l(i'')R_{(k,i),(l,i'')}(-1) \\ \quad +b_l^1(i'')R_{(k,i),(l,i'')}(N-1), & i \geq M - m_p + m_r, \end{cases} \tag{11}$$

corresponds to the paths with delays no greater than that of the current path $m_r \leq m_p$, $1 \leq r \leq p-1$, where $i'' = (i + m_p - m_r) \bmod M$, and

$$\varsigma_{kl}(i,r) = \begin{cases} b_l(i')R_{(k,i),(l,i')}(0), & i \geq m_r + m_p, \\ b_l^{-1}(i')R_{(k,i),(l,i')}(1-N) \\ \quad +b_l(i')R_{(k,i),(l,i')}(1), & i < m_r - m_p, \end{cases} \tag{12}$$

corresponds to the paths with delays no smaller than that of the current path $m_r \geq m_p$, $p+1 \leq r \leq R$, where $i' = (i - m_r) \bmod M$. $R_{(k,i),(l,j)}(m)$ is aperiodic cross-correlation between signatures $\mathbf{s}_k^i$ and $\mathbf{s}_l^j$, $b_l^1(i)$ and $b_l^{-1}(i)$ are, respectively, i-th information bits of the M-blocks following and preceding the current one.

Equation (10) demonstrates that for SI DS CDMA MAI depends on correlations of non-shifted or just one-chip-shifted signatures only, *independently* of the maximal multipath delay. Values of these correlations can be easily retained under control [4], making thus level of MAI much smaller than that of conventional DS CDMA.

4. PERFORMANCE COMPARISON

In order to support abovementioned statement, the performance simulation was carried out for the system with spreading factor $N = 127$, $K = 10$ users under 4-path Rayleigh multipath propagation with path energies $\overline{A_r^2} = [0, -6, -11.9, -17.9]$ dB, random path phases with uniform probability distribution over $[0,\ 2\pi]$ and path delays $m_r = [0, 3, 4, 7]$ chips, which corresponds to the Office Channel B propagation model [6].

For conventional DS CDMA $K = 10$ Gold signatures of length $N = 127$ were used as user signature ensemble, which is a reasonable choice used in practice [1], [7]. For SI DS CDMA user signatures were formed as $L = 7$ even cyclic shifts of Gold signatures as described in [4], since the chosen parameters ($2KL > N$) make usage of ZCZ or low correlation zone ensemble impossible. A random search was performed to find among all possible cyclic shifts the ones minimizing MAI [8]. As an additional reference ensemble of orthogonal Gold signatures [9] was used as user signature ensemble for DS CDMA, and its even cyclic shifts as user signature ensemble for SI DS CDMA.

In both systems 4-finger EGC RAKE receiver was implemented. Dependence of bit error rate (BER) on signal-to-noise ratio $q^2 = 1/\sigma^2$ is shown in Figure 2, where σ^2 is power of the Gaussian noise at the output of the chip matched filter. The results are averaged over users, information bits, ray amplitudes and phases, assuming their independence.

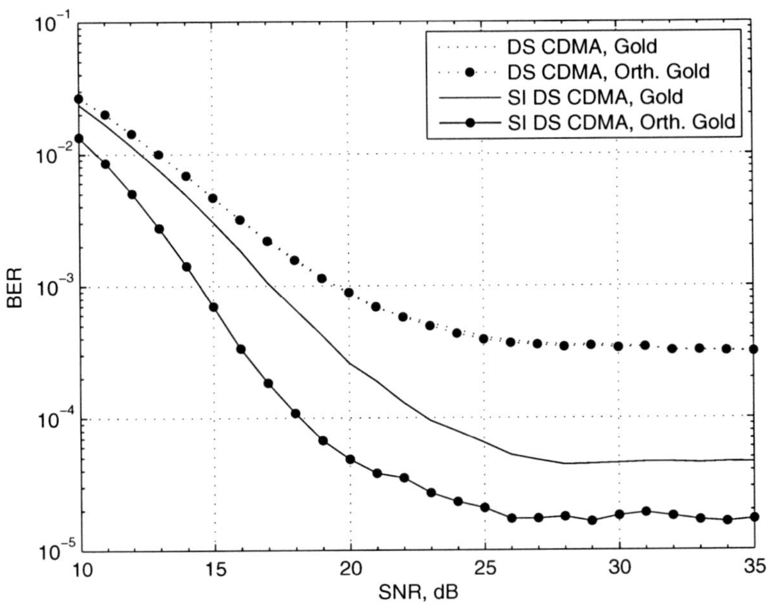

Figure 2. Simulation results

The graphs demonstrate that RAKE receiver for SI DS CDMA system secures smaller probability of error than that of DS CDMA, providing for the Gold ensemble more that 2 dB gain at the BER level 10^{-3} and lowering irreducible BER (error floor). While for conventional DS CDMA switching to orthogonal Gold signature ensemble does not provide any benefit, for SI DS CDMA it increases gain at the BER level 10^{-3} to 5 dB and lowers irreducible BER further. This can be explained by the fact that for conventional DS CDMA MAI level is defined mostly by aperiodic correlations of user signatures of big shifts, which do not differ much for Gold and orthogonal Gold ensembles. At the same time for SI DS CDMA leading role of defining MAI belongs to user signature correlations of zero shift, which for orthogonal Gold ensemble are zero, $R_{kl}(0) = 0$, $1 \leq k, l \leq K$, $k \neq l$.

5. CONCLUSION

Usage of RAKE receiver for Signature-Interleaved DS CDMA in Rayleigh multipath channel was analyzed and its performance was compared with that of conventional DS CDMA system. It was demonstrated that for SI DS CDMA MAI level at the output of the RAKE receiver depends on user signature correlations of zero and one shift only, independently of the multipath delay. Theoretical computations were confirmed by results of the conducted computer simulation, which demonstrated superiority of SI DS CDMA in terms of BER. It is also demonstrated that performance can be further improved by using specifically tailored user signatures.

References

[1] Valery Ipatov. *Spread Spectrum and CDMA: Principles and Applications.* John Wiley and Sons, 2005.

[2] P. Z. Fan, N. Suehiro, N. Kuroyanagi, and X. M. Deng. Class of binary sequences with zero correlation zone. *Electronics Letters*, 35:777–779, 1999.

[3] John Proakis. *Digital Communications.* McGraw-Hill Science/Engineering/Math, August 2000.

[4] A. Dudkov and V. P. Ipatov. Signature-interleaved ds cdma: controlling odd correlation peaks. In *Personal, Indoor and Mobile Radio Communications, 2005. PIMRC 2005. IEEE 16th International Symposium on*, volume 4, pages 2527–2530, 2005.

[5] M. Pursley and D. Sarwate. Performance evaluation for phase-coded spread-spectrum multiple-access communication – part ii: Code sequence analysis communications. *Communications, IEEE Transactions on [legacy, pre - 1988]*, 25:800–803, 1977.

[6] P. A. Bello. Generic channel simulator. Report BTR 103, final report under contract MDA904-95-C-2078, BELLO, Inc., 1997.

[7] Marvin K. Simon, Jim K. Omura, Robert A. Scholtz, and Barry K. Levitt. *Spread spectrum communications handbook (revised ed.).* McGraw-Hill, Inc., New York, NY, USA, 1994.

[8] A. Dudkov. Optimization of signature ensembles for SI DS CDMA using random search. In *Electromagnetic Compatibility and Electromagnetic Ecology, 2007. IEEE 7th International Symposium on*, Saint-Petersburg, Russia, June 2007. To be published.

[9] B. M. Popović. Efficient despreaders for multi-code CDMA systems. In *Universal Personal Communications Record, 1997. Conference Record., 1997 IEEE 6th International Conference on*, volume 2, pages 516–520, 1997.

ACCURATE BER OF MC-DS-CDMA OVER RAYLEIGH FADING CHANNELS

Besma Smida
Division of Engineering and Applied Sciences,
Harvard University, Cambridge, USA
bsmida@deas.harvard.edu

Lajos Hanzo
School of Electronics and Computer Science,
University of Southampton, UK
lh@ecs.soton.ac.uk

Sofiène Affes
INRS-EMT,
Université du Québec, Canada
affes@emt.inrs.ca

Abstract In this contribution an accurate average bit error rate (BER) formula is derived for Rayleigh-faded MC-DS-CDMA in the context of asynchronous transmissions and random spreading sequences. Our analysis is based on the characteristic function and does not rely on any assumption concerning the statistical behavior of the interference. We develop a new closed-form expression for the conditional characteristic function of the inter-carrier interference and require only a single integration for the associated BER calculation. The accuracy of the standard Gaussian approximation (SGA) techniques is also evaluated.

Keywords: multi-carrier direct-sequence CDMA (MC-DS-CDMA), performance evaluation, characteristic function

Introduction

Multi-Carrier Code Division Multiple Access (MC-DS-CDMA) [1, 2] constitutes a particularly attractive design alternative for next-generation wireless communications, since it has numerous reconfigurable parameters, which may be adjusted for the sake of satisfying diverse design goals. The Bit-Error Rate (BER) is considered to be one of the most important performance measures for

S. Plass et al. (eds.), Multi-Carrier Spread Spectrum 2007, 147–156.

communication systems and hence it has been extensively studied. The multiple access interference (MAI) and the inter-symbol interference (ISI) also affect the attainable performance of MC-DS-CDMA systems. Additionally, the MC-DS-CDMA system's performance is also affected by the inter-carrier interference (ICI). When analyzing the BER performance of MC-DS-CDMA systems, the interference source, namely, the MAI, the ISI and the ICI are commonly assumed to be Gaussian distributed [1, 4]. However, the accuracy of the Gaussian approximation technique depends on the specific configuration of the system. It is widely recognized that the Gaussian approximation techniques become less accurate, when a low number of users is supported or when there is a dominant interferer, creating an near-far-scenario [5].

Therefore the employment of accurate BER analysis dispensing with the previous assumptions concerning the distribution of the interfering sources is desirable. In order to avoid the limited accuracy of the Gaussian approximation of the interference, the BER can be calculated in the transform domain. More specifically. two widely used transforms of the decision variable's Probability Density Function (PDF) are its Fourier and Laplace transforms, corresponding to the characteristic function (CF) and the moment generating function (MGF), respectively. The basic philosophy of exact BER calculation is that first the CF or MGF of the decision variable is derived, then an associated inverse transform is performed for calculating the BER, as detailed for example in [6]. Since the decision variable can be exactly characterized by its CF or MGF, the BER can be accurately evaluated via numerical integration techniques. For this reason, CF- and MGF-based methods have received considerable attention. The CF method was used to study the BER performance of DS-CDMA using random sequences in both Rayleigh [6] and Nakagami-fading [7] channels. A saddle point integration-based MGF approach was proposed for computing the error probability of DS-CDMA systems communicating over both Rician [8] and Nakagami-fading [9] channels. This approach has been applied for studying the performance of MC-DS-CDMA systems using deterministic spreading sequences [10]. However, to the best of the authors' knowledge, the accurate BER analysis of asynchronous Rayleigh-faded MC-DS-CDMA using *random sequences* is still an open problem. Hence in this paper, we will derive an accurate BER formula for Rayleigh-faded MC-DS-CDMA in the context of asynchronous transmissions and random spreading sequences. Our analysis is based on the CF, and does not rely on any simplifying assumptions concerning the statistical behavior of the interference. A new closed-form expression, rather than an integral [10] is derived for the conditional characteristic function of the inter-carrier interference.

1. System and Channel Model

We consider an asynchronous Rayleigh-faded MC-DS-CDMA system using BPSK modulation, random spreading sequences and rectangular chip waveforms. The input information sequence of the k-th user is first converted into U parallel data sequences $b_u^k(t)$ for $u = 1, 2, \ldots, U$. The data sequence $b_u^k(t) = \sum_{i=-\infty}^{\infty} b_{u,i}^k P_{T_s}(t - iT_s)$ consists of a sequence of mutually independent rectangular pulses of duration T_s and of amplitude +1 or -1 with equal probability. After serial-to-parallel conversion, the u-th substream BPSK-modulates a subcarrier frequency f_u. Then, the U modulated subcarriers are superimposed, in order to form the complex-valued modulated signal. Finally, spectral spreading is imposed on the complex signal by multiplying it with a spreading code. Therefore, the transmitted signal of the k-th user is given by:

$$s^k(t) = \sum_{u=1}^{U} \sqrt{2P} b_u^k(t) a^k(t) \cos(j2\pi f_u t + \phi_u^k), \tag{1}$$

where P represents the transmitted power per subcarrier, while $a^k(t)$ and ϕ_u^k represent the spreading-code segment and the phase angle introduced in the carrier modulation process. The spreading sequence can be expressed as $a^k(t) = \sum_{l=-\infty}^{\infty} a_l^k P_{T_c}(t - iT_c)$, where a_l^k assumes values of +1 or -1 with equal probability, while $P_{T_c}(t)$ is the rectangular chip waveform that is defined over the interval $[0, T_c)$. $T_c = \frac{T_s}{L}$ is the chip duration, while L is the spreading factor.

For MC-DS-CDMA, the modulated subcarriers are orthogonal over the chip duration. Hence, the frequency corresponding to the u-th subcarrier is $f_u = f_p + u/T_c$, where f_p is the fundamental carrier frequency.

We assume that the channel between the k-th transmitter and the corresponding receiver is a flat slowly Rayleigh fading channel over each subcarrier. The channel impulse response for the k-th transmitted signal over the u-th subcarrier is given by

$$H_u^k(t) = h_u^k \delta(t - \tau_k) exp(-j\varphi_u^k) \tag{2}$$

where h_u^k, τ_k and φ_u^k represent the attenuation factor, delay and phase-shift, respectively. The attenuations h_u^k are independent Rayleigh variables, while φ_u^k are i.i.d. random variables uniformly distributed in the interval $[0, 2\pi)$. The delay τ_k of the k-th user is assumed to be uniformly distributed over $[0, T_s)$.

We consider K asynchronous MC-DS-CDMA users, all of whom have the same number of subcarriers U and the same spreading factor L. The average power received from each user is also assumed to be the same. Consequently, the received signal may be written as:

$$r(t) = \sum_{k=1}^{K} \sum_{u=1}^{U} \sqrt{2P} h_u^k b_u^k(t - \tau_k) a^k(t - \tau_k) \cos(j2\pi f_u t + \varphi_u^k) + N(t), \tag{3}$$

where $\varphi_u^k = \phi_u^k - \varphi_u^k - 2\pi f_u \tau_k$, which is assumed to be an i.i.d. random variable having a uniform distribution in $[0, 2\pi)$, while $N(t)$ represents the additive white Gaussian noise (AWGN) with zero mean and double-sided power spectral density of $N_o/2$.

2. BER Analysis

Decision variable statistics

Consider using a conventional single-user matched filter for coherently demodulating the desired user's signal. We assume, without loss of generality that the reference user's index is $k = 1$ and that we have $\tau_1 = 0$, $P = 2$ and $T_c = 1$. The decision variable of the first user over the v-th subcarrier is:

$$Z_v = \int_0^{T_s} r(t).a^1(t)\cos(j2\pi f_v t + \varphi_v^1)dt \tag{4}$$

$$= D_v + \sum_{k=2}^{K} I_1^k + \sum_{k=2}^{K} \sum_{u=1,u\neq v}^{U} I_2^k + N_v, \tag{5}$$

where

- $D_v = h_v^1 b_{v,0}^1 L$ is the desired output.
- $I_1^k = h_v^k cos(\theta_v^k) W^k$ is the interference imposed by the other users activating the same carrier v, where $\theta_v^k = \varphi_v^1 - \varphi_v^k$. In their impressive study of random spreading sequences used for DS-CDMA, Lehnert and Pursley [11] simplified the expression of the random variable W^k. For a rectangular chip waveform, the random variable W^k was further simplified by Cheng and Beaulieu in [6] as follows:

$$W^k = P^k \nu_k + Q^k(1-\nu_k) + X^k + Y^k(1-\nu_k), \tag{6}$$

where ν_k is a random variable (RV) uniformly distributed over $[0, 1)$, while P^k and Q^k are symmetric Bernoulli RVs. Furthermore, X^k is a discrete RV that represents the sum of A independent symmetric Bernoulli RVs and Y^k is a discrete RV that represents the sum of B independent symmetric Bernoulli RVs. Note that $A + B = L - 1$ and the marginal pdf's of X^k and Y^k are given by:

$$P_{X^k}(j) = \binom{A}{\frac{j+A}{2}} 2^{-A}, \tag{7}$$

$$j \in \mathcal{A} = \{-A, -A+2, \ldots, A-2, A\} \tag{8}$$

and

$$P_{Y^k}(j) = \binom{B}{\frac{j+B}{2}} 2^{-B}, \tag{9}$$

$$j \in \mathcal{B} = \{-B, -B+2, \ldots, B-2, B\}. \tag{10}$$

- $I_2^k = h_u^k W_{u-v}^k$ is the inter-carrier interference imposed by the adjacent carriers ($u \neq v$) of the other users. We derive the random variable W_{u-v}^k as follows:

$$W_{u-v}^k = b_{u,-1}^k R_{k,1,u-v}(\tau_k, \theta_u^k) + b_{u,0}^k \hat{R}_{k,1,u-v}(\tau_k, \theta_u^k), \tag{11}$$

where $R_{k,1,u-v}(\tau_k, \theta_u^k)$ and $\hat{R}_{k,1,u-v}(\tau_k, \theta_u^k)$ are the partial cross-correlation functions defined by:

$$R_{k,1,u-v}(\tau_k, \theta_u^k) = \int_0^{\tau_k} a^k(t-\tau_k)a^1(t)\cos(2\pi(f_u - f_v)t + \theta_u^k), \tag{12}$$

$$\hat{R}_{k,1,u-v}(\tau_k, \theta_u^k) = \int_{\tau_k}^{T_s} a^k(t-\tau_k)a^1(t)\cos(2\pi(f_u - f_v)t + \theta_u^k). \tag{13}$$

In this paper, we follow the same methodology used in [11] and use the formula (23) of [1] to simplify W_{u-v}^k as

$$\begin{aligned} W_{u-v}^k &= \cos(\theta_u^k + \pi\nu_k|u-v|) \times \\ &\quad [(P^k - Q^k)\nu_k \text{sinc}(\nu_k|u-v|) - 2Y^k \nu_k \text{sinc}(\nu_k|u-v|)], \\ &= \cos(\theta_u^k + \pi\nu_k|u-v|)\mathcal{W}_{u-v}^k, \end{aligned} \tag{14}$$

where $\theta_u^k = \varphi_v^1 - \varphi_u^k$ and $(\theta_u^k + \pi\nu_k|u-v|)$ is a uniform RV over $[0, 2\pi)$. Note that X^k does not feature in the formulation of W_{u-v}^k, which will reduce the complexity of the BER calculation.

- N_v is the noise, which is a zero-mean Gaussian random variable with a variance $\sigma_n = N_o L/4$.

Exact BER Analysis

In this section, we will use the CF-based techniques of evaluating the average BER under asynchronous transmission conditions over flat Rayleigh fading. The characteristic function of the interference I_1^k imposed by the other users on the same carrier, $\Phi_{I_1^k|B}$, was derived by Eqs (31)-(32) in [6] as follows:

$$\begin{aligned} \Phi_{I_1^k|B}(w) =& \frac{2^{-(N-1)}}{4} \sum_{i \in A} \sum_{j \in B} \binom{A}{\frac{i+A}{2}} \binom{B}{\frac{j+B}{2}} \\ & \times [J_1(i+1, j) + J_1(i, j-1) + J_1(i, j+1) + J_1(i-1, j)] \end{aligned} \tag{15}$$

where

$$J_1(i,j) = \left\{ \begin{array}{ll} \exp\left(-\frac{i^2}{2}w^2\right), & j=0 \\ \frac{\sqrt{\frac{\pi}{2}}}{j|w|}\{Q(|w|(i-j)) - Q(|w|(i+j))\}, & (j,w \neq 0) \end{array} , \right\} \tag{16}$$

where Q is the standard Q function. In this paper we will derive the characteristic function of the inter-carrier interference I_2^k. We first examine the statistics of each interferer I_2^k. From the previous section, we have $I_2^k = h_u^k \cos(\theta_u^k + \pi\nu_k|u-v|)\mathcal{W}_{u-v}^k$, where $h_u^k \cos(\theta_u^k + \pi\nu_k|u-v|)$ is a zero-mean unit-variance Gaussian RV and $\mathcal{W}_{u-v}^k$ is defined in Eq. (14). This implies that I_2^k given P_k, Q_k, Y_k, ν_k, B is Gaussian with a zero mean and a variance of $\sigma^2_{I_2^k|\mathcal{W}_{u-v}^k} = (\mathcal{W}_{u-v}^k)^2$. Averaging over P_k, Q_k, Y_k (which is equivalent to averaging over all interferes' spreading sequences and data sequences), yields the characteristic function of the inter-carrier interference I_2^k given ν_k and B as:

$$\Phi_{I_2^k|\nu_k,B}(w) = \frac{2^{-B}}{4}\sum_{j\in B}\left(\begin{array}{c} B \\ \frac{J+B}{2} \end{array}\right)\left\{\sum_{l=1,2,3,4} \exp\left\{\frac{1}{2}\sigma_l^2(j,\nu_k,u-v)w^2\right\}\right\}, \tag{17}$$

where

$$\sigma_1^2(j,\nu_k,u-v) = [2j\nu_k \text{sinc}(\nu_k|u-v|)]^2 \tag{18}$$
$$\sigma_2^2(j,\nu_k,u-v) = [2j\nu_k \text{sinc}(\nu_k|u-v|)]^2 \tag{19}$$
$$\sigma_3^2(j,\nu_k,u-v) = [2(1-j)\nu_k \text{sinc}(\nu_k|u-v|)]^2 \tag{20}$$
$$\sigma_4^2(j,\nu_k,u-v) = [2(1+j)\nu_k \text{sinc}(\nu_k|u-v|)]^2. \tag{21}$$

The variables ν_k now appear in the numerator of the exponential function and averaging can be carried out by using Eq. (3.339) of [12], yielding the characteristic function of I_2^k given B as:

$$\Phi_{I_2^k|B}(w) = \int_0^1 \Phi_{I_2^k|\nu_k,B}(w)d\nu_k \tag{22}$$
$$= \frac{2^{-B}}{4}\sum_{j\in B}\left(\begin{array}{c} B \\ \frac{j+B}{2} \end{array}\right) \times [2J(j) + J(j-1) + J(j+1)] \tag{23}$$

where B varies from 0 to $L-1$, and

$$J(j) = \exp\left(-\frac{j^2}{\pi^2(u-v)^2}w^2\right) I_0\left(-\frac{j^2}{\pi^2(u-v)^2}w^2\right). \tag{24}$$

I_0 is the modified Bessel function of order 0. Hence the characteristic function of the total interference is given by:

$$\Phi_{I|B}(w) = \Pi_{k=2}^{K} \left[\Phi_{I_1^k|B}(w)\Pi_{u=1,u\neq v}^{U}\Phi_{I_2^k|B}(w)\right]. \tag{25}$$

The conditional BER evaluated for transmission over a Rayleigh fading channel can be expressed as:

$$\begin{aligned} P_{e|B} &= \frac{1}{2}\left[1 - \frac{L}{\sqrt{L^2+\sigma_n^2}}\right] + \frac{L}{\sqrt{2\pi}}\int_0^{+\infty}[1-\Phi_{I|B}(w)]\Phi_n(w) \\ &\quad \times exp\left\{-\frac{1}{2}w^2L^2\right\}dw, \end{aligned} \tag{26}$$

where σ_n^2 is the variance of the background noise. Finally, the overall average BER is obtained by averaging $P_{e|B}$ over all spreading sequences, yielding:

$$P_e = 2^{-(L-1)}\sum_{B=0}^{L-1}\binom{L-1}{B}P_{e|B}. \tag{27}$$

Standard Gaussian Approximation

Using the derivations found in [3], the average BER P_e in Rayleigh fading approximated by the SGA can be shown to be:

$$P_e \approx \frac{1}{2}\left[1 - \frac{1}{\sqrt{1+\frac{N_o}{4L}+\gamma}}\right], \tag{28}$$

where

$$\gamma = \frac{2(K-1)}{3L} + (K-1)\left[\frac{1}{U}\sum_{u=1}^{U}\sum_{u=1,u\neq v}^{U}\frac{1}{\pi^2(u-v)^2L}\right]. \tag{29}$$

3. Numerical Results

In this section we will compare our simulation results to those obtained from our accurate BER analysis as well as to those generated by the SGA. Since the evaluation of the effect of the ICI on the performance of MC-DS-CDMA is the main objective of our analysis, we assumed first that the effect of noise is negligible. Fig. 1 shows the average BER performance against the number of users. The accuracy of our BER expression was confirmed by simulations. On the other hand, the SGA slightly over-estimates the BER. The inaccuracy of the SGA becomes prevalent when the number of users, the spreading factor, and/or the number of subcarriers decreases. Fig. 2 illustrates the average BER performance versus the per-bit SNR, when the number of users is $K = 4$. It compares the results obtained by the SGA to our accurate BER. Fig. 2 confirms, not surprisingly, that the presence of strong background noise improves the accuracy of the SGA method.

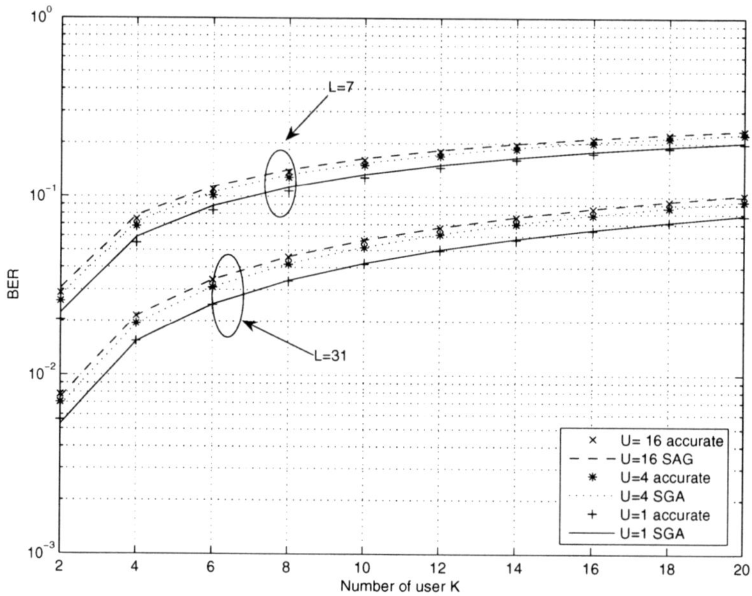

Figure 1. BER versus the number of users K in an asynchronous MC-DS-CDMA system exposed to Rayleigh fading, using random spreading sequences and BPSK modulatior The length of the spreading sequence is $L = 7$ and $L = 31$. The number of subcarriers is $U = 1, 4$, and 9. The average power of all subcarriers and users is equal and the background noise is ignored.

4. Conclusion

We studied the accurate BER calculation of an asynchronous MC-DS-CDMA system exposed to flat Rayleigh fading using random spreading sequences and BPSK modulation. Using the CF approach, we derived a new closed-form expression for the conditional characteristic function of the inter-carrier interference and only a single integration was required for the BER calculation. The inaccuracy of the SGA becomes prevalent when the number of users, the spreading factor, and/or the number of subcarriers decreases.

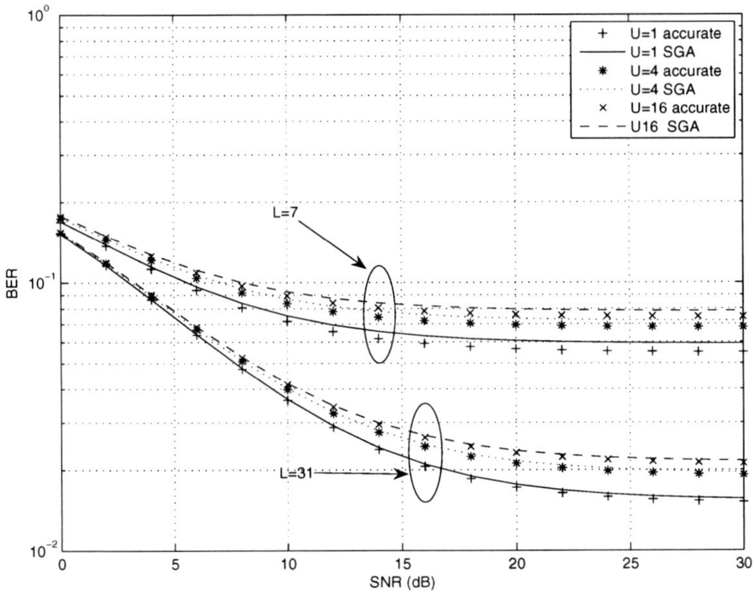

Figure 2. BER versus the per-bit SNR in an asynchronous MC-DS-CDMA exposed to Rayleigh fading, when using random spreading sequences and BPSK modulation. The length of the spreading sequence is $L = 7$ and $L = 31$. The number of subcarriers is $U = 1, 4$, and 9. The average power of all subcarriers and users is equal. The number of users is $K = 4$.

References

[1] E.A. Sourour and M. Nakagawa, "Performance of orthogonal multicarrier CDMA in a multipath fading channel," *IEEE Transactions on Communications*, vol. 44, no. 3, pp. 356-367, Mar. 1996.

[2] L. Hanzo, T. Keller, M. Muenster and B.J. Choi, *OFDM and MC-CDMA for Broadband Multiuser Communications, WLANs and Broadcasting*, John Wiley & Sons Inc, 2003.

[3] Y. L-L. Yang and L. Hanzo, "Performance of generalized multicarrier DS-CDMA over Nakagami-m fading channels," *IEEE Transactions on Communications*, vol. 50, no. 6, pp. 956-966, June 2002.

[4] S. Kondo and L.B. Milstein, "Performance of multicarrier DS CDMA systems," *IEEE Transactions on Communications*, vol. 44, no. 2, pp. 238-246, Feb. 1996.

[5] M. O. Sunay and P. J. McLane, " Caculating error probabilities for DS-CDMA systems: when not to use the Gaussian approximation," *Proc. IEEE GLOBECOM*, vol. 3, London, UK, Nov. 1996, pp. 1744-1749.

[6] J. Cheng and N. C. Beaulieu,"Accurate DS-CDMA bit-error probability calculation in Rayleigh fading," *IEEE Transactions on Wireless Communications*, vol. 1, no. 1, pp. 3-15, Jan. 2002.

[7] J. Cheng and N. C. Beaulieu, "Precise bit error rate calculation for asynchronous DS-CDMA in Nakagami fading," in *Proc. IEEE GLOBECOM*, San Francisco, CA, Nov. 2000, pp. 980-984.

[8] D. Liu, C. L. Despins and W. A. Krzymien, "Low-complexity performance evaluation of binary and quaternary DS-SSMA over Rician fading channels via the characteristic function method," *Wireless Personal Commun.*, vol. 7, pp. 257-273, Aug. 1998.

[9] S. W. Oh and K. H. Li, "Performance evaluation for forward-link cellular DS-CDMA over frequency-selective Nakagami multipath fading channels," *Wireless Personal Commun.*, vol. 18, pp. 275-287, Sep. 2001.

[10] B. Smida, C. L. Despins and G. Y. Delisle, "MC-CDMA performance evaluation over a multipath fading channel using the characteristic function method," *IEEE Transactions on Communications*, vol. 49, no. 8, pp. 1325-1328, Aug. 2001.

[11] J. S. Lehnert and M. B. Pursley, "Error probabilities for binary direct-sequence spread spectrum communications with random signature sequences," *IEEE Transactions on Communications*, vol. 35, pp. 87-98, Jan. 1987.

[12] I. S. Gradshteyn and I.M. Ryzhik, I. M. *Table of Integrals, Series, and Products*, New York: Academic Press, Sixth edition, July 2000.

OFDM/OQAM FOR SPREAD-SPECTRUM TRANSMISSION

Chrislin Lélé, Pierre Siohan, Rodolphe Legouable and Maurice Bellanger

Abstract We propose an alternative to the well-known MC-CDMA technique for downlink transmission by replacing the conventional cyclic-prefix OFDM modulation by an advanced filterbank-based multicarrier system (OFDM/OQAM). Indeed, in one hand, MC-CDMA has already proved its ability to fight against frequency selective channels thanks to the use of the OFDM modulation and its high flexibility in multiple access thanks to the CDMA component. In the other hand, OFDM/OQAM modulation confers a theoretically optimal spectral efficiency as it operates without guard interval. In this paper, we study OFDM/OQAM with CDMA combination and we compare it to the conventional MC-CDMA scheme in terms of Bit Error Rate (BER).

1. Introduction

Multicarrier Code-Division Multiple Access (MC-CDMA) system has been initially proposed in [1], [2]. This technique constitutes a popular way to combine CDMA and Orthogonal Frequency Division Multiplexing (OFDM) with Cyclic Prefix (CP). Nowadays, MC-CDMA is considered as one of the possible candidates for the downlink of B3G communication systems. Indeed, firstly, this technique proposes a good way to fight against frequency selective channels thanks to the OFDM modulation and, secondly, it has a high flexibility in the multiple access scheme thanks to the CDMA component. However, the insertion of the CP leads to spectral efficiency loss since this "redundant" symbol part does not carry useful data information. In addition, the conventional OFDM modulation is based on a rectangular windowing in the time domain which leads to a poor ($sinc(x)$) behavior in the frequency domain. Thus CP-OFDM gives rise to 2 drawbacks: loss of spectral efficiency and sensitivity to frequency dispersion, e.g. Doppler spread. Both of them are counteracted using the OFDM/OQAM modulation.

OFDM/OQAM [3], also known as O-QAM (Orthogonally multiplexed Quadrature Amplitude Modulation) [4], has many common features with

S. Plass et al. (eds.), Multi-Carrier Spread Spectrum 2007, 157–166.

OFDM. Indeed in OQAM the basic principle is also to divide the total transmission bandwidth into a large number of uniform sub-bands. As for OFDM systems, the transmitter and receiver implementations can also benefit of fast Fourier transform (FFT) algorithms. However, instead of a single FFT (or IFFT), a uniform filter bank is used. So one can get a better frequency separation between sub-channels, reducing the Inter-Carrier Interference (ICI) in the presence of frequency shifts. Note however that for OQAM the orthogonality is restricted to the real field. In the same manner as for MC-CDMA system, we have chosen to combine OQAM with spreading applied in the frequency domain. The objective of the present paper is to explore this association in the context of a downlink multiuser transmission. The mathematical foundations of the QAM scheme with spread spectrum is presented in section 2. An analysis of the imaginary component, in the single user case, is provided in section 3. In section 4, the impact of the OQAM pulse shape is discussed in the multiuser case and simulation results are given showing the pulse shape impact on the interfering term. Finally, in section 5, OQAM/CDMA and MC-CDMA are compared in terms of BER.

2. OQAM/CDMA transmission

The baseband equivalent of a continuous-time OFDM/OQAM signal can be written as follows [3]:

$$s(t) = \sum_{m=0}^{M-1} \sum_{n \in \mathbf{Z}} a_{m,n} \underbrace{g(t - n\tau_0) e^{j2\pi m F_0 t} e^{j\phi_{m,n}}}_{g_{m,n}(t)} \tag{1}$$

with $M = 2N$ an even number of sub-carriers, $F_0 = 1/T_0 = 1/2\tau_0$ the subcarrier spacing, g the pulse shape and $\phi_{m,n}$ an additional phase term. The transmitted symbols $a_{m,n}$ are real-valued. They are obtained from a 2^{2K}-QAM constellation, taking the real and imaginary parts of these complex-valued symbols of duration $T_0 = 2\tau_0$, where τ_0 denotes the time offset between the two parts [5, 3, 6].

Assuming a distortion-free channel, perfect reconstruction of real symbols is obtained owing to the following real orthogonality condition:

$$\Re\{\langle g_{m,n} | g_{p,q} \rangle\} = \Re\{\int g_{m,n}(t) g^*_{p,q}(t) dt\} = \delta_{m,p}\delta_{n,q}, \tag{2}$$

$\delta_{m,p} = 1$ if $m = p$ and $\delta_{m,p} = 0$ if $m \neq p$. For concision purpose we set $\langle g \rangle_{m,n}^{p,q} = -j \langle g_{m,n} | g_{p,q} \rangle$, with $\langle g_{m,n} | g_{p,q} \rangle$ a pure imaginary term for $(m, n) \neq (p, q)$.

We denote by N_c the length of the CDMA code used and assume that $N_0 = M/N_c$ is an integer number. Let denote, with $[.]^T$ the transposition and by:

$$\underline{c_u} = [c_{0,u} \dots c_{N_c-1,u}]^T \tag{3}$$

the code used by the uth user. Then, for a user u_0 and given symbol n_0, N_0 different data are transmitted denoted by: $d_{u_0,n_0,0}, d_{u_0,n_0,1}, \dots$, d_{u_0,n_0,N_0-1}. Then by *spreading*, using the set of codes in (3), we get for frequency and time indexes m_0 and n_0 respectively, the real symbol a_{m_0,n_0}:

$$a_{m_0,n_0} = \sum_{u=0}^{U-1} c_{m_0/N_c,u} d_{u,n_0,\lfloor \frac{m_0}{N_c} \rfloor}$$

where U is the number of users, $/$ the modulo operator and $\lfloor\rfloor$ the floor operator. From the a_{m_0,n_0} term, the reconstruction of $d_{u,n_0,p}$ (for $p \in [0, N_0 - 1]$) is insured thanks to the orthogonality of the codes, i.e. $\underline{c_{u1}}^T \underline{c_{u2}} = \delta_{u1,u2}$ (see [8] for more details). Therefore, the *despreading* operator leads to:

$$d_{u,n_0,p} = \sum_{m=0}^{N_c-1} c_{m,u} a_{pN_C+m,n_0}.$$

In [9], we show that in the presence of a propagation channel the demodulated signal for a given sub-carrier m_0 and time n_0 is given by:

$$y^{(c)}_{m_0,n_0} \approx H^{(c)}_{m_0,n_0}(a_{m_0,n_0} + ja^{(i)}_{m_0,n_0}) + \eta_{m_0,n_0}, \tag{4}$$

where $H^{(c)}_{m_0,n_0}$ is the channel frequency response for sub-carrier m_0 and time n_0, η_{m_0,n_0} is the noise contribution and $a^{(i)}_{m_0,n_0}$ is given by:

$$a^{(i)}_{m_0,n_0} = \sum_{(p,q)\in\Omega^*_{\Delta m,\Delta n}} a_{m_0+p,n_0+q} \langle g \rangle^{m_0,n_0}_{m_0+p,n_0+q}, \tag{5}$$

with $\Omega_{\Delta m,\Delta n} = \{(p,q), |p| \leq \Delta m, |q| \leq \Delta n | H^{(c)}_{m_0+p,n_0+q} \approx H^{(c)}_{m_0,n_0}\}$, that defines a neighborhood of the time-frequency location (p,q) where the channel remains approximately constant and $\Omega^*_{\Delta m,\Delta n} = \Omega_{\Delta m,\Delta n} - (0,0)$.

From (4), the estimation of $a_{m_0,n_0} + ja^{(i)}_{m_0,n_0}$ can be obtained by applying a linear *equalizer* to y_{m_0,n_0}. With Zero-Forcing (ZF) equalization the estimation is given by:

$$\hat{a}_{m_0,n_0} + j\hat{a}^{(i)}_{m_0,n_0} = \frac{y^{(c)}_{m_0,n_0}}{H^{(c)}_{m_0,n_0}} \approx a_{m_0,n_0} + a^{(i)}_{m_0,n_0} + \frac{\eta_{m_0,n_0}}{H^{(c)}_{m_0,n_0}}, \tag{6}$$

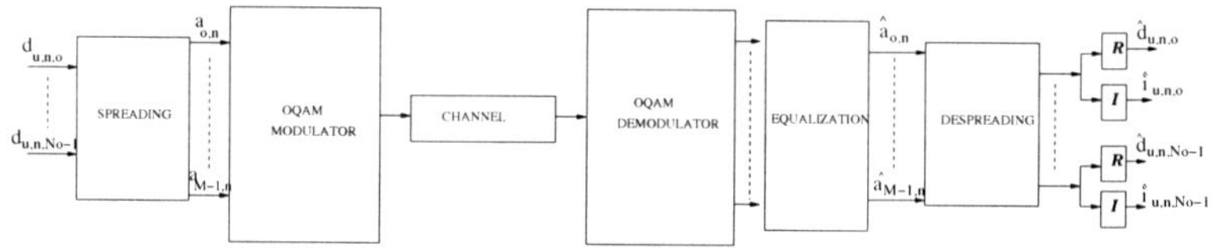

Figure 1. OQAM/CDMA transmission chain.

and with a Minimum-Mean-Square (MMSE) equalization it is:

$$\hat{a}_{m_0,n_0} + j\hat{a}^{(i)}_{m_0,n_0} = \frac{H^{(c)*}_{m_0,n_0} y^{(c)}_{m_0,n_0}}{| H^{(c)}_{m_0,n_0} |^2 + 2\frac{\sigma_a^2}{\sigma_v^2}},$$

where σ_a^2 and σ_n^2 are the real data and noise variance respectively. Then the estimate of $d_{u,n_0,p}$ is given by:

$$\hat{d}_{u,n_0,p} = \sum_{m=0}^{N_c-1} c_{m,u} \hat{a}_{pN_C,n_0}.$$

Similarly, the imaginary part is denoted as follows:

$$\hat{i}_{u,n_0,p} = \sum_{m=0}^{N_c-1} c_{m,u} \hat{a}^{(i)}_{pN_C,n_0}. \quad (7)$$

The transmission chain diagram corresponding to OQAM/CDMA is shown in Fig. 1. Let us now examine some properties of the imaginary component defined in (7).

3. Imaginary component (single user case)

In this section, we assume a single user transmission either in ideal channel or with time de-synchronization. We assume the prototype filter g to be real-valued with unit energy. We adopt the following choice for the parameters: $\phi_{m,n} = \frac{\pi}{2}(m + n + 2mn)$, $N_c = M = 2N$ and Walsh-Hadamard (WH) spreading codes. For simplicity and concision, we suppose one user ($U = 1$), a maximum frequency diversity ($N_c = M$) and we denote $i_{u,n_0,p}$ by i_{n_0}, $d_{u,n_0,p}$ by d_{n_0} and $c_{m,u}$ by c_m. So we can write: $a_{m,n} = c_m d_n$. Assuming that g is well-localized, we can get a good approximation of $a^{(i)}_{m_0,n_0}$ in any given 3×3 neighborhood area [9]:

$$a^{(i)}_{m_0,n_0} \approx \sum_{(p,q)\in\Omega^*_{\Delta 1,\Delta 1}} a_{m_0+p,n_0+q} \langle g \rangle^{m_0,n_0}_{m_0+p,n_0+q}. \quad (8)$$

let us now express more in details, the imaginary component, in the single user case:

$$\begin{aligned}
i_{n_0} &= \sum_{m=0}^{M-1} c_m a_{m,n_0}^{(i)} \\
&\approx \sum_{m=0}^{M-1} c_m \sum_{(p,q)\in\Omega^*_{\Delta 1,\Delta 1}} a_{m+p,n_0+q} \langle g\rangle^{m,n_0}_{m+p,n_0+q} \\
&\approx \sum_{m=0}^{M-1} c_m \sum_{(p,q)\in\Omega^*_{\Delta 1,\Delta 1}} c_{m+p} d_{n_0+q} \langle g\rangle^{m,n_0}_{m+p,n_0+q} \\
&\approx \sum_{m=0}^{M-1} [d_{n_0-1}(c_m c_{m-1} \langle g\rangle^{m,n_0}_{m-1,n_0-1} + c_m c_m \langle g\rangle^{m,n_0}_{m,n_0-1} + c_m c_{m+1} \langle g\rangle^{m,n_0}_{m+1,n_0-1}) \\
&+ d_{n_0}(c_m c_{m-1} \langle g\rangle^{m,n_0}_{m-1,n_0} + c_m c_{m+1} \langle g\rangle^{m,n_0}_{m+1,n_0}) \\
&+ d_{n_0+1}(c_m c_{m-1} \langle g\rangle^{m,n_0}_{m-1,n_0+1} + c_m c_m \langle g\rangle^{m,n_0}_{m,n_0+1} + c_m c_{m+1} \langle g\rangle^{m,n_0}_{m+1,n_0+1})]
\end{aligned}$$

This last expression can be simplified by using the following relations between the different sets of scalar products:

$$\langle g\rangle^{m,n_0}_{m+1,n_0} = -\langle g\rangle^{m,n_0}_{m-1,n_0} = \langle g\rangle^{0,0}_{1,0}$$

$$\langle g\rangle^{m,n_0}_{m,n_0+1} = -\langle g\rangle^{m,n_0}_{m,n_0-1} = (-1)^m \langle g\rangle^{0,0}_{0,1}$$

$$\langle g\rangle^{m,n_0}_{m+1,n_0+1} = \langle g\rangle^{m,n_0}_{m+1,n_0-1} = \langle g\rangle^{m,n_0}_{m-1,n_0+1} = \langle g\rangle^{m,n_0}_{m-1,n_0-1} = (-1)^m \langle g\rangle^{0,0}_{1,1}$$

.

Then, we get:

$$\begin{aligned}
i_{n_0} &= d_{n_0-1}[\langle g\rangle^{0,0}_{1,1}(\sum_{m=0}^{M-1} c_m c_{m-1}(-1)^m + c_m c_{m+1}(-1)^m) - \langle g\rangle^{0,0}_{0,1} \sum_{m=0}^{M-1} c_m^2(-1)^m] \\
&+ d_{n_0}[\langle g\rangle^{0,0}_{1,0}(\sum_{m=0}^{M-1} -c_m c_{m-1} + c_m c_{m+1})] \\
&+ d_{n_0+1}[\langle g\rangle^{0,0}_{1,1}(\sum_{m=0}^{M-1} c_m c_{m-1}(-1)^m + c_m c_{m+1}(-1)^m) + \langle g\rangle^{0,0}_{0,1}] \sum_{m=0}^{M-1} c_m^2(-1)^m
\end{aligned}$$

Note that for any spreading code we have:

$$\sum_{m=0}^{M-1} -c_m c_{m-1} + c_m c_{m+1} = 0$$

and

$$\sum_{m=0}^{M-1} c_m c_{m-1}(-1)^m + c_m c_{m+1}(-1)^m = 0.$$

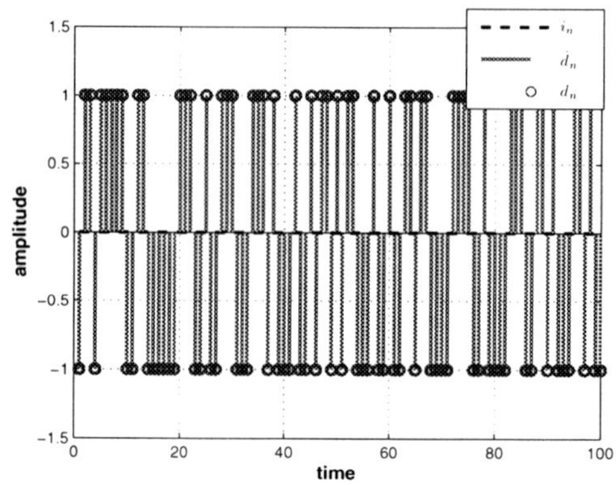

Figure 2. Transmitted (d_n) and received ($\hat{d}_n$) real data and imaginary interference (i_n) for an ideal channel.

As furthermore as for, the set of WH codes, we have: $c_m = \pm\frac{1}{\sqrt{M}}$, $m = 0, ..., M-1$. So, we get:

$$i_{n_0} = 0$$

To confirm this result, for the Isotropic Orthogonal Transform Algorithm (IOTA) [3] prototype, with WH codes of length 32 and $U = 1$, we plotted in Fig. 2 the result obtained with distortion-less and noise free channel. In this figure we show: the transmitted data d_{n0} which are ± 1 values, the estimated data at the reception $\hat{d}_{n0}$ and the value of i_{n0}. We find that the $\hat{d}_{n0}$ and d_{n0} are identical, and the value of i_{n0} is equal to zero.
Now let assume that the transmitter and receiver are not well synchronized, τ being the de-synchronization offset. Plotting the same curves as before, we notice in Fig. 3 that in this case $\hat{d}_{n0}$ and d_{n0} are not the same and i_{n0} is no more equal to zero. In theory, we can write, $\hat{d}_{n0} = d_{n0}\cos\alpha(\tau)$ and $i_{n0} = d_{n0}\sin\alpha(\tau)$, where $\alpha(\tau)$ is the phase rotation due to de-synchronization offset. Therefore, the ratio: $\frac{i_{n0}}{\hat{d}_{n0}} = \tan\alpha(\tau)$ enables us to access the value of $\alpha(\tau)$ and therefore correct the phase rotation and then finally perfectly recover our real data. This principle can be, of course, applied in realistic channel. In the current section we considered only one user ($U = 1$). Let us now examine the multiuser case.

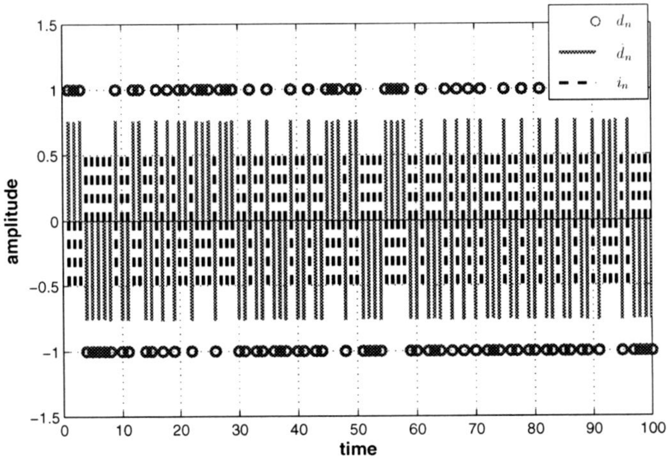

Figure 3. Transmitted (d_n) and received ($\hat{d}_n$) real data and imaginary interference (i_n) for a synchronization offset.

4. Multiuser transmission

We keep the same parameters as in the previous section except that we now consider $U > 1$. We denote for concision, $i_{u,n_0,p}$ by i_{u,n_0} and $d_{u,n_0,p}$ by d_{u,n_0} for multiuser system definition. In this case, $a_{m_0,n_0} = \sum_{u=0}^{U-1} c_{m_0,u} d_{u,n_0}$ and we have:

$$\begin{aligned} i_{n_0,u_0} &= \sum_{m=0}^{M-1} c_{m,u_0} a_{m,n_0}^{(i)} \\ &\approx \sum_{m=0}^{M-1} c_{m,u_0} \sum_{(p,q)\in\Omega^*_{\Delta 1,\Delta 1}} a_{m+p,n_0+q} \langle g \rangle_{m+p,n_0+q}^{m,n_0} \\ &\approx \sum_{m=0}^{M-1} c_{m,u_0} \sum_{(p,q)\in\Omega^*_{\Delta 1,\Delta 1}} \sum_{u=0}^{U-1} c_{m+p,u} d_{u,n_0+q} \langle g \rangle_{m+p,n_0+q}^{m,n_0} \end{aligned}$$

The previous $i_{n0,u0}$ term is not always null since it depends on an appropriate choice of the WH codes. Considering u_0 and u_1, two of the WH spreading codes ($U = 2$) of length 32, we have plotted the mean values of $i_{n0,u0}$, in presence of distortion-less and noise free channel. The values of $i_{n0,u0}$ have been determined either with the IOTA (Fig. 4) or the rectangular window (Fig. 5) prototype function, where m is an integer defining the prototype length $L = mM$, by variation of the couple of the u_0 and u_1 spreading codes chosen among the 32 possible sequences. These curves show that the mean values of $i_{n0,u0}$ depend on the choice

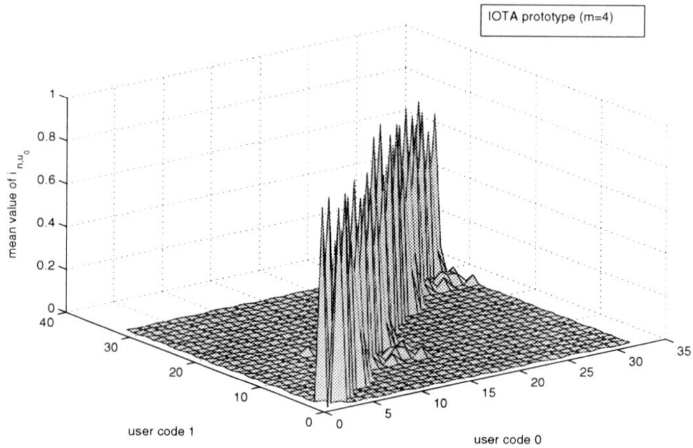

Figure 4. Mean value of i_{n0,u_0} for the IOTA prototype for $m = 4$.

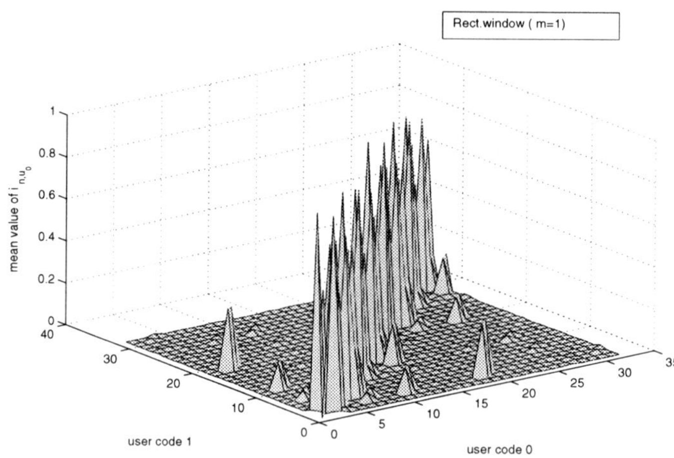

Figure 5. Mean value of i_{n0,u_0} for a rectangular window.

of the codes and on the prototype function. Indeed, for some couples of WH codes, $i_{n0,u0}$ is null showing the importance of the WH sequence choice regarding to the prototype function, if we want to use the $i_{n0,u0}$ term for synchronization process. So specific selection among the different codes have to be carried out. We have mentioned the possibility to use i_{n0,u_0} for synchronization process.

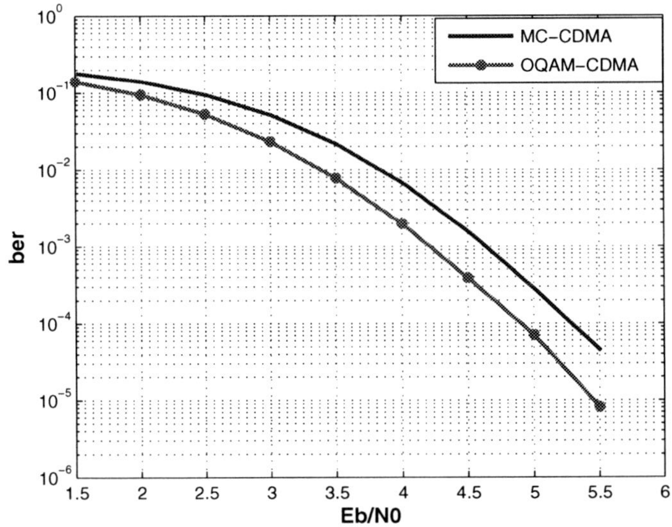

Figure 6. Comparison between MC-CDMA and OQAM/CDMA in full load

5. Simulation results

For the simulation, we have considered the following channel profile and parameters: sampling frequency of 10 MHz, FFT size of 32 points, CP duration = 0.4 μs, QPSK modulation and convolutional channel coding ($K = 7$ with $g_1 = (133)_o$, $g_2 = (171)_o$ and code rate=$\frac{1}{2}$), MMSE equalization, WH codes of length 32. The channel profile is composed of 3 paths having the following powers and delays:

- Power profile (in dB): -0 -3 -2.2204 (dB),
- Delay profile (μs): 0 0.2527 0.32 (μs),

For OQAM we used the IOTA prototype filter with a pulse length of $4T_0$ that contains $L = 4M = 128$ taps.

In Fig. 6, we compare the performance of OQAM/CDMA versus MC-CDMA in full load system, i.e 32 active users and perfect channel estimation. We notice that OQAM/CDMA outperforms MC-CDMA with a gain of about 0.5 dB corresponding to the CP spectral efficiency gain.

6. Conclusion

We have proposed a system combining OQAM modulation with spread spectrum that under some assumptions can guarantee null imaginary component values at the reception side in single user case. We have analyzed the impact of the prototype filter for an OFDM/OQAM spread-spectrum system with 32 sub-carriers occupied by 2 users. It has been shown that different trade-offs have to be studied to find an appropriate pulse shape to preserve the real orthogonality and limit the imaginary interference in a multiuser scenario. In future work, we will investigate the possibility to use this interference for equalization and/or synchronization purpose. Beyond these first results obtained with WH codes we will also look for other spread spectrum codes satisfying our interference cancellation requirement.

References

[1] N. Yee, J.P. Linnartz, and G. Fettweis. "Multicarrier cdma in Indoor Wireless Radio Networks". In *Proc. IEEE PIMRC Conference, Yokohama, Japan*, 1993.

[2] K. Fazel and L. Papke. "On the Performance of Convolutionally-Coded cdma/ofdm for Mobile Communication System". In *Proc. IEEE PIMRC Conference, Yokohama, Japan*, sept. 1993.

[3] B. Le Floch, M. Alard, and C. Berrou. Coded Orthogonal Frequency Division Multiplex. *Proceedings of the IEEE*, 83:982–996, June 1995.

[4] Botaro Hirosaki. An analysis of automatic equalizers for orthogonally multiplexed QAM systems. *IEEE Transactions on Computers*, COM-28(1):73–83, January 1980.

[5] B. Hirosaki. An orthogonally multiplexed QAM system using the discrete Fourier transform. *IEEE Trans. on Communications*, 29(7):982–989, July 1981.

[6] H. Blcskei. Orthogonal frequency division multiplexing based on offset QAM. In *Advances in Gabor theory*. Birkhuser, 2003.

[7] P. Siohan, C. Siclet, and N. Lacaille. Analysis and design of OFDM/OQAM systems based on filterbank theory. *IEEE Transactions on Signal Processing*, 50(5):1170–1183, May 2002.

[8] E.H. Dinan and B. Jabbari. Spreading codes for direct sequence cdma and wideband cdma cellular networks. *IEEE Communications Magazine*, 36(9):48–54, 1998.

[9] C. Lele, J.-P. Javaudin, R. Legouable, A. Skrzypczak, and P. Siohan. Channel estimation methods for preamble-based ofdm/oqam modulations. *to appear in european wireless*, April 2007.

A MULTI-CARRIER DOWNLINK TRANSMITTER FOR LEO SATELLITE

Roman M. Vitenberg
Wavetone Technologies Ltd.

roman@wavetonetech.com

Abstract — This paper discusses the performance of the Satellite Downlink Transmitter that uses a Wavelet based Filtered Multi-Tone (WFMT) modulation. The novel WFMT modulation is proposed to improve the characteristics of the OFDM multicarrier systems. In addition, it can also be used successfully in satellite communications. In this paper, we discuss the advantages of WFMT modulation such as low sensitivity to narrowband RF interference and lower Peak-to-Average Power Ratio (PAR) in comparison with OFDM. It is shown that modern GaAs Solid-State Power Amplifiers are well suitable for transmission of multicarrier signals. An effect of non-linear distortion on power added efficiency (PAE) of the RF Power Amplifier is discussed. The effect of clipping of the WFMT signal on the error performance is described. It is shown that the error correction block-codes can significantly improve PAR characteristics of the communication system. The Power Efficiency of WFMT Satellite Transmitter is significantly higher than in case of OFDM.

Keywords — FMT, WFMT, filter-bank, Satellite, transponder, PAR, Reed Solomon.

1. INTRODUCTION

The Low Earth Orbit (LEO) satellites are used for such applications like Earth Observation, Personal Communication and Internet Access. The major advantage of the LEO Satellites is a low output power (about 1~10 W) of a downlink transmitter and as result the low cost and weight of the satellite.

During last years, the multicarrier modulation methods such as OFDM and MC-CDMA for the satellite communication have been proposed. The multicarrier modulation is well suitable for multi-user communication system that may produce services to many users simultaneously. A main drawback of a multicarrier modulation is a low power efficiency of the satellite downlink transmitter, which has to work with very low non-linear distortions. A downlink transmitter for the LEO satellite may be developed only based on modern semiconductor devices. In the last years, a big progress in the technology of Solid-State-Power Amplifiers (SSPA) has been made. The modern SSPA uses semiconductor devices on the basis of

S. Plass et al. (eds.), Multi-Carrier Spread Spectrum 2007, 167–176.

Gallium Arsenide, which can operate at higher frequencies than silicon devices. The maximum continuous output power of a single GaAs FET ranges from a few watts to several tens of watts. The limiting factor is the generation of heat. The thermal limit for the maximum power is inversely proportional to the square of the frequency. A typical GaAs FET at C-band might have a maximum output power between 100 and 150 W, while at X-band it is about 10 W [13]. Higher power may be obtained by assembling modules using standard combining techniques. However, the number of parallel modules is limited by combination losses. A modern solid-state technology is the microwave monolithic integral circuits (MMIC). This device combines active GaAs FETs and passive elements that are placed on a small chip. The modern SSPA provides very low non-linear distortions. A great progress has been made in research and developing of Solid-State Power Amplifiers for multicarrier communication systems like Wi-Fi and WiMAX. These amplifiers have low non-linear distortion and work with output back off about 7dB. The efficiency of SSPA was improved by using several new techniques. One of these techniques is described in [14]. This paper discusses a GaN WCDMA amplifier, which uses an envelope tracking bias system to achieve high linearity and efficiency. A measured overall power added efficiency (PAE) reached 50% for average output power 37 W. Several new semiconductor technologies were proposed [13], [14]. Wide bandgap transistors such as the GaN HEMTS were recently introduced commercially. The GaN devices wide bandgap increases the breakdown field by five times and the power density by a factor of 10 to 20, compared with GaAs-based amplifiers. The GaN devices are also highly efficient because they can operate at a higher voltage 24-35V, compared to 5-8V for GaAs based devices. In addition, GaN devices built on SiC substrate have a thermal conductivity 10 times higher than those fabricated from GaAs. The GaN HEMTs can also work at higher temperatures, which reduces the need for cooling and allows for more compact module design.

Apart from offering smaller size, GaN transistors have low capacitances, very high gain as well as capability of being operated over wide bandwidths.

The commercial available linear GaN HEMT power amplifiers for WiMAX applications were described in [14]. This article discusses the design and implementation of two linear Class-A/B amplifiers using high voltage Cree GaN transistors. Both designs, operating with peak powers of 15 and 100 watts, demonstrate the multicarrier capability of wide bandgap transistors whilst simultaneously providing high linearity over greater than 16 dB dynamic range. In addition, the transistors exhibit the excellent drain efficiency. Figure 3 illustrates a good linearity and an efficiency of the SSPA for 12-watt average output power on 3.6 GHz frequency [14]. As we see, the modern Solid State Power Amplifiers may be successfully used in the multicarrier downstream transmitters of the LEO Satellites.

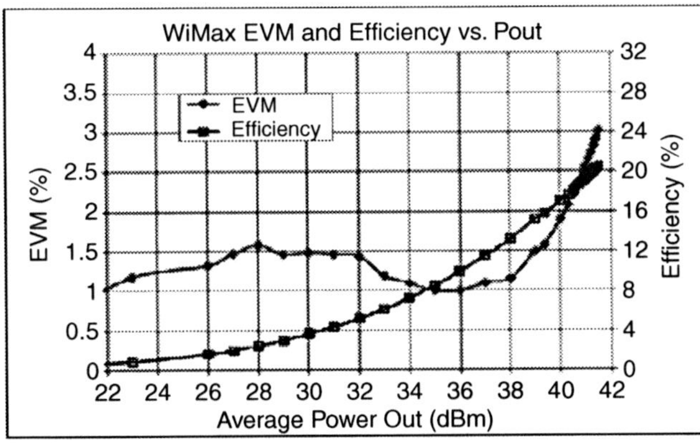

Figure 1. Error Vector Magnitude (EVM) and drain efficiency of the 12 watt WiMAX amplifier

The Power Efficiency of the Downlink Transmitter may be increased by correct choosing of parameters of multicarrier signal and a type of the multi-carrier modulation. A number of carriers and type of constellation diagram has significant impact on power efficiency of the Downlink Transmitter.

2. MULTICARRIER MODULATION: OFDM, FMT, AND WFMT

An Orthogonal Frequency Division Multiplexing (OFDM) Modulation was recently proposed for local area mobile wireless networks, such as WiFi and HIPERLAN-2. By implementing an Inverse Fast Fourier transform (IFFT) at the transmitter and a Fast Fourier Transform (FFT) at the receiver, OFDM transmits data over a number of ISI free sub-channels placed on orthogonal frequency carriers [1]. Although OFDM enables simple equalization, it introduces the following well-known problems.

- The peak-to-average power ratio (PAR) of the transmitted multicarrier signal is large (about 15 dB), which requires using components with very low nonlinear distortions.
- Because information symbols are transmitted on orthogonal sub-carriers, OFDM is very sensitive to the frequency offset between transmitter and receiver oscillators, to the Doppler effect, and to the phase noise.

The Filtered Multi-Tone (FMT) Modulation was proposed by a group of researches from IBM as an alternative technology for xDSL [2]. This technology is based on a Wavelet theory and uses complex filter-banks for synthesis and analysis of a multi-channel signal. The theoretical aspects of FMT Modulation for Wireless Application were developed by Cambridge University [3] and Udine University [4]. It was shown that in many cases FMT modulation guarantees better performance than OFDM. A major drawback of FMT is a high realization complexity. No practical FMT system was used until now.

The novel Wavelet based Filtered Multi-Tone Modulation (WFMT) was developed by Roman M. Vitenberg in 2002 –2003 [5]. The main advantage of the WFMT modulation is its significantly less complexity in comparison with FMT. Therefore, several FPGA prototypes of WFMT system were developed and tested by Wavetone Technology Ltd in 2003-2006 [8], [9].

The WFMT is a version of the Filter-bank modulated communication system described in detail in [8], [9], [10], in which the synthesis and analysis of sub-channel wavelets have been corresponding produced by IFFT and FFT cores. Unlike OFDM, the number M of sub-channels in WFMT system is significantly less than N - the number of IFFT/FFT points. Each sub-channel wavelet is generated in this case from K harmonics (IFFT/FFT points) so there is a simply expression for the number of sub-channels:

$$M \leq \frac{N}{K} \qquad (1).$$

As was shown in [7], the number of harmonics K defines a quality of generated wavelets, in particular the Inter-Symbol Interference (ISI) between wavelets and a bandwidth of sub-channels.

3. PAR REDUCTION, CLIPPING, AND ERROR-CORRECTION CODING IN WFMT SYSTEMS

One of the major advantages of the WFMT technology is the low level of the Power-to-Average Power Ratio (PAR), which limits the transmitter power efficiency. As was shown in [10] the PAR of the WFMT signal is significantly less than the PAR of the OFDM signal. Figure 4 illustrates the PAR of a multicarrier system as a function of the number of sub-channels M. As it is clear from the expression (1), WFMT system has a significant lower number of sub-channels M than the OFDM System, which uses the same IFFT core. Therefore, each WFMT sub-channel has wide bandwidth $W = KF$, where $F = 1/T$ is the distance between OFDM carriers.

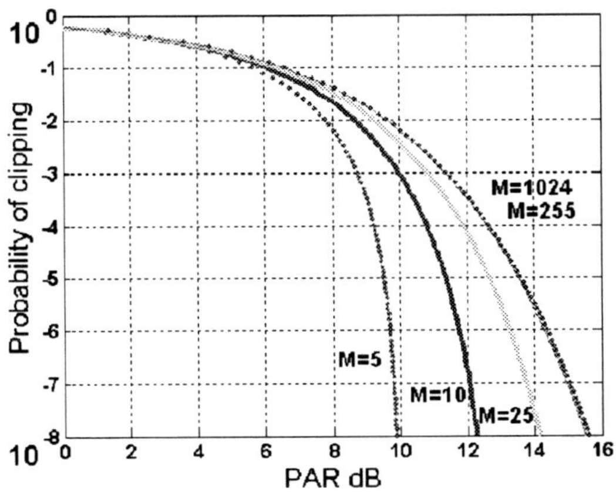

Figure 2. Peak-to-Average Power Ratio of WFMT system with circular constellation diagram

A Clipping of high amplitude peaks is a simple and effective way to reduce PAR in multicarrier systems. [11]. However, clipping causes distortion of the transmitted signal. In general, clipping will cause in-band and out-of-band distortion. It has been shown that out-of-band distortion can be removed using a simple filter [11]. The remaining in-band distortion has an effect on the transmitted signal, such as adding a noise-like component- i.e. clipping noise. This clipping noise can be modelled as an additive noise at the transmitter. Figure 5 shows a block diagram of a multicarrier system with clipping. It is clear that the transmitted signal $S(t)$ can be written as the sum of non-clipped signal $S_0(t)$ and the clipping noise $\varepsilon(t)$:

$$S(t) = S_0(t) + \varepsilon(t). \qquad (3)$$

The clipping process cuts the peaks of the original multicarrier signal. Therefore the clipping noise is a sequence of short pulses followed by different time intervals. These pulses after passing through the communication channel $C(f)$ are processed by a receiver of a multicarrier system. The multicarrier receiver comprises M sub-channel filters $G(f)$. The clipping noise generates on the output of each sub-channel filter $G(t)$ a sequence of impulses $\psi(t)$, which have the length τ, which are significantly longer than the length τ_0 of clipping noise pulses $\varepsilon(t)$.

$$\tau = \tau_0 + \Delta T, \qquad (4)$$

where ΔT is a length of the impulse response $g(t)$.

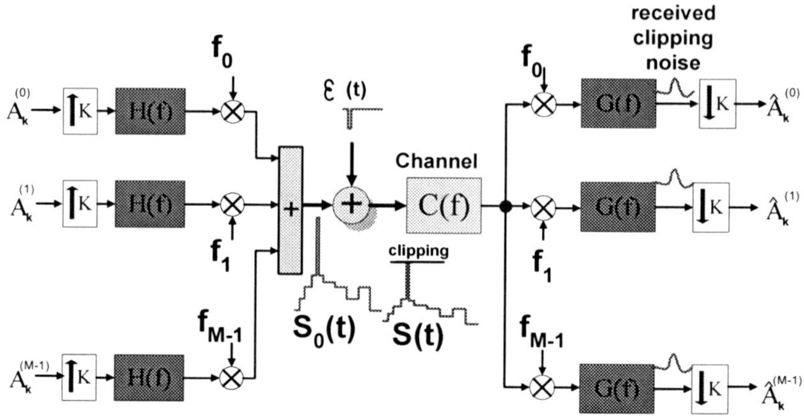

Figure 3. Multicarrier system with clipping

Suppose that $G(f) = 1$ for $0 < f < W$ and $G(f) = 0$ for $f > W$, where W is a bandwidth of the low-pass sub-channel filter $G(f)$, then:

$$g(t) = \sin(2\pi Ft)/2\pi Ft, \quad (5)$$

The most of the $g(x)$ energy is placed in an interval $-2\pi < x < 2\pi$ and ΔT may be approximately defined as:

$$\Delta T \approx 2/W. \qquad (6)$$

The length of the clipping noise pulses τ_0 can be defined correspondingly by using a full bandwidth B of the multicarrier system:

$$\tau_0 = 2/B = 2/MW. \quad (7)$$

Now we can define the relation Ω between the lengths of the transmitted τ_0 and the received τ clipping noise pulses.

$$\Omega = \frac{\tau}{\tau_0} = M + 1, \qquad (8)$$

Note that for the OFDM System the number of sub-channels M is the same as the number of IFFT points: $M_{OFDM} = N$, and the expression (8) can be rewritten as:

$$\Omega_{OFDM} = N + 1 \qquad (9), \qquad\qquad \Omega_{WFMT} = \frac{N}{K} + 1 \quad (10);$$

Figure 4 illustrates a timing diagram of the clipping pulses in the WFMT and OFDM systems. The short pulses of the clipping noise $\varepsilon(t)$ are generated by the limiter schematics in the multicarrier transmitter. These pulses have length τ_0 and a probability P_{CN}.

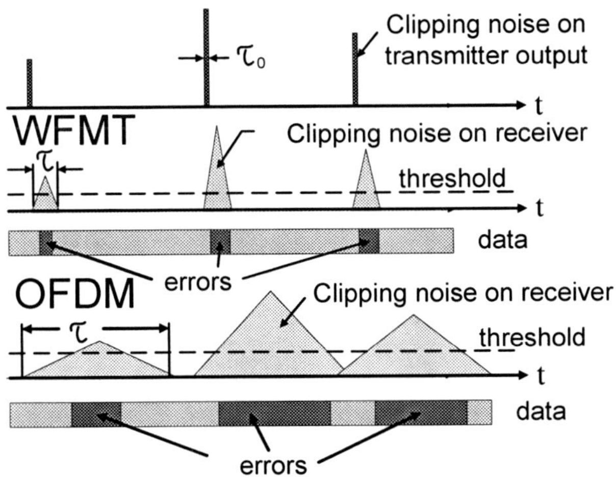

Figure 4. Time diagram of the clipping noise in the OFDM and WFMT systems.

Each pulse of the clipping noise produces an impulse in the receiver of the multicarrier system. The length of the received impulse depends on the bandwidth of the sub-channel. Therefore, the same clipping pulse provides significantly more errors in the OFDM system than in the WFMT system. We start the investigation of this effect from a simple hypothesis.

Suppose that a non-clipped signal $S_0(t)$ is a normal process, then:

$$P_{CN} = 1 - \frac{1}{\sqrt{2\pi}} \int_{-CR}^{+CR} e^{-x^2/2} dx, \quad (11)$$

where CR is a clipping ratio, $CR = CL/\sigma$, CL -clipping level and σ^2 -variance (power) of $S_0(t)$.

It is clear that the peak-to average power ratio of the clipped signal $S(t)$ is:

$$PAR = 10\log(CR). \quad (12)$$

Now we can define an expression for probability of data error P_E in the multicarrier system:

$$P_E = \Omega P_{CN}. \quad (13)$$

Let us use an error correction code to improve the performance of the multicarrier system i.e. the Reed-Solomon Code $RS(n,m,r)$. It generates a sequence of $n-$ coded bytes from $m-$ information bytes, and can correct $r-$ of an erroneous bytes. For this code we define a code rate DE such as:

$$DE = m/n \ , \qquad (14)$$

and the maximal probability of error that may be corrected - P_{MAX} :

$$P_{MAX} = r/n = \frac{1}{2}(1-DE); \ (15)$$

The number of clipping noise pulses that may be corrected by one block of the $RS(n,m,r)$ code Z is:

$$Z = P_{MAX} / P_E = \frac{1-DE}{2P_{CN}\Omega}; \quad (16)$$

Now we can calculate the probability P_Z of the appearance of Z pulses during one block of the $RS(n,m,r)$ code. Of course, this probability is the same as the probability of the non-corrected data error on the output of the Reed-Solomon decoder. Since the number of clipping pulses is the same in the transmitter and receiver of the multicarrier system we can write:

$$P_Z = (P_{CN})^Z = (P_{CN})^{\frac{1-DE}{2\Omega P_{CN}}} \ . (17)$$

Note, that P_{CN} depends only on a clipping ratio CR, and DE is constant, which is defined by the Reed-Solomon Code. A parameter Ω depends only on the bandwidth of the sub-channel (8). Therefore the expression (17) gives relations between probability of an error in the coded and clipped signal and Peak-to-Average Power Ratio (PAR) in the multicarrier system.

Figure 5 illustrates this dependence for a different number of the sub-channels of a multicarrier system. In this case a circular constellation diagram was used for QAM symbols.

As for the case of non-coded transmission, the form of constellation diagram is very important for PAR characteristics of the multicarrier system. The square constellation diagram of QAM modulator increases PAR by ~2 dB both in the case of the non-coded system and in the case of the error-correction coding and clipping of the multicarrier signal.

It is interesting that the code rate DE of the error-correcting RS code has a small impact on the PAR performance of a multicarrier system. Figure 9 demonstrates a PAR characteristic of the WFMT system with $M = 5$ for different code densities of the Reed Solomon Code $DE = 1/2$, 2/3, 3/4, 7/8, 9/10. As we can see, the difference in PAR for different codes does not exceed 1 dB.

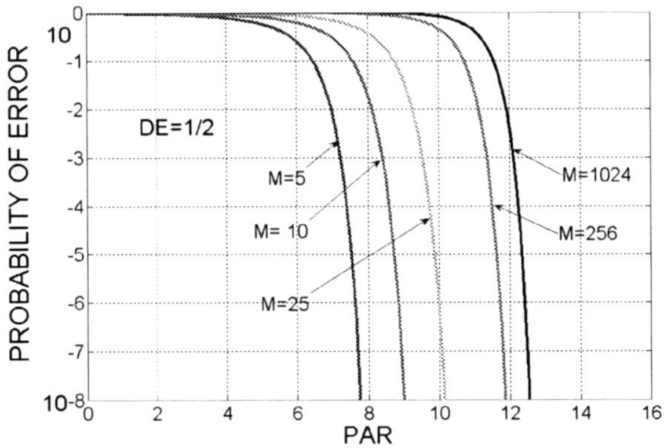

Figure 5. The Peak-to-Average Power Ratio and Probability of error in a multicarrier system with the Reed Solomon correction coding and circular constellation of QAM symbols

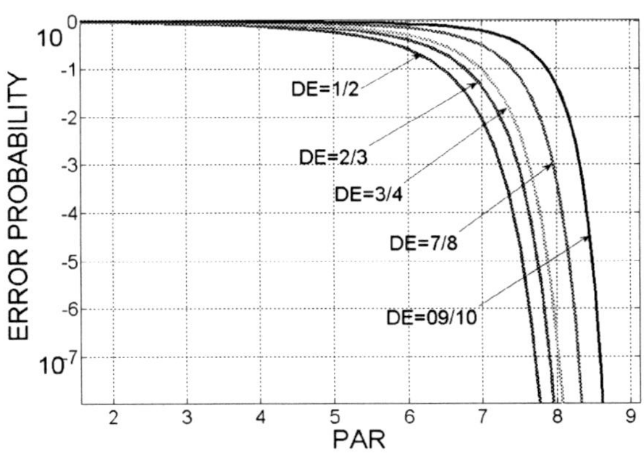

Figure 6. Peak-to-Average Power Ratio of WFMT system with circular constellation diagram and different error-correcting codes

4. CONCLUSIONS

In the last years a big progress in developing new technologies in the fields of wireless communications and the solid-state power amplifiers has been made. The modern GaAs AB-class Solid State Power Amplifiers have a power efficiency of 15~20% in the case of the WiMAX signal transmission in contrast to A-class SSPA, which has power efficiency of no more than 5%. This progress allows using a novel method of multicarrier transmissions based on WFMT modulation in Downstream Transmitters of LEO Satellites.

The main disadvantage of the multicarrier modulation is the high time domain Peak-to-Average Power Ratio (PAR), which limits Downstream Transmitter power efficiency. The Downstream Transmitter power efficiency is one of the main parameters of the satellite communication system, which defines the size and the cost of the satellite.

The novel WFMT modulation has significant fewer PAR than the existing OFDM systems. A using the WFMT modulation increases the downstream transmitter power efficiency by 4~4.5 dB in comparison with OFDM modulation. The efficient method of PAR reduction in multicarrier WFMT systems comprises the clipping of the transmitted signal and using error-correcting block codes.

References

[1] R. Nee, R Prsad :" OFDM for Wireless Multimedia Communications", Norwell, MA; Artech House, 2000.

[2] G. Cherubini, E. Eleftheriou, S. Oelser : "Filtered Multitone Modulation for Very High-Speed Digital Subscribe Lines", IEEE Journal of Selected Areas in Communications, VOL.20, NO. 5, 2002.

[3] Inaki Berenguer, Ian J. Wassell, "Efficient FMT equalization in outdoor broadband wireless systems", Proc. IEEE International Symposium on Advances in Wireless Communications, Victoria, Canada, Sept. 2002.

[4] Andrea Tonello "Discrete Multi-Tone and Filtered Multi-Tone Architectures forBroadband Asynchronous Multi-User Communications" Wireless personal Multimedia Communications Symposium - Aalborg, Denmark - September 9-12, 2001

[5] Roman M. Vitenberg "Method and system for Transmission of information data over communication line" US Patent Application 20050047513

[6] Andrea Tonello , Roman M. Vitenberg "An Efficient Wavelet Based Filtered Multitone Modulation Scheme WPMC 2004 –Albano Terme , Italy , September 2004

[7] Andrea Tonello, Roman M. Vitenberg "An Efficient Implementation of a Wavelet Based Filtered Multitone Modulation Scheme" ISSPIT 04 ITALY, ROME, December 2004.

[8] Roman Vitenberg: "A practical implementation of a Wavelet Based Filtered Multitone Modulation", WOSM'05 Conference . Tenerife, December 2005

[9] Roman Vitenberg: " A WFMT Transmitter for Cable TV Applications", WSEAS Conference, Madrid, February 15-17, 2006.

[10] Roman Vitenberg: "Peak-to-Average Ratio in WFMT System", AIC WSEAS Conference, Elounda, Crete, Greese, August 2006.

[11] Seung Hee Han, Jae Hong Lee : "An overview of Peak-to-Average Power Ratio Reduction Techniques for Multicarrier Transmission", IEEE Wireless Communicatioms, April 2005.

[12] William L. Pribble, Jim M Miligan, and Raimond S. Pengelly: "High efficiency Class-E Amplifier Utilizing GaN HEMT technology", Presentation, Cree Inc. 2005.

[13] S. Wood, P. Smilh, W. Pribble....:" High Efficiency, High Linearity GaN HEMT Amplifiers for WiMAX Applications", High Frequency Electronics, May, 2006.

[14] Don Kimball, Paul Draxler, Jinho Jeong....:."50% PAE WCDNA Basestantion Amplifier Implementation with GaN HFETs ", CSIC'05, IEEE, 30 Okt.-2 Nov. 2005

A NEW DYNAMIC PARTIAL PRECODING TECHNIQUE FOR MC-CDMA SYSTEMS EMPLOYING PSK MODULATION

C. Masouros and E. Alsusa
School of Electrical and Electronic Engineering, The University of Manchester, PO. Box 88, Manchester, M60 1QD, UK

Abstract: A new precoding technique for MC-CDMA is presented, that outperforms conventional precoding by exploiting useful interference. This is done by selectively pre-decorrelating users to destructive interference while allowing interference when it is expected to contribute to their signal. The resulting SINR improvement is achieved by making use of energy existent in the system so performance enhancement is attained without the need for increased transmitted power-per-user. This, however, comes at the expense of additional processing at the transmitter for the allocation of the dynamic precoding matrix. The proposed technique applies to the downlink of cellular MC-CDMA systems. Comparative simulations of this and other precoding methods provide evidence that the proposed method offers significant performance benefits.

Key words: MC-CDMA, precoding, adaptive signal processing, interference multiuser channels, suppression

1. INTRODUCTION

Precoding techniques are gaining a prominent role in modern wireless communications as they offer the best potential for the simplification of Mobile Unites' (MU) receivers. In multiple carrier code division multiple access (MC-CDMA) these techniques are made more complexity-efficient since by use of guard intervals and InterSymbol Interference (ISI) elimination there is no need for blockwise processing. A variety of pre-decorrelating techniques for direct sequence CDMA (DS-CDMA) has been introduced but their application to MC-CDMA has not yet been thoroughly investigated. In [1] the authors propose transferring the channel equalization processing to the BS which yields the pre-equalization technique. This technique's main advantage is that the equalization processing is removed from the MU. However, without the use of multiuser detection (MUD) performance is poor in a multiuser scenario. The authors in [2] propose a system similar to the conventional receiver-based decorellator-detector where the decorrelation procedure happens at the BS prior to transmission. The Transmitter Precoding (TP) method presented in [2] –investigated for DS-CDMA systems – performs a complete orthogonalization amongst all users.

S. Plass et al. (eds.), Multi-Carrier Spread Spectrum 2007, 177–186.

This results in increased transmitted energy which calls for scaling of the signal to be transmitted. An improvement is attained by applying the Joint Transmission (JT) decorrelating procedure in [3]. This optimization – again presented for DS-CDMA – leads to the use of a decorrelation scheme that also employs Pre-Rake processing [4]. This method offers both the benefits of pre-decorrelation as well as the advantages of Pre-Rake over the Rake technique as explained in [4]. Equivalently, in MC-CDMA, JT would apply pre-decorellation processing on a system using pre-equalizaton while TP would utilize post-equalization. Both decorrelating methods introduced in [2,3] involve the inversion of a square matrix which imposes a significant computational burden when blockwise processing is required. Evidently, these techniques could benefit in MC-CDMA from the fact that ISI and consequently Multiple Access Interference (MAI) from symbols of adjacent symbol periods is eliminated and there is no need for blockwise decorrelation.

In this paper, besides introducing the application of pre-decorrelating techniques on MC-CDMA, an improvement of the TP and JT techniques in [2,3] respectively is suggested which takes advantage of the constructive interference concept applicable in PSK modulation that will be analyzed below. The system analysis follows the one presented in [2]. In contrast to the TP and JT techniques where the users are fully orthogonalized, we propose a dynamic partial orthogonalization by means of pre-decorrelation, targeted at the users that are expected to suffer from destructive MAI. This decreases the amount of décorrelating applied, limits the increase in transmitted energy and reduces the scaling required resulting in improved performance. However, since the MAI profile changes between symbol periods, the precoding matrix needs to be dynamically updated according to the current transmitted symbols. This imposes a complexity increase at the transmitter. In cases of fast fading channels, where channel estimation and calculation of the precoding matrix are required at each OFDM symbol period anyway, this complexity penalty diminishes. Since, as proven in [3], JT outperforms TP in multipath scenarios, pre-equalization is considered in all simulations and therefore the proposed technique is compared to the equivalent of JT in MC-CDMA.

2. CONSTRUCTIVE MAI DERIVAITON FOR PSK MODULATION

2.1 Downlink Signal Model and Constructive MAI Definition

Consider the downlink transmission in a discrete-time synchronous frequency selective MC-CDMA system of K equal power users, where all codes and channels are assumed normalized to unit energy and the spreading gain is equal to L. For simplicity we assume that the number of OFDM subcarriers $N_c = L$. The use of cyclic prefix is presupposed so that the ISI is completely suppressed. Assuming pre-equalization at the transmitter the received signal at the u-th MU can be expressed in matrix form as:

$$\mathbf{r}_{\mathrm{iu}} = [\mathbf{x}_{\mathrm{i}} \cdot \mathbf{A} \cdot (\mathbf{C} \circ \mathbf{E})] \circ \mathbf{H}_{\mathrm{u}} + \mathbf{N}_{\mathrm{u}} \tag{1}$$

where $\mathbf{x}_i=[\, x_{1i}\, x_{2i} \ldots x_{Ki}\,]$ is the $1\times K$ matrix containing all users' data for the i-th symbol period, $\mathbf{A}=diag(a_k)$ is the $K\times K$ diagonal matrix of amplitudes, $\mathbf{C}=[\mathrm{C}_1\, \mathrm{C}_2 \ldots \mathrm{C}_K]^{\mathrm{T}}$ and $\mathbf{E}=[\mathrm{E}_1\, \mathrm{E}_2 \ldots \mathrm{E}_K]^{\mathrm{T}}$ are the $K\times L$ matrices containing the users' codes and equalization coefficients for each subcarrier while $\mathbf{H}_{\mathbf{u}}$ and $\mathbf{N}_{\mathbf{u}}$ are the u-th MU's channel and noise matrix of size $1\times L$. In (1) the notation $\circ$ is used to denote element-by-element matrix multiplication. From the received signal the data is extracted as:

$$\mathbf{d}_{\mathbf{iu}} = \mathbf{r}_{\mathbf{iu}} \cdot \mathrm{C}_{\mathrm{u}}^{\mathrm{H}} = a_u \cdot \rho_{uu} \cdot x_{iu} + \sum_{k=1,k\neq u}^{K} a_k \cdot \rho_{ku} \cdot x_{ik} + n_{iu} \tag{2}$$

where

$$\rho_{\mathrm{pq}} = (\mathbf{C}_{\mathrm{p}} \circ \mathbf{E}_{\mathrm{p}} \circ \mathbf{H}_{\mathrm{p}}) \cdot \mathbf{C}_{\mathrm{q}}^{\mathrm{H}}, \tag{3}$$

$$\sum_{k=1,k\neq u}^{K} a_k \cdot \rho_{ku} \cdot x_{ik} = MAI_{iu} \tag{4}$$

is the MAI caused by the other K-1 users and n_{iu} is the noise component. The MAI is constructive when it adds to the desired user's signal energy. It is therefore reasonable to separate MAI in two components, one being the interference that contributes to the useful signal and the other being the one that subtracts from the useful energy. Having the above in mind, the instantaneous SINR (for PSK modulation) can effectively be written as:

$$SINR_e = \frac{S + J_c}{J_d + N} \tag{5}$$

where J_c, J_d are the power of constructive and destructive MAI respectively. It is obvious from this equation that if the J_d component is decreased or completely eliminated while elements of J_c are preserved through selective precoding, the instantaneous $SINR_e$ will be improved and so will the system average performance. This is the objective of the proposed technique for which we present three methods as discussed in the subsequent sections. In the following, constructive MAI is derived for BPSK and generalized to MPSK modulation.

2.2 Constructive MAI Derivation for BPSK Modulation

For BPSK modulation the desired user's signal $x_{iu} \in \{ -1, +1 \}$, so constructive is the MAI_{iu} that has the same sign as x_{iu}, for which $x_{iu} \cdot MAI_{iu} > 0$, as depicted in the shadowed part of the BPSK constellation diagram in Fig. 1a. When unit energy signature waveforms and channels are assumed and $\rho_{uu} = 1$ this, using (4), leads to:

$$a_u \cdot x_{iu} \cdot \sum_{k=1,k\neq u}^{K} a_k \cdot x_{ik} \cdot \rho_{ku} > 0 \tag{6}$$

A matrices representation of the above could be attained as follows. Define **R** as the $K \times K$ crosscorrelation matrix of the multipath corrupted codes with elements as in (3):

$$\mathbf{R} = \begin{pmatrix} 1 & \rho_{12} \cdots\cdots & \rho_{1K} \\ \rho_{21} & 1 \cdots\cdots & \rho_{2K} \\ \vdots & \ddots & \vdots \\ \rho_{K1} & \rho_{K2} \cdots & 1 \end{pmatrix} \tag{7}$$

By observing the matrix

$$\mathbf{M}_i = diag(\mathbf{x}_i \cdot \mathbf{A}) \cdot \mathbf{R} \cdot diag(\mathbf{x}_i \cdot \mathbf{A})$$

$$= \begin{pmatrix} 1 & a_1 a_2 x_{i1} x_{i2} \rho_{12} \cdots\cdots & a_1 a_K x_{i1} x_{iK} \rho_{1K} \\ a_2 a_1 x_{i2} x_{i1} \rho_{21} & 1 \cdots\cdots & a_2 a_K x_{i2} x_{iK} \rho_{2K} \\ \vdots & \ddots & \vdots \\ a_K a_1 x_{iK} x_{i1} \rho_{K1} & a_K a_2 x_{iK} x_{i2} \rho_{K2} \cdots & 1 \end{pmatrix} \tag{8}$$

which is the crosscorrelation matrix of the multipath corrupted codes, modulated by the data, the instantaneous MAI can be investigated. Each element of $\mathbf{M}_i$ provides information on the interference between two users. The criteria in (6) could therefore be written as:

$$sum\left(\left[\mathbf{M}_i - diag(\mathbf{M}_i)\right]_u\right) > 0 \tag{9}$$

where $[.]_u$ denotes the u-th column of a matrix.

From the above analysis and by observing the constellation diagrams in Fig. 1, it can be derived that constructive is the MAI that causes the noiseless part of the received signal to fall in the part of the constellation diagram where the distance from the decision thresholds is increased. Applying this rule to QPSK and 8PSK yields the shadowed part in Fig 1.b,c. The generalization to MPSK is straightforward.

3. DYNAMIC PARTIAL PRECODING METHOD ANALYSIS

It should be clear so far, that the system can benefit from the existence of constructive interference. Consequently, there is no need for it to be removed by applying full pre-decorrelation. This is the main principle of the proposed system which is depicted in Fig. 2. Using Channel State Information (CSI), knowledge of all users' codes and data, readily available at the BS, and with the help of (6) – (9), the interference to each user can be estimated at the BS prior to transmission. By observation of the matrix $\mathbf{M}_i$ the elements of the cross correlation matrix $\mathbf{R}$ to be removed via decorrelation can be determined. Hence the transmitted signal is given as:

$$\mathbf{s} = f \cdot \mathbf{x}_i \cdot \mathbf{A} \cdot \mathbf{T} \cdot (\mathbf{C} \circ \mathbf{E}) \tag{10}$$

where

$$f = \sqrt{\left(\sum_{k=1}^{K} a_k^2\right) \Big/ \left(\sum_{k=1}^{K} a_k^2 T_{k,k}\right)} \tag{11}$$

is the scaling factor and $\mathbf{T}$ is the precoding matrix excluding the spreading operation. Instead of $\mathbf{T}$ being derived by the MMSE optimization proposed in [2,3] we propose the following MMSE optimization:

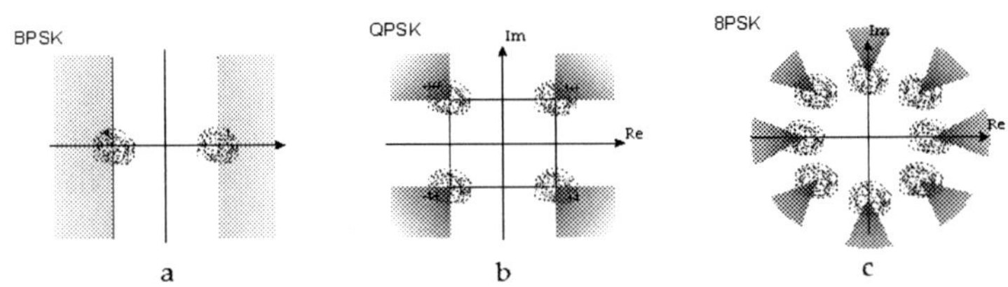

Figure 1. The constellation diagrams of BPSK, QPSK and 8PSK modulation depicting possible noiseless received signals and the constructive MAI sectors

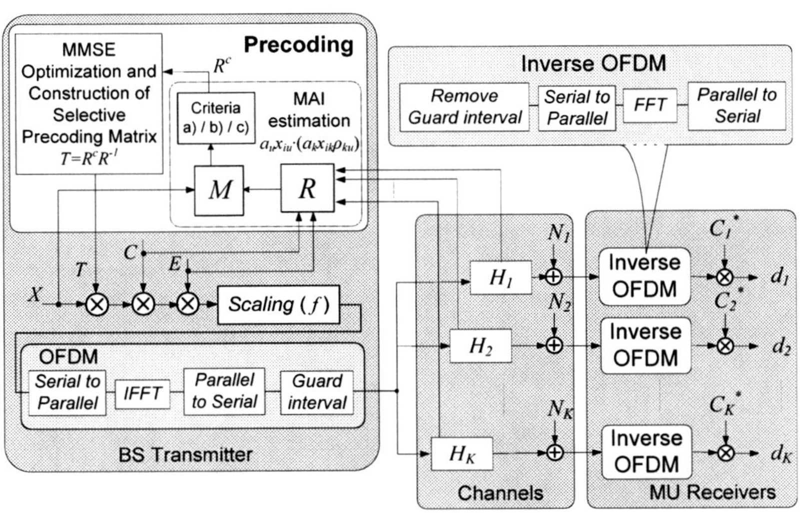

Figure 2. The proposed selective precoding in the DS/CDMA downlink

$$J = E_{x,n}\left\{\left\|\mathbf{xAR}^{C} - \mathbf{d}\right\|^{2}\right\} = E_{x,n}\left\{\left\|\mathbf{xR}^{C} - \left(\mathbf{xTR} + \mathbf{n}\right)\right\|^{2}\right\} \quad (12)$$

where $\mathbf{R}^c$ is the constructive cross correlation matrix that contains the ρ_{uk} elements of $\mathbf{R}$ that yield constructive interference according to the observation of $\mathbf{M}_i$ at every symbol period. Matrix $\mathbf{R}^c$ can be formed according to the three criteria that will be presented in the following sections. The solution to the above optimization is $\mathbf{T}=\mathbf{R}^c\mathbf{R}^{-1}$. For TP and JT in [2,3], $\mathbf{R}^c=\mathbf{I}$ which derives full orthogonalization. Since for the proposed method the constructive cross correlation matrix $\mathbf{R}^c \neq \mathbf{I}$ contains elements of $\mathbf{R}$, selective precoding needs less manipulation. This is because the non-zero elements of $\mathbf{R}^c$ cancel out elements of $\mathbf{R}^{-1}$ and hence the elements of matrix $\mathbf{T}=\mathbf{R}^c\mathbf{R}^{-1}$ are smaller than $\mathbf{T}=\mathbf{R}^{-1}$ for conventional precoding. This means that the scalar f in (11) and consequently the effective SINR will be larger than in [2,3]. Moreover, useful interfering energy is allowed in the system that further enhances the received SINR as shown in eq. (5). The matrix $\mathbf{R}^c$ can be formed using three different criteria as described below.

3.1 Selective Precoding Method a) (SJT A)

The simplest method would be to fully orthogonalize the users that experience destructive cumulative MAI and leave the users that expect constructive cumulative MAI correlated to interference. In mathematical terms for the *i-th* period of interest this could be expressed as:

$$\text{If } a_u x_{iu} \cdot \sum_{k=1,k\neq u}^{K} a_k x_{ik} \rho_{ku} < 0 \text{ then } R^c_{k,u} = 0 \text{ for all } k \neq u$$

$$\text{Else } R^c_{k,u} = \rho_{ku} \text{ for all } k \neq u$$

where $R^c_{k,u}$ is the (k,u)-th element of matrix $\mathbf{R}^c$.

3.2 Selective Precoding Method b) (SJT B)

An alternative to the above method would be to orthogonalize every user but only to the users that impose destructive interference to the useful signal at each symbol period. This would completely remove all destructive while allowing all constructive MAI

Loop for k=1 to K , $k \neq u$

If $a_u a_k x_{iu} x_{ik} \rho_{ku} < 0$ then $R^c_{k,u} = 0$

Else $R^c_{k,u} = \rho_{ku}$

Obviously, this method aims at enhanced performance by retaining purely constructive MAI, but it requires a higher amount of decorrelation and hence increased transmitted power and scaling compared SJT A. The tradeoff to useful MAI energy, though, is better than for SJT A which yields better performance.

3.3 Selective Precoding Method c) (SJT C)

Here an optimization between the complexity, the required scaling and the constructive interference held in the system is attempted. This is done by orthogonalizing the users experiencing destructive cumulative MAI only to the users that impose destructive MAI on them, while leaving the remaining users completely un-decorrelated. This can be expressed as

If $a_u x_{iu} \cdot \sum_{k=1,k\neq u}^{K} a_k x_{ik} \rho_{ku} < 0$ then loop for k=1 to K, $k \neq u$

$$\begin{cases} \text{If} \quad a_u a_k x_{iu} x_{ik} \rho_{ku} < 0 \quad \text{then} \quad R^c_{k,u} = 0 \\ \text{Else} \quad R^c_{k,u} = \rho_{ku} \end{cases}$$

Else $R^c_{k,u} = \rho_{ku}$ for all $k \neq u$.

Evidently, this method requires the least decorrelating and scaling, leaving a larger portion of the useful signal to be exploited at the receiver. Due to the existence of destructive MAI in the system, however, performance is expected to be worse than for SJT B, but the advantage of this method is the reduced complexity compared to the previously proposed techniques.

4. NUMERICAL AND SIMULATION RESULTS

Monte Carlo simulations of the JT and SJT techniques have been performed for a chip-spaced *P*-path decentralized Rayleigh multipath with unity gain and equal average power per channel's path (uniform channel power-profile). BPSK modulation has been employed and orthogonal codes with L=16 have been used. A number of N_c=L=16 OFDM subcarriers is assumed and MRC pre-equalization has been utilized. The use of guard intervals for ISI elimination is supposed. Unless stated otherwise the average transmitted SNR=E_b/N_o per bit per user is considered in the performance results.

In Fig. 3 the BER vs SNR performance for the case of K=12 users in a multipath of P=11 paths is depicted for the JT method applied in MC-CDMA and the three proposed selective precoding (SJT) techniques. It can be seen that all three SJT methods outperform conventional precoding, due to the benefit from the existence of constructive MAI. Since SJT B allows pure constructive MAI it offers the best performance, with an SNR benefit reaching 6dB.

In Fig. 4 the BER vs K performance for SNR=7dB for the same multipath is shown. It is evident that beyond a certain value of K, performance for all techniques severely deteriorates. However, for SJT B and SJT C it can be seen that up to a certain K, performance improves as the users increase. This results from the fact that users are let to interfere constructively, which enhances the SINR and superimposes the effect of scaling and therefore it surpasses single user performance. It can be seen that selective precoding yields significant capacity improvement for all techniques as for the same BER level more users are allowed in the system. For SJT B capacity is almost doubled compared to conventional JT for a BER level of 10^{-2}.

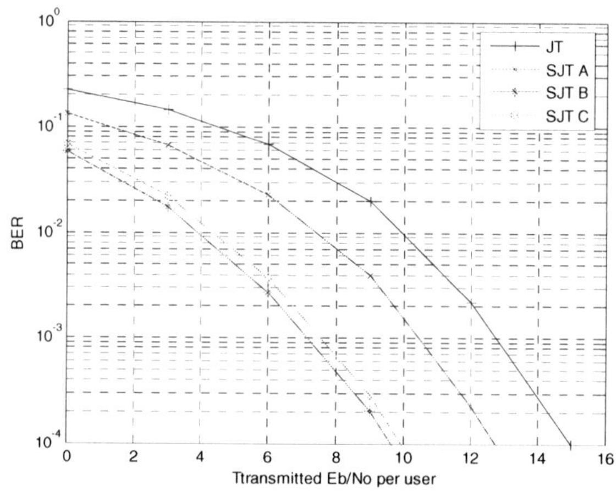

Figure 3. BER vs SNR Performance of conventional precoding (JT) and the proposed precoding (SJT) methods in a Rayleigh fading channel of P=11 paths for K=12, L=16

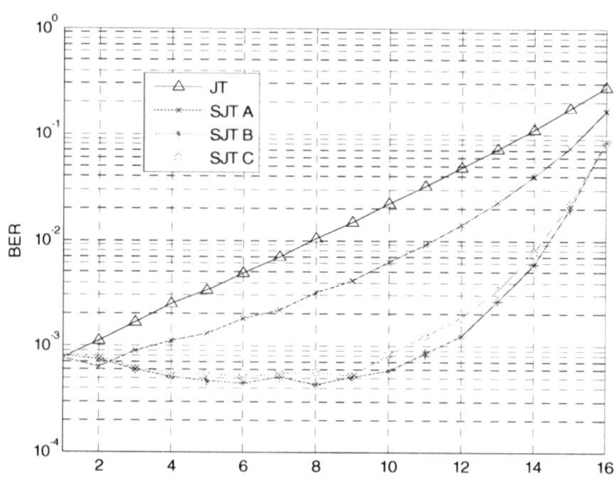

Figure 4. BER vs K Performance of JT and SJT methods in a Rayleigh fading channel of P=11 paths for K=12, L=16

In Fig. 5 the unequal power users scenario is considered. The same values of K=12, L=16 and same channel of P=11 paths is assumed. The users are assumed to have amplitudes of **A**=[K/2-1 K/2-2 . . . 0 . . . –K/2+1 –K/2] in dB with respect to the symbol energy. The average performance of all users is considered with reference to the SNR=0dB level. Conventional JT is afflicted by the fluctuation in amplitude and performance is poor. This is due to the fact that the removal of high power users' interference on the weak users by orthogonalization dramatically increases the transmitted power and therefore a large amount of scaling is required. This in turn results in reduced SNR at the receiver. With SJT, however, this can be avoided. The resulting BER reduction reaches a factor of 100 and the performance gain reaches 7dB for SJT B compared to JT. It can be concluded that selective precoding makes the system more near-far resistant.

Finally, in Fig. 6, complementary to performance comparison, a complexity comparison between the three proposed methods is depicted for P=11, L=16, SNR=7dB. The evaluation is based on the number of elements that are removed from the cross correlation matrix **R** to form $\mathbf{R}^c$ for each technique. The ratio of these over the total number of off-diagonal elements of **R** is then calculated to form the relative decorrelation ratio to conventional JT. The results show that, as explained above, SJT C offers the least complexity with a ratio of 14% to full orthogonalization decorreltaing, where all off-diagonal elements of **R** are removed. SJT A and SJT B provide ratios of about 23%. Combined with the BER performance graphs, the results show that, if an optimization of the performance to complexity is required, SJT C is the preferred method, while when complexity is not an issue, SJT B provides the best alternative.

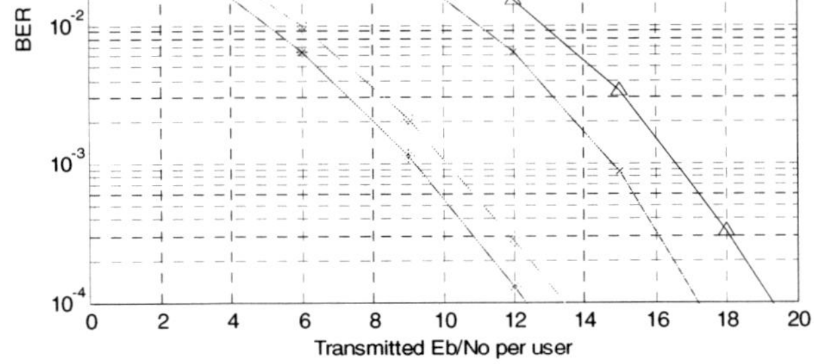

Figure 5. BER vs SNR Performance of JT and the SJT methods in a Rayleigh fading channel of P=11 paths for K=12 unequal power users, L=16

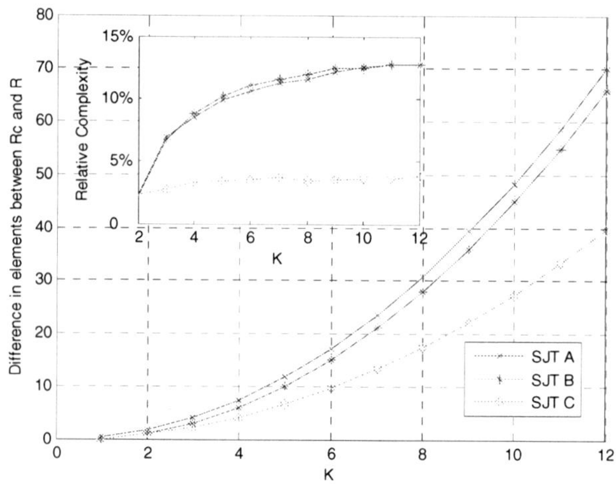

Figure 6. Complexity comparison for increasing number of users of the SJT methods in a Rayleigh fading channel of P=11 for L=16 , SNR=7dB

5. CONCLUSIONS AND FUTURE WORK

We have shown that by allowing the existence of constructive interference, selective precoding can improve BER performance and capacity with no need for additional power-per-user investment. Further work can be carried out towards applying the proposed method to other existent precoding schemes combined with PSK modulation.

REFERENCES

1. P. Bisaglia, N. Benvenuto, and S. Quitadamo, "Performance comparison of singleuser pre-equalization techniques for uplink MC-CDMA systems," in Proceedings IEEE Global Telecommunications Conference (GLOBECOM'03), San Francisco, USA, pp. 3402-3406, December 2003.
2. B. R. Vojcic and W. M. Jang, "Transmitter precoding in synchronous multiuser communications," *IEEE Trans. on Communications*, vol. 46, no. 10, pp. 1346–1355, Oct. 1998.
3. A. Silva and A. Gameiro, "Pre-Filtering Antenna Array for Downlink TDD MC-CDMA Systems", in Proceedings Vehicular Technology Conference, VTC 2003-Spring, pp. 641-645, Apr. 2003.
4. R. Esmailzadeh, E. Sourour, and M. Nakagawa, "Pre-rake diversity combining in time division duplex CDMA mobile communications," *IEEE Trans. on Vehicular Technology*, vol. 48, no. 3, pp. 795–801, May 1999.

THE INFLUENCE OF LINK ADAPTATION IN MULTI-USER OFDM

Stephan Pfletschinger
Centre Tecnolgic de Telecomunicacions de Catalunya (CTTC)
Parc Mediterrani de la Tecnologia (PMT), Av. del Canal Olmpic, s/n
08860 Castelldefels (Barcelona), Spain
stephan.pfletschinger@cttc.es

Abstract We examine the performance gains from two variants of adaptive coding and modulation (ACM) in an OFDM-based cellular system with realistic system parameters. In this system a channel-aware scheduler allocates to each active user a set of subchannels, upon which the users may perform link adaptation with ACM. We investigate the achievable performance gain by frequency-selective ACM, in comparison to non frequency-selective adaptation. Since the scheduler, which is aware of the channel and the traffic conditions, already exploits the multi-user diversity and tends to allocate similar subchannels to a given user, it is not obvious that frequency-selective link adaptation achieves a *significant* performance gain, especially when realistic ACM with limited granularity is considered. In this paper, we evaluate the gain of frequency-adaptive vs. non frequency-adaptive transmission for the uplink under realistic conditions for a 4G system proposal.

1. Introduction and System Model

One of the prime objectives of future wireless communication systems is to provide the user with a variety of services in a wide range of possible environments and deployment scenarios. The radio interface for such a ubiquitous mobile communication system is currently under investigation in the European WINNER project [1]. Due to its well-known benefits and its proven practicability, OFDM is the selected scheme for most transmission modes. The main challenge faced by the physical layer of the air interface is the mobile environment which leads to a channel which is varying on different time scales. While adaptation to slow changes, mainly caused by path loss and shadow fading, is already adopted in current systems, adaptation to the fast, frequency-selective fading is much more challenging, but also promises considerable gains

S. Plass et al. (eds.), Multi-Carrier Spread Spectrum 2007, 187–196.

due to multi-user diversity. Since the transmission bandwidth of 4G systems is expected to extend up to 100 MHz, it seems natural to exploit the available frequency diversity by adapting to it rather than averaging over it. In a multi-user system, there are mainly two mechanisms for adaptation: first, a channel-aware scheduler may assign users to the most appropriate subchannels, and then the users may perform frequency-selective link adaptation. Obviously, the gains from both mechanisms do not sum up and there remains the question if the gain from the second adaptation step is still significant in case of channel-aware scheduling.

1.1 OFDMA Uplink

We consider the uplink in an OFDMA system with N_c active subcarriers, which are grouped into blocks of n_f subcarriers to form $N = N_\mathrm{c}/n_\mathrm{f}$ *subchannels*. The basic unit for adaptation is a *chunk*, which is formed by n_t consecutive OFDM symbols of the same subchannel. The duration of a chunk defines the duration of TDD subframe. The channel is assumed to be constant during one frame, which comprises one uplink and one downlink subframe, and is independent from the previous frame (block fading). The channel transfer coefficients $\mathbf{h}'_k = (h'_{k1}, \ldots, h'_{kN})^\mathrm{T}$ are drawn independently for each user and frame according to a given power delay profile.

The K users are distributed uniformly in a cell of radius r_c, which is modeled by the deterministic distance distribution [2]

$$d_k = \left(1 + 9\sqrt{\frac{k-1}{K-1}}\right)\frac{r_\mathrm{c}}{10}, \qquad k = 1, \ldots, K \tag{1}$$

The path loss is given by $\alpha_k^{(\mathrm{dB})} = 10\beta \lg(d_k) + \alpha_0$, which leads to the transfer coefficients

$$h_{kn} = 10^{-\alpha_k^{(\mathrm{dB})}/20} \cdot h'_{kn} \tag{2}$$

The received signal in the uplink is thus given by

$$y_n(i) = \sum_{k=1}^{K} h_{kn} x_{kn}(i) + w_n(i), \qquad \begin{array}{l} n = 1, \ldots, N \\ i = 1, \ldots, n_\mathrm{t} \end{array} \tag{3}$$

where $w_n(i) \sim \mathcal{CN}(0, N_0)$ is AWGN and $p_{kn} = \mathbb{E}[|x_{kn}|^2]$ is the power on subchannel (chunk) n of user k. The power of each user is limited to $\bar{P}_k$, which leads to the power constraints $\sum_{n=1}^{N} p_{kn} \leq \bar{P}_k$, $\forall k = 1, \ldots, K$. For all following adaptive algorithms, the knowledge of the *channel to noise ratio* (CNR) $T_{kn} \triangleq |h_{kn}|^2/N_0$ is sufficient, i.e. the phases of the transfer coefficients are not required.

1.2 Multiple-Access Capacity Region

The OFDMA uplink constitutes a frequency-selective multiple-access channel (MAC), whose capacity region is given by

$$\mathcal{C}_{\text{MAC}} = \bigcup_{p_{kn}} \left\{ \mathbf{R} \in \mathbb{R}_+^K : \sum_{k \in \mathcal{K}} R_k \leq \frac{1}{N} \sum_{n=1}^{N} \text{ld} \left(1 + \sum_{k \in \mathcal{K}} p_{kn} T_{kn} \right) \right\} \quad (4)$$

where the union is over all powers which fulfill the power constraint and for all non-empty user sets $\mathcal{K} \subseteq \{1, \ldots, K\}$.

The boundary of the region $\mathcal{C}_{\text{MAC}}$ coincides with the solution to the weighted sum rate maximization problem

$$\begin{aligned} \max \quad & \sum_{k=1}^{K} \mu_k R_k \\ \text{s.t.} \quad & \sum_{n=1}^{N} p_{kn} \leq \bar{P}_k, \; p_{kn} \geq 0 \; \forall n, k \end{aligned} \quad (5)$$

which can be found by an iterative algorithm [3, 4].

It is known that a scheduling policy which is based on sum-rate maximization achieves to stabilize the queue lengths as long as the arrival rates are within the region $\mathcal{C}_{\text{MAC}}$ [5, 6]. Note that any other constraint may be considered as long as it can be expressed via the weight vector $\boldsymbol{\mu} = (\mu_1, \ldots, \mu_K)$. Equations (3–5) hold for the uplink, but very similar expressions and an algorithm for sum-rate maximization exist for the downlink as well [7]. Additionally, both cases are linked via the downlink-uplink duality [8].

2. Scheduling and Link Adapation

The optimum solution considers the problem of ACM and user scheduling jointly, taking into account the channel state and all constraints on user rates, priorities, etc. However, for practical implementations this is far too complex and therefore the whole adaptation process is separated into two steps:

1 The *scheduler* allocates a set of subchannels (i.e. chunks) to each active user.

2 The *link adaptation* algorithm selects a suitable modulation and coding scheme (MCS) for each chunk.

The first step hence separates the multi-user problem into K single-user problems which can be solved independently from each other. In this section, before focussing on link adaptation, we provide an overview of the applied scheduling policies. The reason for having more than one scheduling policy is that there is no unique optimization criterion and due to the complexity of the problem, several sub-optimum solutions exist.

2.1 Scheduling Policies

Maximum weighted sum rate. The weighted sum rate criterion according to (5) is well justified since it corresponds to rate vectors which lie on the boundary of the capacity region (4). However, it is not straightforward to express additional constraints via the weight vector $\boldsymbol{\mu}$ and, although iterative algorithms exist [4, 3], their complexity is still considerable, especially for real-time implementations in systems with a high number of subchannels. For a simplified scheduling policy, the weights are chosen to compensate for the path loss:

$$\mu_k = d_k^{\beta} \tag{6}$$

The solution to the weighted sum rate maximization (5) is given in the form of a power allocation p_{kn}, which uniquely defines the rates per user R_k. In order to separate the problem into the two steps described above, we have to derive the subchannel allocation from the power allocation, which can be done by assigning to each subchannel the user with the highest power:

$$a_n = \arg\max_k \{p_{kn}\}, \quad n = 1, \ldots, N \tag{7}$$

The loss in accuracy introduced by this step is rather low as long as the number of subchannels is much higher than the number of simultaneously scheduled users. Based on (7), each user disposes of a set of dedicated chunks over which ACM can be done individually.

Maximum sum rate, considering queue lengths. The uplink of an OFDMA system is a special case of a vector multiple-access channel and thus the iterative water-filling algorithm (IWF) [9] can be applied, which leads to particularly simple implementation for the case of single antennas [10]. For this case, we include a simple model to consider the queue lengths: we assume that in each frame a constant number of B_k bits arrives at each user's queue. First, an estimation $\hat{b}_k$ of each user's rate, expressed in bits per frame, is calculated based on the IWF algorithm. Then a set $\mathcal{K} \subseteq \{1, \ldots, K\}$ of scheduled users is formed, which only includes users whose queue length is higher than the estimated rate per frame:

$$\mathcal{K} = \left\{ k \in \{1, \ldots, K\} : \hat{b}_k \le q_k \right\} \tag{8}$$

This criterion excludes users which could transmit more bits per frame than they have in their queue and, hence, makes efficient use of the available resources. For the users in the set $\mathcal{K}$, the power allocation is computed with the IWF algorithm, from which the subchannel allocation is derived by (7). It has to be noted that this scheduling policy can be regarded as overly simplified for any realistic purpose since it does not consider neither delays nor minimum rate requirements.

Based on the subchannel allocation a_n, the link adaptation scheme computes the actual rate per frame b_k and updates the queue lengths:

$$q_k(t+1) = \max\left(q_k(t) + B_k - b_k, 0\right) \tag{9}$$

Proportional fair. Another policy is proportional fair scheduling, which is often employed as a baseline case due to its simplicity. This strategy is based solely on the channel transfer coefficients and assigns to each subchannel the user which has the best subchannel with respect to his own average:

$$a_n = \arg\max_k \left\{ \frac{T_{kn}}{\frac{1}{N}\sum_{i=1}^{N} T_{ki}} \right\}, \quad n = 1, \ldots, N \tag{10}$$

Since we assume that all users have the same channel statistics, this criterion allocates in the average the same number of subchannels to the users while it nevertheless takes advantage of multi-user diversity.

Queue-proportional. The most complete scheduling policy considered in this paper is based on weighted sum rate maximization with weights proportional to the queue lengths. While the queue model is the same as above, the weights are chosen proportional to the current queue length:

$$\mu_k = q_k \tag{11}$$

2.2 Link Adaptation

In the following, we focus on link adaptation in order to evaluate the impact of ACM on the achievable rates of all users. For this single-user adaptation, we distinguish two modes:

- **Frequency adaptive**. The modulation is adapted per subchannel (chunk). The same code rate (by puncturing of a low-rate mother code) is chosen for all chunks belonging to the same user.

- **Non-frequency adaptive**. The same MCS is selected for all chunks of the same user.

It is immediately clear that the first option will achieve higher throughput. However, the system complexity and the signalling overhead are also higher. It is thus of high practical relevance to evaluate in which environments the gain of frequency-adaptive transmission justifies the additional complexity.

Fig. 1 shows the transmitter and receiver structure for one user. One common encoder/decoder is employed for all subchannels of the same user. This allows to take full advantage of strong coding schemes, which require a certain codeword length, while keeping the system complexity low [11]. The data bits are first encoded with a duo-binary turbo code (DBTC) and interleaved, then bit-loading is performed and the QAM symbols are mapped to the chunks which have been assigned to this user by the scheduler. More details on this scheme and its performance can be found in [12].

For the bit-loading and the adaptation of the coding rate, we employ the Stiglmayr algorithm [13], which has been shown to achieve a throughput comparable to the optimum Hughes-Hartogs algorithm [14], even without power-loading and with a considerably lower computational complexity [12]. Due to these favorable properties, the Stiglmayr algorithm has been selected as a reference for frequency-adaptive transmission in the WINNER project [1]. Fig. 2 shows the achieved throughput with the Stiglmayr and the Hughes-Hartogs algorithms in comparison to the capacity of the Rayleigh channel with constraint inputs and the Rayleigh channel with perfect power adaptation [8].

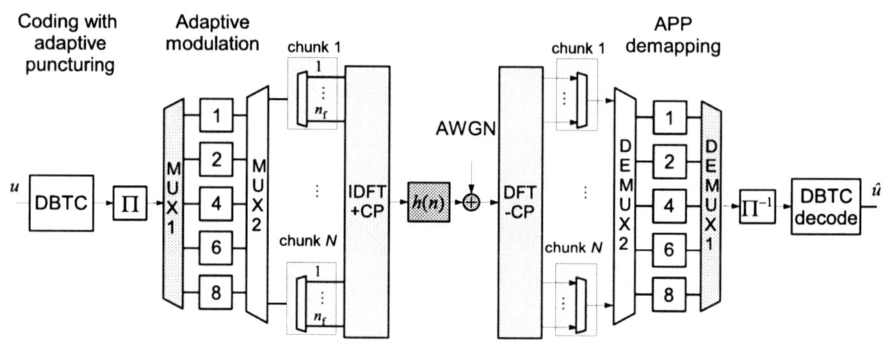

Figure 1. System model for link adaptation: after encoding and QAM modulation, the data of the considered user is mapped to the set of subchannels given by the scheduler.

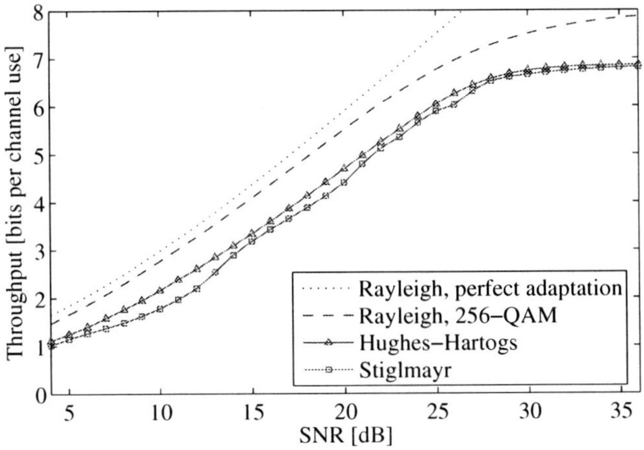

Figure 2. Achieved throughput for one user. $N = 21$ subchannels were allocated randomly to the user under consideration.

3. Simulation Results and Discussion

We consider the TDD mode of the 4G system concept according to the EU project WINNER [1, 15]. The OFDM system uses $N = 208$ subchannels, which are formed by $n_\mathrm{f} = 8$ subcarriers. In each subframe, $n_\mathrm{t} = 15$ OFDM symbols are transmitted. Thus, a chunk, which is the smallest unit for adaptive modulation, holds $n_\mathrm{f} \cdot n_\mathrm{t} = 120$ QAM symbols.

The simulations in this paper are based on a DBTC [15] with code rates $R_2 \in \{\frac{1}{3}, \frac{2}{5}, \frac{1}{2}, \frac{2}{3}, \frac{3}{4}, \frac{4}{5}, \frac{6}{7}\}$ and QAM modulation with $R_1 \in \{1, 2, 4, 6, 8\}$ bits per symbol. Very similar results are achieved with LDPC coding [15]. The results in Fig. 3 have been obtained with the 'B1 NLoS' model according to [16], a receiver noise figure of 5 dB and a cell radius of $r_\mathrm{c} = 100\,\mathrm{m}$. The user rates R_k are expressed in bits per channel use, which is the same as bps/Hz for an anti-aliasing filter with roll-off zero. The rate in bit/s are given by

$$R'_k = \frac{N \cdot n_\mathrm{f} \cdot n_\mathrm{t}}{T_\mathrm{frame}} R_k \tag{12}$$

where the duration of a frame is $T_\mathrm{frame} = 691.2\mu\mathrm{s}$.

From the obtained user rates, several important points can be observed: the achieved rates with frequency-adaptive transmission are significantly higher than for non frequency-adaptive transmission, except for the nearest user. For all three scheduling policies a similar gain is achieved by frequency-selective ACM. Note that while only the second scheduling policy considers the queue lengths, all strategies take the channel variations into account and thus exploit the multi-user diversity

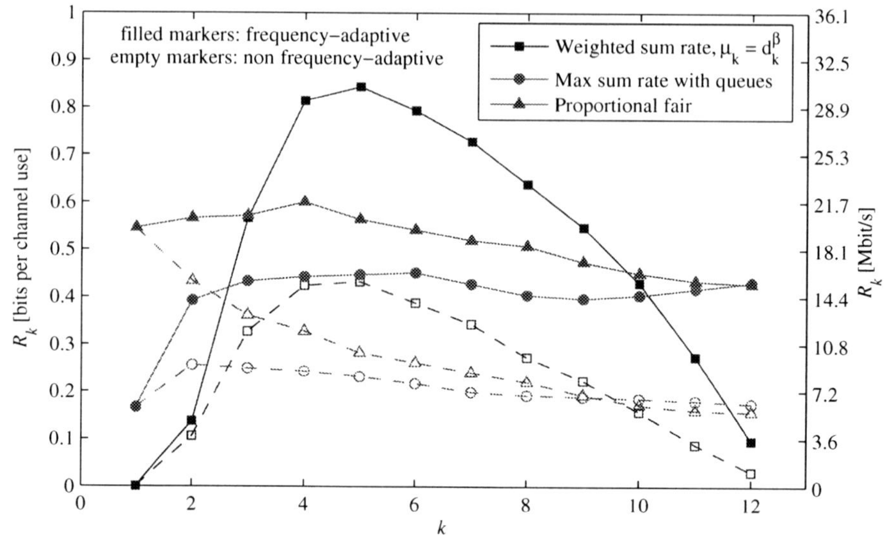

Figure 3. Rates per user for frequency-adaptive and non frequency-adaptive transmission and three scheduling policies.

given by frequency-selective fading. The results also indicate that one has to be cautious when comparing the two ACM modes for a single user. Especially when observing the achievable rates for the nearest user, the results are misleading.

Naturally, the most relevant comparison would be based on the actually chosen scheduler. However, since the scheduling strategy is not fixed at the time of deployment, comparisons have to be performed on the basis of reasonable assumptions about scheduling.

The most complex and complete policy, the queue-proportional scheduler, has been tested in a WLAN environment with $N_c = N = 48$ subchannels, the channel model 'A' according to [17] and path loss function $\alpha_k^{(\mathrm{dB})} = 31\lg(d) + 56$. The achieved user rates in Fig. 4 also show a significant gain from frequency-selective adaptation, even for the nearest user.

4. Conclusion

In this contribution, we evaluate the gain from adapting the modulation per subchannel as opposed to choosing the same modulation scheme for all subchannels in a multi-user environment. The results obtained with four different scheduling policies clearly indicate that frequency-selective adaptation provides substantial gains. Although the scheduler already

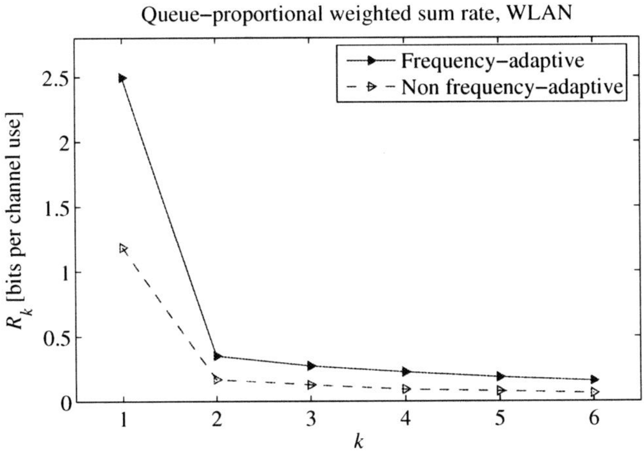

Figure 4. Rates per user for frequency-adaptive and non frequency-adaptive transmission. Queue proportional weighted sum rate scheduling for a WLAN environment is considered

exploits part of the multi-user diversity, which is due to the frequency-selective fading, it is still advantageous to employ frequency-selective link adaptation. Only in some special cases, both link adaptation modes show similar performance. These cases can be recognized by the scheduler, which is located in the base station, and thus unnecessary signalling overhead can be avoided. From the simulation results, it can also be observed that considering only a single link might be misleading since the performance gap between frequency-selective and non frequency-selective adaptation is highly varying for different users.

Acknowledgments

This work has been performed in the framework of the IST project IST-4-027756 WINNER II, which is partly funded by the European Union. The authors would like to acknowledge the contributions of their colleagues. The views expressed in this paper do not necessarily represent those of all partner organizations. This work has also been partially funded by the Catalan Government under grant SGR2005-00690 and the Spanish Ministry of Industry, Tourism and Trade under grant FIT-330220-2005-108.

References

[1] *EU project WINNER II – Wireless World Initiative New Radio, IST-4-027756.* https://www.ist-winner.org/.

[2] Nilo Casimiro Ericsson. *Revenue Maximization in Resource Allocation: Applications in Wireless Communication Networks.* PhD thesis, Uppsala University, Sept. 2004.

[3] Mari Kobayashi and Giuseppe Caire. Iterative waterfilling for weighted rate sum maximization in MIMO-MAC. In *IEEE Workshop on Signal Processing Advances in Wireless Communications (SPAWC)*, Cannes, France, July 2006.

[4] David N. C. Tse and Stephen V. Hanly. Multiaccess fading channels–part I: Polymatroid structure, optimal resource allocation and throughput capacities. *IEEE Trans. Inform. Theory*, 44(7):2796–2815, Nov. 1998.

[5] Edmund M. Yeh and Aaron S. Cohen. Information theory, queueing, and resource allocation in multi-user fading communications. In *Conference on Information Science and Systems (CISS)*, Princeton, USA, March 2004.

[6] Holger Boche and Marcin Wiczanowski. Stability-optimal transmission policy for the multiple antenna multiple access channel in the geometric view. *Signal Processing*, 86(8):1815–1833, Aug. 2006.

[7] Mari Kobayashi and Giuseppe Caire. An iterative water-filling algorithm for maximum weighted sum-rate of gaussian MIMO-BC. *IEEE J. Selected Areas Commun.*, 24(8):1640–1646, Aug. 2006.

[8] Andrea Goldsmith. *Wireless Communications.* Cambridge University Press, New York, 2005.

[9] Wei Yu, Wonjong Rhee, Stephen Boyd, and John M. Cioffi. Iterative water-filling for Gaussian vector multiple access channels. *IEEE Trans. Inform. Theory*, 50(1):145–152, Jan. 2004.

[10] Stephan Pfletschinger. From cell capacity to subcarrier allocation in multi-user OFDM. In *IST Mobile & Wireless Communications Summit*, Dresden, Germany, June 2005.

[11] Stephan Pfletschinger. Multicarrier BICM with adaptive bit-loading and iterative decoding. In *International OFDM-Workshop*, Hamburg, Germany, Aug. 2005.

[12] Stephan Pfletschinger, Adam Piątyszek, and Stephan Stiglmayr. Frequency-selective link adaptation using duo-binary turbo codes in OFDM systems. In *IST Mobile & Wireless Communications Summit*, Budapest, Hungary, July 2007. submitted.

[13] Stephan Stiglmayr, Martin Bossert, and Elena Costa. Adaptive coding and modulation in OFDM systems using BICM and rate-compatible punctured codes. In *European Wireless*, Paris, France, April 2007.

[14] Dirk Hughes-Hartogs. Ensemble modem structure for imperfect transmission media, July 1987.

[15] Johan Axnäs et al. D2.10: Final report an identified RI key technologies, system concept, and their assessment. Technical report, IST-2003-50707581 WINNER, Dec. 2005.

[16] Daniel S. Baum et al. D5.4: Final report on link level and system level channel models. Technical report, IST-2003-50707581 WINNER, Dec. 2005.

[17] ETSI Normalization Committee, Sophia-Antipolis, France. *Channel models for HIPERLAN/2 in different indoor scenarios*, 1998.

ADAPTIVE MULTI-CARRIER SPREAD-SPECTRUM WITH DYNAMIC TIME-FREQUENCY CODES FOR UWB APPLICATIONS

Antoine Stephan, Jean-Yves Baudais and Jean-François Hélard
Institute of Electronics and Telecommunications of Rennes (IETR) – UMR CNRS 6164
INSA IETR, 20 Avenue des Buttes de Coesmes, 35043 Rennes, France
antoine.stephan@ens.insa-rennes.fr, {jean-yves.baudais,jean-francois.helard}@insa-rennes.fr

Abstract: In this paper, we propose a spread spectrum multi-carrier multiple-access (SS-MC-MA) waveform for high data rate UWB applications, taking into consideration the European UWB context. This new UWB scheme respects the parameters of the multiband orthogonal frequency division multiplexing (MB-OFDM) technique which is one of the candidates for wireless personal area networks (WPAN) standardization. We optimize the spreading code length and the number of codes in our proposed scheme in order to maximize the system range for a given target throughput. Furthermore, we dynamically distribute the time-frequency codes that provide frequency hopping between users in order to improve our system range. We show that our adaptive system transmits information at much higher attenuation levels and with larger throughput than the ones of the MB-OFDM proposal. Hence, we conclude that our proposed system can be advantageously exploited for UWB applications.

Key words: Multi-carrier spread-spectrum, resource allocation, UWB.

1. Introduction

Ultra-wideband (UWB) has emerged as an exciting technology for short range, high data rate wireless communications since 2002 when the Federal Communications Commission (FCC) agreed on the allocation of a 3.1–10.6 GHz spectrum for unlicensed use of UWB devices [1]. The FCC imposed a power spectral density (PSD) limit of -41.25 dBm/MHz in order to reduce interference with existing spectrum allocations.

S. Plass et al. (eds.), Multi-Carrier Spread Spectrum 2007, 197–206.

One of the main multiple-access techniques considered by the IEEE 802.15.3a standardization group is a multiband orthogonal frequency division multiplexing (MB-OFDM) supported by the Multiband OFDM Alliance (MBOA) [2], [3]. In this paper, we propose a new UWB waveform based on a spread spectrum multi-carrier multiple-access (SS-MC-MA) scheme while respecting the OFDM parameters of the MBOA solution and taking into consideration the European UWB context. Then, we optimize the spreading code length and the number of codes of the proposed scheme in order to maximize the system range. These optimizations do not take into account the channel coding scheme. Furthermore, we propose an allocation algorithm that maximizes the system range by dynamically distributing the time-frequency codes that provide frequency hopping between users.

This paper is organized as follows. Section 2 presents briefly the multiband OFDM solution, followed by a description of our proposed SS-MC-MA scheme. Section 3 studies the SS-MC-MA system optimization for range maximization. Section 4 describes the dynamic TFC algorithm applied to the SS-MC-MA scheme. Simulation results showing the interest of the proposed adaptive scheme in UWB applications are given in Section 5, followed by the conclusion in Section 6.

2. System model

2.1 Multiband OFDM in the European context

The MBOA solution is based on the combination of an OFDM modulation with a multibanding approach, which divides the 7.5 GHz UWB spectrum into 14 sub-bands of 528 MHz each. The OFDM scheme consists of 128 subcarriers, out of which 100 are assigned to data tones. The modulation used is a quadrature phase-shift keying (QPSK), which leads to the transmission of 200 bits per OFDM symbol. The MBOA solution offers potential advantages for UWB applications, e.g., the signal robustness against channel selectivity and the efficient exploitation of the energy of every signal received within the prefix margin.

Initially, most of the studies have been performed on the first 3 sub-bands (3.1–4.8 GHz). The FCC PSD limit of -41.3 dBm/MHz was imposed on the whole 14 sub-bands, whereas with the European Electronic Communications Committee (ECC) regulations of March 2006 much lower PSD limits were imposed on the UWB spectrum, except on the 6–8.5 GHz range where a similar -41.3 dBm/MHz limit was considered [4]. Hence, we perform our studies on sub-bands 7, 8 and 9 (6.33–7.92 GHz) of the MBOA solution.

2.2 Proposed SS-MC-MA scheme

To improve the system performance, we propose a SS-MC-MA scheme which consists in assigning to each user a specific block of subcarriers [5]. This scheme is applied to UWB while respecting the OFDM parameters of the MBOA solution [6]. The spreading is in the frequency domain in order to improve the signal robustness against the frequency selectivity of the UWB channel and against narrowband interference. If we consider 3 users transmitting simultaneously, at a given time each user is allocated a group of 100 subcarriers equivalent to one of the 3 sub-bands (7, 8 and 9) of 528 MHz bandwidth. Each sub-band can be divided into several blocks, each of them including a number of subcarriers equal to the spreading code length. In addition, the only modulation used is the QPSK as with the MBOA solution.

2.3 Channel model

The channel model used for our study is the one adopted by the IEEE 802.15.3a committee for the evaluation of UWB physical layer proposals [7]. It is a modified version of the Saleh-Valenzuela model for indoor channels, fitting the properties of measured UWB channels. A lognormal distribution is used for the multipath gain magnitude. In addition, independent fading is assumed for each cluster and each ray within the cluster. Moreover, four different channel models (CM1 to CM4) are defined for the UWB system modeling, each with arrival rates and decay factors chosen to match different usage scenarios and to fit line-of-sight (LOS) and non-line-of-sight (NLOS) cases.

3. SS-MC-MA system optimization

The SS-MC-MA system optimization is divided into 2 steps. First, we will find the optimal number of blocks that maximizes the system margin for a target throughput, and then we will optimize the code length and the number of codes in order to maximize the system range with lower variable throughput.

The throughput of an OFDM system in bits per symbol is derived from Shannon theorem by

$$R_{OFDM} = \sum_{k \in S} \log_2 \left(1 + \frac{1}{\Gamma} |h_k|^2 \frac{E_k}{N_0} \right), \tag{1}$$

where S is the group of used subcarriers, Γ the signal-to-noise ratio gap (normalized SNR), h_k and E_k the frequency-domain response and the transmitted power density of the kth subcarrier respectively, and N_0 the noise density. The total throughput in bits per symbol of a SS-MC-MA system using a zero-forcing detection is given by [8]

$$R_{SS-MC-MA} = \sum_{b=1}^{B} R_b = \sum_{b=1}^{B}\sum_{c=1}^{C_b} R_{c,b} = \sum_{b=1}^{B}\sum_{c=1}^{C_b} \log_2\left(1+\frac{1}{\Gamma}\frac{L_b^2}{\sum_{i=1}^{L_b}\frac{1}{|h_{i,b}|^2}}\frac{E_{c,b}}{N_0}\right), \quad (2)$$

where B is the number of blocks, C_b the number of codes used for block b, L_b the code length of block b, $E_{c,b}$ the power allocated to code c of block b, with the constraint

$$\sum_{c=1}^{C_b} E_{c,b} \le E, \quad \forall b. \quad (3)$$

Using the Lagrange multipliers in (2) and (3), we find that the optimal solution which maximizes the system throughput would be to have $E_{c,b} = E/C_b$ and $C_b = L_b, \forall b$. Moreover, in the UWB context, the modulations used are limited to a QPSK. Thus, the UWB throughput in bits per symbol becomes

$$R_{UWB} = 2\sum_{b=1}^{B} L_b \le \sum_{b=1}^{B} L_b \log_2\left(1+\frac{1}{\Gamma}\frac{L_b}{\sum_{i=1}^{L_b}\frac{1}{|h_{i,b}|^2}}\frac{E}{N_0}\right). \quad (4)$$

Property 1: The throughput R_{UWB} is only reachable if the expected value

$$\mathrm{E}\left[\frac{1}{|\hat{h}_{i,b}|^2}\right] \le \frac{1}{3}, \quad (5)$$

with $|\hat{h}_{i,b}|^2 = |h_{i,b}|^2\, E/\Gamma N_0$.

Proof: From (4) we can write

$$\sum_{b=1}^{B}\sum_{i=1}^{L_b}\frac{1}{|\hat{h}_{i,b}|^2} \le \sum_{b=1}^{B}\frac{L_b}{3}. \quad (6)$$

If n is the total number of used subcarriers, then $\sum_b L_b = n$ and (6) becomes

$$\frac{1}{n}\sum_{b=1}^{B}\sum_{i=1}^{L_b}\frac{1}{|\hat{h}_{i,b}|^2}\leq\frac{1}{3}. \tag{7}$$

■

3.1 Optimization of the number of blocks

In this study, we want to find the optimal number of blocks B that maximizes the SS-MC-MA system margin with a fixed modulation and a fixed target throughput (200 bit/symbol). Let γ_b be this margin per block. From (4), we can write

$$R_{UWB} = 2n = 2\sum_{b=1}^{B} L_b = \sum_{b=1}^{B} L_b \log_2\left(1+\frac{1}{\gamma_b}\frac{L_b}{\sum_{i=1}^{L_b}\left(1/|\hat{h}_{i,b}|^2\right)}\right) \tag{8}$$

and

$$\gamma_b = \frac{1}{3}\frac{L_b}{\sum_{i=1}^{L_b}\left(1/|\hat{h}_{i,b}|^2\right)}. \tag{9}$$

Theorem 1: To maximize the noise margin of the SS-MC-MA system, and consequently the system range, a code length equal to the number of useful subcarriers should be used, i.e. one single SS-MC-MA block should be used.

Proof: We want to maximize the minimum value of γ_b.

Let $$\sum_{i=1}^{L_b}\frac{1}{|\hat{h}_{i,b}|^2} = \frac{L_b}{n}\sum_{b=1}^{B}\sum_{i=1}^{L_b}\frac{1}{|\hat{h}_{i,b}|^2}+\alpha_b = \frac{L_b}{n}\alpha+\alpha_b,$$

then $$\gamma_b = \frac{1}{3}\frac{L_b}{\frac{L_b}{n}\alpha+\alpha_b}.$$

We have $$\alpha = \sum_{b=1}^{B}\sum_{i=1}^{L_b}\frac{1}{|\hat{h}_{i,b}|^2} = \sum_{b=1}^{B}\left(\frac{L_b}{n}\alpha+\alpha_b\right) = \alpha+\sum_{b=1}^{B}\alpha_b.$$

Thus, we find that $$\sum_{b=1}^{B}\alpha_b = 0.$$

Let γ be the noise margin of the SS-MC-MA system with one block, and let b' be such that $\gamma_{b'} > \gamma$, then $\alpha_{b'} < 0$. Hence, $\exists b''$ such that $\alpha_{b''} > 0$, i.e. $\gamma_{b''} < \gamma$, and $\min_b \gamma_b < \gamma$.

Thus, $L = n$ maximizes the margin, i.e. the optimal choice for a given throughput is to use a spreading code length equal to the total number of useful subcarriers which is equal to 100 in the UWB context. ■

Consequently, it is not necessary to know the channel coefficients at the transmitter side to distribute the subcarriers between the blocks, since all these subcarriers are used within the same single block. Furthermore, this theorem shows that the SS-MC-MA noise margin can never be lower than the OFDM noise margin. This is due to the energy gathering capability of SS-MC-MA which can exploit, contrarily to OFDM, the residual energy conveyed by each subcarrier. The SS-MC-MA system range is therefore larger than the OFDM system range.

3.2 Range improvement with variable throughput

Now, we optimize the code length L and the number of codes N in order to maximize the system range when the UWB throughput of 200 bit/symbol is not reachable at high attenuation levels.

In a general approach with variable throughput, the number of codes N can be lower than the code length L and *Theorem 1* is not applicable anymore. In this case, a multiple blocks configuration has to be considered and each block can exploits its own code length L_b. But finding the optimal block sizes amounts to resolving a complex combinational optimization problem that can not be reduced to an equivalent convex problem. Then, no analytical solution exists and optimal solution can only be obtained following exhaustive search [8]. In order to avoid prohibitive computations, we assume a single block configuration system.

Maximizing the system range is equivalent to maximizing the system throughput. The optimal non-integer throughput for the single block system is given by

$$R = L \log_2 \left(1 + \frac{L}{\sum_{i=1}^{L} \left(1 / |\hat{h}_i|^2 \right)} \right). \tag{10}$$

Theorem 2: With a QPSK modulation and a PSD constraint of $\sum_{c=1}^{L} E_c \leq E$, the optimal number of codes that can be used for a given spreading code length L is $N = \lfloor L(2^{R/L} - 1)/3 \rfloor$, with R given by (10).

Proof: With the optimal number of codes N, the PSD constraint should be respected, whereas with $N+1$ codes, it shouldn't. Hence, N should satisfy the following 2 conditions

$$\begin{cases} E-\sum_{c=1}^{N} E_c = \frac{L}{\alpha}\left(2^{R/L}-1\right)-\frac{N}{\alpha}\left(2^2-1\right) \geq 0, \\ E-\sum_{c=1}^{N+1} E_c = \frac{L}{\alpha}\left(2^{R/L}-1\right)-\frac{N+1}{\alpha}\left(2^2-1\right) < 0, \end{cases} \tag{11}$$

with
$$\alpha = L^2 / \sum_{i=1}^{L} \frac{N_0 \Gamma}{|h_i|^2}.$$

From (11), $N \leq L\left(2^{R/L}-1\right)/3$ and $N > L\left(2^{R/L}-1\right)/3-1$. Hence, $N = \left\lfloor L\left(2^{R/L}-1\right)/3 \right\rfloor$. ■

From (10) and *Theorem 2*, since the number of codes cannot be larger than the code length, the maximum reachable throughput for a given L becomes

$$R(L) = 2 \times \min\left\{\left\lfloor \frac{L}{3}\left(2^{R/L}-1\right)\right\rfloor, L\right\} = 2 \times \min\left\{\left\lfloor \frac{L^2}{3\sum_{i=1}^{L}\left(1/|\hat{h}_i|^2\right)}\right\rfloor, L\right\}. \tag{12}$$

Finally, the maximum reachable throughput with the optimal code length becomes

$$R_{\max} = \max_{1 \leq L \leq N} \left\{R(L)\right\}. \tag{13}$$

4. Time-frequency codes exploitation

In the MBOA solution, unique logical channels corresponding to different piconets are defined by using different time-frequency codes (TFC) for each sub-band group. These codes provide frequency hopping from a sub-band to another at the end of each OFDM symbol. The configuration proposed by the MBOA solution consists in choosing the TFC regularly without taking into consideration the channel response state of each user for each sub-band. In our study, we use a dynamic TFC (DTFC) distribution, different from the distribution of the MBOA solution, in order to maximize the SS-MC-MA system range.

We consider 3 users, and consequently 3 TFC codes, distributed on sub-bands 7, 8 and 9 of the MBOA solution. Over a given period T equivalent to the duration of an OFDM symbol, each user occupies one of the 3 available sub-bands. To find the number of unique possible distributions of the TFC over 3 successive periods T, we consider a combinational problem: over one period, we have a permutation with order and without repetition of 3 users on 3 sub-bands. Moreover, within each sub-band, the users' order in time is not taken into consideration. We find that the number of unique permutations is equal to 55. In addition, we have a total of 9 different channel responses in the system, since each user has 3 different channel responses corresponding to 3 sub-bands.

5. Simulation results

In this section, we present the simulations performed on sub-bands 7, 8 and 9 of the MBOA solution using the proposed SS-MC-MA scheme and taking into consideration the European context.

Fig. 1 represents the total throughput per OFDM symbol of a single user over CM1 channel model for different channel attenuation levels, before applying the dynamic TFC algorithm. With the MBOA solution, the total throughput of 200 bit/symbol is not reachable at attenuation levels higher than 38 dB, whereas with the proposed SS-MC-MA scheme using a single block of length $L = 100$, we are able to transmit 200 bit/symbol at an attenuation level of 53 dB (15 dB larger range). Moreover, when we optimize the code length L and the number of codes N of the SS-MC-MA system, we are able to transmit data at much higher attenuations (81 dB), and the reachable range and throughput with adaptive SS-MC-MA are always larger than the ones of an adaptive OFDM system. The number of QPSK modulated subcarriers can vary from 100 to 0 with the called adaptive OFDM scheme, whereas with the MBOA solution the number of active subcarriers is always equal to 100. The optimal values of L and N that maximize the range of the adaptive SS-MC-MA system for different attenuation levels are given in Fig. 2. We can notice that at high attenuations, N is lower than L.

Fig. 3 and Fig. 4 represent the total system throughput per 3 OFDM symbols for 3 users over CM1 channel model. In Fig. 3, with the MBOA system using the TFC defined by the solution, the 3 users are able to transmit ($3 \times 3 \times 200 = 1800$ bit/3×symbol) at attenuation levels lower than 41 dB. At attenuation levels from 41 to 43 dB, only 2 users are able to transmit (1200 bit/3×symbol), at levels from 44 to 49 dB only 1 user is able to transmit, and at levels higher than 49 dB no one is able to transmit. When applying the

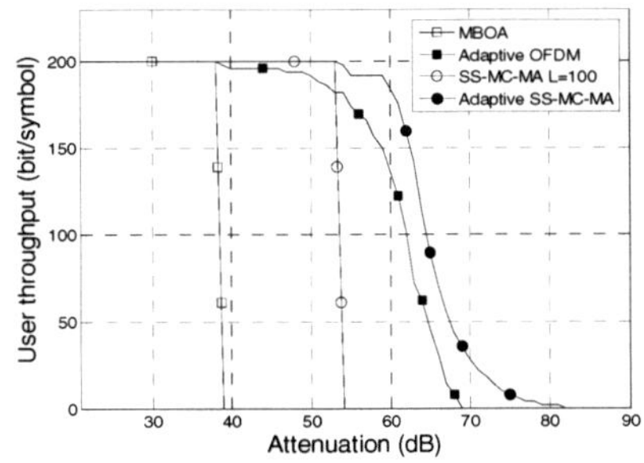

Figure 1. Total throughput of a single user system.

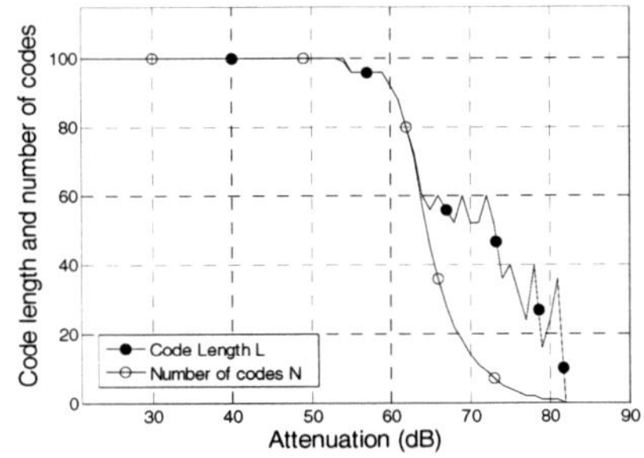

Figure 2. Optimal adaptive SS-MC-MA configuration.

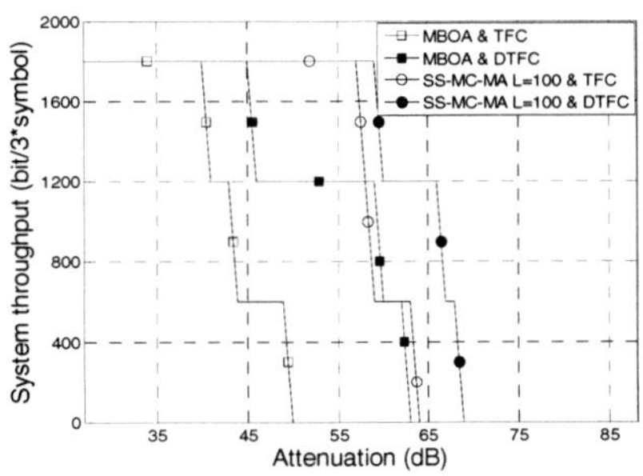

Figure 3. Total throughput of a 3-users system without adaptive schemes.

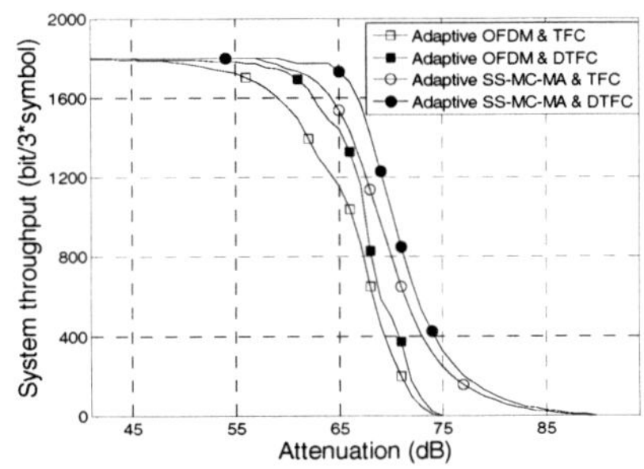

Figure 4. Total throughput of a 3-users system with adaptive schemes.

dynamic TFC (DTFC) algorithm to the MBOA system, the system range increases more than 5 dB, and when applying it to the SS-MC-MA scheme, the system range becomes around 20 dB larger than the MBOA system range. In Fig. 4, the same TFC algorithm is applied to the adaptive OFDM system and to the optimized SS-MC-MA system with variable code length L and variable number of codes N. We notice that the adaptive SS-MC-MA system is able to reach an 89 dB attenuation level. Moreover, with the DTFC applied to the SS-MC-MA system, the reachable range becomes larger, and for a given attenuation level, the throughput is improved. Similar results are obtained with CM2, CM3 and CM4 channel models.

6. Conclusion

In this paper, we proposed a SS-MC-MA waveform which is new for high data rate UWB applications and which respects the OFDM parameters of the MBOA solution, taking into account to the European UWB context. Then, we optimized the spreading code length and the number of codes in order to maximize the system range. Furthermore, we proposed an allocation algorithm which maximizes the system range by distributing the time-frequency codes dynamically, contrarily to the MBOA solution where the time-frequency codes are distributed regularly without taking into consideration the channel response of each user for each sub-band. We showed that the SS-MC-MA system is able to transmit information at attenuation levels much higher than the attenuation limits of the OFDM solution. These optimizations did not take into account the channel coding scheme. However, the performance comparison of the final systems with channel coding shows that our proposed adaptive SS-MC-MA scheme and proposed TFC algorithm can be advantageously exploited for high data rate UWB applications. These improvements can be obtained without changing the radio-frequency front-end compared to the MBOA solution.

Acknowledgment

The authors would like to thank France Télécom R&D/RESA/BWA which supports this study within the contract 46136582.

References

[1] "First report and order, revision of part 15 of the commission's rules regarding ultra-wideband transmission systems," FCC, ET Docket 98-153, Feb. 14, 2002.

[2] IEEE P802.15 Working Group for Wireless Personal Area Networks (WPANs), "Multi-band OFDM physical layer proposal for IEEE 802.15 task group 3a," Sept. 2004.

[3] A. Batra *et al.*, "Design of multiband OFDM system for realistic UWB channel environments," *IEEE Trans. on Microwave Theory and Techniques*, vol. 52, no. 9, Sept. 2004.

[4] "ECC Decision of 24 March 2006 on the harmonised conditions for devices using Ultra-Wideband (UWB) technology in bands below 10.6 GHz," ECC/DEC(06)04, March 2006.

[5] S. Kaiser and K. Fazel, "A flexible spread-spectrum multi-carrier multiple-access system for multi-media applications," in *Proc. IEEE Intern. Symposium on Personal, Indoor and Mobile Radio Commun. (PIMRC'97)*, pp. 100-104, Helsinki, Finland, Sept. 1997.

[6] A. Stephan, J-Y. Baudais and J-F. Hélard, "Resource allocation for multicarrier CDMA systems in ultra-wideband communications," in *Proc. IEEE Intern. Telecommunications Symposium (ITS2006)*, pp. 1002-1007, Fortaleza, Brazil, Sept. 2006.

[7] J. Foerster *et al.*, "Channel modeling sub-committee report final," IEEE802.15-02/490, Nov. 2003.

[8] M. Crussière, J-Y. Baudais and J-F. Hélard, "Adaptive linear precoded DMT as an efficient resource allocation scheme for power line communications," in *Proc. IEEE Global Commun. Conference*, San Francisco, California, USA, Dec. 2006.

RADIO RESOURCE ALLOCATION IN MC-CDMA UNDER QOS REQUIREMENTS

Ismael Gutierrez, Faouzi Bader and Joan L. Pijoan

Ramon Llull University, La Salle School of Engineering. Ps. Bonanova 8. 08022-Barcelona, Spain. E-mail: {igutierrez, joanp}@salle.url.edu

Centre Tecnològic de Telecomunicacions de Catalunya-CTTC, Parc Mediterrani de la Tecnología, Av.Canal Olimpia. 08860-,Barcelona, Spain.e-mail: faouzi.bader@cttc.es

Abstract: Most of the traditional channel-aware schedulers produce very unfair situations, whereas schedulers based on input buffers state and quality of service requirements lead to low spectrally efficient systems. In this paper a novel radio resource allocation algorithm based on combination of the CSI information as well as the input buffers is presented and applied to a multicarrier system[1]. In fact, the Group Orthogonal MC-CDMA and the Spread Spectrum Multicarrier Multiple Access schemes are analyzed and performances compared when applying the proposed algorithm in a typical low-mobility environment.

Key words: SS-MC-MA, GO-MC-CDMA, Radio Resource Allocation, QoS, scheduling

1. INTRODUCTION

Most of the proposed schemes for future communication systems make use of multicarrier transmission due to its high spectral efficiency and its ability to combat multi-path fading effects. During this last decade several combinations based on the Orthogonal Frequency Division Multiplexing (OFDM), and Code Division Multiple Access (CDMA) have been deeply

[1] This work has been partially carried out in the framework of the Celtic project WISQUAS (CP-2-035), and in the national Spanish code project: FIT-330220- 2005-108.

S. Plass et al. (eds.), Multi-Carrier Spread Spectrum 2007, 207–216.

studied[1]. In particular, the Multicarrier (MC) – CDMA architecture which has been proposed in European research projects such as IST-MATRICE[2] or IST-4MORE[3] for the downlink. Also, a modification of the MC-CDMA named Variable Spreading Factor Orthogonal Frequency Code Division Multiplexing (VSF-OFCDM) is becoming of great interest with the MC-CDMA and considered a potential candidate for the 4G downlink commnications [1].

The combination of the sub-carrier allocation and the power loading in OFDM systems by the water-filling algorithm aims to achieve a higher spectral efficiency. In the multiple access case, while the power is assigned following the water-filling principle, the sub-carriers are assigned to the user with best channel state information (CSI) leading to the Orthogonal Frequency Division Multiple Access (OFDMA) scheme [2]. However, most of the adaptive schemes take advantage from the multiuser diversity at the expense of the fairness [3]. By contrast, in multimedia communications, parameters such as the delay or the guarantee of a certain bit rate must be considered, hence the above mentioned schemes must be adapted since different service classes must be guaranteed. M. A. Enright et al. have used in their research work in [4] fixed scaling factors applied to the different service classes in order to allow a modified version of the water-filling algorithm to prioritize the different service classes. Moreover, a cross-layer strategy implemented by X. Guoxing et al. in [5] has demonstrated that the information from the input buffers is very useful to determine the actual transmission rate and to predict the transmission delays.

In this paper, a novel low-complexity adaptive sub-carrier allocation with power and bit allocation algorithm is developed for a MC-CDMA system for the forward transmission case, where mixed QoS requirements are guaranteed for each active user. The proposed scheme makes use of a cross-layer strategy as in [5] where two processes are in charge of guaranteeing the Quality of Service (QoS) and optimizing the spectral efficiency respectively. The first process is applied at the Medium Access Control (MAC) layer and is in charge of assigning priorities to the active users based on their QoS requirements. The second process assigns the active users into their best corresponding sub-carriers according to their respective service application priorities and their CSI. During this second process a modified version of the water-filling algorithm is also introduced which deals with the specific use of the CDMA component by applying the sub-carrier equivalent concept [6]. To the best of the authors' knowledge, this is the first work for MC-CDMA

[2] Multicarrier CDMA Transmission Techniques for Integrated Broadband Cellular Systems. European project IST 2001-32620

[3] European project 4MORE IST-**2002 507039**

systems that combines the adaptive power assignation, bit loading, and sub-carrier allocation while different classes of services are offeredSystem

2. SYSTEM MODEL

In this paper, the system model depicted in Fig. 1 is considered, where the Base Station (BS) transmitter and the receiver blocks of the Mobile Terminal (MT) diagrams are represented. Following this scheme the bandwidth (formed by N_c sub-carriers) is divided into Q groups (or sub-bands) of L sub-carriers each, where K symbols (up to L in the full-loading case) are transmitted with a spreading code factor of length L. Having $L \leq 32$, the receiver is simplified and the Multiple Access Interference (MAI) effect is decreased [1]. However applying a sufficiently large sub-carrier scramblers the frequency diversity can be equally achieved.

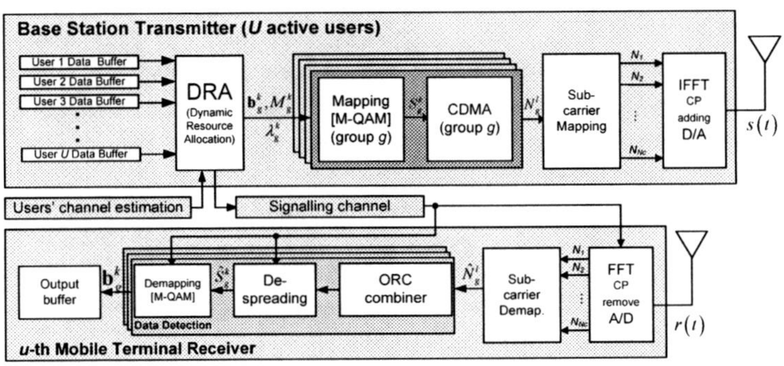

Figure 1. Schematic of the proposed downlink MC-CDMA system.

In order to increase the system capacity, a Dynamic Resource Allocation (*DRA*) block is introduced at the BS. This entity multiplexes the users′ data (available at the input buffers) over the different groups according to two principles; i) maximizing the spectral efficiency , and ii_) accomplishing the different QoS requirements. Since the performance of this block will be described in Section 3, at this point we will assume that somehow a group of K symbols are assigned to each group g. Furthermore, the *DRA* determines the power of each k-th symbol $(\lambda_g^{(k)})$ and the number of bits that are loaded in. In the case where the transmitted symbols over each group belong to different users the resulting transmission scheme is a "Q" MC-CDMA transmission structure [1][6], (named Group Orthogonal -MC-CDMA). On the other hand, whether the transmitted symbols over each g group belong to the same user, the transmission scheme has an SS-MC-MA structure [1]. Anyway, in both cases the data and the users can be multiplexed in the frequency domain (OFDMA) while the Code Division Multiplexing (CDM) brings a frequency diversity, minimize the errors produced by imperfect channel estimation or the inter/intra-cell interference.

Therefore, as a result of the DRA decisions, during each transmitted OFDM symbol $j \in \aleph$, each group g will be assigned a set of K complex symbols where the binary data from the selected user(s) are mapped according to a predefined constellation . Then, the symbol stream assigned to the group g, $S_g^{(k)}$ with $g=\{1,...,Q\}$ and $k=\{1,...,K\}$ is multiplied in the frequency domain by their respective spreading codes, and later all the symbols in each group are added synchronously yielding

$$s(t) = \sqrt{\varepsilon_c} \sum_{j \in \aleph} \sum_{g=1}^{Q} \sum_{k=1}^{L} \sqrt{\lambda_g^{(k)}} S_{j.g}^{(k)} \sum_{l=1}^{L} c_l^{(k)} \cdot cos\left(\omega_{g.l} t\right) \cdot \Pi\left(\frac{t - jT_S}{T_S}\right) \tag{1}$$

$$\text{with } \Pi\left(\frac{t-T_1}{T_2}\right) = \begin{cases} 1, \text{for } T_1 < t < T_1 + T_2 \\ 0, \text{otherwise} \end{cases},$$

where T_S is the OFDM symbol duration , ε_c means the energy of the chips of the spreading code $\boldsymbol{c}$.

At the u-th MT, assuming the channel is flat and slow time variant over T_S, the channel impulse response is assumed shorter than the cyclic prefix (CP) duration. Then the received signal can be expressed as follows

$$r^{(u)}(t) = \left(\sqrt{\varepsilon_c} \sum_{j \in \aleph} \sum_{g=1}^{Q} \sum_{u=1}^{L} \sqrt{\lambda_g^{(k)}} S_{j.g}^{(k)} \sum_{l=1}^{L} \alpha_{j.g.l}^{(u)} c_l^{(k)} \cdot cos\left(\omega_{g.l} t + \varphi_{j.g.l}^{(u)}\right) \cdot p\left(t - jT_S\right) \right) + \eta(t) \tag{2}$$

where $H_{j,g,l}{}^{(u)} = \alpha_{j,g,l}{}^{(u)} \cdot e^{j\varphi_{j,g,l}(u)}$ is the channel transfer function on each sub-carrier, and η the additive white Gaussian noise (AWGN). For brevity but without loss of generality, the the OFDM symbol index, j can be omitted. Then the received signal $r^{(u)}(t)$ during one OFDM symbol is demodulated using the FFT block, and the channel is equalized according to the Orthogonality Restoring principle (also named Zero Forcing). Thus at the output of the combiner the received symbol over the k-th spreading code on the g-th group of sub-carriers (assigned by the *DRA* to the u-th user) is expressed as

$$\hat{S}_g^{(k)} = \underbrace{\frac{S_g^{(k)}}{\sqrt{\varepsilon_c} \cdot L} \sum_{l=1}^{L} \left| c_{g.l}^{(k)} \right|^2}_{\text{desired symbol}} + \underbrace{\frac{1}{\sqrt{\varepsilon_c \lambda_g^{(k)}} \cdot L} \sum_{\substack{i=1 \\ i \neq k}}^{L} \sqrt{\lambda_g^{(i)}} S_g^{(i)} \sum_{l=1}^{L} c_l^{(i)} c_l^{(k)*}}_{\text{MAI}}, \tag{3}$$

$$+ \underbrace{\frac{1}{\sqrt{\varepsilon_c \lambda_g^{(k)}} \cdot L} \sum_{l=1}^{L} \left(\frac{c_l^{(k)*} N_{g.l}}{H_{g.l}^{(u)}} \right)}_{\text{noise}}$$

$(\cdot)^*$ means the complex conjugate , $N_{g,l}$ is the noise contribution on the l-th sub-carrier of the g-th group. Assuming a Walsh-Hadamard spreading codes with unitary power, and a perfect CSI estimation. Thereceived symbol $k$in the g-th group is,

$$\hat{S}_g^{(k)} = \underbrace{S_g^{(k)}}_{\text{desired symbol}} + \underbrace{\frac{1}{\sqrt{\varepsilon_c \lambda_g^{(k)} \cdot L}} \sum_{l=1}^{L} \left(\frac{c_l^{(k)*} r_{g,l}}{H_{g,l}^{(u)}} \right)}_{\text{noise}}. \tag{4}$$

Then the following signal to interference and noise ratio (SNIR) $\gamma_g^{(k)}$, can be obtained from [6]

$$\gamma_g^{(k)} = \frac{\lambda_g^{(k)} P_S}{\sigma_N^{\ 2}} \cdot \frac{L}{\sum_{l=1}^{L} \frac{1}{\left(\alpha_{g,l}^{(u)}\right)^2}}, \tag{5}$$

where $\lambda_g^{(k)} \cdot P_s$ and $\sigma_N^{\ 2}$ are the symbol power and the noise power respectively. Reordering (5) we obtain,

$$\gamma_g^{(k)} = \lambda_g^{(k)} \cdot \underline{SNR} \left| H_{eff,g}^{\ \ (u)} \right|^2 \tag{6}$$

the $\underline{SNR} = P_S / \sigma_N^{\ 2}$ means the average SNR and the term $|H_{eff,\ g}^{\ (u)}|^2$ is the effective channel gain [6] it represents the gain of the channel across all the assigned sub-carriers for the *u-th* user within the *g*-th group.

3. DYNAMIC RESOURCE ALLOCATION

The DRA block aims to maximize the system capacity while guaranteeing the users´ QoS requirements. A common approach is to use a complex cost functions or a set of QoS restrictions that balance the system capacity and the user satisfaction. Alternatively, using a cross-layering implementations this process can be split on two successive sub-processes/approaches. One is in charge of the user scheduling policy, and the second is responsible of the optimization of the system capacity. This paper deals with the second approach..

3.1 User scheduling

The Round Robin user scheduling algorithm gives access to the channel periodically according to the user requirements or on fair principles without considering the actual Channel State Information (CSI), therefore the system capacity is very far away from the limits of the channel capacity. On the other hand, the channel-aware algorithms that maximize the system capacity (i.e. the water-filling algorithm) produce very unfair situations where users with favorable CSI conditions (i.e. proximity to the BS) are assigned with most of the resources.

The algorithm described in Fig.2 is based on a combination of the both above mentioned concepts is proposed. In this algorithm, the users are assigned on base of a certain priority level over each band $PL_g^{(u)}=\{1,\ldots,N_{PL}\}$, according to two principles, the experienced *satisfaction level* and the *expected spectral efficiency*.

$$
\begin{array}{l}
\underline{\text{if}}\left(L^{(u)}>0\right)\\
\quad \Delta_{PL\_QoS}=sign\left(\tau^{(u)}-\tau_{\max}^{(u)}\right)\\
\quad PL^{(u)}{}_{QOS}=PL^{(u)}{}_{QOS}+\Delta_{PL\_QoS}\\
\quad PL_{g}^{(u)}{}_{CSI}=\log_2\left(1+\frac{P_T}{Q\cdot\sigma_N{}^2}\cdot\left|H_{eff\cdot g}{}^{(u)}\right|^2\right)\\
\quad PL_{g}^{(u)}=K_{QoS}\cdot PL^{(u)}{}_{QOS}+K_{CSI}\cdot PL_{g}^{(u)}{}_{CSI}\\
\quad PL_{g}^{(u)}=\max\left(\min\left(PL_{g}^{(u)},N_{PL}\right),1\right)\\
\quad \underline{\text{if}}\left(L_{class}\geq\left(L_{\max}-L^{(u)}\right)\right)\\
\quad\quad PL_{g}^{(u)}=N_{PL}\\
\quad \underline{\text{end}}\\
\underline{\text{else}}\\
\quad PL_{g}^{(u)}=0,\ \text{ for } g=1..Q\\
\underline{\text{endif}}
\end{array}
$$

Figure 2. Priority level assigning algorithm.

The satisfaction level, $PL_{QoS}{}^{(u)}$ is measured on base of the expected delay with regard to the maximum allowed delay , $\tau_{max}{}^{(u)}$ according to each specific service class. The term $\tau^{(u)}$ means the expected delay and is computed as $\tau^{(u)}= L^{(u)}/R^{(u)}$, where $R^{(u)}$ means the instantaneous transmission rate of the *u*-th user. The value $PL_{QoS}{}^{(u)}$ is measured differentially and the term Δ_{PL_QoS} is an indicator of the satisfaction level, whether is negative it means that the network is providing a QoS above the required and then lower channel resources can be assigned. On the other hand the expected spectral efficiency $PL_g{}^{(u)}{}_{CSI}$, is obtained considering the overall available power P_T. Both measurements; the QoS and the CSIs values form a unique priority level as a join linear combination function. The constants K_{QoS} and K_{CSI} can be used to move from an only QoS priority policy base to an only CSI base as long as the experienced satisfaction grade and the *expected spectral efficiency* have the same dynamic range. Finally, the overflow probability has been also considered forcing the priority level to its maximum value when an overflow occur. The term $L^{(u)}$ means the number of buffered bits of the user *u*, L_{max} means the buffers size in bits, and L_{class} is the maximum number of bits for each class can generate between two consecutive evaluation periods (i.e. one frame).

3.2 System capacity optimization

Once each user has been assigned based on its priority level $PL_g{}^{(u)}$ over each sub-band a second routine is used in order to increase the system capacity. Given the structure of the transmission scheme (GO-MC-CDMA or SS-MC-MA) the users with highest priority level are assigned to each sub-band where a total of L users can be assigned (full loading case) in case of a GO-MC-CDMA . Thus, in order to increase the system capacity and given that the users have been already assigned to sub-bands, only the transmission power of each symbol $\lambda_g^{(k)}$ and the modulation are adapted. The expression of the overall system capacity is expressed as,

$$C = \sum_{g=1}^{Q}\sum_{k=1}^{L}\log_2\left(1+\lambda_g^{(k)}\frac{P_G}{\sigma_N^2}\cdot\left|H_{eff\,g}^{(u)}\right|^2\right) \quad [bps/Hz] \tag{7}$$

where the maximization problem must be performed subject to the following restriction,

$$\sum_{g=1}^{Q}\sum_{k=1}^{L}\lambda_g^{(k)} = P_T \tag{8}$$

P_T is the total transmission power. All the symbols in each sub-band should be transmitted with equal power in order to reduce the MAI effect and thus

$$\lambda_g^{(k_1)} = \lambda_g^{(k_2)} \text{ for any } k_1, k_2. \tag{9}$$

The solution of this maximization problem leads to the *Water filling* algorithm [6]. However, when integer bit loading is considered some modifications must be performed to maximize the capacity, i.e. those proposed by Hughes-Hartogs or by Chow et. al. in [8]. Based on previous estimation in [6]the required power to transmit a symbol with integer bits b allocation for M-QAM modulation with Gray mapping is expressed as,

$$\lambda_g^{(k)}\cdot P_S = -\frac{\ln\left(\frac{BER}{0.2}\right)\cdot\left(2^b-1\right)}{1.6\cdot\left|H_{eff\,g}^{(k)}\right|^2}\cdot\sigma_N^2 \tag{10}$$

where **BER** means the Bit Error Rate required without codification (specified by the type of class service). Applying the iterative resolution in [8] the power is estimated given to the symbol that requires less power to increase a specific modulation level (from $b_g^{(k)}$ up to $b_{UP,g}^{(k)}$) until all the power is assigned. This step increment on the power level can be estimated as follows,

$$\Delta P_g^{(k)} = -\frac{\ln\left(BER/0.2\right)}{1.6\cdot\left|H_{eff.g}^{(k)}\right|}\cdot\left(2^{b_{UP,g}^{(k)}} - 2^{b_g^{(k)}}\right)\sigma_N^2. \tag{11}$$

4. SIMULATION RESULTS

The used simulation parameters are based on that depicted in Table I. Moreover, the simulation results are performed for indoor environment, where the BS is located at through an office having an omni directional antenna, and the mobile terminals are distributed uniformly over the office. The adaptation process is here applied within each frame. Note thatthe required protocol signalling is out of the scope of this issue.

Firstly, the resource allocation without QoS requirements is considered, $PL_{QoS}^{(u)}=0$. For this case, the spectral efficiency and the fairness are evaluated,both depicted in Fig. 2. The spectral efficiency is measured as the number of bits transmitted per second over the whole bandwidth. If $\mathbf{x}^{Nx1}$ represents the allocated resources for N users, the fairness of $\mathbf{x}$ is measured according to ,

$$F(\mathbf{x}) = \frac{\left(\sum_{i=1}^{N}\mathbf{x}[i]\right)^2}{N\sum_{i=1}^{N}\mathbf{x}[i]^2}. \quad (12)$$

It is shown in all the figures below that having an SS-MC-MA is is possible to achieve much higher spectral efficiencies, since each sub-band is allocated exclusively for that user with the higher channel gain, those users with bad channel conditions do not penalize the system capacity. It can be also observed that the performance gap between the GO-MC-CDMA transmission scheme, and the SS-MC-MA cheme is reduced when the number of active users increasing. Furthermore it is obtained that for low number of users the spectral efficiency is really low in both schemes since many sub-bands are not in use. On the other hand, fairness is greater for the GO-MC-CDMA scheme since although best users are given more resources In Fig. 4, the assigned power as function of the distance between BS and MS is represented. It is shown that for the GO-MC-CDMA scheme more power is assigned to distant users, and thus a greater cell range could be achieved. The system performance considering the QoS requirement is analyzed with $K_{QoS}=K_{CSI}=0.5$. Two real-time classes of services have been considered: the "*large*" and "*low delay real time*" services under a guaranteed bit rate of 2Mbps and 384kbps, and with a maximum tolerated delay of 300ms and 50ms respectively having the 75% of the simulated users witha. The input buffers size is 2Mbits per user. The plotted results in Fig.5 reflects the Cumulative Density Function of the transmission delay. The performances of the GO-MC-CDMA and SS-MC-MA schemes are compared with a Non-QoS aware scheme where resources are allocated according only to the CSI parameters. It is shown in Fig.5 that when the number of active users is low any scheme achieves good performance in terms of delay, and the performances are quite similar using the different transmission schemes, since the system is able to serve all the users as soon as data is received at the input buffers. When the number of the active users is increased, the GO-MC-CDMA gives the worst performance, whereas the SS-MC-MA scheme performs quite close to the Non-QoS aware scheme.

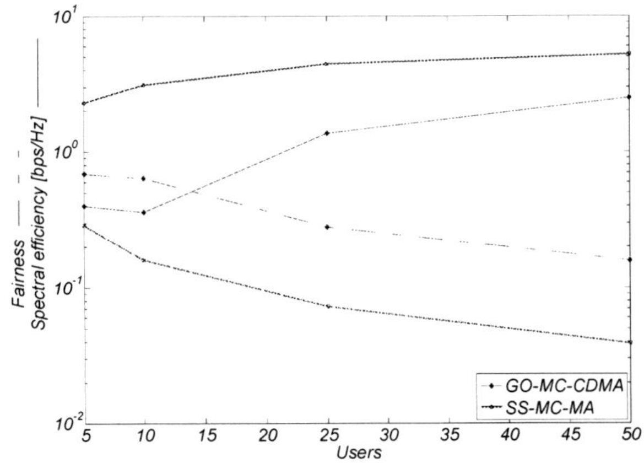

Figure 3. Spectral efficiency and fairness of the SS-MC-MA and GO-MC-CDMA schemes.

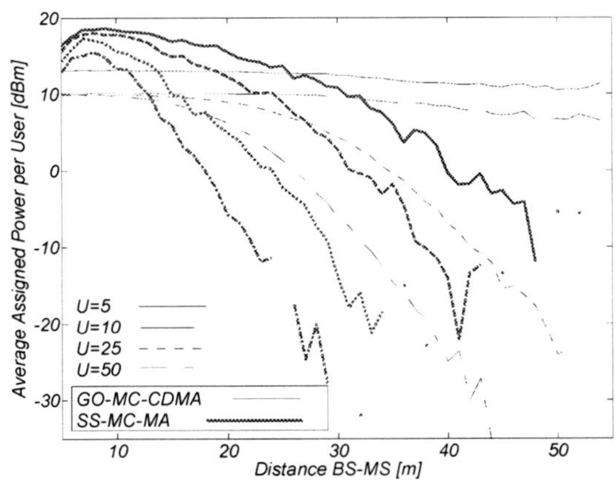

Figure 4. Average power assigned per user as a function of the distance BS-MS.

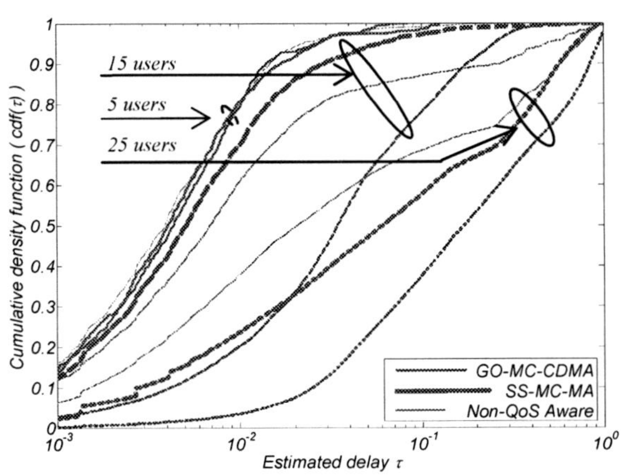

Figure 5. Cumulative density function of the transmission delay for the proposed GO-MC-CDMA and SS-MC-MA resource allocation schemes.

Table I. Simulation parameters

Cell configuration	Square 75m x 75m (regular office)
Total power transmission	200mW (80% for data)
Noise power spectral density	-174dBm/Hz
BS+MS antenna gains, MS Noise Figure, Cables and connectors losses	2dB, 9dB, 2dB *(respectively)*
Transmission mode	Down-link TDD mode, f_c =5 GHz
Number of Carriers ; Modulation	N_c = 1024 ; M-QAM {0-64}
Multipath Channel Model; Terminal Velocity	BRAN channel A (Perf. Estim. & sync.) ; v = 3Km/h
Sampling freq.; Symbol and Cyclic Prefix Time	57.6 MHz; 17.78 μs; 3.75 μs
Frame duration ; UL/DL Guard Time	0.667 mseg ; 20.83 μs
Spreading Factor ; Detection type ; BER	*L*=8 ; *Single user detection, ORC ;* 10^{-3}

5. CONCLUSIONS

The SS-MC-MA outperforms the GO-MC-CDMA scheme when QoS requirements are demanded. Although the grade of fairness is weak compared with than achieved with the GO-MC-CDMA scheme. However, the GO-MC-CDMA performs better in terms of cell-range (*at the expense of system capacity*) when no QoS requirements are included. The main conclusion is that having an accurate design based on a join consideration of K_{QoS} and K_{CSI} parameters is a key point for guaranteeing the QoS demand.

REFERENCES

[1] K. Fazel, S.Kaiser. *Multi-Carrier and Spread Spectrum Systems*. John Wiley & Sons Ltd, 2003.

[2] C. Wong, R. Cheng, K. Letaief, and R. Murch. *Multiuser OFDM with adaptive subcarrier, bit and power allocation*. IEEE Journal on Select. Areas Commun., vol. 17, pp. 1747–1758, Oct. 1999.

[3] F. Brah, L. Vandendorpe. *Proportional Fair Power Allocation for OFDM-CDM Systems in Frequency-Selective Rayleigh Fading Channels.* In Proc. IEEE Workshop on Signal Processing Advances in Wireless Communications (SPAWC'06), July 2006. France.

[4] M.A. Enright, C.J. Kuo. *Energy Allocation for Multicarrier Systems with Mixed QoS Classes. In Proc. 61st IEEE Vehicular Technology Conference. VTC- 2005-Spring.,Vol.2, pp. 1125- 1129.* Stockholm, Sweden. .

[5] X. Guoxin, J. Yang, Z. Jianhua, Z. Ping. *Adaptive OFDMA Subcarrier Assignment for QoS Guaranteed Services. In Proc.* IEEE Vehicular Technology Conference (*VTC' 2005Spring*). pp1817 - 1820 Vol. 3. Stockholm, Sweden

[6] I. Gutierrez, F. Bader, J.L. Pijoan, M. Deumal. *Adaptive Bit Loading with Multi-User Diversity in MC-CDMA*. In Proc. of European Wireless (EW'2006), April 2006, Athens.

[7] X. Cai, S. Zhou, G. B. Giannakis. *Group-Orthogonal Multi-carrier CDMA*. IEEE Transactions on Communications, Vol. 52, No. 1, Jan. 2004.

[8] T. Lestable. *Mécanismes d'adaptation de lien pour des systèmes multi-porteuses de future génération a étalement de spectre.* PhD Thesis dissertation. Paris-SUD University, Oct. 2003.

PERFORMANCE OF CYCLIC DELAY DIVERSITY IN RICEAN CHANNELS

Armin Dammann, Ronald Raulefs, Simon Plass
German Aerospace Center (DLR)
Institute of Communications and Navigation
Oberpfaffenhofen, 82234 Wessling, Germany
{Armin.Dammann,Ronald.Raulefs,Simon.Plass} @DLR.de

Abstract Cyclic delay diversity (CDD) provides additional diversity in Rayleigh fading channels, and therefore, improves system performance. For line-of-sight (LOS) propagation, e.g. the additive white Gaussian noise channel, the implementation of CDD yields to a performance loss. Therefore, we investigate CDD for Ricean channels with different Ricean factors. To combat the SNR loss for significant LOS propagation we propose the principle of antenna power weighting. The idea is to feed different power levels into the multiple transmit (TX) antenna branches rather than distributing the TX power uniformly among the TX-antennas. We exemplarily implement CDD with antenna power weighting to the terrestrial digital video broadcasting system (DVB-T). Simulation results show that antenna power weighting significantly reduces the SNR loss in LOS propagation by the cost of a slight degradation of the SNR gain in non-LOS scenarios.

1. Introduction

Multiple antenna transmission schemes have gained a high attraction since they offer a capacity which rises proportional to the minimum of the number of transmit (TX) and receive (RX) antennas from information theory point of view [1]. In the recent years, several approaches have been proposed, which take advantage of multiple TX- respectively RX-antennas. One representative of multiple antenna schemes is cyclic delay diversity (CDD)[2]. CDD is a variant of delay diversity (DD) [3] and adapted to communications systems with cyclic extensions as guard intervals such as orthogonal frequency division multiplexing (OFDM) for instance. Signal delays in DD may cause intersymbol interference. In contrast, CDD prevents such additional intersymbol interference by using cyclic signal shifts. Typically, multi TX/RX-antenna techniques

S. Plass et al. (eds.), Multi-Carrier Spread Spectrum 2007, 217–226.

like Space-Time coding [4, 5] require signal processing in both the transmitter and the receiver. However, CDD as well as DD can be implemented solely at the transmitter, the receiver or both sides. The fact that the counterpart — e.g. the RX in case of a TX sided implementation — need not be aware of the implementation makes these techniques standard compatible. I.e. they can be implemented as an extension for already existing systems without changing the standard.

CDD has extensively been investigated for Rayleigh fading channels. It can be shown, that CDD can be considered as channel transformation, which increases the number of propagation paths and the delay spread of the channel, observed at the transmitter. This improves the system performance in multipath Rayleigh fading scenarios. A line-of-sight (LOS) propagation, e.g. an AWGN channel, however, would be transformed into a static multipath channel, which definitely decreases performance. Therefore, it is of high interest to investigate the performance of CDD for propagation environments where both LOS and non-LOS (Rayleigh) components occur. In this paper, we investigate CDD for the terrestrial digital video broadcasting standard (DVB-T, [6]) in Ricean multipath channels with different Ricean factors and propose antenna power weighting as a method to combat the system degradation of CDD for extreme LOS propagation scenarios.

2. System Description

Compared to wireless communications systems, LOS propagation in TV broadcasting is more distinct. Transmit (TX) antennas are typically located on high masts and for fixed reception users often install rooftop antennas. Therefore, the propagation conditions for TV broadcasting cover a large variety of scenarios, described by multipath Rayleigh fading channel models with non-LOS, Ricean fading channels (mixed LOS and non-LOS) and, as pure LOS propagation, the AWGN channel. Subsequently, we show the application of CDD to DVB-T and define the channel model, which is used for our investiga tions.

2.1 DVB-T and Cyclic Delay Diversity

The physical layer of a DVB-T transmitter comprises three main parts, which are (i) MPEG-2 source coding and multiplexing, (ii) outer coding with interleaving and (iii) inner coding, interleaving, framing and modulation. The DVB-T standard defines a target bit error rate (BER) of 2×10^{-4} after decoding of the inner channel code, which yields to a quasi error free data stream after decoding of the outer Reed-Solomon code. Therefore, we are interested in the inner DVB-T and model the data stream at the input of the inner system as pseudo random binary sequence. Figure 1(a) shows the block diagram of a

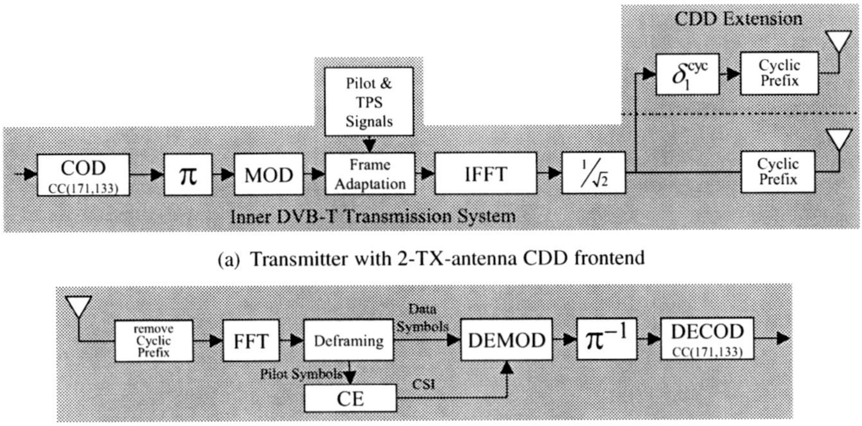

(a) Transmitter with 2-TX-antenna CDD frontend

(b) Receiver

Figure 1. Block diagram of the inner DVB-T system

DVB-T transmitter with 2-TX-antenna CDD. The binary data is encoded, using a convolutional code with generator polynomials $(171, 133)$ in octal form. The mother code of rate $R = 1/2$ may be punctured in order to achieve higher code rates. After interleaving and modulation, the complex valued data symbols together with scattered pilot symbols, continuous pilot carriers (CPC) and transmission parameter signalling (TPS) data are arranged in an OFDM frame, which consists of 68 OFDM symbols. The OFDM symbols are transformed into a time domain signal by an inverse fast Fourier transformation (IFFT). The time domain signal is normalized and split into the TX-antenna branches in such a way, that the overall transmitted power is independent of the number of TX-antennas. In each TX-antenna branch the signal is shifted cyclically by δ_i^{cyc} before the guard interval as cyclic prefix is added. Considering one OFDM symbol, the antenna specific TX signals are

$$
\begin{aligned}
s_i(k) &= \frac{1}{\sqrt{N_{\mathrm{T}}}} \cdot s(k - \delta_i^{\text{cyc}} \bmod N_{\mathrm{FFT}}) \\
&= \frac{1}{\sqrt{N_{\mathrm{T}} \cdot N_{\mathrm{FFT}}}} \sum_{\ell=0}^{N_{\mathrm{FFT}}-1} e^{-\frac{j2\pi \cdot \delta_i^{\text{cyc}} \cdot \ell}{N_{\mathrm{FFT}}}} \cdot S(\ell) \cdot e^{-\frac{j2\pi \cdot k \cdot \ell}{N_{\mathrm{FFT}}}} \qquad (1)
\end{aligned}
$$

For the time interval $k = -N_{\mathrm{G}}, \ldots, N_{\mathrm{FFT}} - 1$ we get the OFDM symbol together with the cyclic prefix. $S(\ell)$ are the complex valued frequency domain symbols, carrying data or pilots.

Table 1. Main DVB-T coding and modulation parameters

Parameter		Specified Values
FFT length	N_{FFT}	2048 (2k), 8192 (8k)
Relative guard interval lengths	$N_{\mathrm{G}}/N_{\mathrm{FFT}}$	1/4, 1/8, 1/16, 1/32
Inner convolutional code rates	R	1/2, 2/3, 3/4, 5/6, 7/8
Modulation		4-QAM, 16-QAM, 64-QAM

The receiver is shown in Figure 1(b). First, the guard interval is removed from the received time domain baseband signal

$$r(k) = \sum_{i=0}^{N_{\mathrm{T}}-1} \sum_{m=0}^{N_{\max}} h_i(m) \cdot s_i(k-m) + n(k). \tag{2}$$

$n(k)$ denotes complex valued additive white Gaussian noise (AWGN) with variance σ^2 and $N_{\max}$ is the maximum channel delay spread. The remaining OFDM time domain symbol is transformed into frequency domain by an FFT, which yields to

$$\begin{aligned} R(\ell) &= \frac{1}{\sqrt{N_{\mathrm{FFT}}}} \sum_{k=0}^{N_{\mathrm{FFT}}-1} r(k) \cdot e^{\frac{j2\pi \cdot k \cdot \ell}{N_{\mathrm{FFT}}}} \\ &= S(\ell) \cdot \underbrace{\frac{1}{\sqrt{N_{\mathrm{T}}}} \sum_{i=0}^{N_{\mathrm{T}}-1} H_i(\ell) \cdot e^{-\frac{j2\pi \cdot \delta_i^{\mathrm{cyc}} \cdot \ell}{N_{\mathrm{FFT}}}}}_{=\tilde{H}(\ell)} + N(\ell). \end{aligned} \tag{3}$$

with $H_i(\ell) = \sum_{k=0}^{N_{\mathrm{FFT}}-1} h_i(k) \cdot e^{\frac{j2\pi \cdot k \cdot \ell}{N_{\mathrm{FFT}}}}$ and the AWGN term $N(\ell)$ again with variance σ^2. Eq. (3) shows that CDD can be described as an *equivalent channel transfer function* $\tilde{H}(\ell)$. For this reason, a receiver cannot distinguish whether a propagation path results from CDD or the channel itself. The deframing unit separates data- and pilot symbols. From the pilots the complex valued channel fading coefficients for each subcarrier are estimated. With this estimation and the received data symbols, the demodulator provides soft information in form of log-likelihood (LL) ratios by applying the MAX-Log-MAP algorithm [7]. The deiterleaved LL values are used as soft input for the Viterbi algorithm [8], which provides a maximum likelihood estimation of the information bits. Table 1 shows possible coding and modulation parameters according to the DVB-T standard.

2.2 Channel Model

For our investigations, we use *Indoor Commercial – Channel B* model as defined in [9]. This 7-path multipath Rayleigh fading channel models large open

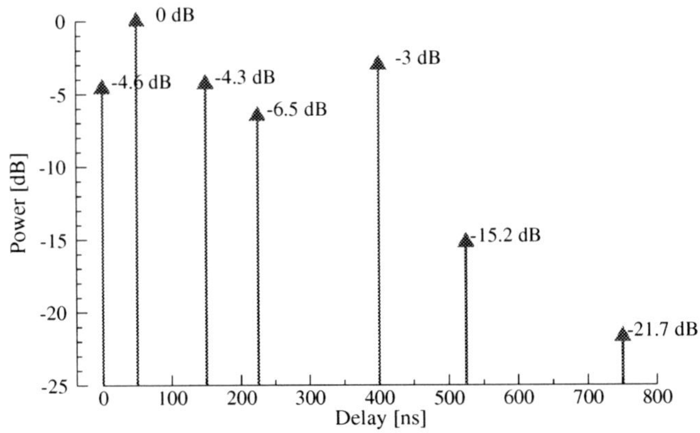

Figure 2. Power-delay profile of the Indoor Commercial – Channel B

centers, such as shopping malls and airports. Its power-delay profile is shown in Figure 2. The Doppler spectrum of the fading processes is uniform with bandwidths in the range of $f_{D_{\max}} = 2 \ldots 5$ Hz.

Subsequently, we are interested in propagation scenarios, which contain LOS, therefore we define an additional propagation path at delay zero with power $|h|^2$. The average channel impulse response power of the resulting a multipath Riceean fading channel model is normalized to one, i.e.,

$$|h|^2 + \sum_{p=0}^{N_P-1} E\{|h_p|^2\} \stackrel{!}{=} 1, \tag{4}$$

where $E\{|h_p|^2\}$ is the average power of the fading process h_p for propagation path p. We define the power ratio of LOS and non-LOS propagation paths as the Ricean factor

$$K = 10 \cdot \log \frac{|h|^2}{\sum_{p=0}^{N_P-1} E\{|h_p|^2\}} \stackrel{(4)}{=} 10 \cdot \log \frac{|h|^2}{1-|h|^2} \quad \text{[dB]} \tag{5}$$

Note that $K = -\infty$ results in the original indoor Rayleigh fading model (Fig. 2), whereas $K = +\infty$ yields to an AWGN channel.

3. Antenna Power Weighting

LOS propagation is a severe problem for CDD since the constant (LOS) paths of the channel are transformed into a static frequency selective one. Using 2-TX-antenna CDD, for instance, transforms an AWGN channel with CTF $H_i(\ell) = 1$ into an equivalent channel with an absolute square CTF $|\tilde{H}(\ell)|^2 = 1 + \cos(2\pi \cdot \delta_1^{\text{cyc}} \cdot \ell / N_{\text{FFT}})$ according to (3). This is depicted for $\delta_1^{\text{cyc}} = 10$ samples in Figure 3 with graph $\Delta P = 0$ dB, where frequency $f = \Delta f \cdot \ell$ can be

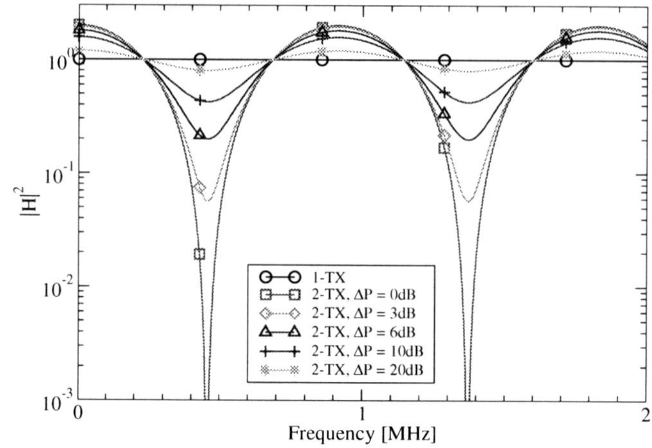

Figure 3. Equivalent CTF $|H(f)|^2$ for pure LOS (AWGN), 2-TX CDD, $\delta_1^{\mathrm{cyc}} = 10$ samples

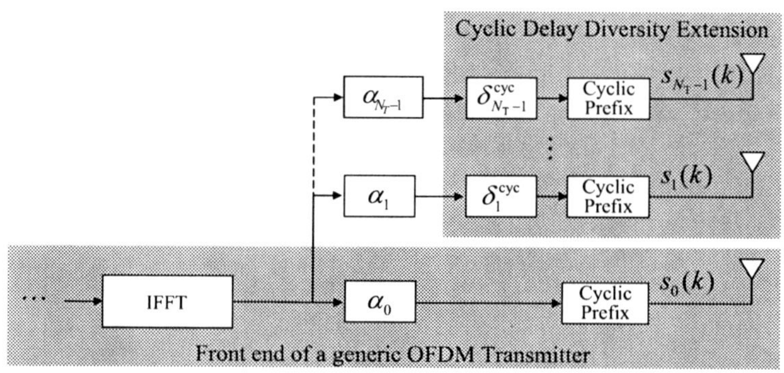

Figure 4. Principle of antenna power weighting

calculated from subcarrier index ℓ and the subcarrier spacing $\Delta f = 4464$ Hz of the considered DVB-T 2k mode. We can clearly observe deep fades, which degrade the system performance compared to the 1-TX antenna case. The reason for these deep fades is the equal power distribution among the TX-antennas.

A solution to overcome this problem is to weight the signal at each TX-antenna branch by different factors α_i. The implementation principle is shown in Figure 4. To keep the transmitted power independent of the numer of TX-antennas yields to the normalization

$$\sum_{i=0}^{N_{\mathrm{T}}-1} E\{|\alpha_i|^2\} \stackrel{!}{=} 1. \tag{6}$$

First of all, the implementation shown in Fig. 4 allows a flexible allocation of power to the different TX-antenna branches with several degrees of freedom. In order to describe the power distribution by one parameter, we define

$$\Delta P = 10 \cdot \log \frac{|\alpha_0|^2}{1 - |\alpha_0|^2} \quad \text{[dB]} \tag{7}$$

as the TX power ratio between the first TX antenna and the CDD extension, i.e., TX-antennas $1 \dots N_\mathrm{T} - 1$. The parameter ΔP allows to switch on/off CDD softly. Note that $\Delta P = +\infty$ completely switches off the CDD extension. Subsequently we investigate 2-TX antenna CDD by simulation. For that, definition (7) provides a unique description of the power distribution. For pure LOS (AWGN) and 2-TX antenna CDD, the equivalent CTF is

$$|H(\ell)|^2 = 1 + \frac{2 \cdot \sqrt{\Delta P_\mathrm{lin}}}{1 + \Delta P_\mathrm{lin}} \cdot \cos\left(\frac{2\pi \cdot \delta_1^\mathrm{cyc} \cdot \ell}{N_\mathrm{FFT}}\right), \tag{8}$$

which is shown in Figure 3 for $\delta_1^\mathrm{cyc} = 10$ and different TX antenna power ratios. $\Delta P_\mathrm{lin} = \frac{|\alpha_0|^2}{1-|\alpha_0|^2}$ is the linear representation of ΔP. For more than 2-TX antennas, the power distribution within the CDD extension part is still ambiguous, and therefore, provides room for optimization.

4. Results

In this section, we provide simulation results for CDD with antenna power weighting applied to the inner DVB-T system as introduced in Section 2.1. We use the 2k mode with a subcarrier spacing of $\Delta f = 4464$ Hz, 16-QAM modulation and an inner code rate of $R = 3/4$. The guard interval length is $N_\mathrm{G} = 1/32 \cdot N_\mathrm{FFT} = 64$ samples, which equals 7 μs for 8 MHz channels. This parameter set results in a net bit rate after the outer Reed-Solomon decoder of 18.1 Mbit/s for 8 MHz channels. We consider 1-TX and 2-TX antenna CDD with $\delta_1^\mathrm{cyc} = 10$. Results in [2] have shown that no further gain is achievable for the considered channel model if we further increase the cyclic delay δ_1^cyc. The Doppler spectrum of the Rayleigh (non-LOS) components is uniform with a bandwidth of $f_{\mathrm{D}_\mathrm{max}} = 4.464$ Hz, which is 0.1% of the subcarrier spacing and thus negligible in terms of intercarrier interference.

In Rayleigh fading channels, CDD provides additional propagation paths, which increases the available diversity. In pure LOS (AWGN), however, these additional propagation paths are static, and thus, transform the AWGN channel into a static frequency selective one, which degrades the system performance. So, the SNR gain turns into a loss if we increase the LOS component in a Ricean channel. This can clearly be seen in Figure 5. For the indoor Rayleigh fading environment, we get an SNR gain of 3.5 dB at BER $= 2 \cdot 10^{-4}$ for 2-TX antenna CDD compared to the 1-TX antenna case (Fig. 5(a)). For the

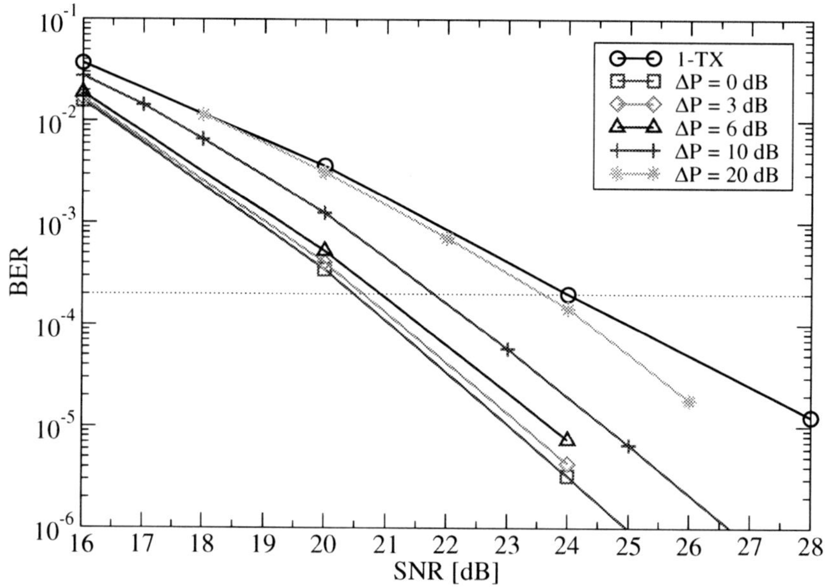

(a) $K = -\infty$ dB (Indoor Commercial – Channel B)

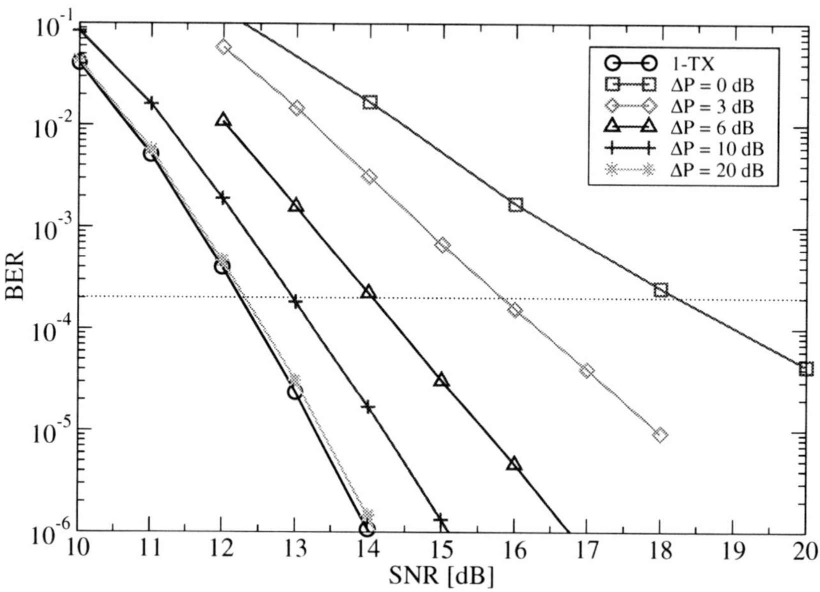

(b) $K = +\infty$ dB (AWGN channel)

Figure 5. BER after the inner Viterbi decoding of 1-TX and 2-TX CDD, DVB-T 2k-mode, 16-QAM, R=3/4, perfect CE

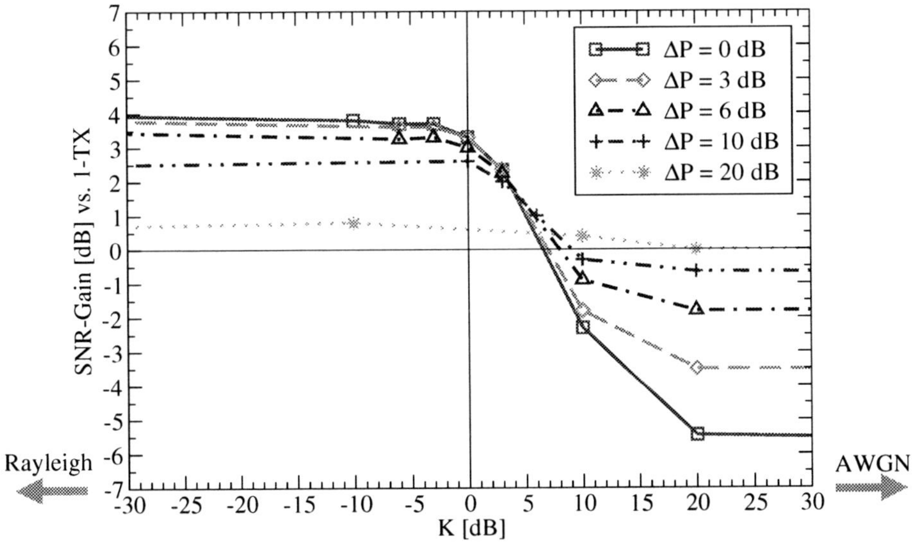

Figure 6. SNR gain of 2-TX CDD compared to 1-TX at BER $= 2 \cdot 10^{-4}$ versus the Ricean factor K, DVB-T 2k-mode, 16-QAM, R=3/4, perfect CE, Ricean channel

AWGN channel (Fig. 5(b)), however, we observe an SNR loss of 6 dB. As we change the power distribution among the TX-antennas the SNR loss decreases significantly an almost vanishes for $\Delta P = 20$ dB. For the Rayleigh fading channel the SNR gain decreases. For $\Delta P = 10$ dB the SNR loss reduces by 5.3 dB to 0.7 dB, whereas the SNR gain for the Rayleigh channel reduced by 1.2 dB only. Antenna power weighting allows to find a compromise between SNR gains and losses in non-LOS respectively LOS scenarios. Figure 5 has shown the extremal cases ($K = \pm\infty$) of the Ricean channel, introduced in Section 2.2. Our interest now is in on the SNR gain/loss in Ricean channels for different Ricean factors K. These results are shown in Figure 6. As the LOS propagation path increases in power, the SNR gain vanishes and turns over into an SNR loss for $K = 6.5$ dB. It is interesting to note that this turnover point shifts to higher Ricean factors with increasing power weighting ratio ΔP.

5. Conclusions

CDD is a standard conformable antenna diversity technique, which improves system performance of OFDM systems in Rayleigh fading (non-LOS) propagation environments. For LOS propagation, however, there are severe problems in terms of performance losses. We have introduced the idea of antenna power weighting in order to combat these losses. We have seen that antenna power weighting drastically reduces the SNR loss for Ricean channel with a significant LOS propagation part by the cost of a slightly decreased SNR gain for non-LOS scenarios. With the approach of antenna power weighting it is possible to switch on/off CDD softly, and therefore, to find good system performance compromises dependent on environment conditions, i.e. the ratio between LOS and non-LOS propagation. Investigations by simulation have been done for a DVB-T system with 2-TX antenna CDD in an indoor propagation environment. Simulation results indicate that increasing the number of TX-antennas further increases the SNR gain for non-LOS propagation. Therefore, it will be worth to investigate the system performance with an increased number of TX-antennas for LOS channels with different Ricean factors.

Acknowledgment

This work have been performed in the context of the IST project PLUTO (FP6-2004-IST-4-026902), which is partly funded by the European Union.

References

[1] Gerard J. Foschini and Michael J. Gans. On limits of wireless communications in a fading environment when using multiple antennas. *Wireless Personal Communications*, 6(3):311–335, March 1998.

[2] Armin Dammann and Stefan Kaiser. Performance of low complex antenna diversity techniques for mobile OFDM systems. In *Proceedings 3rd International Workshop on Multi-Carrier Spread-Spectrum & Related Topics (MC-SS 2001), Oberpfaffenhofen, Germany*, pages 53–64, September 2001. ISBN 0-7923-7653-6.

[3] Armin Wittneben. A new bandwidth efficient transmit antenna modulation diversity scheme for linear digital modulation. In *Proceedings IEEE International Conference on Communications (ICC 1993), Geneva, Switzerland*, pages 1630–1634, May 1993.

[4] Vahid Tarokh, Nambi Seshadri, and A. Robert Calderbank. Space-time codes for high data rate wireless communication: Performance criterion and code construction. *IEEE Transactions on Information Theory*, 44(2):744–764, March 1998.

[5] Siavash M. Alamouti. A simple transmit diversity technique for wireless communications. *IEEE Journal on Selected Areas in Communications*, 16(8):1451–1458, October 1998.

[6] European Telecommunications Standard Institute (ETSI). *Digital Video Broadcasting (DVB); Framing structure, channel coding and modulation for digital terrestrial television*, July 1999. EN 300 744 V1.2.1.

[7] Patrick Robertson, Emmanuelle Villebrun, and Peter Höher. A comparison of optimal and sub-optimal map decoding algorithms operating in the log domain. In *Proceedings IEEE International Conference on Communications (ICC 1995), Seattle, USA*, volume 2, pages 1009–1013, June 1995.

[8] Andrew J. Viterbi. A personal history of the Viterbi algorithm. *IEEE Signal Processing Magazine*, 23(4):120–142, July 2006.

[9] Joint Technical Committee on Wireless Access. *Final Report on RF Channel Characterization*, September 1993. JTC(AIR)/93.09.23-238R2.

DOWNLOWNLINK PERFORMANCE OF MC-CDMA SYSTEMS WITH SPATIAL PHASE CODES IN FADING CHANNELS

Stefan Kaiser
DoCoMo Communications Laboratories Europe GmbH
Landsberger Str. 312, 80687 Munich, Germany
kaiser@docomolab-euro.com

Abstract: Spatial phase coding (SPC) can closely approach the performance of optimum pre-coding schemes, whereas SPC requires less complexity and overhead with respect to channel estimation and feedback. This paper analyses the effects of SPC in MC-CDMA multi-user systems. The downlink performance and complexity is evaluated and compared to that of known pre-coding schemes. It can be shown that MC-CDMA with SPC applying single as well as multi-user detection can outperform OFDM(A) with SPC.

Key words: MC-CDMA, OFDM, spreading, spatial pre-coding, phase flipping, SPC, transmit diversity.

1. INTRODUCTION

MC-CDMA is a powerful multiple access scheme since it enables multi-user detection at low complexity [1][2]. Depending on the amount of channel knowledge available at the transmitter and receiver, different combinations of pre- and post-equalization are available [3]. The trend in mobile communications is to go for multiple antenna schemes where benefits even at high mobility are given by additional spatial diversity exploitation. This paper analyses the impact of spatial phase coding (SPC) [4] on MC-CDMA systems with single and multi-user detection in the downlink.

SPC transmits the signal from several antennas such that the probability of destructive superposition of the signals at the receiver antenna is reduced.

S. Plass et al. (eds.), Multi-Carrier Spread Spectrum 2007, 227–236.

The principle is based on coordinated phase flipping. The promising feature of SPC with respect to complexity is that only the superimposed channel has to be estimated at the receiver and not all channels from the different transmit (Tx) antennas. The performance of MC-CDMA with SPC is compared to that of the known pre-coding schemes maximum ratio transmission (MRT), equal gain transmission (EGT), quantized equal gain transmission (QEGT) and transmitter antenna selection diversity (SD) [5][6]. Moreover, a performance comparison with SPC applied in OFDM(A) is shown where it can be observed that SPC MC-CDMA outperforms SPC OFDM(A).

2. MULTI-CARRIER CDMA SYSTEM

2.1 Transmitter

An MC-CDMA system in the downlink using 2 Tx antennas is investigated. The MC-CDMA transmitter applying SPC is shown in Fig. 1. After channel encoding and symbol mapping the data symbols are spread with the user specific spreading code of length L. Since the downlink is considered, the data of all K active users are superimposed (multiplexed) before spatial pre-coding, i.e.,

$$s_l = \sum_{k=1}^{K} d^{(k)} z_l^{(k)} \tag{1}$$

where $d^{(k)}$ is the data symbol of user k and $z_l^{(k)}$ is the l-th chip of the spreading sequence assigned to user k. The resulting chip sequence is $\mathbf{s}=(s_1,s_2,\ldots,s_L)^T$. The symbol $(.)^T$ denotes the transposition of a vector. In order to reduce the complexity of the mobile receiver the *M&Q* Modification is applied which is in detail explained in [1]. The *M&Q* Modification achieves a spreading code length L much smaller than the total number of sub-carriers N_c. This reduces the complexity especially of multi-user

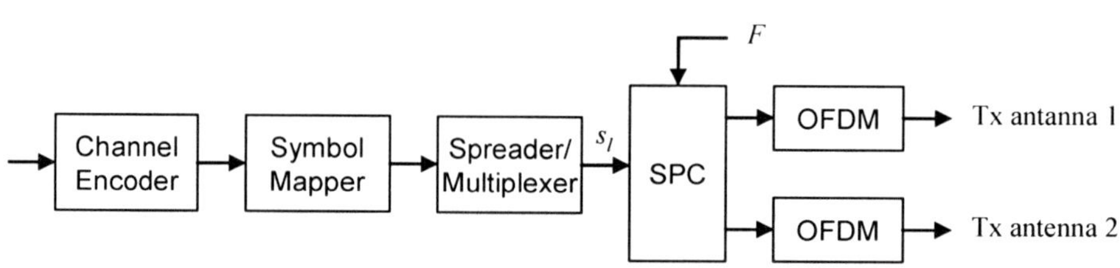

Fig. 1. MC-CDMA transmitter with spatial phase coding (SPC).

detectors. The symbols transmitted on the 2 Tx antennas after pre-coding are $s_l^{(1)}$ and $s_l^{(2)}$, where $^{(m)}$ is the Tx antenna index m=1, 2. The pre-coding is given by

$$\mathbf{s}_l = s_l \mathbf{c}_l = \left(s_l^{(1)}, s_l^{(2)}\right)^T \quad (2)$$

where $\mathbf{c}_l=(c_l^{(1)},c_l^{(2)})^T$ is the spatial pre-coding vector. According to (1), the l-th chip to be transmitted is given by s_l. After pre-coding the symbols on each antenna are modulated on sub-carriers by applying OFDM. The OFDM operation also includes the insertion of a cyclic extension as guard interval. F is the feedback information required for pre-coding.

2.2 Receiver

The MC-CDMA receiver with SPC is shown in Fig. 2. The received signal after inverse OFDM is given by

$$\begin{aligned} r_l &= \left(c_l^{(1)} H_l^{(1)} + c_l^{(2)} H_l^{(2)}\right) s_l + N_l \\ &= H_l s_l + N_l \end{aligned} \quad (3)$$

where H_l represents the superimposed pre-coded channel and N_l the additive noise affecting s_l. The channels from the 2 Tx antennas to the receiver (Rx) antenna are given by the complex-valued fading coefficients $H_l^{(1)}$ and $H_l^{(2)}$ respectively.

Channel estimation is required for pre-coding as well as for data detection. The feedback F to the transmitter contains the information required for pre-coding. The detection comprises user specific despreading, data symbol demapping and channel decoding. For data detection two different schemes are applied in this paper.

2.2.1 Single User Detection with MMSE Equalization

Equalization according to the minimum mean square error (MMSE) criterion

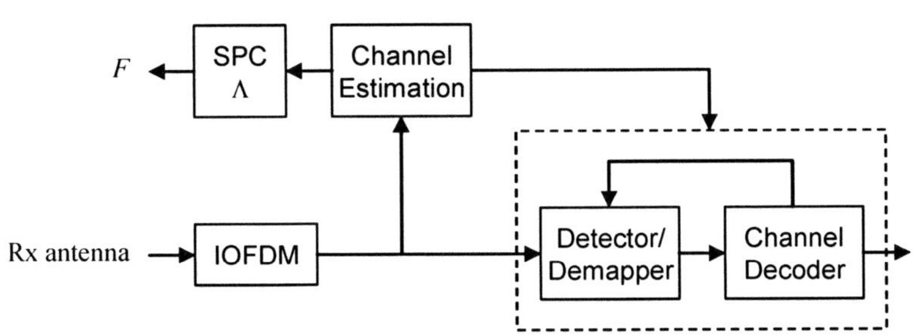

Fig. 2. MC-CDMA receiver with spatial phase coding (SPC).

minimizes the mean square value of the error between the transmitted signal and the output of the equalizer. The equalization coefficient based on the MMSE criterion results in

$$G_l = \frac{H_l}{|H_l|^2 + \sigma^2}. \tag{4}$$

The variance of the noise affecting s_l is given by σ^2.

2.2.2 Multi-user Detection with Soft Interference Cancellation

The principle of soft interference cancellation is to take reliability information about the detected interference into account in the interference cancellation process. Soft values instead of hard values are subtracted. The channel decoding is included in the iterative process and reliability information about the interference is obtained from a soft-in/soft-out channel decoder [7].

3. SPATIAL PRE-CODING

3.1 Spatial Phase Codes (SPC)

The pre-coding vector $\mathbf{c}_l$ with SPC is defined as

$$\mathbf{c}_l = \begin{cases} \frac{1}{\sqrt{2}}(1,1)^T & \text{State A} \\ \frac{1}{\sqrt{2}}(1,e^{-j\pi})^T & \text{State B} \end{cases}. \tag{5}$$

The factor $1/\sqrt{2}$ normalizes the overall Tx power. The pre-coder can be either in State A or in State B. Important for SPC is that the phase relation between the signals from the 2 Tx antennas is changed by π with a state transition. This can be achieved by any appropriate modification of the phases of both signals. The realization presented in (5) and Fig. 3 is one possible implementation. The decision for a state transition from A to B or vice versa is made by the receiver and indicated to the transmitter via the feedback F.

The criterion to change the state and with that to flip the phase relation between the signals from the 2 Tx antennas by π is

$$F = \begin{cases} \text{no state trans.} & |H_l| \geq \Lambda \\ \text{state trans.} & |H_l| < \Lambda \end{cases}. \tag{6}$$

The receiver has to estimate the absolute value of the superimposed pre-coded channel H_l and not $H_l^{(1)}$ and $H_l^{(2)}$. If $|H_l|$ is above a predefined threshold Λ, the spatial phase pre-coder remains in its current state. If the absolute value of the superimposed channel H_l is below the threshold Λ, the spatial phase pre-coder performs a state transition. This state transition from State A to B or vice versa, depending on the previous state, causes that the phase relation between the signals on the 2 Tx antennas changes by π compared to the previous state. To indicate a state transition, 1 bit is sufficient for the feedback signal F.

The principle of SPC is shown in Fig. 3. Fig. 3a) shows $H_l^{(1)}$ and $H_l^{(2)}$ with a relative angle α_l such that $|\alpha_l|<\pi/2$ where they superimpose constructively. In Fig. 3b) the relative angle α_l between both channels is such that $\pi/2<|\alpha_l|<\pi$ where $H_l^{(1)}$ and $H_l^{(2)}$ superimpose destructively before spatial phase pre-coding. By comparing the absolute value of the superimposed channel $|H_l|$ with a predefined threshold Λ, SPC can get an indication for a destructive superposition at the receiver and signals to the transmitter over a feedback channel that the phase relation of the transmitted signals should be changed by π. One possible realization is to flip the phase of the signal at one Tx antenna by π. This is indicated in Fig. 3b).

3.2 Selection Diversity (SD)

With SD the signal is transmitted over the antenna where the absolute value of the individual channel is greater, i.e.,

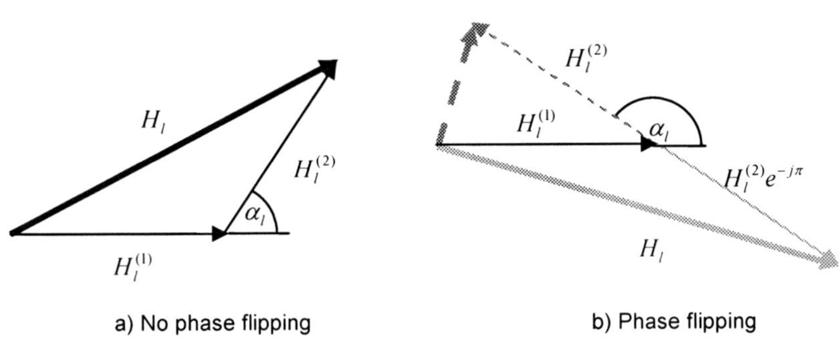

Fig. 3. Principle of spatial phase coding (SPC).

$$\mathbf{c}_l = \begin{cases} (1,0)^T & \text{if } \left|H_l^{(1)}\right| \geq \left|H_l^{(2)}\right| \\ (0,1)^T & \text{otherwise} \end{cases}. \tag{7}$$

With SD the receiver has to estimate both channels in order to determine which channel has a larger absolute value and has to indicate this back to the transmitter via a feedback channel. As with SPC, 1 bit is sufficient for the feedback. However, SD needs twice the overhead for channel estimation than SPC since SD needs the estimation of $|H_l^{(1)}|$ and $|H_l^{(2)}|$ instead of $|H_l|$ only.

3.3 Equal Gain Transmission (EGT)

EGT transmits the signals at the Tx antennas with a phase shift such that they superimpose constructively at the receiver antenna (α_l=0). The pre-coding vector is

$$\mathbf{c}_l = \frac{1}{\sqrt{2}}\left(1, e^{j\alpha_l}\right)^T \tag{8}$$

where α_l is the relative phase difference between $H_l^{(1)}$ and $H_l^{(2)}$ (see Fig. 3). The phases of channel $H_l^{(1)}$ and $H_l^{(2)}$ have to be estimated at the receiver and the relative phase difference α_l between both channels has to be fed back to the transmitter.

3.4 Quantized Equal Gain Transmission (QEGT)

QEGT is EGT with 1 bit quantized feedback. It is defined as

$$F = \begin{cases} \text{no state trans.} & \left|\alpha_l\right| \leq \frac{\pi}{2} \\ \text{state trans.} & \frac{\pi}{2} < \left|\alpha_l\right| \leq \pi \end{cases} \tag{9}$$

where $\pi/2$ is the threshold for the decision on a phase flip. The variable F is the feedback from the receiver to the transmitter to indicate a state change. Since it is only a binary feedback, 1 bit is sufficient for the feedback signal. The receiver has to estimate the relative angle α_l (see Fig. 3) between the vectors $H_l^{(1)}$ and $H_l^{(2)}$. Thus, a channel estimation of both channels $H_l^{(1)}$ and $H_l^{(2)}$ is necessary. QEGT is a lower bound for SPC.

3.5 Maximum Ratio Transmission (MRT)

The optimum pre-coding scheme is MRT. The pre-coding coefficient of MRT is

$$\mathbf{c}_l = w_l \left(H_l^{(1)*}, H_l^{(2)*} \right)^T . \qquad (10)$$

The symbol $(.)^*$ denotes complex conjugation. Compared to the previous schemes, MRT additionally allows for an unequal distribution of the transmit power between both Tx antennas. The power normalization factor w_l results in

$$w_l = \frac{1}{\sqrt{\sum_{m=1}^{2} \left| H_l^{(m)} \right|^2}} \qquad (11)$$

The full channel information about $H_l^{(1)}$ and $H_l^{(2)}$ has to be estimated at the receiver. Thus, MRT requires the most complex channel estimation with the highest overhead of all schemes investigated in this paper. Moreover, MRT requires full feedback of amplitude and phase of each channel, which also exceeds the feedback required by the previous schemes.

4. SIMULATION RESULTS

An MC-CDMA system with 2 Tx antennas and 1 Rx antenna is investigated. The transmission bandwidth is 2MHz and the carrier frequency is 2GHz. The multicarrier modulation is realized by OFDM occupying N_c=512 subcarriers. An OFDM frame consists of 24 OFDM symbols. The guard interval duration is 5μs for the outdoor environment. Results are presented for a rate 1/2 coded transmission using convolutional codes with memory 6. Soft decision decoding with log-likelihood ratios optimized for MC-CDMA are applied [8]. QPSK is chosen for symbol mapping. The threshold Λ is optimized for SPC as shown in [4]. In order to compare the performance differences between SPC and other spatial pre-coding schemes without side effects, effects due to channel estimation, synchronization and feedback errors are omitted by assuming that these components are perfect. The effects of channel estimation errors on SPC are investigated in [9].

As propagation channel the COST 207 Typical Urban (TU) channel model is taken [10]. The maximum velocity of the mobile user is 30km/h which corresponds to a Doppler frequency of 55.6Hz. A classical Doppler spectrum is assumed [11]. The channels from different Tx antennas to the Rx antenna are assumed to be uncorrelated. The signal-to-noise power ratio (SNR) given in the following results always refers to E_t/N_0 where E_t is the total energy per bit at the transmitter and N_0 is the one-sided noise power spectral density.

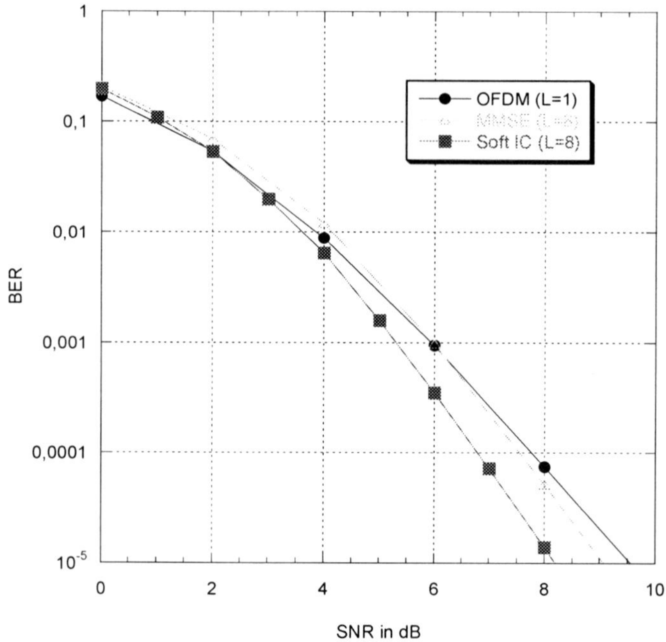

Fig. 4 BER vs. SNR with SPC applied in an MC-CDMA system with single and multi-user detection and in an OFDM(A) system. Fully loaded system.

Fig. 4 shows the BER versus the SNR of an MC-CDMA system and an OFDM(A) system where both apply SPC. Both systems are fully loaded and have the same throughput. The performance of MC-CDMA is shown with MMSE single user detection and multi-user detection applying soft interference cancellation (soft IC). It can be observed that MC-CDMA outperforms OFDM(A) at BERs less than 10^{-2} with multi-user detection and at BERs less than 10^{-3} with single user detection.

The BER versus SNR for SPC, SD, QEGT, EGT and MRT applied in an MC-CDMA system is shown in Fig. 5. The MC-CDMA system applies soft interference cancellation. The system is fully loaded. As reference, the performance of an MC-CDMA system with 1 transmit antenna (1Tx) is given. The results in Fig. 5 show increasing gains with spatial pre-coding when increasing channel state information is available at the transmitter. It can be observed that the least complex pre-coding scheme SPC achieves a similar performance than the more complex schemes. SPC requires a 0.7dB higher SNR at a BER of 10^{-4} than the lower bound given with QEGT. Since the SPC scheme requires only the estimation of the superimposed channel H_l instead of $H_l^{(1)}$ and $H_l^{(2)}$, the required overhead due to pilots for channel estimation is reduced by a factor of 2 compared to the other schemes. The additionally required pilot overhead and with that pilot power for QEGT,

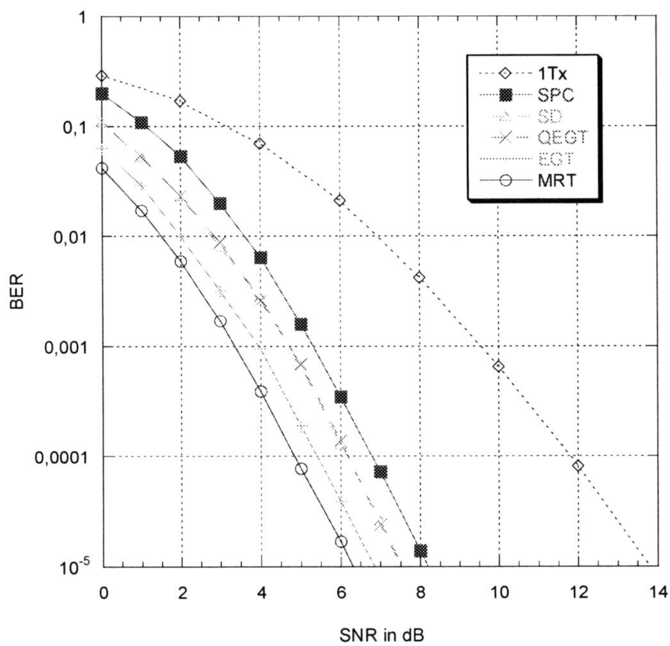

Fig. 5. BER vs. SNR for an MC-CDMA system with different spatial pre-coding schemes. Fully loaded system.

SD, EGT and MRT (not taken into account in this paper) further reduces the performance advantage of these schemes compared to SPC [9]. Moreover, while MRT has to feed back the full channel state information of both channels $H_l^{(1)}$ and $H_l^{(2)}$ and EGT the phase difference α_l, SPC only has to feed back 1 bit to indicate if a state change is necessary or not.

5. CONCLUSIONS

Spatial phase coding (SPC) has been applied to an MC-CDMA multi-user system in the downlink. The principle of SPC is to avoid at the receiver antenna destructive superposition of the signal transmitted from multiple antennas. Compared to optimum solutions like maximum ratio transmission (MRT), SPC requires less overhead for channel estimation at the receiver since only the superimposed channel has to be estimated and not all channels from the different transmit antennas. Moreover, while MRT requires a feedback providing full channel knowledge at the transmitter about all channels from the transmit antennas to the receiver antenna, SPC requires only a 1 bit feedback indicating if the spatial phase encoder has to perform a state transition or not. The achievable performance of the proposed system can be summarized as follows. The performance of a fully loaded MC-CDMA system with channel coding in the downlink can be improved by several dB with SPC compared to a system with only 1 Tx antenna at a BER of 10^{-4}. The degradation compared to the lower bound given by QEGT is about 0.7dB in SNR. Moreover, an SPC MC-CDMA system with single as well as multi-user detection outperforms an SPC OFDM(A) systems at BERs < 10^{-3}.

6. REFERENCES

[1] K. Fazel and S. Kaiser, *Multi-Carrier and Spread Spectrum Systems*. John Wiley & Sons, 2003.

[2] K. Higuchi, H. Kawai, N. Maeda, H. Taoka, and M. Sawahashi, "Experiments on real-time 1-Gb/s packet transmission using MLD-based signal detection in MIMO-OFDM broadband radio access," *IEEE Journal on Sel. Areas in Com.*, vol. 24, pp. 1141-1153, June 2006.

[3] I. Cosovic and S. Kaiser, "A unified analysis of diversity exploitation in multi-carrier CDMA," accepted for *IEEE Trans. Vehicular Techn.*, 2007.

[4] S. Kaiser, "Spatial pre-coding with phase flipping for wireless communications," in Proc. *IEEE Wireless Commun. & Networking Conf. (WCNC 2007)*, Hong Kong, China, March 2007.

[5] R.W. Heath and A. Paulraj, "A simple scheme for transmit diversity using partial channel feedback," in *Proc. 32nd Asilomar Conf. on Signals, Systems, Computers*, Pacific Grove, USA, pp. 1073-1078, Nov. 1998.

[6] T.K.Y. Lo, "Maximum ratio transmission," *IEEE Trans. Commun.*, vol. 47, pp.1458-1461, Oct. 1999.

[7] S. Kaiser and J. Hagenauer, "Multi-Carrier CDMA with iterative decoding and soft-interference cancellation," in *Proc. IEEE Global Telecom. Conf. (GLOBECOM'97)*, Phoenix, USA, pp. 6-10, Nov. 1997.

[8] S. Kaiser, "OFDM code division multiplexing in fading channels," *IEEE Trans. Commun.*, vol. 50, pp. 1266-1273, Aug. 2002.

[9] S. Kaiser, "Effects of channel estimation errors on spatial pre-coding schemes with phase flipping," in Proc. *IEEE Vehicular Techn. Conf. (VTC 2007 Spring)*, Dublin, Ireland, April 2007.

[10] COST 207, *Digital Land Mobile Radio Communications*. Final Report, Luxembourg: Office for Official Publications of the European Communities, 1989.

[11] W.C. Jakes, *Microwave Mobile Communications*. John Wiley & Sons, 1974.

DOPPLER-COMPENSATION FOR OFDM-TRANSMISSION BY SECTORIZED ANTENNA RECEPTION

Peter Klenner and Karl-Dirk Kammeyer
University of Bremen
Department of Communications Engineering
28359 Bremen, Germany
Email:{klenner,kammeyer}@ant.uni-bremen.de

Abstract Rapidly time-varying channels are a major obstacle for successful data-transmission via OFDM. The resulting loss of orthogonality among neighboring subcarriers leads to intercarrier-interference, which affects channel estimation. This in turn impedes the subsequent data detection. Literature contains numerous approaches to cope with this problem working either in time- or frequency-domain. Our concern in this paper is a novel time-domain method which relies on the use of multiple directional receive antennas. Each of these antennas experiences only a fraction of the total Doppler spread of a comparable omnidirectional antenna. This not only eases channel estimation but also allows for a diversity gain due to maximum-ratio-combining. We will demonstrate that our scheme copes well with large maximum Doppler frequencies.

1. Introduction

Transmitting data over frequency-selective channels is easily accomplished by the use of OFDM. Applying a suitable cyclic prefix not only avoids inter-symbol interference between successive OFDM symbols but also performs the transformation of the linear convolution with the channel's impulse response into a circular convolution. In frequency-domain the subcarriers will then be received via flat-fading channels. However, rapidly time-varying channels with large Doppler spread will introduce intercarrier interference (ICI) in frequency domain since subcarriers loose their orthogonality.
A popular approach in literature to deal with rapid channel fluctuations is the assumption of a linear model. Besides linear channel variation the authors of [1] additionally assume a block diagonal structure for the channel matrix neglecting off-diagonal elements, whose inversion demands less computional complexity but fails to capture the effect of ICI sufficiently for larger Doppler

S. Plass et al. (eds.), Multi-Carrier Spread Spectrum 2007, 237–246.

and delay spreads. Noticing that actually the FFT introduces ICI the authors of [2] propose to estimate the time-variant channel impulse from pilot symbols in time-domain. Depending on the channel parameters this approach demands a large number of training symbols and becomes infeasible with larger delay spread. A pilot-based method was proposed in [3] based on an initial Least-Square estimate of the mean channel impulse response, followed by a reconstruction of the channel matrix by assuming again a linearly time-variant channel model. This scheme requires a rather high amount of computational cost, i.e., setting up and inverting the channel matrix, and will degrade as soon as the Doppler spread leads to channel variations which are no longer linear. Our method relies on avoiding large Doppler spread in the first place by applying directional antennas which divide the horizontal reception space into sectors. This allows for a separation of incoming paths according to their angle of incidence and, thereby, their position in the Doppler spectrum. A suitable arrangement of the sector limits splits the Doppler spectrum into evenly spaced subspectra with reduced Doppler spread [4, 5]. Each sector/antenna (these terms will be used interchangeably in the following) is associated with a certain Doppler range corresponding with its comprised range of angles of incidence. We present a coherent receiver which exploits this structure and demonstrate that in fact channel estimation is alleviated. This eventually allows for the application of higher-level modulation at maximum Doppler frequencies which could not be handled in the case of traditional omnidirectional reception.

2. System Model

Fig. 1 depicts the assumed OFDM-transmitter in the equivalent complex baseband. Information bits $b(\xi)$ (IID) are convolutionally encoded (CC) and ran-

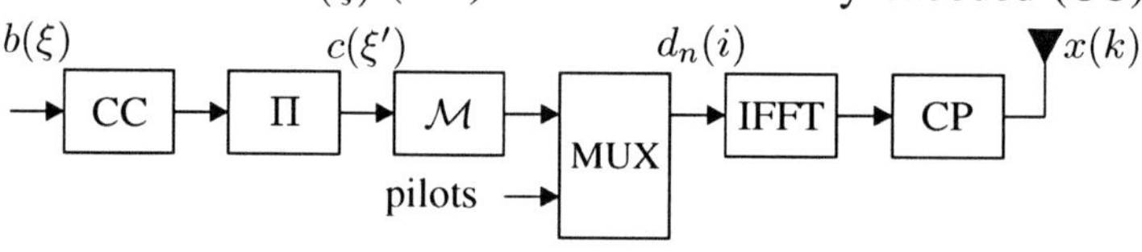

Figure 1. OFDM transmitter

domly bit-interleaved (Π). The coded bits $c(\xi')$ are then mapped by the mapping function $\mathcal{M}$ to an M-ary QAM/PSK signal constellation. The data symbols are multiplexed (MUX) with pilot symbols which enable channel estimation at the receiver side. We denote the symbol on the n-th subcarrier for the i-th OFDM symbol by $d_n(i)$. The OFDM time-domain signal is computed by an N-point IFFT followed by prepending the cyclic prefix (CP) consisting of N_g symbols. The transmitted signal thus reads

$$x(k) = \frac{1}{\sqrt{N}} \sum_{i=-\infty}^{\infty} \sum_{\nu=0}^{N-1} d_\nu(i) \cdot e^{j2\pi\nu(k-i(N+N_g))/N} . \tag{1}$$

2.1 Sectorized Receive Antenna

The impulse response according to the widely accepted WSSUS channel model [6] reads

$$h_\ell(k) = \frac{1}{\sqrt{N_\mathrm{e}}} \sum_{\mu=0}^{N_\mathrm{e}-1} a(\mu) \mathrm{e}^{\mathrm{j}2\pi f(\mu)Tk} \delta(\ell - \ell(\mu)) \tag{2}$$

with delay index $\ell = 0, \ldots, L-1$, time index k, sampling period t, and the length of the channel impulse response L. The N_e path components are characterized by their path amplitude $a(\mu)$, Doppler frequency $f(\mu)$ and delay $\ell(\mu)$. Jakes' Doppler spectrum is generated by choosing $f(\mu) = f_{\mathrm{D,max}} \cos(\theta(\mu))$ with maximum Doppler frequency $f_{\mathrm{D,max}}$ and uniformly distributed phases $\theta(\mu)$ in the interval $[0, 2\pi]$. As an example Fig. 2 illustrates a sectorized antenna and its effect on Jakes' Doppler spectrum

$$P(f) = \left(\pi\sqrt{f_{\mathrm{D,max}}^2 - f^2}\right)^{-1}, \quad -f_{\mathrm{D,max}} < f < f_{\mathrm{D,max}}\,. \tag{3}$$

Fig. 2a shows its division into $S = 4$ sectors. The angles are chosen such

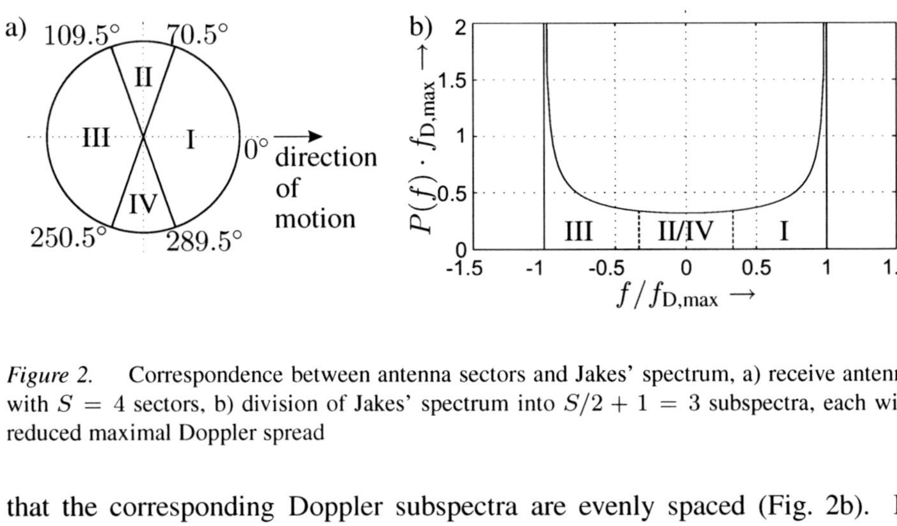

Figure 2. Correspondence between antenna sectors and Jakes' spectrum, a) receive antenna with $S = 4$ sectors, b) division of Jakes' spectrum into $S/2 + 1 = 3$ subspectra, each with reduced maximal Doppler spread

that the corresponding Doppler subspectra are evenly spaced (Fig. 2b). In Tab. 1 we have summarized those angles which lead to equisized subspectra for up to eight antennas. In the current case of $S = 4$ antennas the original Doppler spread has been reduced by $1/3$ per sector. Each sector $s = 1, \ldots, S$ corresponds with a time-variant channel impulse response $h_{\ell,s}(k)$. The key point here is that the impulse responses are accompanied by a reduced Doppler spread. We assume perfect sectorization and find the impulse responses $h_{\ell,s}(k)$ by generating (2) and by summation of those paths whose angle of incidence fall into the angle range comprised by the s-th antenna (cf. Tab.1).

S	ϕ_1	ϕ_2	ϕ_3	ϕ_4	ϕ_5	ϕ_6	ϕ_7	ϕ_8
2	90	270						
4	70.5	109.5	250.5	289.5				
6	60	90	120	240	270	300		
8	53.1	78.5	101.5	126.9	233.1	258.5	281.5	306.9

Table 1. Sector angles for equal Doppler partitioning (values in degree)

2.2 Receiver

Fig. 3 depicts our coherent receiver. The receive signal at the s-th antenna

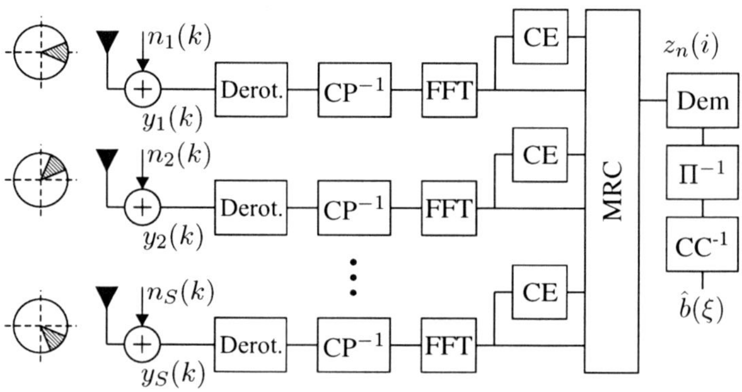

Figure 3. OFDM-receiver with maximum-ratio-combining

reads

$$y_s(k) = \sum_{\ell=0}^{L-1} h_{\ell,s}(k) \cdot x(k-\ell) + n_s(k), \quad 1 \le s \le S \tag{4}$$

where $n_s(k)$ denotes additive white Gaussian noise (AWGN). The Doppler spectrum is divided into a set of subspectra. These exhibit no longer a pure Doppler spread but additionally a frequency shift (cf. Fig. 2) which is compensated in the block "Derot.". In [4] we find the approximation for the compensating frequency to be

$$f_c(s) = \begin{cases} f_{\mathrm{D,max}} \cos(\phi_s/2)\,, & s = 1 \\ -f_{\mathrm{D,max}} \sin(\phi_s/2)\,, & s = S/2+1 \\ f_{\mathrm{D,max}} \cos((\phi_s + \phi_{s-1})/2)\,, & \text{else}\,. \end{cases} \tag{5}$$

Frequency compensation is simply accomplished by

$$\tilde{y}_s(k) = \mathrm{e}^{-\mathrm{j}2\pi f_c(s)Tk} y_s(k)\,. \tag{6}$$

Please note that $f_c(s)$ is not the center frequency of the s-th subspectrum. It is an approximation for the center of gravity of the corresponding Doppler frequencies, i.e., $f_c(s)$ considers a bias towards large Doppler frequencies which occur with higher probability than small Doppler frequencies. Subsequently, after removing the cyclic prefix (CP^{-1}) the frequency-domain signal is computed by the DFT. The receive signal, $r_{n,s}(i)$, on the n-th subcarrier for the i-th OFDM symbol and the s-th sector thereby reads

$$r_{n,s}(i) = \frac{1}{\sqrt{N}} \sum_{\mu=0}^{N-1} \tilde{y}_s(\mu + i(N + N_g))e^{-j2\pi\mu n/N} . \tag{7}$$

2.2.1 Channel Estimation. We consider two channel estimation approaches. The first follows [7] where the authors describe a two-dimensional MMSE filter. Pilot symbols are multiplexed in the OFDM time-frequency grid. For simplicity we assume a rectangular scattered pilot grid throughout the paper. Here pilot symbols are spaced at a distance of Δ_f symbols in frequency direction and at a distance of Δ_t symbols in time direction. It is crucial that the pilot symbol spacing adheres to the sampling theorem. To capture the maximal Doppler frequency the pilot spacing in time direction should fulfil

$$\Delta_t \leq \left\lfloor \frac{N}{2\gamma(N + N_g)} \right\rfloor \tag{8}$$

with the normalized maximum Doppler frequency $\gamma = f_{D,\max}/\Delta f$ and subcarrier spacing Δf. To sample the channel transfer function correctly the pilot spacing in frequency domain should follow

$$\Delta_f \leq \lfloor N/L \rfloor . \tag{9}$$

We find an estimate for the channel coefficients at all pilot symbol positions through division by the respective pilot symbol

$$\hat{H}^{\text{LS}}_{n\Delta_f}(i) = \tilde{r}_{n\Delta_f}(i)/d_{n\Delta_f}(i) . \tag{10}$$

Based on (10) MMSE filtering allows for computation of all remaining channel coefficients associated with data carrying subcarriers denoted by $\hat{H}^{\text{MMSE}}_n(i)$. In [7] omnidirectional reception is assumed. Hence, time correlation of neighboring receive samples is governed by the Bessel function of the first kind, $J_0(\cdot)$. Time correlation in each branch of our sectorized receiver is different from the omnidirectional case. For the latter the Bessel function holds, for the former one can assume that a sufficient number of sectors S leads to correlations described by a Sinc function. Although this is not strictly true, we found from simulations that the resulting performance penalty is negligible. Furthermore, this assumption allows the use of the same estimator for all sectors.

Our second CE algorithm performs linear interpolation in both directions, frequency and time, based on the same rectangular grid as the MMSE approach. In a first step we determine the remaining channel coefficients for the i-th OFDM symbol, i.e., in frequency direction. The linearly interpolated channel coefficients between the $(n\Delta_f)$-th and the $((n+1)\Delta_f)$-th pilot carrying subcarrier is given by

$$\hat{H}^{\mathsf{LIN}}_{n\Delta_f+n'}(i) = \frac{\hat{H}^{\mathsf{LS}}_{(n+1)\Delta_f}(i) - \hat{H}^{\mathsf{LS}}_{n\Delta_f}(i)}{\Delta_f} n' + \hat{H}^{\mathsf{LS}}_{n\Delta_f}(i), \;\; 1 \leq n' < \Delta_f \quad (11)$$

Upon interpolation in frequency domain we perform interpolation in time direction, i.e.,

$$\hat{H}^{\mathsf{LIN}}_{n}(i\Delta_t + i') = \frac{\hat{H}^{\mathsf{LS}}_{n}((i+1)\Delta_t) - \hat{H}^{\mathsf{LS}}_{n}(i\Delta_t)}{\Delta_t} i' + \hat{H}^{\mathsf{LS}}_{n}(i\Delta_t), \;\; 1 \leq i' < \Delta_t. \quad (12)$$

This scheme follows no optimality criterion like the MMSE approach. Nevertheless, its simplicity is appealing since no knowledge about channel correlation or noise variance is required. But usually linear interpolation can handle only a limited Doppler range since rapid channel fluctuations between OFDM symbols will violate the necessity of linear channel variations. However, dividing the Doppler spectrum into subspectra avoids large Doppler spread in the first place. Hence, linear interpolation becomes possible again even for large Doppler spread, if the number of sectors is chosen suitably large.

To improve the performance of linear interpolation, especially for channels with long impulse responses, i.e., which exhibit strong frequency selectivity, we incorporate the findings of [8] into our subsequent simulation results. The respective authors observe that as a side effect convolution with a long impulse response introduces a timing offset which acts as a frequency offset in frequency domain. In [8] it is proposed to remove this frequency offset prior to linear interpolation and attaching it afterwards.

The final steps of our receiver after channel estimation are maximum-ratio-combining

$$z_n(i) = \frac{\sum_{s=1}^{S} \left(\hat{H}^{\mathsf{LIN/MMSE}}_{n,s}(i)\right)^* r_{n,s}(i)}{\sum_{s=1}^{S} \left|\hat{H}^{\mathsf{LIN/MMSE}}_{n,s}(i)\right|^2} \quad (13)$$

followed by APP demodulation ("Dem"), deinterleaving (Π^{-1}) and Viterbi decoding (CC^{-1}).

3. Simulation Results

We present results for an OFDM system with $N = 64$ subcarriers and $N_{\mathrm{g}} = 16$ guards taps. The channel is characterised by an uniformly distributed power-delay profile with either $L = 3$ or $L = 10$ taps corresponding to the weak and strong frequency-selective case, respectively. We apply the non-systematic convolutional code $(133, 171)_8$ and QPSK as well as 16-QAM. MMSE-CE is performed according to [7]. The MMSE filters had 20 coefficients. Perfect knowledge about channel length, maximum Doppler frequency and direction of motion at the receiver side was assumed.

3.1 Single Antenna Performance

We start to investigate the performance limits for single antenna reception, i.e., $S = 1$. Fig. 4 depicts simulation results for a QPSK transmission over weakly

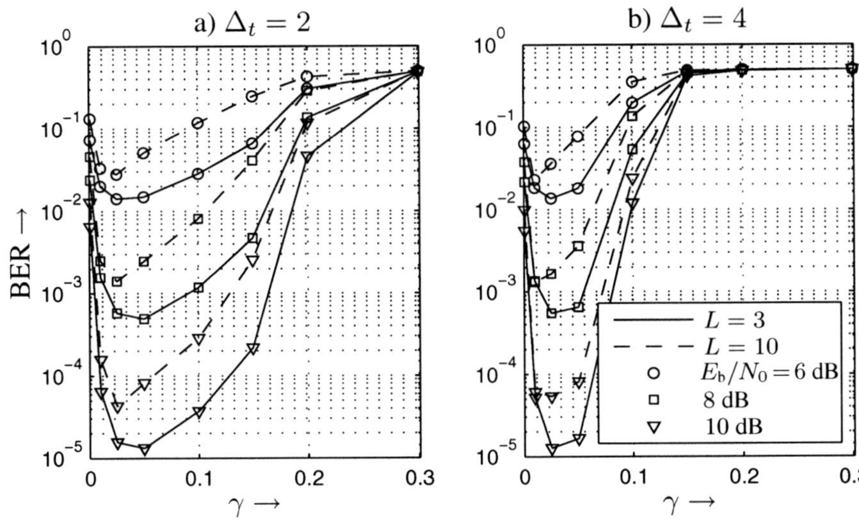

Figure 4. Single antenna bit error performance vs. normalized Doppler frequency γ for different signal-to-noise power ratios, QPSK, $S = 1$, $\Delta_f = 5$, MMSE channel estimation

and strongly frequency-selective channels for different signal-to-noise ratios. In Fig. 4a the pilot spacing in time direction was $\Delta_t = 2$, in Fig. 4b $\Delta_t = 4$. According to (8) the maximal Doppler frequency which can be captured in a critical sense is $\gamma \approx 0.19$ for $\Delta_t = 2$ and $\gamma \approx 0.094$ for $\Delta_t = 4$. However, we see that the BER deteriorates already before the normalized Doppler frequency closes in on these critical values. If the Doppler frequency surpasses the critical Doppler frequencies the sampling theorem is evidently violated and channel estimation fails.

3.2 Multiple Antenna Performance

To facilitate channel estimation for large Doppler environments we could choose to decrease the pilot spacing to $\Delta_t = 1$. But since then every OFDM symbol carries pilot data, bandwidth effciency is decreased. Besides, the largest Doppler frequency which we could capture according to (8) for this case amounts to 0.375. The channel fluctuations accompanied by such large Doppler spreads lead to intercarrier interference which further impedes channel estimation. Instead we apply our sectorization approach to decrease the time-selectivity of the channel. This allows us to keep a scattered pilot scheme with reasonable bandwith efficiency.

3.2.1 Sectorized vs. Single Antenna Performance. In Fig. 5 we have fixed the maximal Doppler frequency to $\gamma = 0.2$ which is critical for the pilot spacing $\Delta_t = 2$. Again results are given for weak and strong frequency selectivity. The number of sectors ranges from $S = 1, \cdots, 8$. Fig. 5a depicts

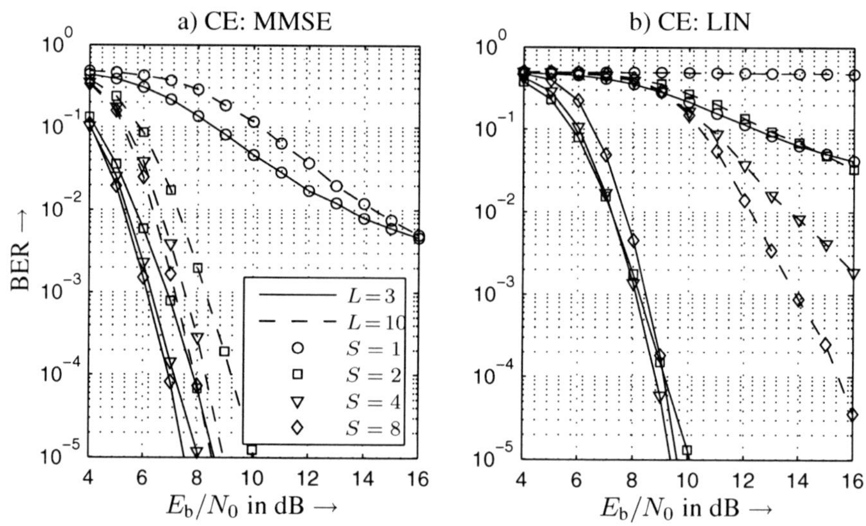

Figure 5. $\gamma = 0.2$, QPSK, $\Delta_f = 5$, $\Delta_t = 2$

results for MMSE channel estimation and Fig. 5b for linear interpolation. Single antenna reception ($S = 1$) leads to an error-floor in all cases due to critical sampling of the channel. Usage of two sectors ($S = 2$) already avoids an error-floor.

We see that MMSE channel estimation copes better with the strongly frequency selctive channel than linear interpolation does. This problem was mentioned at the end of Sec. 2.2.1 and is rooted in the timing offset accompanied by long channel impulse responses. For the results in Fig. 6 we perform phase compensation prior to linear interpolation as proposed in [8]. The performance improvement is impressive. However, the receiver has to have knowledge about the channel length.

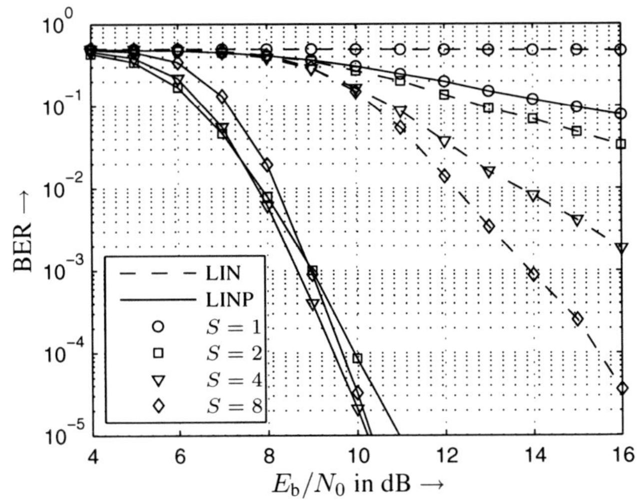

Figure 6. $\gamma = 0.2$, QPSK, $\Delta_f = 5$, $\Delta_t = 2$; LIN=linear interpolation, LINP=linear interpolation with phase correction

3.2.2 Diversity vs. Sectorization. In the previous section we have seen the benefits of sectorization. Reducing the time-selectivity of the channel greatly alleviates channel estimation allowing for reliable data detection even for severe Doppler conditions. We compare now the performance of sectorization against diversity reception. The latter means we apply the same number of *omnidirectional* antennas which are spaced far enough to yield uncorrelated signals. However, each of these signals experiences the full Doppler spread.
Fig. 7 shows that diversity reception is superior to sectorization for small Doppler frequencies. The sectorized receiver suffers from a lack of time diversity since the already slowly fluctuating channel becomes quasi-static after sectorized reception. It is however evident that large Doppler frequencies can not be handled by diversity reception. Here, sectorized reception performs much more robust.

4. Conclusions

We have described a coherent OFDM-receiver with Doppler compensation based on directional antennas. These devide the Doppler spectrum into a set of subspectra with smaller Doppler spread. Since these subspectra correspond with slowly time-varying channels, channel estimation is alleviated. We have presented simulation results for two CE algorithms based on the MMSE criterion and on linear interpolation. We have shown that diversity reception is superior over sectorization for small Doppler frequencies, however, rapidly fluctuating channels are successfully coped with only by sectorization.

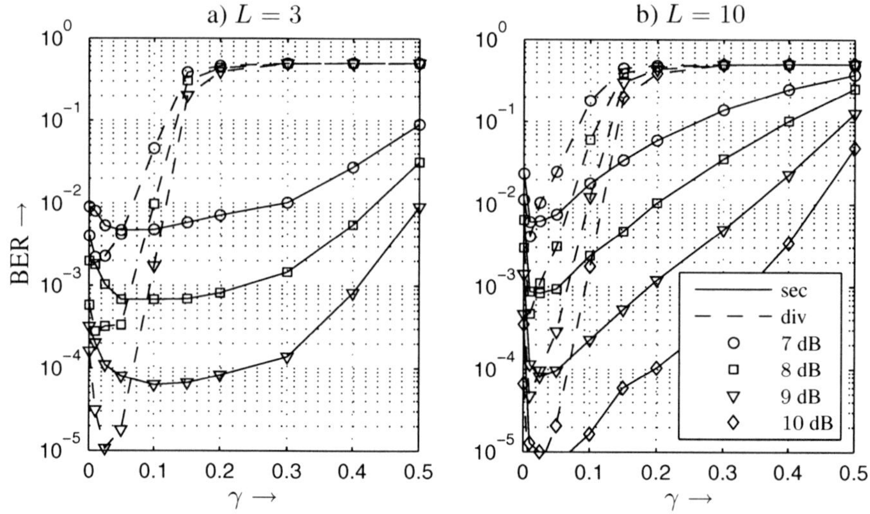

Figure 7. $S = 8$, $\Delta_f = 5$, $\Delta_t = 4$, 16QAM, MMSE channel estimation

References

[1] W. G. Jeon, K. H. Chang. "An equalization technique for orthogonal frequency-division multiplexing systems in time-variant multipath channels". *IEEE Trans. Comm.*, pp. 27-32, Vol. 47, Jan. 1999

[2] Y.-S. Choi, P. J. Voltz, F. A. Cassara. "On Channel Estimation and Detection for Multicarrier Signals in Fast and Selective Rayleigh Fading Channels". *IEEE Trans. Comm.*, pp.1375-1387, Vol. 49, No. 8, Aug. 2001

[3] Y. Mostofi, D. C. Cox. "ICI Mitigation for Pilot-Aided OFDM Mobile Systems". *IEEE Trans. on Wireless Comm.*, pp. 765-774, Vol. 4, No. 2, March 2005

[4] O. Norklit, R. G. Vaughan. "Angular Partitioning to Yield Equal Doppler Contributions". *IEEE Trans. on Vehicular Comm.*, pp. 1437-1442, Vol. 48, N0. 5, Sept. 1999

[5] W. T. Ng, V. K. Dubey. "Effect of Employing Directional Antennas on Mobile OFDM System With Time-Varying Channel". *IEEE Communications Letters*, pp. 165-167, Vol. 7, April 2003

[6] P. Höher. "A Statistical Discrete-Time Model for the WSSUS Multipath Channel". *IEEE Trans. on Vehicular Technology*, pp. 461-468, Vol. 41, Nov. 1992

[7] P. Höher, S. Kaiser, and P. Robertson. "Two-Dimensional Pilot-Symbol-Aided Channel Estimation by Wiener Filtering". *ICASSP*, vol.3, pp. 1845-1848, April 1997

[8] C. H. Lim, D. S. Han. "Robust LS Channel Estimation with Phase Rotation for Single Frequency Network in OFDM". *IEEE Consumer Electronics*, pp. 1173-1178, Vol. 52, Nov. 2006

ON THE RECEIVE SIGNAL DYNAMICS IN ENERGY EFFICIENT MIMO OFDM MULTI-USER MOBILE RADIO DOWNLINKS

Faruk Keskin and Andre Egelhof
Research Group for Radio Communications, University of Kaiserslautern,
P.O. Box 3049, D–67653 Kaiserslautern, Germany, Email: fkeskin@rhrk.uni-kl.de

Abstract In future mobile radio concepts, the cost and complexity of the mobile terminals (MT) should be low. Concerning the downlink, this demand can be fulfilled by resorting to receiver orientation (RO). In conventional RO MIMO OFDM multi-user downlinks, to the data elements to be transmitted uniquely defined complex values are assigned by modulation, such that the transmit processing can be characterized by a modulator matrix, with for instance represents the Transmit Zero Forcing (TxZF) filter. Unfortunately, such a unique assignment does not provide any transmit power reductions compared to the equivalent transmitter oriented (TO) schemes. However, dramatic transmit power reductions become possible if the conventional unique mapping of data elements is replaced by a selective assignment within a set of valid representatives and/or representative domains. The benefit of low transmit power - in addition to low receiver complexity - has to be bargained for increased signal amplitude dynamics at the receivers. In this paper we analyze this dynamics increase and discuss a possible counter measure.

1. Introduction

In future mobile radio concepts, the cost and complexity of the mobile terminals (MT) should be low. Concerning the downlink, this demand can be fulfilled by resorting to RO [1]. In the present paper we consider mobile radio DLs, where we, in order to keep the MT complexity low, resort to RO. Conventional RO DL transmission schemes of the type Transmit Zero Forcing (TxZF) described for instance in [1] do not have any transmit power advantage over TO schemes. However, unconventional schemes providing this advantage [2, 3], [4, 5] go along with increased signal dynamics at the receiver side, which are examined in this paper, including their adverse consequences and avoidance by the introduction of additional constraints in the transmit signal generation.

S. Plass et al. (eds.), Multi-Carrier Spread Spectrum 2007, 247–256.

2. MIMO OFDM multi-user system model

Our target system consists of an access point (AP) with Q transmit antennas and N MTs with one receive antenna each, see Fig. 1, where we assume $Q \geq N$. We abstract the self-evident OFDM typical operations of serial-to-parallel conversion, IFFT and insertion of the cyclic prefix in the transmitter, and of FFT in the receiver. Then, in each OFDM symbol slot and for each of the N_{F} subcarriers the complex amplitudes of the Q transmit signals on the one side and the complex amplitudes of the N receive signals on the other side can be quantitatively related by the channel matrix [3]

$$\underline{\mathbf{H}} = \begin{pmatrix} \underline{h}_{1,1} & \cdots & \underline{h}_{1,Q} \\ \vdots & \ddots & \vdots \\ \underline{h}_{N,1} & \cdots & \underline{h}_{N,Q} \end{pmatrix} \in \mathbb{C}^{N \times Q} . \tag{1}$$

We consider two versions of a stationary time variant channel, namely a fair channel with the channel matrix $\underline{\mathbf{H}}_{\mathrm{fair}}$, and an unfair channel with the channel matrix $\underline{\mathbf{H}}_{\mathrm{unfair}}$, see also [3]. In both cases the realizations of the channel matrices are assumed to have independent bivariate Gaussian entries, which remain constant over one OFDM symbol slot, but may vary from symbol slot to symbol slot. The realizations of the channel matrices valid for different subcarriers are assumed to be independent. All elements of $\underline{\mathbf{H}}_{\mathrm{fair}}$ have unit variances σ_{h}^2 of their real and imaginary parts. In the channel matrix $\underline{\mathbf{H}}_{\mathrm{unfair}}$ the variances of the ele-

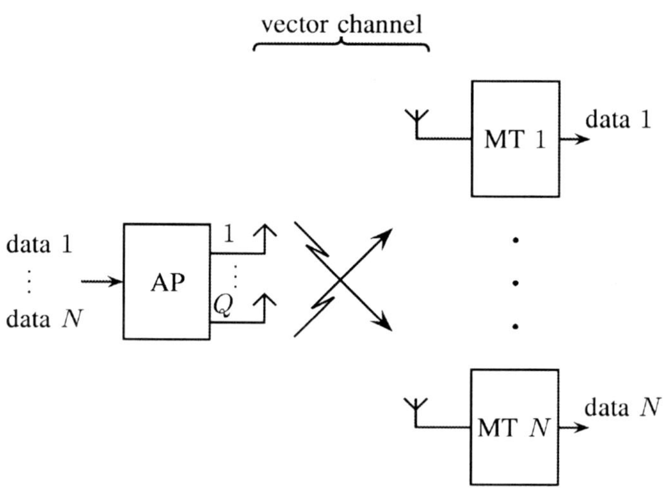

Figure 1. Target MIMO OFDM multi-user DL for each of the N_{F} subcarriers

ments of the first $N/2$ rows are all equal and 10 dB below the variances of the elements of the second $N/2$ rows, which are also all equal. The unfair channel matrix is normalized in such a way that the Euclidian norms $\|\underline{\mathbf{H}}_{\mathrm{unfair}}\|$ and $\|\underline{\mathbf{H}}_{\mathrm{fair}}\|$ become identical. In the case of the fair channel all N MTs experience the same mean channel attenuation - hence the attribute fair -, whereas in the case of the unfair channel the first $N/2$ MTs are subject to mean channel attenuations 10 dB higher than those of the second $N/2$ MTs. Due to fast and slow fading, attenuation-wise unfairness is typical of real world mobile radio scenarios. As the reason for considering here, beside the fair, also an unfair channel, in the case of the latter the decrease of the required transmit energy is particularly large with the energy efficient unconventional schemes as compared to TxZF. In our simulations we choose Q and N equal to four.

We adopt a subcarrierwise approach and, concerning the symbol alphabet on each of the N_{F} subcarriers, we choose the cardinality four in the sense that, in the case of the reference system TxZF, we have the modulation scheme 4-PSK with the representatives $(\pm 1 \pm \mathrm{j})/2$, and in the case of the unconventional schemes the correspondingly extended versions of this scheme [2, 3]. The transmitted data $\mathfrak{a}$ consist of N data elements, each mapped on a complex subcarrier amplitude, and due to the cardinality four of the data alphabet there are 4^N different realizations of $\mathfrak{a}$ with each of these realizations being mapped on a set of N symbols, that is complex subcarrier amplitudes. We assume that all realizations of $\mathfrak{a}$ occur with the same probability.

3. Transmit signal generation

Fig. 2 shows a generic precoding transmission system [1], which consists of the modulator $\underline{\mathbf{M}}$, the channel $\underline{\mathbf{H}}$, the additive noise $\underline{\tilde{\mathbf{n}}}$ and the demodulator $\underline{\mathbf{D}}$. As mentioned in the introduction, in this paper we resort to RO. As reference scheme for conventional RO we refer to TxZF [1], in which the elements of data vector $\mathfrak{a}$ are encoded corresponding to the utilized modulation scheme of cardinality M, that is non-ambiguously mapped on complex values $\underline{d}_m$, $m = 1 \ldots M$, chosen to be the components of the undisturbed vector $\underline{\mathbf{d}}_{\mathrm{Rx}}$ at the output of the demodulator.

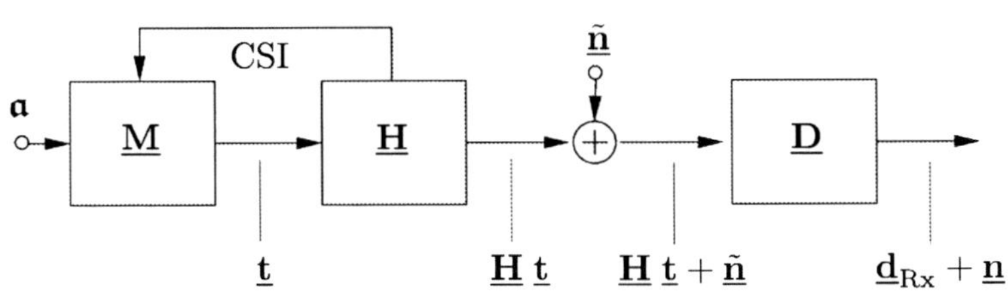

Figure 2. Generic precoding transmission system

In general, according to [1], the transmit vector

$$\underline{\mathbf{t}} = \underbrace{(\underline{\mathbf{D}}\,\underline{\mathbf{H}})^{\mathrm{H}}\left(\underline{\mathbf{D}}\,\underline{\mathbf{H}}\,(\underline{\mathbf{D}}\,\underline{\mathbf{H}})^{\mathrm{H}}\right)^{-1}}_{\underline{\mathbf{M}}}\underline{\mathbf{d}}_{\mathrm{Rx}} = \underline{\mathbf{M}}\,\underline{\mathbf{d}}_{\mathrm{Rx}}, \tag{2}$$

follows from $\underline{\mathbf{d}}_{\mathrm{Rx}}$, that is the Moore-Penrose Pseudo-Inverse of the matrix product $\underline{\mathbf{D}}\,\underline{\mathbf{H}}$ is employed to enforce the desired receiver output $\underline{\mathbf{d}}_{\mathrm{Rx}}$ with minimum required transmit energy

$$T = \frac{1}{2}\underline{\mathbf{t}}^{\mathrm{H}}\underline{\mathbf{t}}. \tag{3}$$

As can be seen in Fig. 2, the undisturbed received vector $\underline{\mathbf{d}}_{\mathrm{Rx}}$ is corrupted by the noise vector $\underline{\mathbf{n}}$ according to

$$\underline{\mathbf{d}}_{\mathrm{Rx}} + \underline{\mathbf{D}}\,\underline{\tilde{\mathbf{n}}} = \underline{\mathbf{d}}_{\mathrm{Rx}} + \underline{\mathbf{n}}. \tag{4}$$

In what follows, we assume that $\underline{\mathbf{n}}$ consists of i.i.d. bivariate Gaussian entries with equal variances $\sigma_{\mathbf{n}}^2$. Then, in order to achieve reliable communication in the presence of noise, that is to statistically ensure that a certain symbol error probability P_{s} is not exceeded, the ratio of the distance among adjacent representatives a and the standard deviation $\sigma_{\mathbf{n}}$ must not fall short of a certain limit given by the Q-function [6], yielding

$$Q\left(\frac{a}{2\sigma_{\mathbf{n}}}\right) \geq 1 - \sqrt{1 - P_{\mathrm{s}}} \tag{5}$$

for the case of 4-PSK modulation, which is conventionally adjusted to fulfill equality in order not to waste transmit power.

Unfortunately, the unique assignment of TxZF does not provide any transmit power reductions compared to the equivalent transmitter oriented (TO) schemes [1]. However, as shown in [2, 3], [4, 5] dramatic transmit power reductions become possible if the conventional unique mapping of data elements is replaced by a selective assignment based on a set of valid representatives and/or representative domains. Hence, the possibility of choosing among multiple equivalent representations of the same data vector $\mathfrak{a}$ can be exploited for optimization through selective assignment. With the required transmit energy per subcarrier T of $\underline{\mathbf{t}}$ of (2), and $\mathbb{C}^N$ the N-dimensional complex space, we quite generally formulate the transmit energy efficient generation of $\underline{\mathbf{t}}$ by precoding as the conditioned minimization of T [2]:

$$\begin{gathered}\underline{\mathbf{t}} = \arg\left\{\min_{\underline{\mathbf{t}}'\in\mathbb{C}^{Q\times 1}}\left(\underline{\mathbf{t}}'^{\mathrm{H}}\underline{\mathbf{t}}'\right)\right\} \\ \text{s.t. } \underline{\mathbf{d}}_{\mathrm{Rx}} \in \mathbb{G}_{\mathfrak{a}} \subset \mathbb{C}^N .\end{gathered} \tag{6}$$

As indicated in (6), the domain $\mathbb{G}_{\mathfrak{a}}$ conditions $\underline{t}$ and depends on the data vector $\mathfrak{a}$. In the case of TxZF [1], $\mathbb{G}_{\mathfrak{a}}$ in (6) is constituted by N discrete complex values, taken from the set $\in (\pm 1 \pm \mathrm{j})\frac{a}{2}$, which we term representatives of the elements of $\mathfrak{a}$. This value discreteness typical of TxZF exerts a rather strong constraint on $\underline{t}$ and hampers the minimization of T. Also in the case of the precoding scheme Transmit Non-linear Zero Forcing (TxNZF) [5], $\mathbb{G}_{\mathfrak{a}}$ consists of discrete complex values. However, now to each of the N data elements correspond a number $P > 1$ of alternative representatives taken from a set as shown in Fig. 3. In this case, the solution of (6) would require an exhaustive N-dimensional search, which would be prohibitively expensive [5]. Therefore, we resort to an approximate solution of (6) in the form of TxNZF which relies on a stepwise procedure based on ordered Tomlinson-Harashima-Precoding (THP) [5]. The liberty to choose from P alternatives loosens the constraint on $\underline{t}$ when performing (6) so that T can be expected to become smaller than in the case of TxZF. In Fig. 3 we also show the Voronoi regions (VR) [7] and the lattice-reducing quantization scheme associated with the discrete valued multiple representatives of TxNZF. In the case of TxZF the VRs are identical with the four quadrants. Due to the impact of the noise vector $\underline{\mathbf{n}}$, the sum of an aimed at representative and the corresponding component of $\underline{\mathbf{n}}$ may land outside the correct VR, and a symbol error occurs.

Also in the case of Minimum Energy Soft Precoding (MESP) [2] the constraint on $\underline{t}$ is loosened as compared to TxZF. This is effected by using, instead of discrete valued representatives, continuous valued representative domains (RD) when establishing $\mathbb{G}_{\mathfrak{a}}$; Fig. 4 shows an example of such RDs, which will be explained in more detail in the following. In this paper, the idea of landing 'softly' somewhere in continuous valued RDs instead of 'hard' spot landings on discrete valued representatives is primarily considered an extension of TxZF. However, it should be mentioned that also TxNZF could be extended in this way, which should be the topic of future research.

The choice of the representatives in the case of TxZF and TxNZF, see Figs. 3, is straightforward. Concerning the choice of the RDs of MESP, some more thought is required. As already explained in [2], we define these domains as follows: A given symbol error probability $P_{\max}$ is not exceeded if and only if a component of $\underline{\mathbf{d}}_{\mathrm{Rx}}$, see (6), lands, instead of a spot landing on the correct discrete valued representative, within the corresponding RD. The shapes of the RDs depend on $P_{\max}$ and σ_{n}^2. They are bounded by Contours of Constant Symbol error Probability (CCSP) $P_{\max}$. The example of Fig. 4 holds for $P_{\max} = 10^{-3}$ and $\sigma_{\mathbf{n}}^2 = 0.01$.

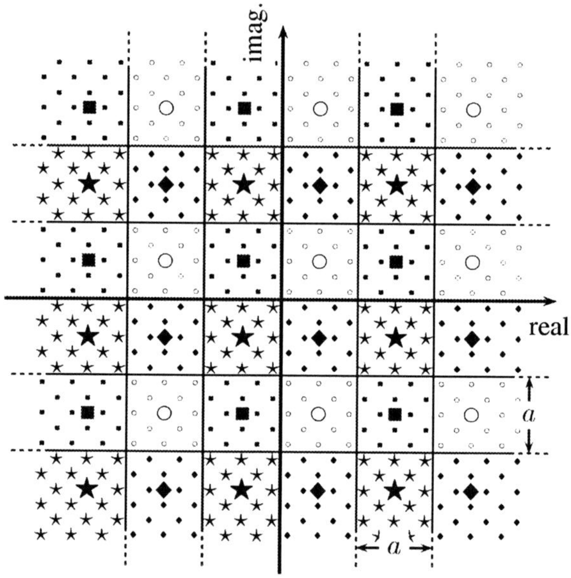

Figure 3. Representatives ○, ■, ★, ♦ and corresponding VRs of a lattice-reducing quantization scheme corresponding to 4-PSK

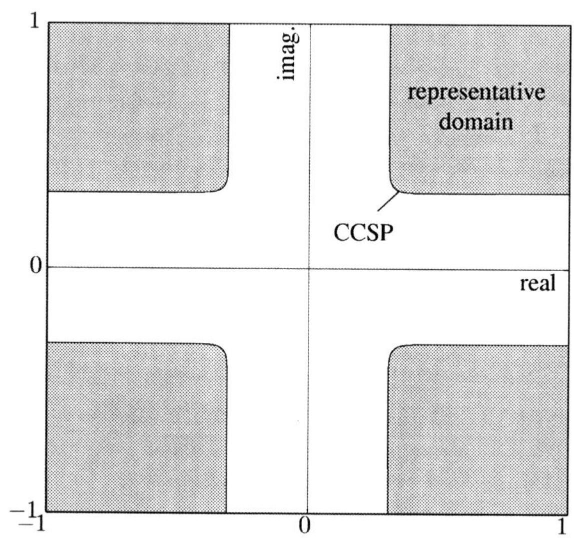

Figure 4. Representative domains (RD) and Contours of Constant Symbol error Probability (CCSP) for $P_{\max} = 10^{-3}$ and $\sigma_{\mathbf{n}}^2 = 0.01$ corresponding to 4-PSK

In the next section we determine, for the channel types $\underline{\mathbf{H}}_{\text{fair}}$ and $\underline{\mathbf{H}}_{\text{unfair}}$ introduced in Section 2, the performance of the schemes discussed above, considering both the transmit energy T and the receive signal dynamics in order to compare the unconventional energy efficient schemes TxNZF and MESP with the conventional TxZF scheme.

4. Performance comparison

As explained in the previous section, MESP and TxNZF enable transmit energy reductions, namely by allowing larger complex amplitudes within the receive vector $\underline{\mathbf{d}}_{\text{Rx}}$. Hence, the benefit of low transmit power of TxNZF and MESP, in addition to their low receiver complexity, has to be bargained for larger dynamics of the receive signals, which requires amplitude resolutions of the receiver AD-converters higher than in the case of TxZF. In this section, we recapitulate the performance advantage of TxNZF and MESP over TxZF by quantifying for all three schemes the required transmit energy T by simulations and then quantitatively assess the corresponding received signal dynamics.

In our simulations we choose Q and N equal to four, and, as mentioned earlier, adopt OFDM as channel access scheme. Then, the demodulator matrix $\underline{\mathbf{D}}$ is the identity matrix of dimensions $N \times Q$ [3]. For the channel matrix $\underline{\mathbf{H}}$ of (1), we consider the two choices $\underline{\mathbf{H}}_{\text{fair}}$ and $\underline{\mathbf{H}}_{\text{unfair}}$ introduced in Section 2. Concerning the performance, in our comparison, with T of (3) we refer to the average radiated transmit energy

$$T_{av} = \frac{1}{2} \cdot \underset{\mathfrak{a},\underline{\mathbf{H}}}{E}\{T\} \tag{7}$$

determined over many realizations of $\mathfrak{a}$ and $\underline{\mathbf{H}}$, satisfying $P_{\text{s}} \leq 10^{-3}$ for all N receivers.

Concerning the receive signal dynamics, typically, the number N_{F} of OFDM subcarriers is large, for instance in the order of 1000. Then, the input signals of the receive antennas can be modelled as Gaussian bandpass signals with certain variances. We define the variance of the n-th receive antenna as

$$\sigma_n^2 = \frac{1}{2} \cdot \underset{\mathfrak{a},\,\underline{\mathbf{H}}}{E}\left\{\underline{\mathbf{d}}_{\text{Rx},n}^{*}\underline{\mathbf{d}}_{\text{Rx},n}\right\} \tag{8}$$

obtained by averaging over many realizations of $\mathfrak{a}$ and $\underline{\mathbf{H}}$. In the case of the fair channel, these variances σ_n^2 are equal for all N receive antennas, and we term them σ^2. In the case of the unfair channel, the variances valid for the second two receive antennas, which are subject to a lower channel attenuation, tend to be larger than those valid for the first two and, therefore, are dynamic-wise more critical. We term these larger variances also σ^2. In Table 1 we list T_{av} of (7) and σ^2 of (8) resulting for TxZF, TxNZF and MESP for the fair and unfair channels, normalized by the number N_{F} of utilized OFDM subcarriers.

	fair channel		unfair channel	
	T/dB	σ^2	T/dB	σ^2
TxZF	10.9	0.125	25.0	0.125
MESP	5.6	0.348	18.1	12.61
TxNZF	4.2	0.306	11.3	11.51

Table 1. Average transmit energies T_{av} and variances σ^2 for $\underline{\mathbf{H}}_{\text{fair}}$ and $\underline{\mathbf{H}}_{\text{unfair}}$

The results in Table 1 show that the receive signal dynamics of the schemes TxNZF and MESP are virtually equal. As can be seen, the required transmit power of TxNZF is smaller than that of MESP, especially in the case of the unfair channel. On the other side, due to the lattice-reducing quantization [3] in the receivers of TxNZF, this scheme, even though the complexity of both TxNZF and MESP is far below the one of TO systems, has a higher receiver complexity than MESP. However, both schemes require higher receive signal dynamics, and, therefore, higher A/D converter resolution compared to TxZF. This additional demand is particularly pronounced in the case of the unfair channel, as the A/D converter needs a seven bit higher resolution, whereas in the case of the fair channel two additional bits would suffice.

5. Dynamics reduction by subcarrierwise receive amplitude restrictions

Now, in order to reduce the receive signal dynamics of MESP and TxNZF, we introduce a maximum amplitude constraint $A_{\max}$ during the transmit signal generation such that

$$\max(\|\Re\{\underline{d}_{\text{Rx},n}\}\|, \|\Im\{\underline{d}_{\text{Rx},n}\}\|) \leq A_{\max},\ n = 1...N, \tag{9}$$

holds for each component of $\underline{\mathbf{d}}_{\text{Rx}}$. Due to this constraint the variances σ^2 become upper bounded by $A_{\max}$ and, consequently, attain smaller values. Unfortunately, the potential transmit energy reduction is limited simultaneously such that as the variances σ^2 attain smaller values, the transmit energies increase. We now let $A_{\max}$ vary and determine the occurring transmit energies T_{av} and variances σ^2. However, because the dynamics increase is especially critical for unfair channels, we only consider this channels in the following. Fig. 5 illustrates the trade-off between the transmit energies T_{av} and receive signal variances σ^2. For convenience, the average transmit energies T_{av} are normalized by the value occurring in case of TxZF, termed T_{ZF}, such that in Fig. 5 the transmit energy reduction with respect to TxZF is shown versus A_{max}. In Fig. 5 we also normalize the variances σ^2 by the value in the case of TxZF, and, to give an immediate impression of the additionally required number of bits in the A/D conversion, we depict $\log_2(\frac{\sigma^2}{0.125})$ versus A_{max}.

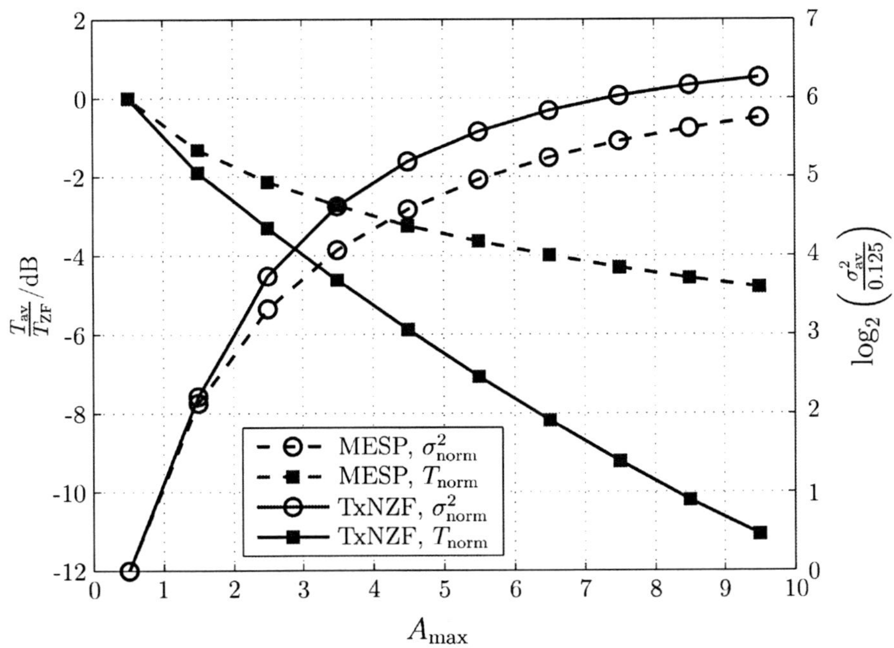

Figure 5. Normalized energies T_{av} and variances σ^2 versus A_{max}, unfair channel

As can be seen from Fig. 5, the transmit energy advantage is counteracted by the requirement of larger word length of the A/D converters in the receivers, which is a drawback for practical implementation. The trade-off presented may help the system designer to find a reasonable compromise between A/D converter expense and performance degradation. This trade-off can be further elaborated by the application of more complicated cost functions rather than amplitude restrictions at the transmitter side, the determination of which would be a topic of further research.

6. Summary

In this paper, it is shown that by means of unconventional precoding schemes the required transmit energies can be significantly reduced. However, simultaneously, the received signal dynamics scale up. In order to reduce this dynamics increase, we propose a subcarrierwise amplitude restriction and illustrate the trade-off between the obtained transmit energy reduction and receive signal dynamics.

7. Acknowledgment

The authors are indebted to Siemens AG for supporting part of this work in the framework of the BMBF project 3GET, and to DFG for enabling basic scientific investigations on energy efficient radio transmission within the focus program AKOM.

References

[1] M. Meurer, P. W. Baier, and W. Qiu. Receiver Orientation versus Transmitter Orientation in Linear MIMO Transmission Systems. *EURASIP Journal on Applied Signal Processing*, 9:1191–1198, 2004.

[2] F. Keskin, J. Hahn, and P. W. Baier. Minimum energy soft precoding. *IET Electronics Letters*, 43(6), 2007, to appear.

[3] J. Hahn, M. Meurer, P. W. Baier, and W. Zirwas. Spectrum and energy efficient MIMO OFDM multi-user downlinks for B3G and 4G mobile radio systems. In *Proc. 13th Int. Conference on Telecommunications (ICT'06)*, Funchal, 2006.

[4] P. W. Baier, M. Meurer, W. Qiu, and T. Weber. Method for operating a transmitting station in a radio communications system and transmitter device. *Patent pending, EP 1538774A1 (EU), WO 2005/055509 (US)*, 2005.

[5] M. Meurer, T. Weber, and W. Qiu. Transmit Nonlinear Zero Forcing: Energy Efficient Receiver Oriented Transmission in MIMO CDMA Mobile Radio Downlinks. In *Proc. IEEE 8th International Symposium on Spread Spectrum Techniques & Applications (ISSSTA'04)*, pages 260–269, Sydney, 2004.

[6] J. G. Proakis. *Digital-Communications*. McGraw-Hill, New York, 1995.

[7] J. H. Conway and N. J. A. Sloane. *Sphere Packings, Lattices and Groups*. Springer, New York/Berlin/Heidelberg, 1999.

DOA ESTIMATION FOR MC-CDMA UPLINK TRANSMISSIONS

Antonio Alberto D'Amico, Michele Morelli and Luca Sanguinetti
University of Pisa
Department of Information Engineering
Pisa, Italy
{antonio.damico,michele.morelli,luca.sanguinetti}@iet.unipi.it

Abstract We consider the uplink of an MC-CDMA network and assume that the base station is endowed with a uniform linear array. Transmission takes place over a multipath channel and the goal is the estimation of the directions of arrival (DOAs) of the signal from the active users. In a multi-user scenario difficulties are due primarily to the large number of parameters involved in the estimation of the DOAs which makes this problem much more challenging than in single-user transmissions. The solution we propose allows estimating the DOAs of different users independently, thereby leading to significant reduction in the system complexity. In the presence of multipath propagation, however, estimating the DOAs of a given user through maximum likelihood methods remains a formidable task since it involves a search over a multidimensional domain. Therefore, we look for simpler solutions and discuss two alternative schemes based on the SAGE and ESPRIT algorithms.

1. Introduction

In the present paper we consider the uplink of a Multi-Carrier Code-Division Multiple-Access (MC-CDMA) network and propose a method for estimating the directions of arrival (DOAs) of the signals of each active user. Transmission takes place over a multipath time-varying channel in which several paths with possibly different DOAs are present for each user. In a multi-user scenario the main obstacle is the large number of parameters involved in the estimation of the DOAs [1]- [2] which makes this problem much more challenging than in single-user transmissions [3]- [5]. A practical solution to this problem consists of separating each user from the others before applying conventional DOA estimation schemes. To this purpose we first estimate the channel response and the data symbols of each active user by resorting to the method

S. Plass et al. (eds.), Multi-Carrier Spread Spectrum 2007, 257–266.

discussed in [6]. Once channel estimates and data decisions are obtained, they are exploited to reconstruct the interfering signals, which are then subtracted from the received waveform. This produces a multiple-access interference (MAI) free signal which is finally used for DOA estimation purposes. In this way the DOAs are estimated independently for each user but, contrarily to [2], no knowledge of the MAI statistics is required.

In spite of the significant simplification achieved by means of users' separation, estimating the DOAs of a given user through maximum likelihood (ML) methods is still difficult as it involves a numerical search over a multidimensional domain. To reduce the system complexity we investigate two alternative schemes. The first is based on the Space-Alternating Generalized Expectation-Maximization (SAGE) algorithm [7], in which the DOAs of a given user are estimated sequentially instead of jointly. The second scheme exploits the ESPRIT (Estimation of Signal Parameters by Rotational Invariance Techniques) algorithm [8] and estimates the DOAs in closed form.

2. Signal Model

We consider the uplink of an MC-CDMA network in which the total number of subcarriers, N, is divided into I smaller groups of Q elements ($I = N/Q$). The base station (BS) is equipped with P antennas and employs the subcarriers of a given group to communicate with K users that are separated through orthogonal Walsh-Hadamard (WH) codes of length $Q \geq K$. Without loss of generality, we concentrate on the ith group and denote $\{i_n; 1 \leq n \leq Q\}$ the subcarrier indexes in the group, with $i_n = i + (n-1)I$. The symbol transmitted by the kth user during the mth MC-CDMA block is denoted $a_k(m)$.

We assume a quasi-synchronous system in which each user is time-aligned to the BS reference in a way similar to that discussed in [9]. In this case, the DFT output at the pth antenna over the nth subcarrier of the considered group during the mth MC-CDMA block takes the form

$$X_p(m,n) = \sum_{k=1}^{K} H_{p,k}(m,i_n)c_k(n)a_k(m) + w_p(m,n) \tag{1}$$

where $H_{p,k}(m,i_n)$ is the kth user's channel frequency response over the i_nth subcarrier at the pth antenna while $c_k(n) \in \{\pm 1/\sqrt{Q}\}$ is the nth element of the kth user's WH code $\mathbf{c}_k = [c_k(1), c_k(2), \ldots, c_k(Q)]^T$. Also, $w_p(n,m)$ is thermal noise modeled as a Gaussian random variable with zero mean and variance σ^2. Assuming that the P receive antennas are arranged in a uniform linear array with inter-element spacing δ, $H_{p,k}(m,i_n)$ takes the form

$$H_{p,k}(m, i_n) = \sum_{\ell=1}^{L} \alpha_{\ell,k}(m) e^{-j2\pi\, i_n \tau_{\ell,k}/T} e^{j(p-1)\,\omega_{\ell,k}} \tag{2}$$

where L is the number of distinct paths, $\tau_{\ell,k}(m)$ is the delay of the ℓth path, $\alpha_{\ell,k}(m)$ the corresponding complex amplitude while $\omega_{\ell,k}(m)$ is defined as

$$\omega_{\ell,k}(m) = \frac{2\pi}{\lambda} \delta \sin[\varphi_{\ell,k}(m)] \tag{3}$$

λ being the free-space wavelength and $\varphi_{\ell,k}(m)$ denoting the DOA of the ℓth path. In the following we assume that the path delays and DOAs do not change significantly with time, i.e., we set $\tau_{\ell,k}(m) \approx \tau_{\ell,k}$ and $\varphi_{\ell,k}(m) \approx \varphi_{\ell,k}$. Vice versa, the path gains $\alpha_{\ell,k}(m)$ are modeled as independent Gaussian processes with zero-mean and average power $\sigma_\ell^2 = \mathrm{E}\{|\alpha_{\ell,k}(m)|^2\}$.

It is worth pointing out that in our multiuser scenario the number of resolvable paths is KL. In practice, KL may be so large to prevent the joint estimation of the DOAs of all active users. To overcome this obstacle, we propose to separate each users from the others using estimates of the channel frequency responses and data decisions. The latter are obtained resorting to the method discussed in [6] where the transmission is organized in frames and some blocks carrying known symbols (training blocks) are placed at the beginning of each frame for channel acquisition. Channel tracking is accomplished by means of the least-mean-square (LMS) algorithm.

3. DOA Estimation

In this Section we show how the channel estimates and data decisions are exploited to perform DOA estimation. To this end, we denote $\{\hat{a}_k(m)\}$ the data decisions and $\{\hat{H}_{p,k}(m, i_n)\}$ the estimates of the channel frequency responses. We begin by computing the following quantities during the mth received block

$$y_{p,k}(m, i_n) = Q\, \hat{a}_k^*(m) c_k(n) \left[X_p(m, n) - \sum_{\substack{\ell=1 \\ \ell \neq k}}^{K} \hat{a}_\ell(m) c_\ell(n) \hat{H}_{p,\ell}(m, i_n) \right]. \tag{4}$$

Substituting (1) into (4) and assuming $\hat{H}_{p,\ell}(m, i_n) \approx H_{p,\ell}(m, i_n)$ and $\hat{a}_\ell(m) \approx a_\ell(m)$ yields

$$y_{p,k}(m, i_n) = \sum_{\ell=1}^{L} d_{\ell,k}(m, i_n) e^{j(p-1)\omega_{\ell,\,k}} + w_p(m, n) \qquad (5)$$

where $d_{\ell,k}(m, i_n) = \alpha_{\ell,k}(m) e^{-j2\pi\, i_n \tau_{\ell,\,k}/T}$ and we have set $|a_k(m)| = 1$ (i.e., PSK constellations). Inspection of the right-hand-side (RHS) of (5) reveals that, apart from thermal noise, only the contribution of the kth user is present. This amounts to saying that the quantities $\{y_{p,k}(m, i_n)\}$ are MAI-free and, therefore, they can be used to estimate the DOAs of the kth user independently from the others. Without loss of generality, we concentrate on the first user and we aim at estimating $\boldsymbol{\omega}_1 = [\omega_{1,1} \;\; \omega_{2,1} \;\; \cdots \;\; \omega_{L,1}]^T$ based on the observation of $\{y_{p,1}(m, i_n)\}$. For notational simplicity we drop the subscript identifier for the first user.

3.1 ML Estimation

To proceed further, we arrange the quantities $\{y_p(m, i_n)\}$ into P-dimensional vectors $\mathbf{y}(m, i_n) = [y_1(m, i_n), y_2(m, i_n), \ldots, y_P(m, i_n)]^T$. We assume slow channel variations so that $d_\ell(m, i_n)$ can be considered practically constant over N_s consecutive blocks. Then, we divide the data section of the frame into adjacent segments, each containing N_s blocks, and compute the following average

$$\mathbf{Y}_r(i_n) = \frac{1}{N_s} \sum_{i=0}^{N_s-1} \mathbf{y}(i + rN_s - N_s + N_T, i_n) \qquad r = 1, 2, \ldots, R \qquad (6)$$

where r is the segment index and R denotes the number of segments within the frame (the number of data blocks in each frame is $N_B = N_s \times R$). Substituting (5) into (6) and bearing in mind that $d_\ell(m, i_n) \approx d_\ell(rN_s - N_s/2,\, i_n)$ over the rth segment (i.e., for $N_T + rN_s - N_s \le m \le N_T + rN_s - 1$), yields

$$\mathbf{Y}_r(i_n) = \sum_{\ell=1}^{L} d_{\ell,r}(i_n) \mathbf{f}(\omega_\ell) + \boldsymbol{\eta}_r(i_n) \qquad (7)$$

where $d_{\ell,r}(i_n) = d_\ell(rN_s - N_s/2,\, i_n)$, $\mathbf{f}(\omega_\ell)$ has entries $[\mathbf{f}(\omega_\ell)]_p = e^{j(p-1)\omega_\ell}$ $(1 \le p \le P)$ and $\{\boldsymbol{\eta}_r(i_n)\}$ are statistically independent Gaussian vectors with zero-mean and covariance matrix $\mathbf{C}_\eta = (\sigma^2/N_s)\,\mathbf{I}_P$.

Skipping mathematical derivations, it can be easily seen that the joint ML estimation of $\boldsymbol{\omega}$ and $\{d_{\ell,r}(i_n)\}$ based on the observation of $\{\mathbf{Y}_r(i_n)\}$ for $1 \le r \le R$ and $1 \le n \le Q$ is given by

$$\hat{\mathbf{d}}_r(i_n) = [\mathbf{F}^H(\tilde{\boldsymbol{\omega}}) \mathbf{F}(\tilde{\boldsymbol{\omega}})]^{-1} \mathbf{F}^H(\tilde{\boldsymbol{\omega}}) \mathbf{Y}_r(i_n) \qquad (8)$$

and

$$\hat{\boldsymbol{\omega}} = \arg\max_{\tilde{\boldsymbol{\omega}}} \left\{ \sum_{r=1}^{R} \sum_{n=1}^{Q} \mathbf{Y}_r^H(i_n)\mathbf{F}(\tilde{\boldsymbol{\omega}})[\mathbf{F}^H(\tilde{\boldsymbol{\omega}})\mathbf{F}(\tilde{\boldsymbol{\omega}})]^{-1}\mathbf{F}^H(\tilde{\boldsymbol{\omega}})\,\mathbf{Y}_r(i_n) \right\}. \tag{9}$$

Unfortunately the maximum of (9) can only be found performing a search over the L-dimensional space spanned by $\tilde{\boldsymbol{\omega}}$. As the computational load would be too intense, in the sequel we discuss two practical DOA estimators based on the SAGE and ESPRIT algorithms. Due to space limitations, they are applied to our problem without further explanation. The reader is referred to [7] and [8] for details.

3.2 SAGE-based Estimation

We apply the SAGE algorithm in such a way that the parameters of a single path are updated at a time. This leads to the following procedure consisting of cycles and steps. A cycle is made of L steps and each step updates the parameters of a single path. In particular, it can be shown that the ℓth step of the ith cycle looks for the minimum of

$$\lambda(\{\tilde{d}_{\ell,r}(i_n)\}, \tilde{\omega}_\ell) = \sum_{r=1}^{R} \sum_{n=1}^{Q} \left\| \mathbf{Y}_{\ell,r}^{(i)}(i_n) - \tilde{d}_{\ell,r}(i_n)\mathbf{f}(\tilde{\omega}_\ell)) \right\|^2 \tag{10}$$

where $\mathbf{Y}_{\ell,r}^{(i)}(i_n) \in \mathbb{C}^P$ is defined as

$$\mathbf{Y}_{\ell,r}^{(i)}(i_n) = \mathbf{Y}_r(i_n) - \sum_{q=1}^{\ell-1} \hat{d}_{q,r}^{(i)}(i_n)\mathbf{f}(\hat{\omega}_q^{(i)}) - \sum_{q=\ell+1}^{L} \hat{d}_{q,r}^{(i-1)}(i_n)\mathbf{f}(\hat{\omega}_q^{(i-1)}) \tag{11}$$

and $\{\hat{d}_{q,r}^{(i)}(i_n), \hat{\omega}_q^{(i)}\}$ denotes the estimate of $\{d_{q,r}(i_n), \omega_q\}$ at the ith cycle. Minimizing (10) with respect to $\{\tilde{d}_{\ell,r}(i_n), \tilde{\omega}_\ell\}$ produces

$$\hat{\omega}_\ell^{(i)} = \arg\max_{\tilde{\omega}} \left\{ \sum_{r=1}^{R} \sum_{n=1}^{Q} \left| \mathbf{f}^H(\tilde{\omega})\mathbf{Y}_{\ell,r}^{(i)}(i_n) \right|^2 \right\} \tag{12}$$

and

$$\hat{d}_{\ell,r}^{(i)}(i_n) = \frac{1}{P}\mathbf{f}^H(\hat{\omega}_\ell^{(i)})\mathbf{Y}_{\ell,r}^{(i)}(i_n) \qquad 1 \le r \le R, \quad 1 \le n \le Q. \tag{13}$$

Note that only one-dimensional searches are involved in (12).

3.3 ESPRIT-based Estimation

To begin, we exploit vectors $\{\mathbf{y}(m, i_n)\}$ to compute the sample correlation matrix

$$\widehat{\mathbf{R}}_y = \frac{1}{N_B Q} \sum_{n=1}^{Q} \sum_{m=N_T}^{N_T+N_B-1} \mathbf{y}(m, i_n)\mathbf{y}^H(m, i_n). \tag{14}$$

Then, based on the forward-backward (FB) approach [10], we obtain the following modified sample correlation matrix

$$\widetilde{\mathbf{R}}_y = \frac{1}{2}(\widehat{\mathbf{R}}_y + \mathbf{J}\widehat{\mathbf{R}}_y^T\mathbf{J}) \tag{15}$$

in which $\mathbf{J}$ is the exchange matrix with 1's on its anti-diagonal and 0's elsewhere. In the ESPRIT method, the eigenvectors associated with the L largest eigenvalues of $\widetilde{\mathbf{R}}_y$ are arranged into a $P \times L$ matrix $\mathbf{V} = [\mathbf{v}_1\ \mathbf{v}_2\ \cdots\ \mathbf{v}_L]$. Next, we consider the matrices $\mathbf{V}_1 = [\mathbf{I}_{P-1}\ \mathbf{0}]\,\mathbf{V}$ and $\mathbf{V}_2 = [\mathbf{0}\ \mathbf{I}_{P-1}]\,\mathbf{V}$, where $\mathbf{0}$ is an L-dimensional column vector with zero entries. The estimate of ω_ℓ is eventually obtained as

$$\hat{\omega}_\ell = \arg\{\lambda_\ell^*\} \qquad \ell = 1, 2, \ldots, L \tag{16}$$

where $\{\lambda_1,\ \lambda_2, \ldots, \lambda_L\}$ are the eigenvalues of

$$\mathbf{S} = (\mathbf{V}_1^H\mathbf{V}_1)^{-1}\mathbf{V}_1^H\mathbf{V}_2 \tag{17}$$

and $\arg\{\lambda_\ell^*\}$ denotes the phase angle of λ_ℓ^* in the interval $[-\pi, \pi)$.

4. Simulation Results

We consider a cellular system operating over a typical urban area with a cell radius of 1 km. The information bits are mapped onto QPSK symbols using a Gray map. The total number of subcarriers is $N = 64$ while WH codes of length $Q = 8$ are used for spreading purposes. The signal bandwidth is $B = 8$ MHz, corresponding to a block duration of $T = N/B = 8\ \mu$s. A cyclic prefix of $T_G = 2\ \mu$s is adopted to eliminate inter-block interference. The number of data blocks in each frame is $N_B = 128$. The carrier frequency is $f_0 = 2$ GHz (corresponding to a wavelength $\lambda = 15$ cm) and the inter-element spacing in the antenna array is $\delta = \lambda/2$. The channel frequency response of the active users are generated as indicated in (2) with $L = 3$. The path delays and DOAs of the desired user are equal to $(\tau_1 = 0,\ \varphi_1 = 0°)$, $(\tau_2 = 1.5\,T_s,\ \varphi_2 = -20°)$ and $(\tau_3 = 3.5\,T_s,\ \varphi_3 = 45°)$. Vice versa, path delays and DOAs of the interfering users are uniformly distributed within $[0,\ 1]\ \mu$s and

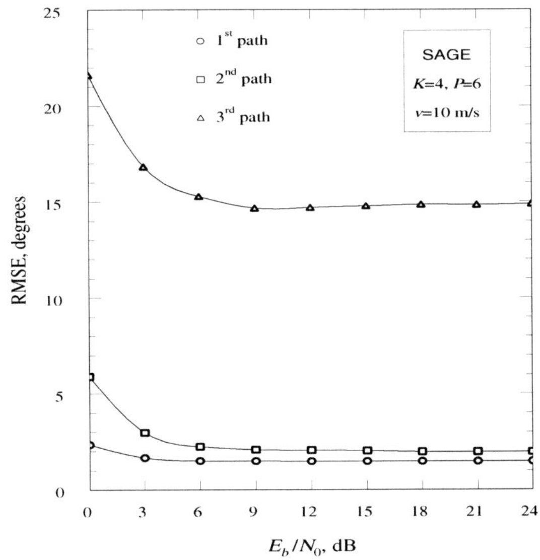

Figure 1. RMSE of the SAGE estimator with $K = 4$, $P = 6$ and $v = 10$ m/s

$[-60°, 60°]$, respectively, and are kept constant over a frame. For all active users (including the desired one) the path gains have powers

$$\sigma_\ell^2 = \sigma_H^2 \exp(-\ell) \qquad \ell = 0,\ 1,\ 2 \tag{18}$$

where σ_H^2 is chosen such that the channel energy is normalized to unity. Each path varies independently of the others within a frame and is generated as in [6].

A simulation run begins with the generation of the channel responses of each user. Channel acquisition and tracking is performed using the method discussed in [6]. Once channel estimates and data decisions are obtained, they are passed to the proposed SAGE- or ESPRIT-based DOA estimators. The SAGE computes the function $\Psi_\ell^{(i)}(\tilde{\omega})$ over a grid of $N_g = 64$ values and it is stopped at the end of the second cycle. Parameter N_s in (6) is fixed to 16, so that $R = N_B/N_s = 8$.

The system performance has been assessed in terms of root mean-square-error (RMSE) of the DOA estimates. For simplicity, the number L of paths is assumed perfectly known at the receiver.

Figure 1 illustrates the performance of the SAGE-based scheme vs. E_b/N_0 (E_b is the average received energy per bit and $N_0/2$ is the two-sided noise spectral density) for $K = 4$. The mobile speed v is 10 m/s while $P = 6$. We see that the curves exhibit a floor. In particular, the

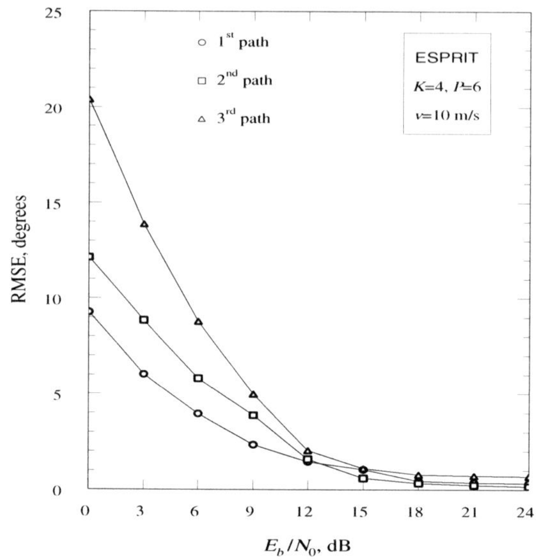

Figure 2. RMSE of the ESPRIT estimator with $K = 4$, $P = 6$ and $v = 10$ m/s

RMSE of the weakest path is approximately 15 degrees for $E_b/N_0 > 10$ dB. The appearance of the floor can be explained as follows. Inspection of (12) reveals that, at the first step of the first cycle, the SAGE looks for the maximum of the periodogram $\psi_1^{(1)}(\tilde{\omega})$. Neglecting the effect of thermal noise, we expect that $\psi_1^{(1)}(\tilde{\omega})$ has three peaks located at the angular frequencies $\omega_i = \pi \sin \varphi_i$ for $i = 1, 2, 3$. As is known [10], in periodogram-based methods the width of the main lobe is approximately $2\pi/P$. It follows that, if a pair of angular frequencies are separated by less than $2\pi/P$, then the corresponding peaks appear as a single broader peak (smearing effect). In these circumstances the two paths cannot be resolved and large estimation errors may occur even in the absence of noise.

Figure 2 shows simulation results as obtained with the ESPRIT estimator in the same operating conditions of Figure 1. As we see, the RMSE curves have no floor. The reason is that ESPRIT is a *high-resolution* technique, meaning that it can resolve angular frequencies separated by less than $2\pi/P$. Comparing with Figure 1, however, it turns out that the SAGE estimator performs better than the ESPRIT at low E_b/N_0.

Figures 3 and 4 illustrate the performance of the proposed schemes for several mobile speeds for $K = 4$ and $P = 6$. For simplicity, only

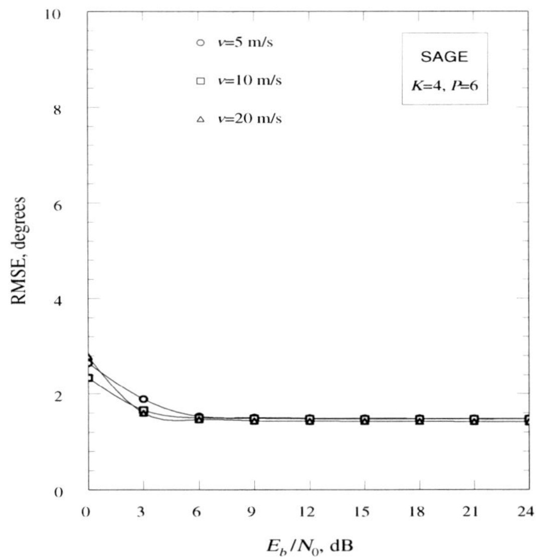

Figure 3. RMSE of the SAGE estimator for $K = 4$, $P = 6$ and some mobile speeds

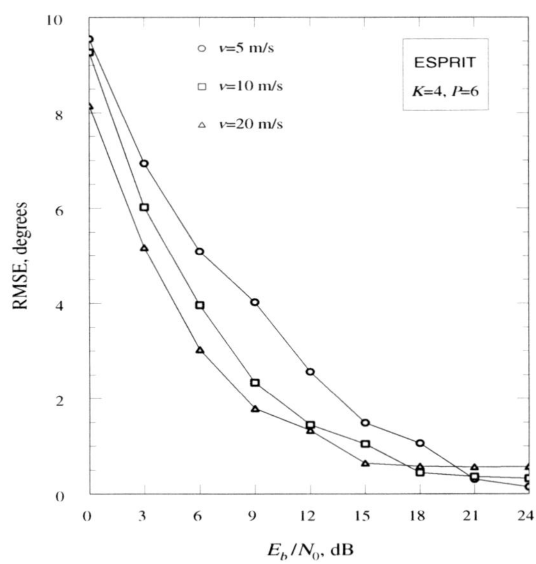

Figure 4. RMSE of the ESPRIT estimator for $K = 4$, $P = 6$ and some mobile speeds

the RMSE of the strongest path is shown. We see that the system performance improves as the mobile speed increases. This is due to the channel variations that provide the system with time diversity.

5. Conclusions

We have discussed a method for estimating independently the DOAs of the each user in the uplink of a multi-user MC-CDMA network. Our solution exploits channel estimates and data decisions to isolate the contribution of each user from the received signal and employs either SAGE or ESPRIT algorithms for estimating the DOAs of each one. Comparisons between the two schemes are not simple because of the different weights that may be given to the various performance indicators, i.e., estimation accuracy and computational complexity. It is likely that the choice will depend on the specific application. For example, the ESPRIT is simpler and has good accuracy. On the other hand, the SAGE outperforms ESPRIT at low SNR values but has limited resolution. Computer simulations indicate that both schemes are robust against channel variations.

References

[1] Z. Lei and T. J. Lim. Estimation of Directions of Arrival of Multipath Signals in CDMA Systems. 48:1022–1028, June 2000.

[2] A. A. D'Amico, U. Mengali, and M. Morelli. Joint DOA and Channel Parameter Estimation for Code-Division Multiple-Access Systems. 2005.

[3] M. Wax and A. Leshem. Joint Estimation of Time Delays and Directions of Arrival of Multiple Reflections of a Known Signal. 45:2477–2484, Oct. 1997.

[4] A.-J. van der Veen, M. C. Vanderveen, and A. Paulraj. Joint Angle and Delay Estimation Using Shift-Invariance Techniques. 46:405–418, Feb. 1998.

[5] L.C. Godara. Application of Antenna Arrays to Mobile Communications, Part II: Beam-Forming and Direction-of-Arrival Considerations. In *Proceedings of IEEE*, volume 85, pages 1195–1245, Aug. 1997.

[6] L. Sanguinetti and M. Morelli. Channel Acquisition and Tracking for MC-CDMA Uplink Transmissions. *IEEE Trans. Vehicular Technology*, 55:956–967, May 2006.

[7] J. A. Fessler and A. O. Hero. Space-Alternating Generalized Expectation-Maximization Algorithm. 42:2664–2677, Oct. 1994.

[8] R. Roy and T. Kailath. ESPRIT-Estimation of Signal Parameters Via Rotational Invariance Techniques. *IEEE Trans. Acoustic, Speech, Signal Processing*, ASSP-37:984–995, July 1989.

[9] M. Morelli. Timing and Frequency Synchronization for the Uplink of an OFDMA System. 52:296–306, Feb. 2004.

[10] P. Stoica and R. Moses. *Introduction to Spectral Analysis*. Englewood Cliffs, NJ: Prentice-Hall.

THE CELLULAR ALAMOUTI TECHNIQUE

Simon Plass and Ronald Raulefs
German Aerospace Center (DLR)
Institute of Communications and Navigation
Oberpfaffenhofen, 82234 Wessling, Germany
{simon.plass, ronald.raulefs}@dlr.de

Abstract This paper handles a transmit diversity technique to provide adequate performances at the cell border in a next generation cellular downlink of a mobile communications system. The used technique is based on the Alamouti scheme, and therefore, we call the scheme cellular Alamouti technique (CAT). The CAT is exemplarily applied to an orthogonal frequency division multiplexing access (OFDMA) which is a promising candidate for future communications systems. For improving the performance at the cell borders, the CAT exploits resource management to apply the Alamouti technique and to avoid inter-cellular interference using both adjacent base stations to transmit the desired signal. The influence of the CAT to a cellular OFDMA system is investigated for different resource management strategies. Simulation results show a significant performance improvement at the cell border by using CAT. The performance depends on the resource allocations of the chosen subcarriers. Nevertheless, CAT is even robust to high-speed scenarios. Furthermore, investigations regarding the system throughput with different modulation alphabets offer a reliable and flexible adaptation to different mobile terminal requests.

1. Introduction

Terrestrial wireless communications systems such as GSM or UMTS provide data transmission services for a large area. Usually, a peer-to-peer connection — *e.g.*, a phone call — between clients or mobile terminals (MT) is established indirectly via a base station (BS). Propagation conditions and system constraints, such as limited transmission power or a required data throughput per area, do not allow covering such large areas by a single base station. Therefore, large areas are split into smaller ones, called cells, where each of them is served by one base station. In order to avoid interference, these cells have to split and cooperatively share the available resources, *i.e.*, frequency bands or time slots. Due to path loss, these resources can be reused if cells are separated far enough

S. Plass et al. (eds.), Multi-Carrier Spread Spectrum 2007, 267–276.

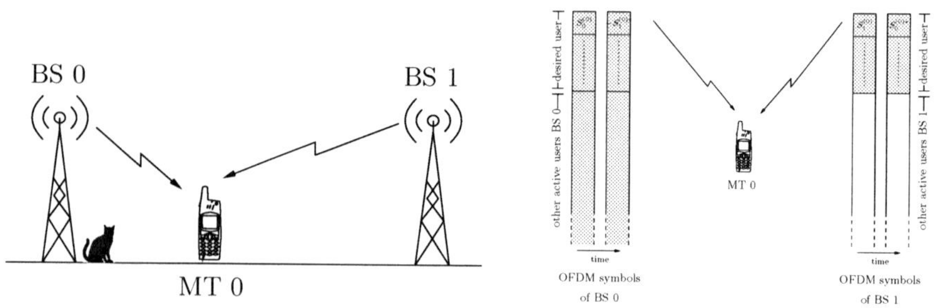

(a) Cellular system design for CAT (b) OFDM symbol design for CAT

Figure 1. Principle of the cellular Alamouti technique

from each other. GSM for instance defines static frequency reuse patterns [1]. For a wireless communications system of the 4th generation (4G), it will be necessary to implement more sophisticated, dynamic radio resource management (RRM) strategies in order to achieve its requirements in terms of frequency reuse or data throughput per cell area. For dynamic RRM, adjacent cells are in principle allowed to allocate the same resources in time-, frequency-, or code-domain. Such systems allocate the available resources dependent on the user's data traffic load and the distribution of the mobile terminals within the managed cells. Furthermore, it becomes possible to transmit data signals for a desired user from BSs of different adjacent cells. This procedure is an essential component for soft handoffs [2].

The Alamouti technique increases the robustness of a transmitted signal over a selective fading channel within a single cell [3]. The two transmit antennas of the Alamouti technique are spatially separated to assure independently faded channels. This essential requirement of two spatially separated transmit antennas was picked up by Inoue *et al.* [4]. We investigate this idea further regarding inter-cellular interference reductions at the cell border and reliable throughputs over the whole cell area. Figure 1(a) shows a single user fed by two adjacent base stations applying the Alamouti technique. The promising candidate orthogonal frequency division multiplexing access (OFDMA) for a 4G downlink transmission technique is used throughout this paper which provides a flexible communications system [5].

RRM offers the avoidance of inter-cellular interference by allocating different time and frequency slots to different base stations. The available resources can be used to perform soft handoff concepts for mobile terminals moving to another cell. Soft handoff concepts [2] have shown that the usage of two base stations at the same time increases the robustness of the received data and avoids interruption and calling resources for re-initiating a call. We will show that the cellular Alamouti technique

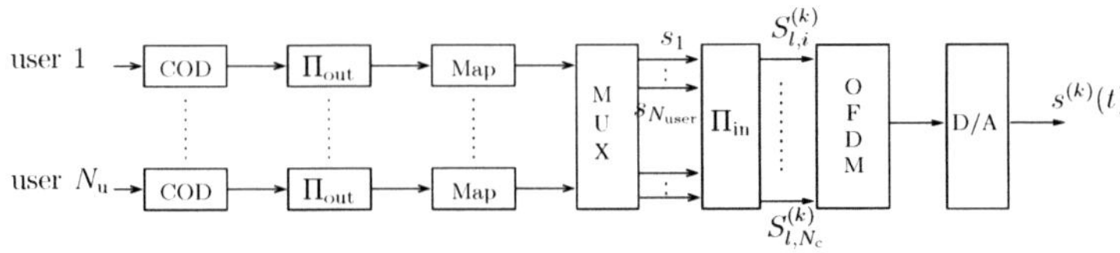

(a) Transmitter of the kth base station

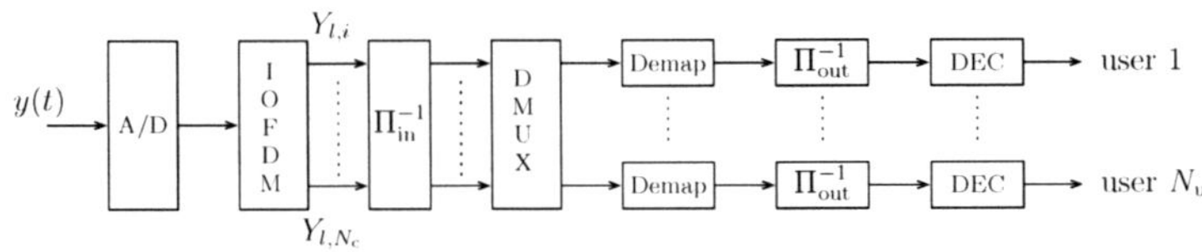

(b) Receiver

Figure 2. Structure of the OFDMA system

offers major advantages to increase the throughput and the robustness in a cellular system.

In the following sections, we first introduce the OFDMA system, channel model, and cellular environment layout under investigation. Then, the cellular Alamouti principle is described and introduced. Finally, simulation results show the performance of the CAT scheme in the cellular structure in comparison to a conventional cellular system.

2. System Model

2.1 OFDMA System

The block diagram of a transmitter using OFDMA is shown in Figure 2(a). The information bit streams of the $N_{\rm u}$ active users are convolutionally encoded and interleaved by the outer interleaver $\Pi_{\rm out}$. With respect to the modulation alphabet, the bits are mapped to complex-valued data symbols. In the sub-carrier allocation block $N_{\rm user}$ symbols per user are arranged for each OFDM symbol.

An inner random sub-carrier interleaver $\Pi_{\rm in}$ allows a better exploitation of diversity. The input block of the interleaver is denoted as one OFDM symbol and $N_{\rm s}$ OFDM symbols describe one OFDM frame. By taking into account a whole OFDM frame, a two-dimensional interleaving in frequency and time direction is possible. Finally, an OFDM modulation is performed which includes an inverse fast Fourier transformation (FFT) and insertion of a guard interval (GI) to avoid inter-symbol interference (ISI).

On the receiver side, the lth received OFDM symbol at sub-carrier i becomes

$$Y_{l,i} = \sum_{k=0}^{K} S_{l,i}^{(k)} H_{l,i}^{(k)} + N_{l,i}\,, \tag{1}$$

where $S_{l,i}^{(k)}$ denotes the value of the ith sub-carrier in the lth OFDM symbol at base station k, $H_{l,i}^{(k)}$ is the channel transfer function (CTF) from BS k to the mobile terminal, and $N_{l,i}$ is an additive white Gaussian noise (AWGN) process with zero mean. At the receiver, see Figure 2(a), the transmitter signal processing is inverted. The symbol demapper calculates the log-likelihood ratio for each bit based on the selected alphabet. The code bits are deinterleaved and finally decoded using soft-decision Viterbi decoding [1].

The resource load (RL) of an OFDMA system can be seen as the ratio of the number of assigned sub-carriers and the total number of available sub-carriers, *i.e.*, $\mathrm{RL} = (N_{\mathrm{user}} \cdot N_{\mathrm{u}})/N_{\mathrm{c}}$.

2.2 Channel Model

The channel model C of the IEEE 802.11n standard is chosen in this paper. This model represents a large open space (indoor and outdoor) with non-line-of-sight conditions of a short range scenario. Therefore, the maximum cell size is chosen to $r = 300$ m. The channel is modelled by a tapped delay-line model with 14 taps and a maximum delay $\tau_{\max} =$ 200 ns. [6]. These channel and propagation model are used throughout all cells.

2.3 Cellular Environment

We consider a synchronized cellular system in time and frequency with two cells, see also Figure 3(a). For example, two mobile users at the cell border are fed together from each base station. The superimposed signal from the kth BSs at the mth MT is given by

$$Y_{l,i}^{(m)} = S_{l,i}^{(0,m)} H_{l,i}^{(0,m)} + S_{l,i}^{(1,m)} H_{l,i}^{(1,m)} + N_{l,i}^{(m)} = R_{l,i}^{(0,m)} + R_{l,i}^{(1,m)} + N_{l,i}^{(m)}\,. \tag{2}$$

Depending on the position of the MT the carrier-to-interference ratio (C/I) varies and is defined by

$$\frac{C}{I} = \frac{E\{|R_{l,i}^{(0,m)}|^2\}}{E\{|R_{l,i}^{(1,m)}|^2\}}\,. \tag{3}$$

3. Cellular Alamouti Technique

In this section, we introduce the concept of transmit diversity by using the space-time block codes (STBCs) from orthogonal designs, namely the Alamouti technique. Then, we will apply this scheme to the aforementioned cellular scenario.

In [7] space time block codes from orthogonal designs were presented. These STBCs are based on the theory of (generalized) orthogonal designs for both real and complex valued signal constellations. The complex valued STBCs, can be described by a matrix

$$\begin{array}{c} \leftarrow \text{ space } \rightarrow \\ \mathbf{B} = \begin{pmatrix} b_{0,0} & \cdots & b_{0,K-1} \\ \vdots & \ddots & \vdots \\ b_{n-1,0} & \cdots & b_{n-1,K-1} \end{pmatrix} \end{array} \begin{array}{c} \uparrow \\ \text{time} \\ \downarrow \end{array}, \tag{4}$$

where n and K is the STBC length and the number of BS (TX-antennas) respectively. The simplest case is the Alamouti code [3],

$$\mathbf{B} = \begin{pmatrix} x_0 & x_1 \\ -x_1^* & x_0^* \end{pmatrix}. \tag{5}$$

The respective assignment for the Alamouti-STBC to the mth MS is

$$\vec{y}^{(m)} = \begin{pmatrix} y_0^{(m)} \\ y_1^{(m)*} \end{pmatrix} = \begin{pmatrix} h^{(0,m)} & h^{(1,m)} \\ h^{(1,m)*} & -h^{(0,m)*} \end{pmatrix} \cdot \begin{pmatrix} x_0 \\ x_1 \end{pmatrix} + \begin{pmatrix} n_0^{(m)} \\ n_1^{(m)*} \end{pmatrix}. \tag{6}$$

$\vec{y}^{(m)}$ is obtained from the received complex values $y_i^{(m)}$ or their conjugate complex $y_i^{(m)*}$ at the receiver. At the receiver, the vector $\vec{y}^{(m)}$ is multiplied from left by the Hermitian of matrix $\mathbf{H}^{(m)}$. The fading between the different fading coefficients is assumed to be quasi static, and then, we obtain the (weighted) STBC information symbols

$$\begin{aligned} \vec{\hat{x}} &= \mathbf{H}^{(m)H} \cdot \vec{y}^{(m)} = \mathbf{H}^{(m)H} \cdot \mathbf{H}^{(m)} \vec{x} + \mathbf{H}^{(m)H} \cdot \vec{n}^{(m)} \\ &= \mathbf{H}^{(m)H} \cdot \vec{n}^{(m)} + \vec{x} \cdot \sum_{i=0}^{1} |h^{(i,m)}|^2, \end{aligned} \tag{7}$$

corrupted by noise. For STBCs from orthogonal designs, MIMO channel estimation at the receiver is mandatory, *i.e.*, $h^{(k,m)}$, $k = 0 \ldots K-1$, $m = 0 \ldots M-1$ must be estimated. Pilot symbols can guarantee a separate channel estimation for each BS [4]. Since the correlation of the subcarrier fading coefficients in time direction is decreasing with increasing Doppler spread — *i.e.*, the quasi stationarity assumption of the fading is incrementally violated — the performance of this STBC class will suffer from higher Doppler frequencies. Later we will see, this is not necessarily true as the stationarity of the fading could also be detrimental in case of burst errors in fading channels.

3.1 Traffic management at the cell border

Figure 3(a) shows two mobile users sojourning at the cell borders. There is no performance difference for each MT if choosing the scenarios of Figure 1(a) or Figure 3(a). The additional freedom is of course paid by a reduced available spectrum at the cell border. On the other hand the power from each base station could be reduced and therefore the cellular interference to the other adjacent cells is lowered.

Radio resource management works perfectly if all information about the mobile users, like the channel state information, is available at the base station. This is especially true if the RRM could be intelligently managed by a single genie manager. As this will be very unlikely the described scheme offers an improved performance especially at the critical cell border without the need of any information about the channel state information on the transmitter side. The main goal is to increase performance by avoiding interference and increasing diversity at the most critical environment. Therefore, the term C/I is misleading, as there is no I (interference). On the other hand it describes the ratio of the power from the desired base station to the temporary base station and indicates where the mobile user is in respect to the base stations. For $C/I = 0$ dB the MT is directly between the two BSs, for $C/I > 0$ dB the MT is closer to the desired BS, and for $C/I < 0$ dB the MT is closer to the other BS. By using CAT the critical cell border area can be also seen as a broadcast scenario with a multiple access channel.

Additionally, the awareness of the MT location can also be exploited. If *e.g.*, a single MT or two MTs are aware that they are at the cell border, they could already ask for the CAT procedure on the first hand to avoid any interference at all. Possible schemes are shown in Figure 1(b) for a single user or in Figure 3(b) for two MTs. In both figures the data $s^{(k)}$ of the mth mobile user allocates the same sub-carriers. In the simulation section, we compare two different scenarios. The first scenario will investigate that the used sub-carriers are not interleaved by Π_{in}, only the data from the remaining users is interleaved for diversity. The second scenario will apply the interleaver for all sub-carriers and data symbols within a whole frame. As the multi-path channel offers frequency diversity latter will violates (7) at the expense of the error performance. It also indicates how robust the system performs for a non-smooth transition.

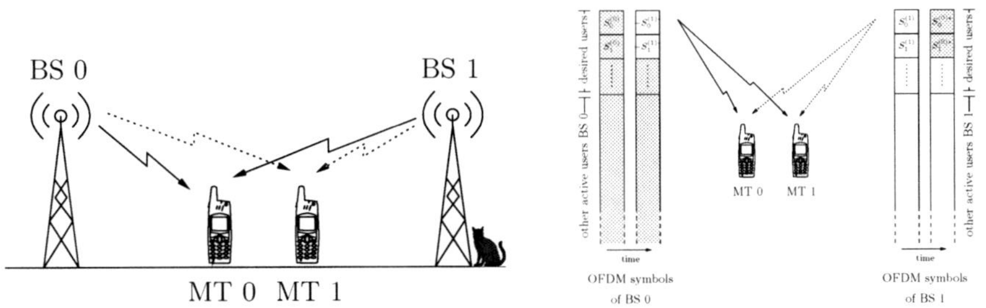

(a) Simultaneous application of CAT

(b) OFDM symbol design with location awareness of 2 MTs

Figure 3. Principle of CAT to two close-by mobile terminals

4. Simulation Results

The used parameters for the simulations are based on the assumptions of the European Union IST-project WINNER for next generation mobile communications system [8]. The transmission system is based on a carrier frequency of 5 GHz, a bandwidth of 100 MHz, and an FFT length of $N_{\mathrm{FFT}} = 2048$. Therefore, one OFDM symbol length (excluding the GI) is 20.48 μs and the GI is set to 0.8 μs (corresponding to 80 samples). Furthermore, a $(561, 753)_{\mathrm{oct}}$ convolutional code with rate $R = 1/2$ was selected as channel code. The MT moves with an average velocity of 40 mph or is static. For all simulations, perfect channel knowledge from all BSs is assumed and each generated signal of each BS is transmitted over independent channels. Two different modulation alphabets are chosen, namely 4-QAM and 16-QAM.

The bit error rate (BER) performances for the conventional cellular OFDM system for different C/I values are given in Figure 4(a). The system is half (RL = 0.5) and fully loaded (RL = 1.0). For the simulations a signal-to-noise ratio (SNR) of 5 dB is assumed. The closer the MT is at the cell border ($C/I = 0$ dB), the more the performance degrades for higher loads due to the increased inter-cellular interference. Figure 4(a) shows also the performances of the applied CAT in the cellular system for different scenarios. In contrast to the conventional system, the BER can be dramatically improved at the cell border. By using the CAT, the MT exploits the additional transmit diversity where the maximum is given at the cell border. If the MT moves with higher velocity (40 mph), the correlation of the sub-carrier fading coefficients in time direction decreases. This incremental violation of the quasi-stationarity assumption of the fading is profitable compensated by the channel code.

The total violation of the aforementioned constraint of CAT is achieved by a fully-interleaved OFDMA frame. There is a large performance

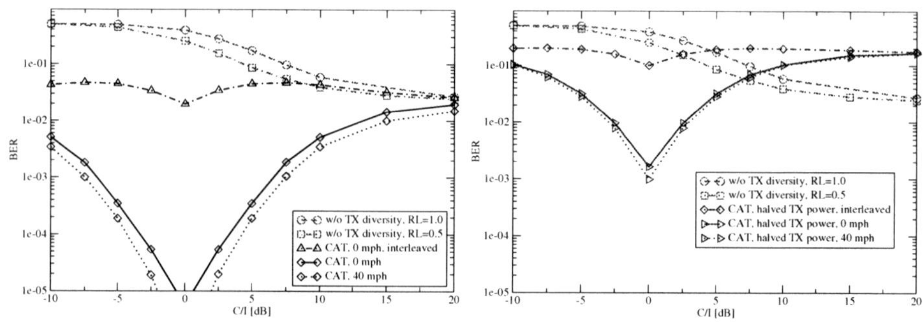

(a) Performance with no transmit diversity and CAT

(b) Performance of using the halved transmit power at each base station

Figure 4. BER versus C/I for an SNR of 5 dB for different scenarios

degradation compared to the CAT performance with a non-interleaved frame. Nevertheless, a residual transmit diversity exists, the MT benefits at the cell border, and the performance is improved. The applied CAT is not only robust for varying MT velocities but also for non-quasi-static channel characteristics.

To establish a more detailed understanding we analyze the CAT with halved transmit power. In contrast to the investigations in Figure 4(a), the total desired received power at the MT is equal to the conventional OFDMA system. Figure 4(b) shows that there is still a performance gain due to the exploited transmit diversity for $C/I < 7.5$ dB. The characteristics of the CAT performances are similar to the performances with full transmit power. Another benefit of the halved transmit power for the used CAT sub-carriers is a reduction of the inter-cellular interference for the neighboring cells.

Simulation results at the cell border regarding the SNR are shown in Figure 5(a) for different modulation alphabets and velocities. For a BER of 10^{-3}, the CAT performances degrades by about 4 dB SNR by using 16-QAM. On the other hand, the CAT performance improves for higher velocities due to the increased time diversity. Since the performances of the conventional system are quite similar for different modulation alphabets and velocities, only the best performance (4-QAM or 16-QAM with 0 mph and RL $= 0.5$) is displayed in the figure. The reference system has a marginal improvement for higher SNR values. In contrast, the CAT — even for higher modulation alphabets — can provide a BER of 10^{-3} or 10^{-4} at the cell border for acceptable SNR values (4-QAM: 2.5 dB|4 dB, 16-QAM: 6.5 dB|9 dB).

Finally, we investigate the throughput of the systems. The conventional OFDMA system with different system loads and modulation alphabets is compared with the CAT. For the CAT different scenarios

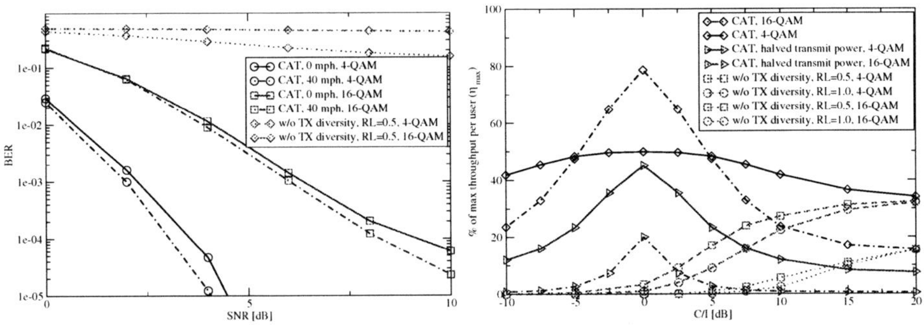

(a) Performances at the cell border (C/I = 0 dB) for different modulation alphabets and velocities

(b) Performances of using no transmit diversity and CAT with full and halved transmit power

Figure 5. BER versus SNR and throughput per user versus C/I performances

are chosen: full and halved transmit power; 4-QAM and 16-QAM. The investigations are limited to the medium access control (MAC) layer automatic-repeat-request (ARQ) scheme [9]. We assume one transmission and no retransmission of an erroneous packet and frame. By using the frame error rate (FER), we can calculate the throughput η by $\eta = \eta_{\max}(1 - \text{FER})$. Since we assume the total number of sub-carriers is equally distributed to the maximum number of users per cell, the systems have a maximum throughput for each user of $\eta_{\max}$. We regard the maximum throughput of 16-QAM as the reference value for $\eta_{\max}$. Therefore, the maximum throughput of the 4-QAM systems is $\eta_{\max}/2$.

Figure 5(b) shows the throughput performances versus C/I for an SNR of 5 dB. The CAT outperforms the conventional system at the cell border for all scenarios. The closer the MT is to the desired BS ($C/I > 15$ dB), the performances merge with the conventional system. An exception represents the CAT with halved transmit power for high C/I as the received power is only obtained from one BS. For CAT with 4-QAM the throughput is nearly constant and achieves its maximum throughput of $\eta_{\max}/2$ for all C/I. Between $C/I = [-5\text{ dB}, 5\text{ dB}]$, the throughput can be increased by using 16-QAM. In contrast, for halved transmit power 4-QAM outperforms 16-QAM. A power and/or modulation adaptation from the BSs opens the possibility for the MT to request a higher throughput in the severe cell border area. Even for the CAT with halved transmit power and 4-QAM or 16-QAM at the cell border the performance surpasses the conventional system without inter-cellular interference ($C/I > 15$ dB). Therefore, it is possible to decrease the inter-cellular interference at the cell border and to provide a quasi-constant throughput throughout the cell. All these characteristics can be utilized by soft handoff concepts.

5. Conclusions

In this paper we investigated the potential of a transmit diversity technique to increase and to offer a stable performance in the most critical area, namely the cell border. The technique is based on the Alamouti scheme which is applied on adjacent base stations in a cellular communications system. Our investigations showed that at the cell border the system: achieves a higher throughput; performs robust to different velocities and to non-stationary fading values; can provide reliable performances simultaneously for one or two mobile terminals; allows to decrease the overall inter-cellular interference.

6. Acknowledgement

This work has been performed in the framework of the IST projects IST-4-027756 WINNER [10] and IST-2002-507039 4MORE [11], which are partly funded by the European Union.

References

[1] Gordon L. Stüber. *Principles of Mobile Communication.* Kluwer Academic Publishers, 2nd edition, 2002.

[2] Daniel Wong and Teng J. Lim. Soft handoffs in CDMA mobile systems. *IEEE Personal Communications*, 36:6–15, December 1997.

[3] S. M. Alamouti. A simple transmit diversity technique for wireless communications. *IEEE Journal of Select Areas Communications*, 16(8):1451–1458, October 1998.

[4] Manabu Inoue, Takeo Fujii, and Masao Nakagawa. Space time transmit site diversity for OFDM multi base station system. In *Proceedings IEEE Mobile and Wireless Communication Networks (MWCN 2002), Stockholm, Sweden*, pages 40–34, September 2002.

[5] H. Sari and G. Karam. Orthogonal frequency-division multiple access and its application to CATV networks. *European Transactions on Telecommunications (ETT)*, 9(6):507–516, November–December 1998.

[6] IEEE 802.11-03/940r2. IEEE P802.11 wireless LANs, TGn channel models, January 2004.

[7] Vahid Tarokh, Hamid Jafarkhani, and A. Robert Calderbank. Space-time block codes from orthogonal designs. *IEEE Transactions on Information Theory*, 45(5):1456–1467, July.

[8] IST-2003-507581 WINNER. Final report on identified RI key technologies, system concept, and their assessment, December 2005.

[9] Shu Lin, Daniel J. Costello Jr., and Michael J. Miller. Automatic-repeat-request error-control schemes. *IEEE Communications Magazine*, 22(12):5–17, December 1984.

[10] http://www.ist-winner.org.

[11] http://www.ist-4more.org.

ITERATIVE INTERCELL INTERFERENCE CANCELLATION FOR DL MC-CDMA SYSTEMS

M. Chacun, M. Hélard and R. Legouable
France Telecom R&D, Broadband Wireless Access Laboratory
4 rue du Clos Courtel, 35512 Cesson-Sévigné, France
(maryline.helard,rodolphe.legouable)@orange-ftgroup.com

Abstract This paper investigates the cancellation of intercellular interference for downlink (DL) MC-CDMA systems. We especially focus on two turbo-iterative receiver schemes that allow cancellation of a part of the intercell interference. After deriving the block diagram receivers and the mathematical expressions at different receiver stages for both receivers, we present several link-level simulation results considering B3G multicellular scenario in presence of one interfering cell. We demonstrate the significant performance improvements of the proposed schemes over a wide signal to interference ratio (SIR) area at half and full loads.

Index Terms— Iterative receiver, intercell interference, MC-CDMA.

1. Introduction

Multicarrier code-division multiple access (MC-CDMA) systems have been initially proposed in [1, 2]. This technique constitutes an efficient way to combine CDMA and orthogonal frequency division multiplexing (OFDM). Nowadays, MC-CDMA is considered as one of the possible candidates for the B3G downlink communication systems. In practice MC-CDMA systems operate over a dispersive channel. However, no intersymbol interference (ISI) is created due to guard interval insertion in terms of cyclic prefix (CP) larger than the longest channel echo.

Degradations due to intercell interference constitute one of the main challenge for modern wireless communication systems, whenever a frequency reuse factor of one is targeted. It has been verified in [3, 4] that it can be modelled as white Gaussian noise and affects the performance in a similar way to Gaussian noise of the same power.

S. Plass et al. (eds.), Multi-Carrier Spread Spectrum 2007, 277–286.

Interference cancellation often investigated in different contexts but mainly in single cell environments. Processing like successive interference cancellation (SIC) [5] and parallel interference cancellation (PIC) detectors [6, 7] are often suggested. Concerning intercell interference, [9] proposes schemes based on PIC detectors that achieve already significant improvements in a multi-cell environment.

Nevertheless, the knowledge of estimated interfering data can efficiently be used to limit the negative effect of interference thanks to iterative receivers. Iterative processes are from now well known in the digital communication systems. Based on turbo principles, [10] proposes interference cancellation detector based on MMSE filter using estimated symbols issued from the channel decoder. We will adapt this detector scheme to cancel the intercell interference for DL MC-CDMA systems.

The sequel of the paper is organized as follows. Section II introduces the signal model for an MC-CDMA system in a multicellular environment. In Section III, the proposed iterative receiver is described and adapted to a MC-CDMA multicellular transmission. Simulation results are then presented in Section IV for both half and full loads considering parameters close to those used within the 3GPP/LTE framework. The behavior of the iterative receiver is analyzed with the help of exit-charts illustrating exchanges of reliability between the parallel structures. Conclusion are drawn in section V.

2. Multi-cell MC-CDMA system

The MC-CDMA technique has the characteristic to multiplex spread data symbols over a large number of orthogonal carriers. Consider a DL MC-CDMA system with the following parameters: N the FFT size, N_{mod} the number of modulated subcarriers, Δ the length of the CP, L_c the processing gain, and $N_u \leq L_c$ the number of active users. We assume that an integer number of m CDMA symbols constitutes the OFDM symbol, with $m = \frac{N_{mod}}{L_c}$. A scrambling matrix $\mathbf{C_s}(n)$, specific to each cell multiplies in the frequency domain the transmit signal before OFDM modulation.

Let $s_{i,k}(n)$ the emitted symbol at the n-th time instant, of the k-th user, and lying in the i-th set inside the OFDM symbol (with $i \in [1, m]$) then we can define $\mathbf{s}(n)$ as follows

$$\mathbf{s}(n) = \begin{bmatrix}\mathbf{s}_1^T(n) & \cdots & \mathbf{s}_m^T(n)\end{bmatrix}^T, \tag{1}$$

where $\mathbf{s}_i(n) = [s_{i,1}(n) \ \cdots \ s_{i,N_u}(n)]^T$. Let $\mathbf{C}$ denote the Walsh-Hadamard matrix and $\mathbf{C}_{N_u}$ the selection of the former's columns that correspond to the N_u active users. Thus, we obtain the spread vector $\mathbf{x}(n)$ of size

$N_{mod} \times 1$ that is equal to

$$\mathbf{x}(n) = \left[\mathbf{C}_{N_u}\mathbf{s}_1(n) \ldots \mathbf{C}_{N_u}\mathbf{s}_m(n)\right]^T . \tag{2}$$

Finally, the transmitted signal is given by

$$\mathbf{z}(n) = \begin{bmatrix} \mathbf{F}^H \\ \tilde{\mathbf{F}}^H \end{bmatrix} \begin{bmatrix} \mathbf{0}_k \\ \mathbf{C}_\mathbf{s}(n)\mathbf{x}(n) \\ \mathbf{0}_k \end{bmatrix}, \tag{3}$$

where $\mathbf{F}$ stands for the normalized DFT matrix and $\tilde{\mathbf{F}}$ corresponds to the first Δ columns of the former. Moreover, $\mathbf{0}_k$ denotes a zero vector of length k, and $\mathbf{C}(n)$ constitutes a diagonal $N_{mod} \times N_{mod}$ matrix containing the scrambling coefficients. For this purpose, we use m-sequences of size $2^{15} - 1$ as it was proposed in [8].

The transmitted signal propagates through a multipath channel with impulse response $\mathbf{h} = [h_0 \ \ldots \ h_\Delta]^t$. Here we have assumed that the channel has a finite impulse response of length at most $\Delta + 1$ not exceeding the CP length (plus one).

At the receiver's end after OFDM demodulation (CP removal, Fourier transform, and null carrier removal) the resulting signal takes the following form

$$\mathbf{y}(n) = \mathcal{H}(\omega)\mathbf{C}_\mathbf{s}(n)\mathbf{x}(n) + \eta(n), \tag{4}$$

where $\mathcal{H}(\omega)$ is an $N_{mod} \times N_{mod}$ diagonal matrix of the channel frequency response at the modulated subcarriers, $\eta(n)$ is the additive white Gaussian noise (AGWN) vector.

Consider now an interfering base station which is synchronized with the one of interest. In such a case, (4) becomes

$$\mathbf{y}(n) = \underbrace{\mathcal{H}(\omega)\mathbf{C}_\mathbf{s}(n)\mathbf{x}(n)}_{\mathbf{g}(n)} + \underbrace{\mathcal{H}^I(\omega)\mathbf{C}_\mathbf{s}{}^I(n)\mathbf{x}^I(n)}_{\mathbf{g}^I(n)} + \eta(n), \tag{5}$$

where the superscript "I" denotes quantities coming from the interfering base station.

3. Proposed iterative receiver

3.1 Cancellation of intercell interferences

The proposed iterative receiver shown in Fig.1 is based on the turbo equalization principle firstly dedicated to intersymbol interference cancellation [10]. Since then, this iterative principle has widely been adapted at receiver side to cancel all kinds of interferences (multi-users, multiple antennas...) and for different systems (single or multi-carrier, CDMA)

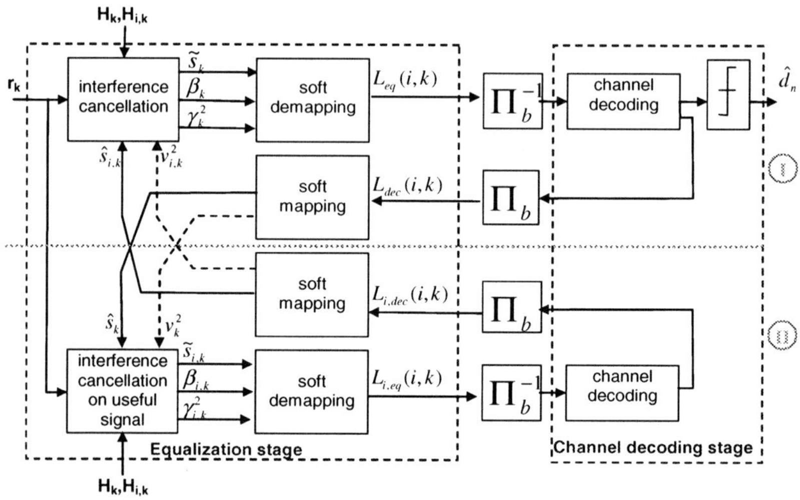

Figure 1. Iterative intercell interference cancellation

[11]. For OFDM based systems with interferences to deal with, such as MC-CDMA with multi-user interference, the iterative receiver is performed after OFDM demodulation in the frequency domain leading simple equalization functions per carrier.

Iterative receivers are commonly composed of 2 main stages: an equalization stage and a channel decoding stage both exchanging soft extrinsic information allowing at each iteration an improvement of the other stage decoding: L_{eq} and L_{dec} are metrics provided respectively by the equalizer and the channel decoder with index k for useful data and (i,k) for interfering data. The two stages are separated by interleavers in order to decorrelate exchanged information. Since interference cancellation is performed at symbol levels whereas channel coding works here on bits, demapping and mapping convert bits into symbols and vice versa. For equalization stage, channel estimations are required, i.e. $\mathbf{H_k}$ and $\mathbf{H_{i,k}}$ for the cell of interest and the interfering one respectively.

Moreover and as it can be observed in Fig.1, two iterative parts (I and II) are operating in parallel aiming at improving the estimation of the useful and of the interfering symbols respectively. In fact, for one iteration, part I required as input an estimation of the interfering symbol $\hat{s}_{i,k}$ and provides an improved estimation of useful symbol $\hat{s}_k$ for next iteration of part II, while part II required as input an estimation of the useful symbol $\hat{s}_k$ and provides an improved estimation of interfering symbol $\hat{s}_{i,k}$ for next iteration of part I.

3.2 Adaptation to MC-CDMA transmission

When considering MC-CDMA transmission in multicell environnment, 2 schemes are considered here to carry out the cancellation function depending on the place to despread signals in the 'interference cancellation process'. The first proposed receiver (IC1) processes the interference cancellation separately from the cell descrambling and despreading functions whilst the second one (IC2) jointly performs them. The 'interference cancellation' presented in general case on Fig.1 is detailed in Fig.2 with the respective equations of each receiver. The structure of the detailed block is the same whatever the part I or II, but represented on this Figure for part I.

IC1: Interference cancellation is thus performed on scrambled and spread symbols $\tilde{u}$ before descrambling, despreading and user demultilexing to be performed. The two weight vectors p_k and q_k are optimized under the MMSE criterion [12] such as

$$(p_k^{opt}, q_k^{opt}) = arg \min_{p_k, q_k} \mathbf{E}\left[|u_k - \tilde{u}_k|^2\right] \tag{6}$$

IC2: Descrambling and despreading are performed before Interference cancellation, thus multi-user interference is also treated in the receiver and p_k, q_k and z_k obtained thanks to

$$(p_k^{opt}, q_k^{opt}, z_k^{opt}) = arg \min_{p_k, q_k, z_k} \mathbf{E}\left[|s_k - \tilde{s}_k|^2\right] \tag{7}$$

4. Performance results

In this section, firstly we list the system parameters used in our DL MC-CDMA simulations. We give some BER (Bit Error Rate) versus SIR (Signal to Interference Ratio) performance results regarding to the system's load and the channel coding rate. To explain the BER performance, we carry out a analysis by exit chart representation.

4.1 System parameters

For performance evaluation, we have carried out simulations with the following parameter values, typical of B3G system: sampling frequency

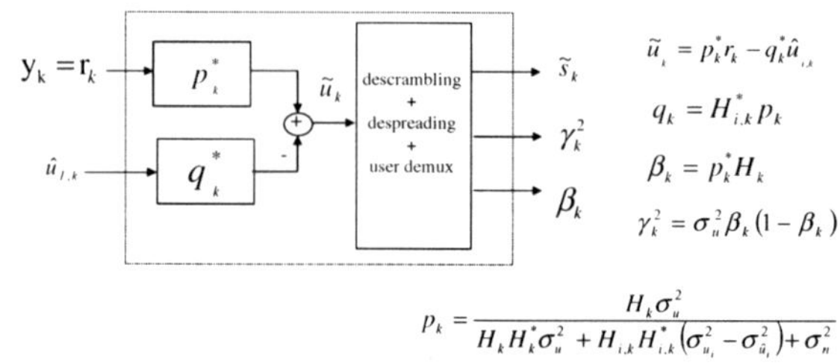

IC1: disjoint interference cancellation and 'descrambling + despreading'

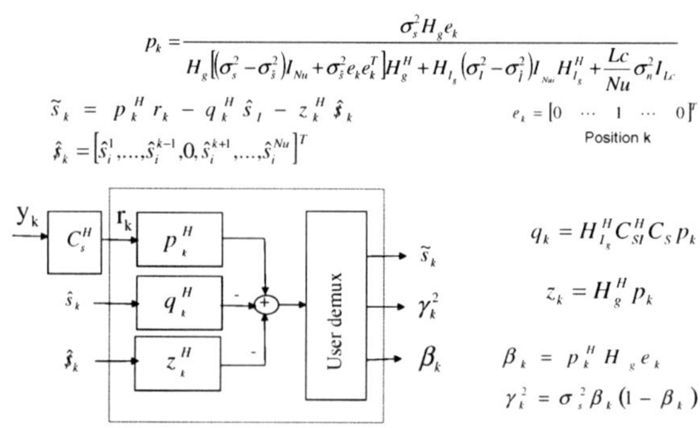

IC 2: joint interference cancellation and 'descrambling + despreading'

Figure 2. Different schemes for intercell interference cancellation for MC-CDMA

$F_e = 15.36MHz$, $N = 1024$, $N_{mod} = 704$, $\Delta = 72$, $L_c = 16$, QPSK modulation, and UMTS turbo-coding scheme of rate $\frac{2}{3}$ or $\frac{1}{3}$, frame length of 12 OFDM symbols, bit interleaving depth of 1056 bits and cell scrambling applied on the whole N_{mod} carriers. The σ_{si}^2 and σ_s^2 values measured at the transmitter before FHT, when considering the IC2 detector, are equal to 1. All the simulations have been carried out by considering the Typical Urban (TU) propagation channel model [13]. Finally, the SNR is fixed to 15dB that corresponds to a BER of 10^{-4} in single cell. The SIR values vary from -20dB to 20dB.

4.2 Performance at half-load

Fig.3 represents the performance obtained with the IC2 detector, for the case where both cells are working at half load, i.e. $K = 8$ users and

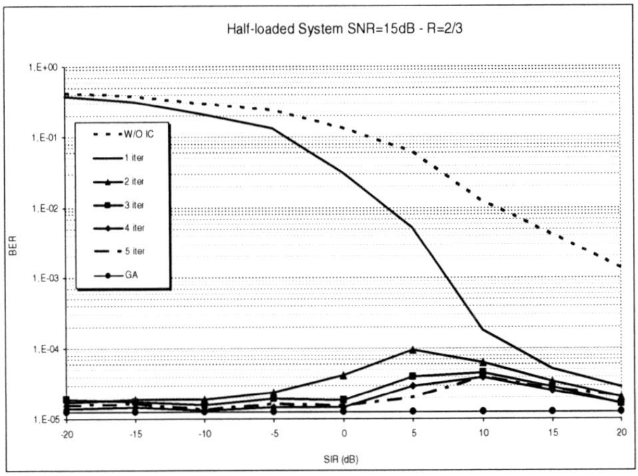

Figure 3. BER of IC2 detector; SNR=15 dB, half-load and $R = \frac{2}{3}$.

with channel coding rate of $\frac{2}{3}$. We can easily observe that the detector improves the performance in the course of iterations compared to none interference cancellation (W/O IC curve), especially in the most critical SIR area ($-10dB < SIR < 10dB$). After the first iteration, corresponding to one single passage into the turbo decoding process that is different to the "W/O IC" curve, by the fact that the interfering channel coefficients are taken into account into the first interference canceller stage, improving the performance. Compared to results obtained with the IC1 detector (not shown here), the IC2 detector gets better convergence and results, showing that carrying out the interference cancellation operation jointly to the despreading-descrambling processes improves the performance of the receiver whatever the SIR value.

Even at the cell border (SIR=0dB), the performance is close to the limit bound (GA curve) leading to more than 25dBs gain compared to a system with classical MMSE equalization. We note a small loss in performance (ascent of the curve) around the SIR=5dB value. This degradation is due to a bad cancellation of the interferer, providing by the approximation of the bias values (β_k and $\beta_{i,k}$) inside the interference cancellation receiver structure.

4.3 Performance at full-load

Fig.4 illustrates the performance obtained with the IC2 detector, for the case where both cells are working at full load, i.e. $K = 16$ users and with channel coding rate of $\frac{1}{3}$. Compared to one full-loaded system implementing channel coding of $\frac{2}{3}$ rate (curve not shown here), a significant

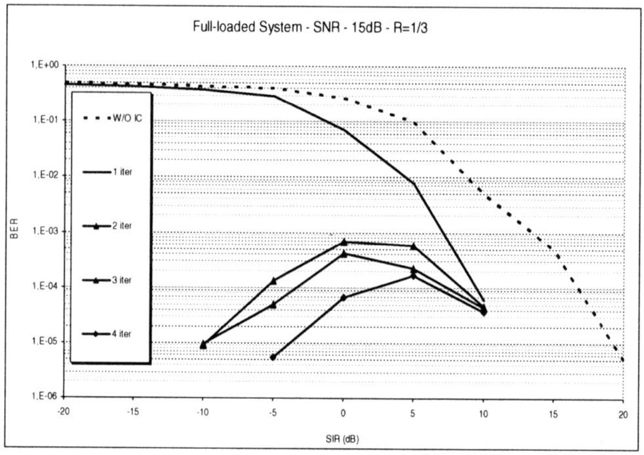

Figure 4. BER of IC2 detector, SNR=15 dB, full-load and $R = \frac{1}{3}$.

gain in performance is obtained even at full load and in a large SIR range (that is not the case with channel coding rate of $\frac{2}{3}$). These results show that the exchange information between the 2 decoders inside the parallel structure is enough reliable to allow the system working. In order to explain the behavior of the receiver according to the system load and the channel coding rate, exit charts are represented in the following subsection.

4.4 Exit-chart representation

The exit chart diagrams is a 2-D representation of the amount of obtained information (decoding output) issued from the amount of given information (decoding input). So, we can plot for each decoder, these informations corresponding either to the exit chart of the useful information bits or of the interfering bits. In this paper, we have only plotted the exit chart about the useful bit since the same behavior is recorded on the interfering bits. Fig.5 represents the exit charts, plotted at $SIR = 0dB$, for a system working either with a channel coding rate of $\frac{2}{3}$ or $\frac{1}{3}$ and/or at half load or full load. The aim of plotting this kind of graphs allows us to see if the system converge or not. Indeed, at half load $R = \frac{2}{3}$ and full load $R = \frac{1}{3}$, starting from one curve, it is possible to attain to the value $Z = 1$, showing that the parallel structure converges as the number of iterations increases. It is not the case at full load $R = \frac{2}{3}$ since the curves cut across.

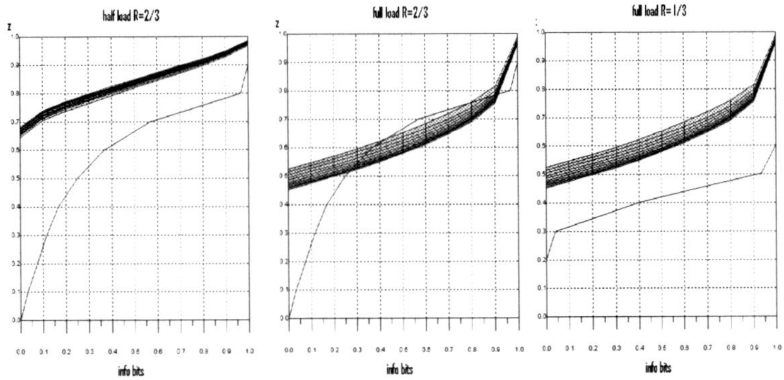

Figure 5. Exit chart of IC2 detector, $R = \frac{2}{3}$ or $R = \frac{1}{3}$, half or full load, SIR=0 dB.

5. Conclusion

The present work focuses on the topic of intercell interference cancellation for downlink MC-CDMA systems. We proposed two iterative receiver schemes that offer significant performance gains in all SIR levels. The best performance has been obtained with the IC2 detector that carries out the interference cancellation operation jointly to the despreading-descrambling processes. The IC2 detector allows to improve the performance even when considering a full loaded system. However, when the load of the system increases, the channel coding rate has to be reduced to allow the convergence of the iterative detector and then significant performance gain compared to one system that doesn't apply interference cancellation. The convergence principle has been explained by exit chart illustration. The main advantage of using such a detector is that it allows to use an unitary frequency reuse factor inside the network, an increase the cell coverage as well as lower transmit powers. Perspective work will focus on complexity evaluation of such kind of receiver with the goal to propose a scheme with good performance/complexity comprise.

Acknowledgment

Part of this work has been carried by M. Chacun when he was a internship student at France Telecom Research and Development.

References

[1] N. Yee, J.P. Linnartz, and G. Fettweis, "Multicarrier CDMA in Indoor Wireless Radio Networks," in Proc. of IEEE PIMRC Conference, Yokohama, Japan, Sep. 1993.

[2] K. Fazel and L. Papke, "On the Performance of Convolutionally-Coded CDMA/OFDM for Mobile Communication System," in Proc. of IEEE PIMRC Conference, Yokohama, Japan, Sep. 1993.

[3] G. Auer, S. Sand, A. Dammann, and S. Kaiser, "Analysis of the Cellular Interference for MC-CDMA and its Impact on Channel Estimation," *European Transactions on Telecommunications*, pp. 173-184, vol. 15, May-June 2004.

[4] X.G. Doukopoulos and R. Legouable, "Impact of the Intercell Interference in DL MC-CDMA Systems," 5th International Workshop on Multi-Carrier Spread Spectrum, MC-SS'2005, Oberpfaffenhofen, Germany, Sep. 2005.

[5] J.G. Andrews and T.H.Y. Meng, "Performance of Multicarrier CDMA with Successive Interference Cancellation in a Multipath Fading Channel," *IEEE Transactions on Communications*, vol. 52, pp. 811-822, May 2004.

[6] R. Hofstad and M.J. Klok, "Performance of DS-CDMA Systems with Optimal Hard-Decision Parallel Interference Cancellation," *IEEE Transactions on Information Theory*, vol. 49, pp. 2918-2940, Nov. 2003.

[7] D. Divsalar, M.K. Simon, and D. Raphaeli, "Improved Parallel Interference Cancellation for CDMA," *IEEE Transactions on Communications*, vol. 46, pp. 258-268, Feb. 1998.

[8] E.H. Dinan and B. Jabbari, "Spreading Codes for Direct Sequence CDMA and Wideband CDMA Cellular Networks," *IEEE Communications Magazine*, vol. 36, no 9, pp. 48-54, Sep. 1998.

[9] X. Doukopoulos and R. Legouable, "Intercell Interference Cancellation for MC-CDMA Systems," *IEEE VTC07*, Dublin, Ireland, Apr. 2007.

[10] A. Glavieux, C. Laot and J. Labat, "Turbo equalization over a frequency selective channel ". *In Proceedings of ISTC'97*, Brest, France, pp 96-102, Sept. 1997.

[11] P.J. Bouvet and M. Helard, "Near optimal performance for high data rate MIMO MC-CDMA scheme", 5th International Workshop on Multi-Carrier Spread Spectrum, MC-SS'2005, Oberpfaffenhofen, Germany, Sep. 2005.

[12] C. Laot, R. Le Bidan, and D. Leroux, "Low complexity linear turbo equalization: A possible solution for EDGE", IEEE Trans. Wireless. Commun., 2005.

[13] COST 207. "Digital Land Mobile Radio Communications". Report, Office for Official Publications of the European Communities, July 1989.

POSITIONING WITH GENERALIZED MULTI-CARRIER COMMUNICATIONS SIGNALS

Christian Mensing, Simon Plass, and Armin Dammann
German Aerospace Center (DLR)
Institute of Communications and Navigation
Oberpfaffenhofen, 82234 Wessling, Germany
{christian.mensing, simon.plass, armin.dammann}@dlr.de

Abstract This paper covers positioning techniques for cellular networks using generalized multi-carrier communications signals. These techniques are based on symbol-timing synchronization algorithms to find the starting point for the underlying orthogonal frequency division multiplexing symbols of the incident signals. Well known algorithms are, e.g., the Schmidl-Cox algorithm and the Minn algorithm. These algorithms are investigated w.r.t. their positioning capabilities. Furthermore, they are extended to improve the performance in multipath environments and to achieve a sufficient accuracy with weak signals from neighboring out-of-cell base stations.

1. Introduction

Positioning in wireless networks became very important in recent years. Services and applications based on accurate knowledge of the location of the mobile station (MS) will play a fundamental role in future wireless systems [1]. In addition to vehicle navigation, fraud detection, automated billing, and further promising user-side applications, it is stated by the United States Federal Communications Commission (FCC) that all wireless service providers have to deliver the location of all enhanced 911 (E-911) callers with specified accuracy [2]. Additionally, system-side applications (e.g., resource management, handover, etc.) can improve their performance by making use of the knowledge of the MS position.

MS localization using global navigation satellite systems (GNSSs) such as the Global Positioning System (GPS) or the future European Galileo system [3] deliver very accurate position information for good environmental conditions but require expensive extensions of the MS. Therefore, in this paper we concentrate on determination of the MS

S. Plass et al. (eds.), Multi-Carrier Spread Spectrum 2007, 287–296.

location by exploiting already available communications signals. Generally, this localization process is based on measurements in terms of time of arrival (TOA), time difference of arrival (TDOA), angle of arrival (AOA), and/or received signal strength (RSS) [4], provided by the base stations (BSs) or the MS, where the achievable accuracy is the highest with the timing measurements. TDOA measurements are also foreseen for positioning in future fourth generation (4G) mobile communications systems as it is proposed, e.g., within the WINNER project [5] [6].

Extracting timing measurements from the communications signals directly results in a synchronization problem where normally the arrival time of the signal transmitted from the BS has to be measured at the MS. For code division multiple access (CDMA) signals these algorithms and their positioning capabilities are well studied [4]. However, the performance in orthogonal frequency division multiplexing (OFDM) based systems, namely generalized multi-carrier (GMC) [7], is of particular interest because of the future relevance of this modulation scheme.

Usually, OFDM with its cyclic extension by the guard interval is designed in a way, that the symbol-timing synchronization is robust against multipath effects disturbing the signal. The fast Fourier transform (FFT) window can be shifted within a certain range of the guard interval, depending on its length and the length of the channel impulse response (CIR). But for positioning purposes, each deviation from the perfect symbol-timing yields a positioning error. Well-known symbol-timing synchronization algorithms for OFDM are the Schmidl-Cox algorithm [8] and the Minn algorithm [9]. In this paper, both are investigated w.r.t. their positioning capabilities in performance limiting multipath channels. Additionally, the influence of carrier-frequency offsets is analyzed. Furthermore, we introduce an offset correction based on the estimated CIR to improve the performance in a fine tuning step and test the positioning performance of the proposed approach in a cellular communications system.

In Section 2, the GMC system model for synchronization is introduced. Section 3 describes the considered synchronization algorithms with main focus on the positioning capabilities. To obtain the position of the MS from the measurements the navigation equation has to be solved which is part of Section 4. Finally, in Section 5 the simulation results are presented.

Throughout this paper, vectors and matrices are denoted by lower and upper case bold letters. The operation '$\otimes$' denotes the Kronecker product, $\mathrm{E}\{\cdot\}$ expectation, $(\cdot)^{\mathrm{T}}$ transpose, $(\cdot)^*$ conjugate, $(\cdot)^{\mathrm{H}}$ conjugate transpose, $\|\cdot\|_2$ the Euclidean norm, and $[\cdot]_{2\pi}$ the modulo 2π operation.

2. System Model

We consider a link level GMC based transmission scheme where the samples of the transmitted baseband OFDM signal with assumed ideal Nyquist pulse shaping can be expressed as

$$s(k) = \frac{1}{\sqrt{N_{\mathrm{FFT}}}} \sum_{n=0}^{N_{\mathrm{u}}-1} S(n) \exp\left(\jmath \frac{2\pi k n}{N_{\mathrm{FFT}}}\right), \qquad (1)$$
$$-N_{\mathrm{GI}} \leq k \leq N_{\mathrm{FFT}} - 1,$$

where $S(n)$ are the modulated data symbols for the N_{u} used subcarriers, N_{FFT} is the FFT length, and N_{GI} is the number of samples for the guard interval. After transmission over a frequency selective fading channel of length L with path gains

$$h(\ell), \quad 0 \leq \ell \leq L-1, \qquad (2)$$

the signal becomes

$$x(k) = \sum_{\ell=0}^{L-1} h(\ell)\, s(k-\ell) + n(k), \qquad (3)$$

with zero-mean additive white Gaussian noise (AWGN) $n(k)$ of variance σ_n^2. At the receiver, there generally exist symbol-timing offset, carrier-frequency offset, and sampling clock offset w.r.t. the transmitter. But in this paper we assume perfect sampling clock offset correction because timing and frequency offset errors are usually more dominant compared to sampling clock errors [9]. Hence, the received samples can be written as

$$r(k) = \exp\left(\jmath \frac{2\pi k v}{N_{\mathrm{FFT}}}\right) x(k), \qquad (4)$$

with assumed carrier-frequency offset of v normalized by the subcarrier spacing. The timing point of the start of the FFT window is determined by the timing synchronization to be at the sample $r(\varepsilon)$, where ε is the symbol-timing offset in terms of OFDM samples.

3. Communications Synchronization Algorithms for Positioning

Aim of the symbol-timing synchronization is to find the start of the OFDM symbols. Due to the guard interval with the cyclic extension, OFDM based systems are quite robust against imperfect symbol-timing, i.e., in a multipath channel of length L the start of the FFT window can be shifted within an interval of $N_{\mathrm{GI}} - L$ samples (range without any intersymbol interference). Contrary, for positioning this timing offset directly results in an estimation error. Usually, in a first step a coarse timing synchronization is performed. After that, the carrier-frequency offset is estimated and integrated in the following channel estimation. Finally, with the knowledge of the CIR a fine symbol-timing is calculated. Clearly, for OFDM communications systems a final fine frequency-offset compensation is required. However, for positioning purposes we are just interested in the correct timing point. Usually, OFDM synchronization algorithms are based on specifically designed training or pilot sequences at the beginning of each OFDM frame. In the following, we describe a commonly used synchronization algorithm for communications by Minn [9] and extend it for positioning purposes. Note that the well-known Schmidl-Cox algorithm can be seen as a special case of the more general Minn algorithm (cf. Section 3.1).

3.1 Coarse Symbol-Timing Offset Estimation

Minn algorithm estimates both symbol-timing offset and carrier-frequency offset where different pilot patterns are considered [9]. Generally, the pilot patterns are given in terms of repeated sequences $\boldsymbol{a}$ in time-domain of equal length M and follow Golay complementary sequences (cf. [10]). Hence, in each OFDM symbol $B = N_{\mathrm{FFT}}/M$ of such sequences can be included, e.g.,

$$\boldsymbol{s} = \left[\boldsymbol{s}^{\mathrm{T}}(0), \ldots, \boldsymbol{s}^{\mathrm{T}}(B-1)\right]^{\mathrm{T}} = \left[-\boldsymbol{a}^{\mathrm{T}}, \boldsymbol{a}^{\mathrm{T}}, -\boldsymbol{a}^{\mathrm{T}}, -\boldsymbol{a}^{\mathrm{T}}\right]^{\mathrm{T}} \tag{5}$$

is an appropriate sequence for $B = 4$. Generally, the coarse timing estimation makes use of the structure of these pilot symbols — without knowledge of the transmitted pilot sequence itself — and calculates the correlation properties between the different sequences. Thus, Minn's metric for this structure is given as

$$\Lambda(d) = \left(\frac{B}{B-1}\frac{|P(d)|}{E(d)}\right)^2, \tag{6}$$

with

$$P(d) = \sum_{b=0}^{B-2} g(b) \sum_{m=0}^{M-1} r^*(d + bM + m) \cdot r(d + (b+1)M + m) \tag{7}$$

and

$$E(d) = \sum_{m=0}^{M-1}\sum_{b=0}^{B-1} |r(d + m + bM)|^2, \tag{8}$$

where

$$g(b) = \frac{\operatorname{sign}\left(\boldsymbol{s}^{\mathrm{T}}(b)\right)\operatorname{sign}\left(\boldsymbol{s}(b+1)\right)}{M}, \quad 0 \le b \le B-2, \tag{9}$$

includes the structure of the sequence $\boldsymbol{s}$. The parameter d describes the time index in samples. The correlation function $\Lambda(d)$ gives a peak at the correct timing point in the multipath-free and noise-free case, and hence, the coarse timing estimate can be found at

$$\hat{\varepsilon}_{\mathrm{Minn}} = \underset{d}{\operatorname{argmax}} \, \Lambda(d) \,. \tag{10}$$

The metric proposed by Schmidl-Cox (SC) can be derived from Minn's metric by using the pattern $\boldsymbol{s} = \left[\boldsymbol{a}^{\mathrm{T}}, \boldsymbol{a}^{\mathrm{T}}\right]^{\mathrm{T}}$ and approximating $E(d) \approx 2\sum_{m=0}^{M-1} |r(d+m+M)|^2$. Contrary to Minn, this metric gives a plateau of length N_{GI} in the multipath-free and noise-free case where the timing point is found by the 90%-method (cf. [8]): The maximum point of $\Lambda(d)$ is first found, and then two points with 90% of the maximum value to the left and to the right of the maximum value. The timing estimate $\hat{\varepsilon}_{\mathrm{SC}}$ is then given as the average of these two 90% points.

3.2 Coarse Carrier-Frequency Offset Estimation

The coarse carrier-frequency offset estimator is an adaptation by Minn [9] of the carrier-frequency offset estimator proposed in [11]. It is given by

$$\hat{v} = \frac{B}{2\pi} \sum_{m=1}^{B/2} w(m)\, \phi(m) \,, \tag{11}$$

with the pre-factors

$$w(m) = 3\frac{4(B-m)(B-m+1) - B^2}{2(B^2-1)}, \qquad 1 \leq m \leq \frac{B}{2}, \tag{12}$$

and the phase differences

$$\phi(m) = \left[\arg\{R_r(m)\} - \arg\{R_r(m-1)\}\right]_{2\pi} \,, \qquad 1 \leq m \leq \frac{B}{2}. \tag{13}$$

The coefficients

$$R_r(m) = \frac{1}{N_{\mathrm{FFT}} - mM} \sum_{k=\hat{\varepsilon}+mM}^{\hat{\varepsilon}+N_{\mathrm{FFT}}-1} r^*(k-mM)\, r(k) \,, \qquad 0 \leq m \leq \frac{B}{2}, \tag{14}$$

describe the correlation between the received samples after the coarse estimated timing point.

3.3 Impact of Channel Estimation

Multipath channel environments result in a bias of the symbol-timing offset estimates. Hence, for fine tuning the estimate of the CIR is required for performance improvement already in the synchronization entity. Clearly, different channel estimation algorithms can be considered with different performance and complexity. However, we will use the least squares based algorithm adapted in [9] that can directly be integrated in the proposed synchronization scheme and gives the channel estimate $\hat{h}(l)$.

3.4 Fine Symbol-Timing Offset Estimation

The multipath channel results in a bias of the symbol-timing offset estimation. Therefore, we propose to use an offset correction method based on the knowledge of the estimated CIR. Exploiting the linearity of the timing estimators, we can calculate the offset as

$$\Delta\varepsilon = \frac{\sum_{\ell=0}^{L-1} (\ell+1) |h(\ell)|^2}{\sum_{\ell=0}^{L-1} |h(\ell)|^2} \tag{15}$$

which is the center of gravity of the CIR. Using this correction term, the corrected symbol-timing offset can be calculated as

$$\hat{\varepsilon}_{\text{Minn,fine}} = \hat{\varepsilon}_{\text{Minn}} - \Delta\varepsilon + 1 \tag{16}$$

for the Minn algorithm, and

$$\hat{\varepsilon}_{\text{SC,fine}} = \hat{\varepsilon}_{\text{SC}} + \frac{N_{\text{GI}} - \Delta\varepsilon + 1}{2} \tag{17}$$

for the Schmidl-Cox algorithm.

3.5 Performance with Weak Signals

Usually, for positioning measurements with at least three BSs have to be taken into account. Clearly, for communications purposes the network is designed in a way that overhead is minimized and only one strong BS should be hearable for the MS for spectral efficiency reasons. This results in the fact that the signal from the serving BS is very strong where the signals from the neighboring out-of-cell BSs are very weak. However, for positioning purposes, we are not interested in the transmitted data itself. Thus, averaging over several received pilots (e.g., over one second) is possible without considerable limitation of the MS's mobility. This procedure results in an equivalent signal-to-noise ratio (SNR) increase. Furthermore, inter- and intracell interference can limit the performance in cellular networks and suitable interference mitigation techniques have to be applied which is not in the focus of this paper.

4. Solution of the Navigation Equation

In the sections before, we have just considered link-level algorithms to obtain the TOA of the incident signal. Nevertheless, state-of-the-art positioning uses TDOA based solutions where the difference of TOA measurements is delivered w.r.t. one reference BS (cf. [1, 12]). At least three BSs are required to get a unique solution in the two-dimensional case.

The time synchronized BSs are organized in a cellular network with cell radius R according to Figure 1(a). The MS is located at $\boldsymbol{x} = [x, y]^{\mathrm{T}}$ and only the N_{BS} nearest BSs at $\boldsymbol{x}_\mu, \mu \in \{1, 2, \ldots, N_{\mathrm{BS}}\}$ are used for positioning. The distance between the BSs and the MS is given by

$$r_\mu(\boldsymbol{x}) = \|\boldsymbol{x}_\mu - \boldsymbol{x}\|_2 = \sqrt{(x_\mu - x)^2 + (y_\mu - y)^2}. \tag{18}$$

We treat distances and propagation times as equivalent, and thus, the TDOAs for BS $\nu \in \{2, 3, \ldots, N_{\mathrm{BS}}\}$ w.r.t. BS 1 can be written as

$$d_{\nu,1}(\boldsymbol{x}) = r_\nu(\boldsymbol{x}) - r_1(\boldsymbol{x}), \tag{19}$$

where — without loss of generality — we use BS 1 as the reference BS. From a geometric point of view the MS lies on hyperbolas with foci at the two related BSs. The at most $N_{\mathrm{BS}} - 1$ linear independent TDOAs compose the vector

$$\boldsymbol{d}(\boldsymbol{x}) = [d_{2,1}(\boldsymbol{x}), d_{3,1}(\boldsymbol{x}), \ldots, d_{N_{\mathrm{BS}},1}(\boldsymbol{x})]^{\mathrm{T}}, \tag{20}$$

and the corresponding TDOA measurements are given by

$$\mathsf{d} = [\mathsf{d}_{2,1}, \mathsf{d}_{3,1}, \ldots, \mathsf{d}_{N_{\mathrm{BS}},1}], \tag{21}$$

based on the measurement model

$$\mathsf{d} = \boldsymbol{d}(\boldsymbol{x}) + \mathsf{n}, \tag{22}$$

where

$$\mathsf{n} = [\mathsf{n}_{2,1}, \ldots, \mathsf{n}_{N_{\mathrm{BS}},1}]^{\mathrm{T}} \tag{23}$$

is assumed zero-mean AWGN with covariance matrix

$$\boldsymbol{\Sigma}_{\mathsf{n}} = \mathrm{E}\left\{\mathsf{n}\mathsf{n}^{\mathrm{T}}\right\}. \tag{24}$$

Note that the TDOA measurements can directly be calculated from the estimated timing-offsets $\hat{\varepsilon}$ as shown in the previous section. For the solution of the estimation problem for the MS location it is a common way to follow the weighted non-linear least squares approach [4] which minimizes the cost function

$$\varepsilon(\boldsymbol{x}) = (\mathsf{d} - \boldsymbol{d}(\boldsymbol{x}))^{\mathrm{T}} \boldsymbol{\Sigma}_{\mathsf{n}}^{-1} (\mathsf{d} - \boldsymbol{d}(\boldsymbol{x})) \tag{25}$$

w.r.t. the unknown MS position $\boldsymbol{x}$ yielding

$$\hat{\boldsymbol{x}} = \underset{\boldsymbol{x}}{\operatorname{argmin}}\, \varepsilon(\boldsymbol{x}). \tag{26}$$

In the general case, there exists no closed-form solution to this non-linear two-dimensional optimization problem, and hence, iterative approaches are necessary. A standard approach to deal with (26) is based on the Gauss-Newton (GN) algorithm [13, 12]. The GN algorithm linearizes the system model in (22) about some initial value $\boldsymbol{x}^{(0)}$ yielding

$$\boldsymbol{d}(\boldsymbol{x}) \approx \boldsymbol{d}\left(\boldsymbol{x}^{(0)}\right) + \boldsymbol{\Phi}(\boldsymbol{x})\Big|_{\boldsymbol{x}=\boldsymbol{x}^{(0)}} \left(\boldsymbol{x} - \boldsymbol{x}^{(0)}\right), \tag{27}$$

with the elements of the $(N_{\mathrm{BS}} - 1) \times 2$ Jacobian matrix

$$\boldsymbol{\Phi}(\boldsymbol{x}) = \left[\frac{\partial}{\partial x}, \frac{\partial}{\partial y}\right] \otimes \boldsymbol{d}(\boldsymbol{x}) . \tag{28}$$

Afterwards, using (27) and (25), the linear least squares procedure is applied resulting in the iterated solution

$$\begin{aligned} \boldsymbol{x}^{(k+1)} = \boldsymbol{x}^{(k)} + \left(\boldsymbol{\Phi}^{\mathrm{T}}\left(\boldsymbol{x}^{(k)}\right) \boldsymbol{\Sigma}_{\mathbf{n}}^{-1} \boldsymbol{\Phi}\left(\boldsymbol{x}^{(k)}\right)\right)^{-1} \\ \cdot \boldsymbol{\Phi}^{\mathrm{T}}\left(\boldsymbol{x}^{(k)}\right) \boldsymbol{\Sigma}_{\mathbf{n}}^{-1}\left(\mathbf{d} - \boldsymbol{d}\left(\boldsymbol{x}^{(k)}\right)\right) . \end{aligned} \tag{29}$$

5. Simulation Results

Simulation results are shown for a system with QPSK modulation, FFT length of $N_{\mathrm{FFT}} = 1024$, used subcarriers of $N_{\mathrm{u}} = 1000$, and $N_{\mathrm{GI}} = 128$ samples for the guard interval. The bandwidth is 50 MHz, which results in a chip duration of $T_{\mathrm{C}} = 20\,\mathrm{ns}$. From a positioning point of view this chip duration yield an equivalent chip length of $6\,m$. For the simulations we use the WINNER C2 wide area channel model [5]. Furthermore, we assume a carrier-frequency offset of $v = 10$ normalized to the subcarrier spacing.

In Figure 1(b) the timing metrics for the algorithms are shown for coarse and fine symbol-timing estimation for the WINNER wide area channel model C2 in the noise-free case. The performance improvement after fine tuning is about 20 samples (equivalent to $120\,m$) for this realization. We further can see the clear peak given by Minn's algorithm where the Schmidl-Cox algorithm with its plateau will result in ambiguities especially under noisy conditions.

Figure 1(c) shows the root mean square error (RMSE) in meters as performance measure for the timing measurement over the SNR for the different algorithms using the WINNER C2 channel model, supplemented by a comparison with the AWGN channel. For instance, at an RMSE of $20\,m$ we lose approximate 7 dB if we compare the multipath channel with the AWGN channel by using Minn's algorithm. The Schmidl-Cox algorithm has a distinctive error floor at large RMSEs, and thus, it is not suitable for high accurate positioning applications.

Finally, in Figure 1(d) the performance of the Minn algorithm is tested in a cellular network environment under multipath conditions. Note that NLOS propagation is not investigated here. For positioning the strongest $N_{\mathrm{BS}} = 3$ BSs are used and an averaging is performed over 100 synchronization symbols which is equivalent to about 0.5 seconds [14]. Furthermore, we assume an SNR of -5 dB at the cell edge where the cell radius is $R = 500\,m$. The cumulative density function (CDF) in Figure 1(d) shows the probability that the RMSE is below a fixed value e_{err}, where several MS positions and noise realizations were performed. We

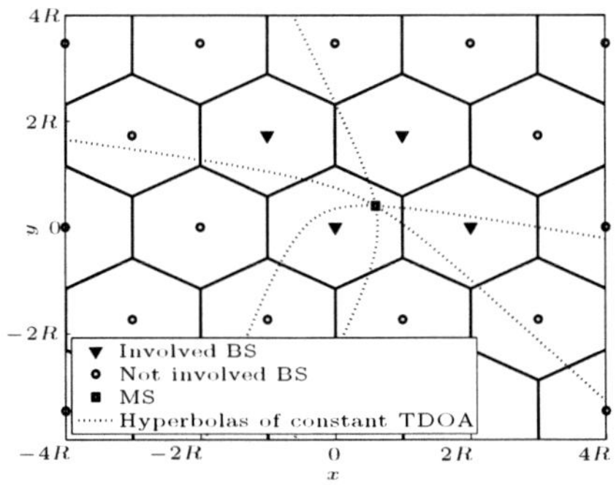

(a) TDOA positioning in cellular networks with $N_{\mathrm{BS}} = 4$ involved BSs

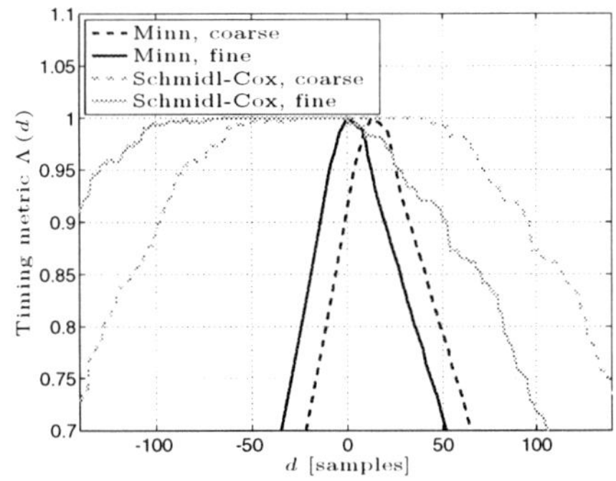

(b) Timing metric for different algorithms using WINNER channel model C2, no noise

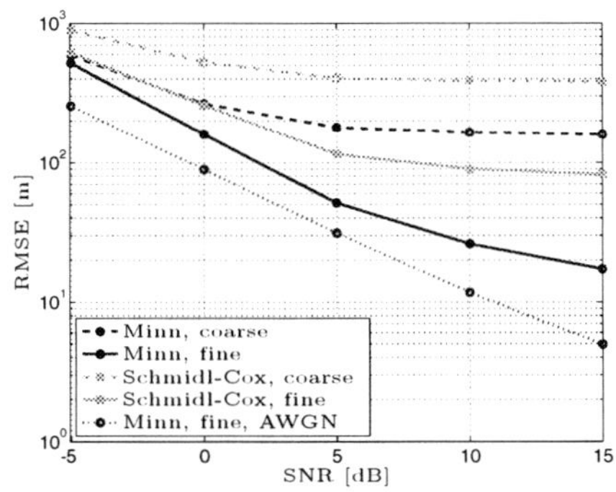

(c) RMSE versus SNR for different algorithms using WINNER channel model C2

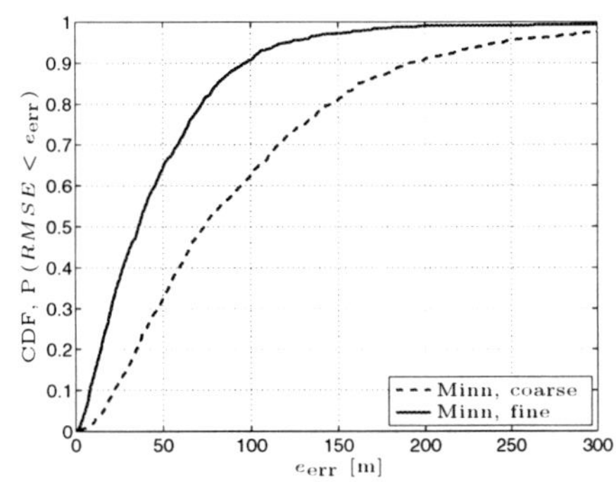

(d) CDF for positioning in a cellular network with $R = 500\,\mathrm{m}$ and $N_{\mathrm{BS}} = 3$

Figure 1. Cell structure and performance evaluation for positioning algorithms

observe that, e.g., in 90% of the cases the RMSE is smaller than 95 m and we see a significant gap between the coarse and fine estimation.

6. Conclusions

In this paper we have investigated synchronization methods for positioning applications in cellular networks with generalized multi-carrier communications signals. These approaches are based on symbol-timing synchronization algorithms to find the starting point of the incident signals where a detailed description of the Minn and the Schmidl-Cox algorithm was given. Furthermore, a symbol-timing offset correction which makes use of the estimated channel impulse response was proposed. Simulations have shown that the performance in terms of the root mean square error is at about 18 m for a signal-to-noise ratio of 15 dB. An averaging over several synchronization symbols can further improve the performance, especially in environments where weak signals from out-of-cell base stations have to be processed. This procedure was used in the overall performance test in a cellular network which has shown that in 90% of the cases the root mean square error is smaller than 95 m for the investigated scenario.

7. Acknowledgments

This work has been partially performed in the framework of the IST project IST-4-027756 WINNER II, which is partly funded by the European Union. The authors would like to acknowledge the contributions of their colleagues.

References

[1] A. H. Sayed, A. Tarighat, and N. Khajehnouri. Network-Based Wireless Location. *IEEE Signal Processing Magazine*, 22(4):24–40, July 2005.

[2] Federal Communications Commission (FCC). FCC 99-245: Third Report and Order. *http://www.fcc.gov/911/enhanced/*, October 1999.

[3] B. W. Parkinson and J. J. Spilker Jr. *Global Positioning System: Theory and Applications, Volume 1.* Progress in Astronautics and Aeronautics, Volume 163, 1996.

[4] J. J. Caffery. *Wireless Location in CDMA Cellular Radio Systems.* Kluwer Academic Publishers, 2000.

[5] IST-2003-507581. WINNER Project. *https://www.ist-winner.org.*

[6] IST-2003-507581. WINNER Deliverable D4.8.1: WINNER II Intramode and Intermode Cooperation Schemes Definition. *https://www.ist-winner.org*, June 2006.

[7] Z. Wang and G. Giannakis. Wireless Multicarrier Communications - Where Fourier Meets Shannon. *IEEE Signal Processing Magazine*, 17(3):29–48, May 2000.

[8] T. M. Schmidl and D. C. Cox. Robust Frequency and Timing Synchronization for OFDM. *IEEE Transactions on Communications*, 45(12):1613–1621, December 1997.

[9] H. Minn, V. K. Bhargava, and K. B. Letaief. A Robust Timing and Frequency Synchronization for OFDM Systems. *IEEE Transactions on Wireless Communications*, 2(4):822–839, July 2003.

[10] M. J. E. Golay. Complementary Series. *IRE Transactions on Information Theory*, IT-7:82–87, April 1961.

[11] M. Morelli and U. Mengali. An Improved Frequency Offset Estimator for OFDM Applications. *IEEE Communications Letters*, 3(3):75–77, March 1999.

[12] C. Mensing and S. Plass. Positioning Algorithms for Cellular Networks Using TDOA. *Proceedings of the IEEE International Conference on Acoustics, Speech, and Signal Processing (ICASSP)*, 4:513–516, May 2006.

[13] W. Foy. Position-Location Solutions by Taylor-Series Estimation. *IEEE Transactions on Aerospace and Electronic Systems*, 12:187–193, March 1976.

[14] IST-2003-507581. WINNER Deliverable D6.13.7: Test Scenarios and Calibration Cases Issues 2. *https://www.ist-winner.org*, December 2006.

JOINT CHANNEL TRACKING AND PHASE NOISE COMPENSATION FOR OFDM IN FAST FADING MULTIPATH CHANNELS*

Roberto Corvaja
University of Padova – Department of Information Engineering
corvaja@dei.unipd.it

Ana García Armada
University Carlos III of Madrid – Department of Signal Theory and Communications
agarcia@tsc.uc3m.es

Abstract OFDM systems such as the schemes proposed for 3G evolution exhibit a great sensitivity to the effects of phase noise and channel estimation, which can introduce inter-channel interference due to the loss of orthogonality among sub-carriers. In this work channel estimation and tracking schemes are investigated, proposing to compensate jointly the effects of phase noise and of the channel. A decision feedback scheme is proposed together with an estimator based on the pilots, to improve the performance also in the presence of fast fading multipath.

Keywords: Multi-carrier systems, channel tracking, phase noise.

1. Introduction

OFDM systems are already being employed in wireless networks for fixed and low-mobility environments (as WLAN) and are considered as candidates for the next generation of mobile communications, for their flexibility, easiness of implementation and equalization [1]. In particular, the 3GPP proposal for the long term evolution (LTE) of UMTS considers again OFDM, but with many more sub-carriers than in WLAN. Moreover, the use of OFDM in a mobile environment gives rise to a degradation due to the changing channel. Phase noise in OFDM has been considered in several works [2]–[7] and schemes to compensate the common phase error (CPE) are considered for example in [6].

*This work has been partly funded by projects "MACAWI" TEC2005-07477-C02-02 and "MAMBO" UC3M-TEC-05-027

S. Plass et al. (eds.), Multi-Carrier Spread Spectrum 2007, 297–306.

In [7] a dual branch estimation is proposed, with an un-modulated version of the phase noise coming from the receiver oscillator. To combat he effect of the inter-channel interference (ICI) a MMSE per-carrier equalization is considered in [10], while schemes to compensate also the ICI are considered in [9, 11]. However in the phase noise compensation schemes, channel estimation is considered almost ideal or separately from the phase noise compensation, which is difficult to accomplish in a practical implementation.

Here the idea is to consider jointly the effect of phase noise and of the channel, proposing an equalization and tracking scheme based on the pilots, jointly with a decision feedback (DF).

2. OFDM system

The complex baseband equivalent of the OFDM transmission chain is depicted in Fig. 1.

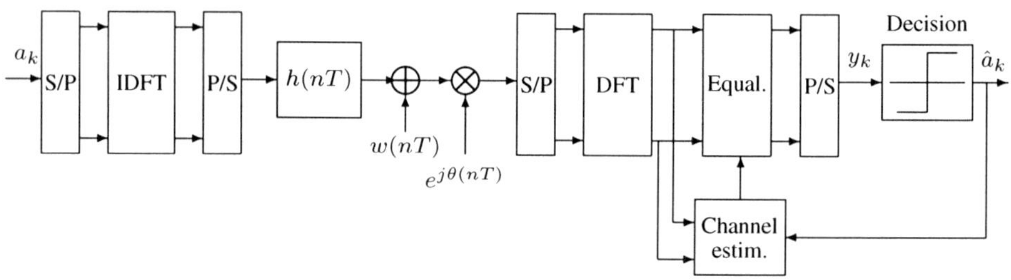

Figure 1. Scheme of the OFDM transmission system.

The data symbols a_k are OFDM modulated by means of the inverse discrete Fourier transform (IDFT) over M points. The phase noise $\theta(t)$ is introduced at the receiver, although it can account for the effects of both the term $\theta_T(t)$ introduced in the up-conversion and $\theta_R(t)$ introduced in the down-conversion at the receiver side. The phase noise is assumed to evolve according to a Wiener random walk process (non-stationary), although the noisy carrier is wide-sense stationary with a Lorentian line. Note however that also a stationary model could be adopted [14], characterized by the power spectral density. Indeed the error probability is referred to the phase noise increment variance σ_θ^2 over the sample period, that can be obtained from the Wiener model or from the phase noise spectrum [14].

After OFDM, the signal is passed to the DAC giving the analog output for the up-conversion to RF and applied to the multipath radio channel, with baseband-equivalent impulse response $h(t)$. The use of a suitable cyclic prefix removes the intersymbol interference. We consider the channel maximum delay contained within the margin of the cyclic prefix. The multipath channel is assumed with independent Rayleigh fading processes on the channel taps, with equivalent impulse response in the k-th OFDM symbol

$$h_k(t) = \sum_{n=0}^{N_{Ch}} h_k[n]\,\delta(t - nT)\ . \tag{1}$$

The Doppler process is characterized by a maximum spread frequency f_{max}, whose value, normalized to the symbol rate, is $f_D = f_{max}T$. The power delay profile considered in the following is the UMTS vehicular channel A [15].

The sampled signal is demodulated by the discrete Fourier transform (DFT) to give the decision variable y_k. The received signal at time k in the i-th sub-carrier, after cyclic prefix removal and OFDM demodulation, is

$$x_k[i] = H_k[i]\,\Theta_0\, a_k[i] + I_k[i] + W_k[i]\ , \tag{2}$$

where $H_k[i]$ is the complex channel coefficient for the sub-carrier i at time k, given by the DFT of the equivalent discrete-time channel impulse response,

$$H_k[i] = \sum_{n=0}^{M-1} h_k[n] e^{-j2\pi\frac{in}{M}}\ , \quad i = 0, \ldots, M-1\ . \tag{3}$$

and $I_k[i]$ is the ICI contribution, originated by the loss of orthogonality due to phase noise

$$I_k[i] = \sum_{\substack{j=0\\ j\neq i}}^{M-1} a_k[j] H_k[j]\,\Theta_{i-j} \tag{4}$$

with

$$\Theta_m = \sum_{n=0}^{M-1} e^{j\theta_n}\, e^{j2\pi n\frac{m}{M}}\ . \tag{5}$$

Note that $\Theta_0 = \sum_{n=0}^{M-1} e^{j\theta_n}$ represents the common phase error. In (2) $W_k[i]$ is the noise on sub-channel i, which can be expressed as

$$W_k[i] = \sum_{n=0}^{M-1} w_k[n] e^{-j2\pi\frac{in}{M}} \qquad i = 0, \ldots, M-1 \tag{6}$$

where w_n is the additive Gaussian noise at the receiver input (see Fig. 1), whose samples have variance σ_w^2, while the effect of phase noise on the additive noise samples is neglected, due to circular symmetry. The received signal can be written in matrix form, from eq. (2) as

$$\boldsymbol{X}_k = \boldsymbol{Q}_k \boldsymbol{a}_k + \boldsymbol{w}_k \tag{7}$$

where the matrix $\boldsymbol{Q}_k$ comprises both the effect of the channel and of the phase noise and is expressed by

$$\boldsymbol{Q}_k = \begin{bmatrix} H_k[0]\Theta_0 & H_k[1]\Theta_{-1} & \cdots & H_k[M-1]\Theta_{-M+1} \\ H_k[0]\Theta_1 & H_k[1]\Theta_0 & \cdots & H_k[M-1]\Theta_{-M+2} \\ \vdots & & & \\ H_k[0]\Theta_{M-1} & H_k[1]\Theta_{M-2} & \cdots & H_k[M-1]\Theta_0 \end{bmatrix} . \quad (8)$$

The decision variable y_k is obtained by a proper equalization scheme, as described in the next Section.

3. Receiver schemes

According to eq. (2) and (8) different choices are possible for detection and equalization: i) To consider a per-sub-carrier approach by means of a one-tap equalizer based on the MMSE criterion, as done in [2] for the phase noise contribution. This corresponds to compensating the CPE only. ii) To partially estimate the matrix $\boldsymbol{Q}_k$ and apply a simple ZF equalizer to remove the main contributions of ICI. iii) To equalize the whole OFDM symbol by inversion of $\boldsymbol{Q}_k$ (de-correlation).

Per sub-carrier equalization. Equalization by a per sub-carrier approach can be obtained by the multiplication of each sub-carrier by a coefficient $K_k[i]$. The optimum coefficient is derived according to the minimum mean square error (MMSE) criterion, which gives

$$K_k[i] = \frac{\tilde{H}_k[i]^*}{\left|\tilde{H}_k[i]\right|^2 + \sigma_{ICI}^2 + \sigma_w^2} . \quad (9)$$

where $\tilde{\boldsymbol{H}}_k$ is the estimated channel frequency response. This corresponds to the choice of considering the ICI as an additive noise term. Estimation of the ICI variance and even a more detailed statistical characterization has been addressed in several works [8]–[10]. This approach corresponds to the estimation of the diagonal terms in the matrix and removes the effect of the CPE only, which can be acceptable for small phase noise and Doppler.

ICI reduction. A simple ZF equalization scheme can be considered, which achieves a partial ICI cancellation. The idea is based on the fact that in the frequency domain, the overall "channel" matrix $\boldsymbol{Q}_k$ can be seen as the convolution of the the vector channel $\boldsymbol{H}_k$ and of the phase noise vector $\boldsymbol{\Theta}_k$. Noting that the phase noise is slow, with respect to the symbol-rate, we can approximate the phase noise Fourier transform vector as different from zero only in few terms around the DC component. This corresponds to considering the matrix as a band matrix and in particular we can consider it "tri-diagonal", that is, with elements different form zero only in the positions (i, j) with $|i - j| \leq 1$. Then the ZF equalizer is obtained in the frequency domain, giving

$$y_k[i] = \sum_{n=i-1}^{i+1} x_k[n]\, K_k[n] \,, \tag{10}$$

where the equalizer $K_k[n]$ is obtained according to the zero forcing criterion and removes the ICI contribution from one precursor and one postcursor.

OFDM symbol equalization. The possibility of performing a block equalization on the whole OFDM symbols leads to a better performance than a simple per carrier scheme in the presence of phase noise only, as shown in [11].

The same approach can be adopted here, considering the equalization performed on the whole symbol by inversion of the matrix $\boldsymbol{Q}_k$.

$$\boldsymbol{y}_k = \boldsymbol{Q}_k^{-1}\, \boldsymbol{x}_k \tag{11}$$

This corresponds to considering as different from zero all the elements of the matrix. Note that this approach leads to a high computational complexity in the determination of the inverse matrix, which is usually quite large for the number of sub-carriers of new mobile standards, such as LTE.

4. Channel estimation and tracking

All the possible schemes presented before need the estimation of the channel matrix $\boldsymbol{Q}_k$ or at least of some of its elements. Moreover, tracking of the estimation is crucial as well, not only in fast fading, but also due to the presence of phase noise, as shown in Fig. 2. The known elements that can be employed for the channel estimation are the first pilot OFDM symbol, or possibly more than one, depending on the frame structure, and the pilots scattered along each symbol, which provide an updated channel frequency response. The initial estimate can be derived on the first pilot OFDM symbol, with a lest squares (LS) or MMSE approach. Here the LS approach is used. Tracking however is a key aspect, also for a static terminal ($f_D = 0$) due to the time variability which is introduced also by the phase noise process. We can consider two approaches for tracking and updating the channel estimation: i) interpolation of the estimate provided by the pilots sub-carriers; ii) a decision feedback channel estimator.

Pilot channel estimation and interpolation. The pilot symbols scattered along the OFDM symbol can provide an updated channel estimation, on the OFDM symbol itself under detection. A frequency interpolation approach is followed here. The channel coefficients obtained by the LS criterion on the pilots are extended to adjacent sub-carriers by an FFT, followed by zero-padding and a subsequent IFFT to have the equivalent interpolated channel.

Decision feedback channel estimation. A DF scheme can provide the information to estimate the channel, as an alternative to the pilots. In other words, the symbols needed for inversions of the relation (8), which can be partial or complete, are derived from the detected symbols themselves. This approach has the disadvantage that the estimated channel can be already "old" in the case of fast Doppler, especially if we need all the data symbols of the previous OFDM symbol, as required by the full matrix inversion.

5. Results

The 3GPP LTE parameters (set 1) are used here: $M = 512$ sub-carriers; symbol-rate $1/T = 7.68$ Msymbols/s; useful sub-carriers 299; OFDM symbols per Transmission Time Interval (TTI) 27. Fig. 2 shows the effect of phase noise after one OFDM symbol and at the end of the TTI. The reference curve is obtained with no Doppler and no phase noise.

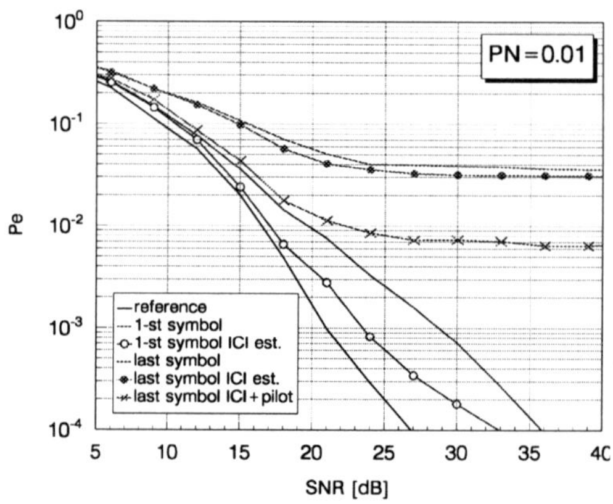

Figure 2. Error probability after one OFDM symbol and at the end of TTI, $f_D = 0$, phase noise variance $\sigma_\theta^2 = 0.01$.

It can be seen the dramatic effect of phase noise at the end of the TTI, due to the introduction of a time-varying characteristic equivalent to a fast fading. Also, the gain given by the estimation of the ICI variance, to evaluate the performance of a MMSE equalization, is limited to the first symbols, where performance is insensitive to the channel variation. On the other hand, at the TTI end, at least a tracking based on the pilots should be employed.

Per-carrier (one-tap) MMSE equalization. In Fig. 3 the error probability at the end of the TTI interval is shown in the presence of phase noise and a quite slow Doppler fading, using the initial estimate only, using only the pilot subcarriers, and using a DF scheme. Equalization is obtained by multiplication of a coefficient.

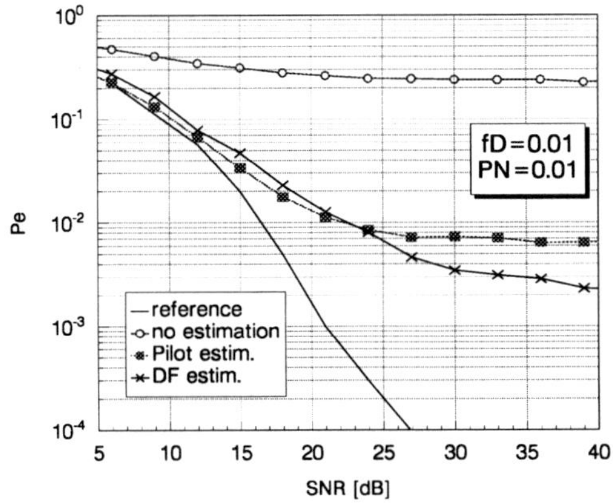

Figure 3. Error probability (end of TTI), with $f_D = 0.01$ and $\sigma_\theta^2 = 0.01$.

It can be seen that only the tracking capabilities provided by a large number of pilots such as in LTE can give a performance comparable to the DF scheme. However at large SNRs DF schemes perform better. Note that the value of the Doppler spread is quite small. If we increase the value of the Doppler spread, the performance of the DF scheme looses some of its effectiveness, as shown in Fig. 4.

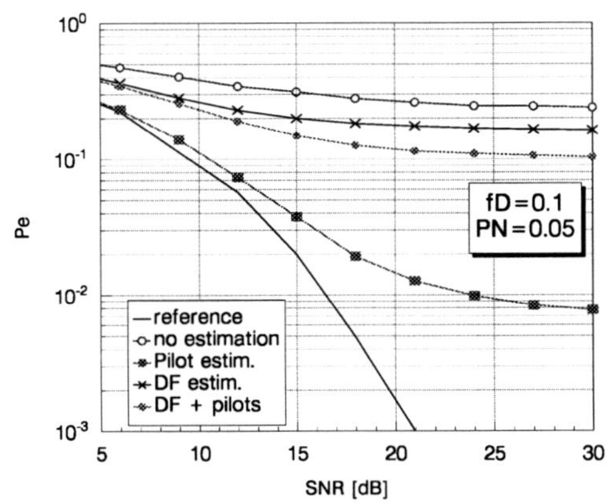

Figure 4. Error probability (end of TTI), with $f_D = 0.1$ and $\sigma_\theta^2 = 0.05$.

This is due to the fact that the DF estimate is not updated, since it refers to the previous OFDM symbol, while the pilots sub-carriers are available in the current symbol. The DF estimation can be combined with the one provided by pilots, but it gives only a slight improvement.

Partial ICI cancellation. In Fig. 5, with a larger amount of phase noise, the error probability is compared using only the pilots for estimation, a combination of DF and pilots, and showing the effect of a partial ICI cancellation.

For the ICI cancellation by the ZF three-tap equalizer, the estimation again is obtained as combination of the pilots and the DF estimations.

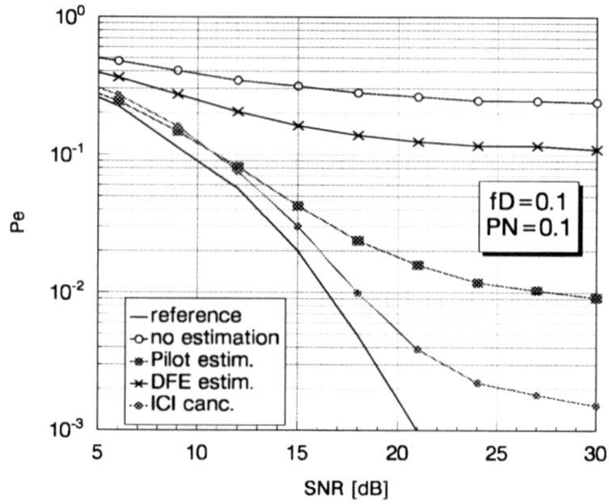

Figure 5. Error probability (end of TTI) with $f_D = 0.1$, and $\sigma_\theta^2 = 0.1$.

ICI cancellation by matrix inversion. In Fig. 6, we present the gain obtained by inversion of the complete matrix, for the same phase noise and the same Doppler as in Fig. 5.

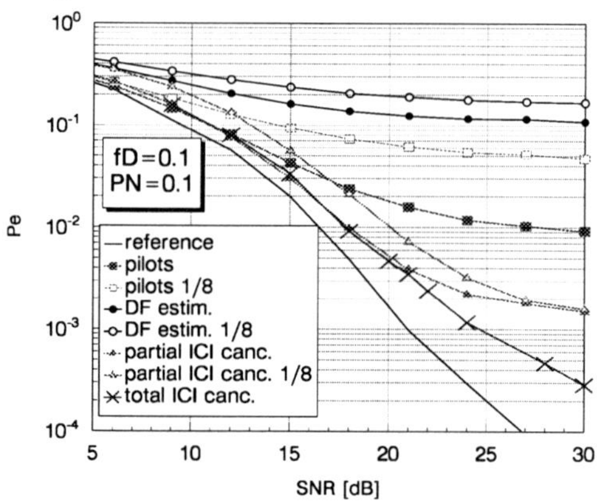

Figure 6. Error probability (end of TTI), $f_D = 0.1$ and $\sigma_\theta^2 = 0.1$.

Up to high values of SNR the performance achieved by the simplified ICI cancellation scheme is quite close. In the same figure the effect of reducing the number of subcarriers can be appreciated, from one every 4 to one every 8. To see the sensitivity of the proposed algorithms to the phase noise and to the Doppler spread frequency f_D, the penalty, that is the SNR increase to achieve $P_e = 10^{-2}$ (with respect to the ideal case), is plotted in Figs. 7–8.

Up to high values of phase noise, the penalty of the ICI cancellation schemes is quite low and the reduced ICI canceller is almost equivalent to the complete

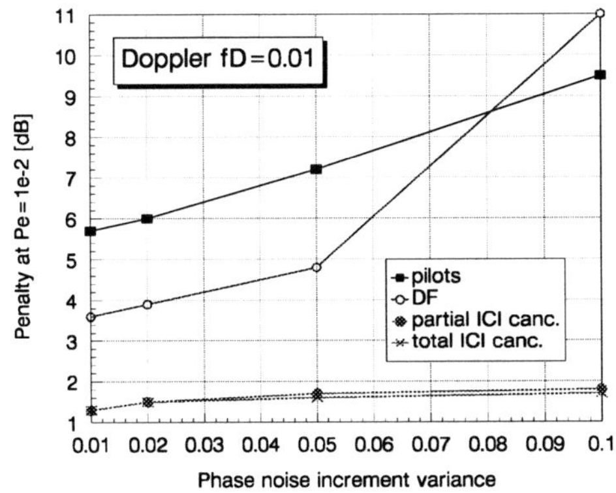

Figure 7. Penalty at $P_e = 10^{-2}$ as a function of σ_θ^2, for $f_D = 0.01$.

ICI cancellation scheme. On the other hand, this occurs for a small Doppler spread. If we consider the sensitivity to the Doppler spread, the penalty is shown in Fig. 8.

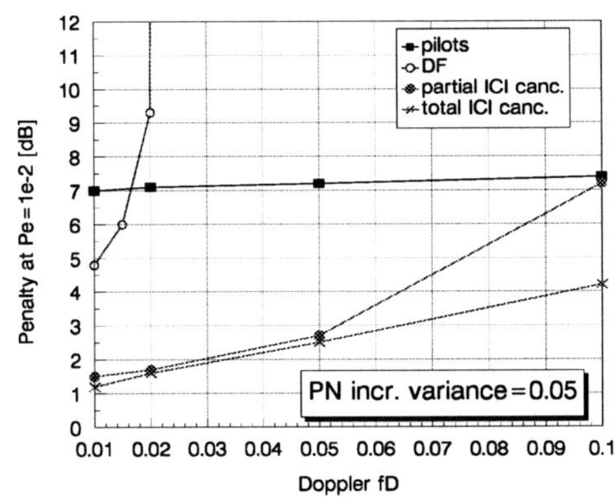

Figure 8. Penalty at $P_e = 10^{-2}$ as a function of f_D, for $\sigma_\theta^2 = 0.05$.

It can be seen that the scheme based on the pilots is almost insensitive to the Doppler spread, since the estimation relies on the actual OFDM symbol under detection. On the other hand, the DF estimation soon degrades as the channel is changing faster, leading to an infinite penalty, due to an error probability floor greater than $P_e = 10^{-2}$, for any value of SNR.

6. Conclusions

We propose a scheme to compensate the channel and the phase noise for OFDM in fast fading, using an estimated channel which combines a pilot-based estimation and a DF one. To combat the ICI a per-subcarrier equalization or the matrix channel inversion are considered. The best performance is obtained by the full channel matrix inversion, but we get very close by considering only the first terms of ICI, that is, a tri-diagonal matrix, with a lower complexity.

References

[1] J. Moon, J.-Y. Ko, Y.-H. Lee, "A framework design for the next-generation radio access system" *IEEE J. on Selec. Areas in Comm.*, Vol. 24, No. 3, pp. 554–564, March 2006.

[2] S. Wu, Y. Bar-Ness,"OFDM Systems in the Presence of Phase Noise: Consequences and Solutions", *IEEE Trans. on Comm.*, Vol. 52, No. 11, pp. 1988–1996, Nov. 2004.

[3] L. Tomba, "On the effect of Wiener phase noise in OFDM systems," *IEEE Trans. on Comm.*, Vol. 46, No. 5, pp. 580–583, May 1998.

[4] L. Tomba, W. A. Krzymien, "Sensitivity of the MC-CDMA access scheme to carrier phase noise and frequency offset," *IEEE Trans. on Vehic. Techn.*, Vol. 48, pp. 1657–1665, Sept. 1999.

[5] A. García Armada, M. Calvo, "Phase noise and sub-carrier spacing effects on the performance of an OFDM communication system", *IEEE Comm. Letters*, Vol. 2, No. 1, pp. 11–13, Jan. 1998.

[6] F. Munier, T. Eriksson, A. Svensson "Receiver Algorithms for OFDM Systems in Phase Noise and AWGN", *Proc. IEEE PIMRC 2004*, pp. 1998–2002, Sept. 2004.

[7] G. Liu and W. Zhu, "Compensation of Phase Noise in OFDM Systems Using an ICI Reduction Scheme", *IEEE Trans. on Broadcasting*, Vol. 50, No. 4, pp. 399–407, Dec. 2004.

[8] D. Petrovic, W. Rave, G. Fattweis, "Properties of the Intercarrier Interference due to Phase Noise in OFDM", *Proc. IEEE ICC 2005*, pp. 2605–2610, May 2005.

[9] Z. Jianhua, H. Rohling, "Analysis of ICI Cancellation Scheme Systems with Phase Noise", *IEEE Trans. on Broadcasting*, Vol. 50, No. 2, pp. 97–106, June 2004.

[10] S. Wu, Y. Bar-Ness,"A Phase Noise Suppression Algorithm for OFDM-Based WLANs", *IEEE Communications Letters*, Vol. 6, No. 12, pp. 535–537, Dec. 2002.

[11] A. García Armada, "Estimation and correction of Phase noise effects in Orthogonal Frequency Division Multiplexing", *Proc. IEEE GLOBECOM*, San Francisco, Nov. 2006.

[12] D. N. Kalofonos, M. Stojanovic, and J. G. Proakis, "Performance of Adaptive MC-CDMA Detectors in Rapidly Fading Rayleigh Channels", *IEEE Trans. on Wireless Comm.*, Vol. 2, No. 2, pp. 229–239, March 2003.

[13] S. Hara and R. Prasad, "Design and Performance of Multicarrier CDMA System in Frequency-Selective Rayleigh Fading Channels", *IEEE Trans. on Vehic. Technol.*, Vol. 48, No. 5, pp. 1584–1595, Sept. 1999.

[14] T. E. Parker, "Characteristics and sources of phase noise in stable oscillators", *Proc. 41st Annual Frequency Control Symposium*, pp. 99–110, 1987.

[15] 3G TS 25.221. Physical Channels and Mapping of Transport Channels onto Physical Channels. 3GPP TSG RAN WG1, version 3.2.0, March 2000.

ESTIMATION OF SAMPLING TIME MISALIGNMENTS IN IFDMA UPLINK

Alexander Arkhipov, Michael Schnell
German Aerospace Center (DLR)
Institute of Communications and Navigation
Oberpfaffenhofen, 82234 Wessling, Germany
{alexander.arkhipov,ronald.raulefs,michael.schnell}@dlr.de

Abstract This article investigates the influence of sampling time offset on the performance of an interleaved frequency-division multiple-access (IFDMA) uplink scheme for mobile radio transmission. It is shown that sampling time offset causes signal-to-noise ratio (SNR) degradation and leads to inter-symbol interference (ISI). In addition, an efficient algorithm is proposed which is capable to estimate sampling time offset by applying oversampling at the receiver. The proposed algorithm requires only three-times oversampling, does not use any pilot symbols, and is independent from the transmission channel and modulation alphabet.

1. Introduction

Compared to conventional OFDMA [1], IFDMA seems more attractive due to its lower complexity, since IFDMA requires no discrete Fourier transform (DFT) at the transmitter. Moreover, IFDMA shows a significantly lower peak-to-average power ratio (PAPR) than conventional OFDMA.

If a transmit filter is applied, the correct sampling at the receiver becomes important. Several studies have been accomplished to estimate the time offset for OFDMA uplink systems [2] [3]. In these works, rectangular pulse shaping is assumed, and the problem of correct sampling is not considered. Standard sampling time offset estimation algorithms for single-carrier systems [4] cannot be applied to an IFDMA receiver. First, in an IFDMA receiver, equalization is preferably performed in the frequency domain. Second, the influence of a time-dispersive mobile radio channel must be taken into account. Third, received uplink signals from different users are distorted by different transmission channels.

S. Plass et al. (eds.), Multi-Carrier Spread Spectrum 2007, 307–316.

In this article, we investigate the performance of an IFDMA uplink system for mobile radio transmission in the presence of sampling time offset misalignment and propose an algorithm for sampling time offset estimation. We assume that sampling time offset is equal or less than the chip duration. The base station receives data from different users simultaneously and, ideally, all users are perfectly aligned in time. However, in our investigation, we consider that all users can have random time delays, which have to be less than the chip duration.

The obtained estimates of the sampling time offsets of each user can be used for the correcting of sampling time instances at the base station receiver, or to transmit time correction values back to the mobile terminals in order to fine-tune users in time.

We use the following notation in this article: variables denoting signals in time domain are written in small letters, variables denoting signals in frequency domain are written in capital letters. Bold variables designate matrices or vectors.

The remainder of this article is organized as follows. In section II IFDMA scheme is briefly renewed. Section 3 describes the proposed sampling time offset estimation algorithm and simulation results are discussed in section 4. Finally, section 5 concludes this article.

2. Interleaved Frequency-Division Multiple Access Scheme

In the following, IFDMA is described as a special case of OFDMA-CDM. For additional descriptions of the IFDMA scheme refer e.g to [5] [6]. The standard OFDMA-CDM [7] uplink transmission system is able to support up to K simultaneously active users. Assume that N_u users ($N_u \leq K$) transmit simultaneously. Each OFDMA-CDM user, $i = 1, \ldots, N_u$, performs a blocked transmission of Q complex-valued symbols $d_q^{(i)}, q = 1, \ldots, Q$, which are assumed to be equally likely and are taken from the underlying modulation alphabet. The average energy of the transmitted data symbol $d_q^{(i)}$ is unity and the symbol duration is T_s. The Q symbols are arranged to build the symbol vector $\mathbf{d}^{(i)}$ of user i

$$\mathbf{d}^{(i)} = (d_1^{(i)}, d_2^{(i)}, \ldots, d_Q^{(i)})^T. \qquad (1)$$

The symbol vector $\mathbf{d}^{(i)}$ is first serial-to-parallel converted. Then, a code-division multiplexing (CDM) component is introduced by spreading each symbol with a different spreading code $\mathbf{c}_q, q = 1, \ldots, Q$, of length $L \geq Q$ and adding together the resulting Q spread vectors. After summation, we obtain vector $\mathbf{s}^{(i)}$ which is referred as a vector of the spread symbols.

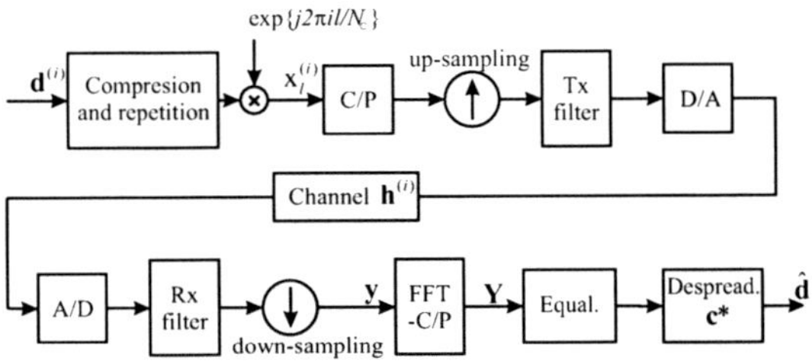

Figure 1. Simulation model for IFDMA uplink

It can be represented by its components $s_n^{(i)}, n = 0, \ldots, L-1$ as

$$s_n^{(i)} = \sum_{q=1}^{Q} c_{n,q} \cdot d_q^{(i)}, n = 0, \ldots, L-1, \tag{2}$$

where $c_{n,q}$ is the nth component of the spreading code $\mathbf{c}_q$. Finally, an inverse DFT (IDFT) with user-specific frequency mapping is performed for the vector $\mathbf{s}^{(i)}$. Block interleaving achieves optimal frequency diversity, since it distributes the L spread symbols $s_n^{(i)}$ of the Q data symbols of user i equally over the whole transmission bandwidth. Note, OFDMA-CDM exclusively assigns to each of the K users a set of L frequencies out of all possible $N_c = KL$ subcarriers. If all subcarriers assigned to user i are uniformly spaced over the whole available transmission bandwidth at a distance N_c/K, the time domain transmit signal $x_l^{(i)}, l = 0, \ldots N_c - 1$, obtained after the IDFT operation in the transmitter, can be represented by

$$x_l^{(i)} = e^{j\frac{2\pi}{N_c}il} \frac{1}{\sqrt{N_c}} \sum_{n=0}^{L-1} s_n^{(i)} e^{j\frac{2\pi}{L}nl}. \tag{3}$$

As in standard OFDMA, a guard interval larger than the maximum channel delay is added as a cyclic prefix (CP) in OFDMA-CDM in order to avoid interference from preceding OFDMA-CDM transmission symbols.

IFDMA as a special case of OFDMA-CDM can be realized if a fully loaded system, i.e. $L = Q$, is considered and Fourier codes with components $c_{n,q} = \{e^{-j\frac{2\pi}{Q}nq}\}, n = 0, \ldots, L-1, q = 1, \ldots, Q$, are used for spreading. In this case, the Fourier codes for spreading and the rotation factors of the Fourier transform cancel out and the transmit signal given in (3) reduces to

$$x_l^{(i)} = \underbrace{e^{\frac{j2\pi}{N_c}il}}_{\text{user specific phase factor}} \cdot \frac{1}{\sqrt{K}} d_{l \bmod L}^{(i)}, l = 0, \ldots, N_c - 1. \quad (4)$$

Comparing (4) and the IFDMA signal generation as described in [5] [6] it follows that IFDMA is equivalent to OFDMA-CDM with block interleaved frequency allocation and can be completely generated in the time domain without using the Fourier transform. The time domain generation of IFDMA uses compression and repetition as described in detail in [5]. Note, the multiplication with the user specific phase factor in (4) ensures the user discrimination by assigning to each user i a set of subcarriers orthogonal to all other users' subcarrier sets. Each sample $x_l^{(i)}$ represents the transmission chip $l, l = 0, \ldots, N_c - 1$. The duration of the resulting chips is $T_c = T_s/K$.

The simulation model for IFDMA is shown in Fig. 1. After addition of CP, the obtained signal is up-sampled. Up-sampling is implemented by adding zero samples between successive samples $x_l^{(i)}$. After up-sampling, pulse-shaping is performed by a transmit filter. After digital-to-analog (D/A) conversion, we assume transmission over a time-dispersive mobile radio channel. At the receiver, we consider oversampling per transmitted chip $x_l^{(i)}$. After sampling, the signal is filtered by a receive filter, which is matched to the transmit pulse-shaping filter. In Fig. 2, an example of the IFDMA signal after the receive filter is represented. The solid lines represent the pulse-shapes corresponding to the different chips and the dotted line is the sum signal. In the down-sampling operation at least one sample per transmitted chip is chosen for performing equalization, despreading and estimation of the transmitted data.

For the sake of simplicity, the convolution of the receive and transmit filters is represented by the function $g(t)$, which is chosen as a Nyquist pulse

$$g(t) = \text{sinc}\left(\frac{t}{T_c}\right) \frac{\cos(\pi\alpha t/T_c)}{1 - 4\alpha^2 t^2/T_c^2}, \quad (5)$$

where α denotes the roll-off factor.

In this section only one receive sample $y_l, l = 0, \ldots, N_c - 1$ per transmitted chip $x_l^{(i)}$ is considered. We assume that the channel has an impulse response described by the vector $\mathbf{h}^{(i)}$ of dimension M with components $h_m^{(i)}, m = 1, \ldots, M$. Without loss of generality, we assume that $N_c > M$, in the following.

Assume that y_l represents the lth element of the received vector $\mathbf{y}$, which can be written as

$$\mathbf{y} = \sum_{i=1}^{N_u} \hat{\mathbf{h}}^{(i)} \mathbf{z}(\tau^{(i)}) + \mathbf{n}, \tag{6}$$

where the vector $\mathbf{n}$ denotes the samples of the additive white Gaussian noise (AWGN). The channel matrix $\hat{\mathbf{h}}^{(i)}$ is the $N_c \times N_c$ right circular matrix whose rows are cyclically shifted versions of the vector $\tilde{\mathbf{h}}^{(i)}$. The vector $\tilde{\mathbf{h}}^{(i)}$ is obtained by appending $N_c - M$ zeros to $\mathbf{h}^{(i)}$.

The elements $z_l(\tau^{(i)})$ of vector $\mathbf{z}(\tau^{(i)})$ in (6) can be represented as

$$z_l(\tau^{(i)}) = e^{j\frac{2\pi}{N_c} i(l + \frac{\tau^{(i)}}{T_c})} \frac{1}{\sqrt{K}} \cdot \tag{7}$$

$$\left[\underbrace{d^{(i)}_{l \bmod L}\, g\Big(\tau^{(i)}\Big)}_{\text{useful signal}} + \underbrace{\sum_{\substack{p=0 \\ p\neq l}}^{N_c-1} d^{(i)}_{p \bmod L}\, g\Big(\tau^{(i)} + (p-l)T_c\Big)}_{\text{ISI}} \right].$$

The normalized time offset $\tau^{(i)}$ can take on values in the range $-T_c/2 \leq \tau^{(i)} \leq T_c/2$. The orthogonality between users is not violated since $\tau^{(i)}$ does not introduce any frequency offset. However, ISI appears if a sampling time offset $\tau^{(i)}$ is present as shown in Fig. 2. The optimum sampling points correspond to a situation where $\tau^{(i)} = 0$. In this case, the received signal of user i is ISI-free.

Defining $\mathbf{W}$ as a $N_c \times N_c$ DFT matrix, the frequency domain representation $\mathbf{Y}$ of the received signal $\mathbf{y}$ can be written as

$$\mathbf{Y} = \sum_{i=1}^{N_u} \mathbf{H}^{(i)} \mathbf{Z}(\tau^{(i)}) + \mathbf{N}, \tag{8}$$

where $\mathbf{H}^{(i)} = \mathbf{W}\hat{\mathbf{h}}^{(i)}$ is a $N_c \times N_c$ diagonal matrix, which is the DFT transformation of the channel impulse response $\hat{\mathbf{h}}^{(i)}$. The received signal of user i in the frequency domain is defined as $\mathbf{Z}(\tau^{(i)}) = \mathbf{W}\mathbf{z}(\tau^{(i)})$ and $\mathbf{N} = \mathbf{W}\mathbf{n}$.

The vector $\mathbf{Y}$ can be used for the estimation of the transmitted data. User i transmits its L spread symbols $s_n^{(i)}$ on subcarriers $\kappa = nK + i, n = 0, \ldots, L-1$. Other users transmit their data on separated orthogonal subcarriers and their influence is not considered in the following. Thus, we omit subscript i for simplicity and focus on the data of user i. After

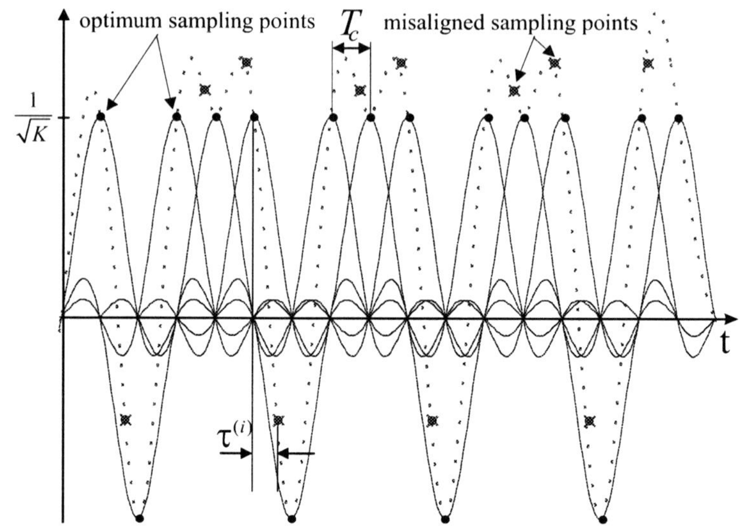

Figure 2. Example of the IFDMA signal after the receive filter. Sampling time offset $\tau^{(i)}$ causes SNR degradation and leads to ISI; $\mathbf{d}^{(i)} = [1, -1, 1, 1]^T$; $L = Q = 4$; $\alpha = 0.25$; only user $i = 0$ is active; AWGN channel with a negligible noise power.

frequency domain equalization, the L received values Y_κ can be despread with the complex-conjugate Fourier codes (IDFT operation of length L) to create an estimation $\hat{\mathbf{d}}$ of the symbol vector $\mathbf{d}$. The considered receiver is equivalent to the conventional OFDMA-CDM receiver [7], only Fourier codes are used instead of Walsh-Hadamard codes.

3. Time offset estimation

Our goal is to estimate the time offset τ without any additional pilot symbols. In the following, we propose an algorithm which allows to estimate τ using tree times oversampling per transmitted chip. Thus, we consider the case where sampling is performed with a period $\frac{T_c}{3}$ and in addition to vector $\mathbf{y}$ we introduce two vectors $\mathbf{y}'$ and $\mathbf{y}''$. If we omit the influence of other users, $\mathbf{y}'$ and $\mathbf{y}''$ can be defined as

$$\mathbf{y}' = \hat{\mathbf{h}}\mathbf{z}\left(\tau - \frac{T_c}{3}\right) + \mathbf{n}' \tag{9}$$

and

$$\mathbf{y}'' = \hat{\mathbf{h}}\mathbf{z}\left(\tau + \frac{T_c}{3}\right) + \mathbf{n}'', \tag{10}$$

respectively. We introduce also two additional frequency domain vectors $\mathbf{Y}'' = \mathbf{W}\mathbf{y}''$ and $\mathbf{Y}' = \mathbf{W}\mathbf{y}'$. Note, only the elements Y'_κ and $Y''_\kappa, \kappa = nK + i, n = 0, \ldots, L-1$ of vectors $\mathbf{Y}'$ and $\mathbf{Y}''$ represent data of user i.

Thus, we have three frequency domain vectors $\mathbf{Y}$, $\mathbf{Y}'$ and $\mathbf{Y}''$. The expectations of the elements of these vectors Y_κ, Y'_κ and Y''_κ are given as

$$E\{|Y_\kappa|^2\} = E\{|H_{\kappa,\kappa}|^2\}E\{|Z_\kappa(\tau)|^2\} + \sigma_n^2, \tag{11}$$

$$E\{|Y'_\kappa|^2\} = E\{|H_{\kappa,\kappa}|^2\}E\left\{\left|Z_\kappa\left(\tau - \frac{T_c}{3}\right)\right|^2\right\} + \sigma_n^2 \tag{12}$$

and

$$E\{|Y''_\kappa|^2\} = E\{|H_{\kappa,\kappa}|^2\}E\left\{\left|Z_\kappa\left(\tau + \frac{T_c}{3}\right)\right|^2\right\} + \sigma_n^2, \tag{13}$$

where σ_n^2 is the variance of the filtered noise, $Z_\kappa(\tau)$, $Z_\kappa(\tau - \frac{T_c}{3})$ and $Z_\kappa(\tau + \frac{T_c}{3})$ are the components of vectors $\mathbf{Z}(\tau)$, $\mathbf{Z}(\tau - \frac{T_c}{3})$ and $\mathbf{Z}(\tau + \frac{T_c}{3})$ taken at $\kappa = nK + i, n = 0, \ldots, L-1$. The fading coefficients $H_{\kappa,\kappa}$, $\kappa = nK + i, n = 0, \ldots, L-1$ are the diagonal elements of $\mathbf{H}$. It can be noted that $H_{\kappa,\kappa}$ are the same in (11), (12) and (13). This assumption can be considered as realistic, since system design rules assume that fading coefficients remain constant over a duration of one IFDMA symbol, i.e, during a time period equal to $N_c T_c$.

The expectations $E\{|Y_\kappa|^2\}$, $E\{|Y'_\kappa|^2\}$ and $E\{|Y''_\kappa|^2\}$ can be approximated as averages of the energies of the received data Y_κ, Y'_κ and Y''_κ. For example, such an averaging can be performed over the whole transmission frame.

The DFT operation does not change the energy of the signal, i.e. $E\{|Z_\kappa(\tau)|^2\} = K \cdot E\{|z_l(\tau)|^2\}$. Therefore, the value $E\{|Z_\kappa(\tau)|^2\}$ can be approximated as

$$E\{|Z_\kappa(\tau)|^2\} = E_s(\tau) = \sum_{p=-N_c/2}^{N_c/2} g^2(\tau + pT_c)\,. \tag{14}$$

Taking into account (11), (12), (13) and (14) and introducing new variables $A = E\{|Y_\kappa|^2\}$-$E\{|Y''_\kappa|^2\}$, $B = E\{|Y'_\kappa|^2\}$-$E\{|Y_\kappa|^2\}$ and $C = E\{|Y''_\kappa|^2\}$-$E\{|Y'_\kappa|^2\}$, the estimate $\hat{\tau}$ of the normalized time offset τ can be found as

$$\hat{\tau} \;= \arg\min_{\tilde{\tau}} \left| AE_s\left(\tilde{\tau} - \frac{T_c}{3}\right) + BE_s\left(\tilde{\tau} + \frac{T_c}{3}\right) + CE_s(\tilde{\tau})\right|. \tag{15}$$

The following remarks are of interest:

- The estimation in (15) does not require the knowledge of the transmission channel and the modulation alphabet.

- The estimation in (15) can be performed only if $\frac{T_c}{2} \leq \tau \leq \frac{T_c}{2}$, since function $E_s(\tau)$ is periodic with a period of T_c.

- The proposed estimation requires only one dimensional search over possible values of τ. Values $E_s(\tau)$ can be precalculated at the receiver according to formula (14).

One of the vectors $\mathbf{Y}$, $\mathbf{Y}'$ and $\mathbf{Y}''$ can be used for despreading and data demodulation.

4. Simulation Results

The transmission system considered for simulations has a bandwidth of 20 MHz and the carrier frequency of 5 GHz. The number of subcarriers is $N_c = 1024$. The spreading length is $L = 64$ and the total number of users is $K = 16$. A minimum mean square error (MMSE) detector is used as described in [7]. For the channel coding, a convolutional code with rate 1/2 and memory 6 is used in our simulations. The transmission frame consists of 24 IFDMA symbols. QPSK is used for modulation. Influence of the receive and the transmit filter is modelled as the Nyquist pulse $g(t)$, which assumed to be non-zero only within $-3T_c < t < 3T_c$. Therefore, if sampling time offset is present, each transmitted chip x_l experiences ISI from three preceding and three following chips. The roll-off factor α varies from 0.25 to 0.75.

A multipath channel model is considered, which consists of a tapped delay line with $M = 11$ statistically independent Rayleigh fading taps. The average power for each channel component $h_m, m = 1, \ldots, M$, is modelled as

$$E\{|h_m|^2\} = -m + 1 \text{ (dB)}. \tag{16}$$

Additionally, the average channel attenuation is normalized to unity in our simulations. We evaluate the system performance in terms of bit-error rate (BER) and mean square error (MSE) of the sampling time offset estimate. In Fig. 3 the system performance versus SNR is present for sampling time offsets $\tau = \frac{4}{18}T_c$ and $\tau = \frac{7}{18}T_c$. As a reference, the ideal case with $\tau = 0$ is presented. It is shown that the performance of the system degrades with increasing τ, due to the increase of ISI and SNR degradation. However, SNR degradation does not play a significant role in the performance loss. According to (2), SNR degradation is -0.74

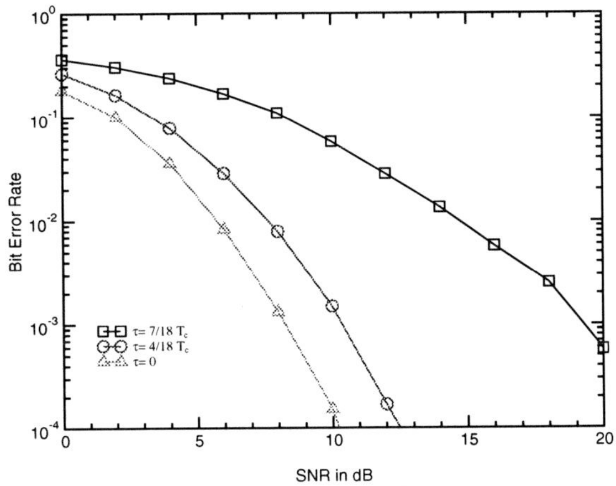

Figure 3. BER versus SNR for the different values of sampling time offset τ; $\alpha = 0.25$

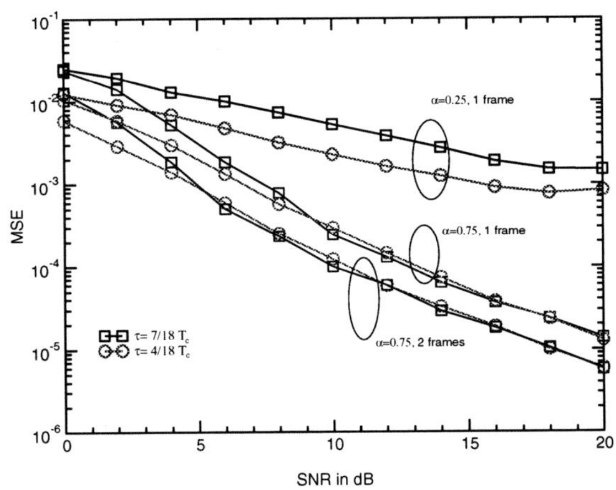

Figure 4. MSE of the sampling time offset estimate versus SNR

dB for $\tau = \frac{4}{18}T_c$ and -2.36 dB for $\tau = \frac{7}{18}T_c$. Therefore, ISI is responsible for the residual performance degradation.

The MSE of the sampling time offset estimation in (15) is presented in Fig. 4 for $\tau = \frac{4}{18}T_c$ and $\tau = \frac{7}{18}T_c$ as well as for different values of α equal to 0.25 and 0.75. For $\alpha = 0.25$, the performance of the proposed algorithm degrades if τ increases. In contrast, the performance of the algorithm is independent from τ if $\alpha = 0.75$. Close inspection of the function $E_s(\tau)$ explains this fact. If $\alpha = 0.25$, the function $E_s(\tau)$ is relatively flat. Hence, the function in (15) does not have a distinct minimum, and the obtained estimation is easily affected by noise. The situation changes completely, if we increase α. The difference between the maximum and minimum value of $E_s(\tau)$ is nearly two times larger for $\alpha = 0.75$ than for $\alpha = 0.25$. If $\alpha = 0.75$, the function $E_s(\tau)$ increases steeply and this steepness remains nearly constant with increasing τ. Therefore, MSE is independent of τ for $\alpha = 0.75$. How it is seen from Fig. 4, the required SNR to achieve a MSE of 10^{-3} is 14 dB less for $\alpha = 0.75$ than for $\alpha = 0.25$.

The precision of the estimate can be improved if averaging in (11), (12) and (13) is performed over the frame with a large size or over many frames. In Fig. 4 this averaging is performed over two frames for the case of $\alpha = 0.75$. It can be seen that the performance is improved by 3 dB. However, such an improved averaging leads to a delay in the sampling time offset estimation and is not desirable.

5. Conclusions

The performance of the IFDMA uplink system is investigated in the presence of sampling time offset. It has been shown that sampling time offset results in ISI and SNR degradation. Moreover, we have proposed an algorithm for sampling time offset estimation using three times over-sampling at the receiver. The proposed algorithm is independent from the transmission channel and modulation alphabet.

References

[1] R.V. Nee and R. Prasad. *OFDMA for Wireless Multimedia Communications.* Artech House, 2002.

[2] M. Morelli. Timing and frequency synchroniozation for the uplink of an OFDMA system. *IEEE Transactions on Communications*, 52, no. 2:286–306, Feb. 2004.

[3] M. Park, K.Ko, H.Yoo, and D.Hong. Performance analysis of OFDMA uplink systems with symbol timing misallignment. *IEEE Communications Letters*, 7:376–379, Aug. 2003.

[4] J. Proakis. *Digital Communications.* McGraw Hill Higher Education, Dec. 2000.

[5] M. Schnell, I. De Broeck, and U. Sorger. A promising new wideband multiple-access scheme for future mobile communications systems. *European Transactions on Telecommunications (ETT)*, 10, No. 4:417–427, Jul./Aug. 1999.

[6] U. Sorger, I.De Broeck, and M. Schnell. Interleaved FDMA - a new spread-spectrum multiple-access scheme. *IEEE International Conference on Communications (ICC'98)*, pages 1013–1017, June.1998.

[7] S. Kaiser. OFDM code division multiplexing in fading channels. *IEEE Transactions on Communications*, 50, no. 8:1266–1273, Aug. 2002.

COMBINED TIME AND FREQUENCY DOMAIN OFDM CHANNEL ESTIMATION

Maik Bevermeier and Reinhold Haeb-Umbach
Department of Communications Engineering
University of Paderborn, Germany
{bevermeier,haeb}@nt.uni-paderborn.de

Abstract In this paper we present a novel channel impulse response estimation technique for block-oriented OFDM transmission based on combining estimators: the estimates provided by a Kalman Filter operating in the time domain and a Wiener Filter in the frequency domain are optimally combined by taking into account their estimated error covariances. The frequency domain Wiener Filter smoothes Maximum-Likelihood estimates of the channel frequency response obtained by the Expectation-Maximization algorithm. Different variants of the proposed scheme are experimentally investigated in an IEEE 802.11a-like system setup. They compare favourably with known approaches from the literature resulting in reduced mean square estimation error and bit error rate. Further, robustness and complexity issues are discussed.

Keywords: Orthogonal frequency division multiplexing, MAP estimation, Kalman filtering, Wiener filtering

1. Introduction

Orthogonal Frequency Division Multiplexing (OFDM) has received a lot of interest in recent years for both wired and wireless communications due to its spectral efficiency and implementation simplicity [1]. By transmitting a high-rate data stream by many low-rate streams in parallel, a frequency selective channel is turned into a set of parallel non-frequency selective narrowband transmission channels, for which a simple one-tap equalization can be carried out.

In this article we are concerned with channel estimation and tracking for wireless packet-oriented coherent OFDM single-input and single-output transmission. Channel estimation is usually carried out on the known symbols of the preamble. Semi-blind techniques have been proposed when the size of the payload or the terminal velocity is large.

S. Plass et al. (eds.), Multi-Carrier Spread Spectrum 2007, 317–326.

Then the channel estimates obtained on the preamble serve as initial values which are refined on the data of the payload [2].

We also carry out channel estimation both on the preamble and on the payload. However, we seek for an optimal combination of the two. First, a Kalman Filter (KF) is used for estimation on the preamble in the time domain. This has been shown to be computationally more efficient and to result in lower variance of the estimates compared to a frequency domain KF [3]. Next, we conduct channel frequency response (CFR) estimation on the symbols of the payload, which consists of two steps. The first step is a Maximum-Likelihood (ML) estimation, either solely at the subcarriers of the interspersed pilot data or also on the data subcarriers using the Expectation-Maximization (EM) algorithm. In a second step the estimates are improved by Wiener filtering. The final step is the optimal combination of the Kalman and Wiener Filter (WF) estimates. The resulting estimation formulae are surprisingly simple and we will give an alternative interpretation of them.

The article is organized as follows: In the next section the OFDM transmission model is outlined. Section 3 describes the proposed channel estimation. In Section 4 we present simulation results and discuss robustness issues, before finishing with conclusions drawn in section 5.

2. System Model

We consider a block-oriented OFDM transmission over a multipath fading channel. In the following, we assume perfect timing and frequency synchronization, and absence of phase noise. Let

$$\tilde{\mathbf{a}}(k) = (\tilde{a}_0(k), \ldots, \tilde{a}_{M-1}(k))^T \tag{1}$$

denote the $(M \times 1)$ symbol vector in the frequency domain at time $k = iB + n$, where B is the number of symbols per block, i counts the blocks (packets) and n the symbols within a block. For simplicity of notation, let us assume that the known preamble consists of one symbol at $n = 0$ and the payload of the remaining $B-1$ symbols ($n = 1, \ldots, B-1$). The OFDM modulated symbol vector is obtained as

$$\mathbf{x}(k) = (\mathbf{W}_{M\times M})^H \tilde{\mathbf{a}}(k). \tag{2}$$

Here, $\mathbf{W}_{M\times M}$ is the DFT matrix of dimension $(M \times M)$ with the (i,l)-th entry $(\mathbf{W})_{i,l} = \frac{1}{\sqrt{M}} \exp(-j2\pi il/M)$, and $(\cdot)^H$ denotes the Hermitian transpose.

The dispersive fading channel is characterized by the channel impulse response (CIR) vector

$$\mathbf{h}^{(0)}(k) = (h_0^{(0)}(k), \ldots, h_{L_h-1}^{(0)}(k))^T \tag{3}$$

of length L_h. The $h_l^{(0)}(k)$, $l = 0, \ldots, L_h - 1$, are independent complex Gaussian random variables with a Jakes' power density spectrum.

Prior to transmission, a cyclic prefix of length $L > L_h$ symbols is prepended and removed after transmission. The received signal can then be written as follows [1]:

$$\mathbf{r}(k) = \mathbf{X}(k)\mathbf{h}^{(0)}(k) + \mathbf{n}(k). \tag{4}$$

$\mathbf{X}(k)$ is a $(M \times L_h)$ circulant matrix formed from the vector $\mathbf{x}(k)$, and the additive white noise vector $\mathbf{n}(k)$ consists of M independent complex Gaussian random variables of variance σ_n^2 per dimension.

3. Channel Estimation

The proposed channel estimator consists of a time domain KF operating on the preamble, a frequency domain WF working on ML estimates of the CFR obtained on the payload, and the combination of the estimates.

3.1 Kalman Filter

For the design of the KF we assume signal propagation along assumed distinct paths $\hat{L}_h$, where the channel response of each path is described by a first-order Markov model [4]:

$$h_l((i+1)B) = f \cdot h_l(iB) + g \cdot w_l(iB); \quad l = 0, \ldots, \hat{L}_h - 1, \tag{5}$$

where $w_l(iB)$ is complex white Gaussian noise of zero mean and unit variance. Further $f = J_0(2\pi \hat{f}_d T_B)$ and $g = \sqrt{(1-f^2)/\hat{L}_h}$. J_0 denotes the modified Bessel function of first kind and 0-th order. $\hat{f}_d$ is the assumed Doppler frequency, and T_B is the duration of a data block. In matrix notation, we have the following state equation:

$$\mathbf{h}(i+1) = \mathbf{F}\mathbf{h}(i) + \mathbf{G}\mathbf{w}(i), \tag{6}$$

where $\mathbf{F} = f \cdot \mathbf{I}_{\hat{L}_h \times \hat{L}_h}$, $\mathbf{G} = g \cdot \mathbf{I}_{\hat{L}_h \times \hat{L}_h}$, and

$$E[\mathbf{h}(i)\mathbf{h}^H(i)] = (1/\hat{L}_h)\mathbf{I}_{\hat{L}_h \times \hat{L}_h}. \tag{7}$$

$\mathbf{I}_{\hat{L}_h \times \hat{L}_h}$ denotes the identity matrix of dimension $(\hat{L}_h \times \hat{L}_h)$. Note in this model we assume for simplicity that each propagation path has the same power, whereas in practice the power profile is unknown at the receiver. Since the data of the preamble are known, eq. (4) may serve as measurement equation.

The KF computes the a posteriori probability density of the CIR vector, given all past observations, as a Gaussian with mean $\mathbf{h}^{(KF)}(i|i)$, the MMSE estimate of the CIR, and covariance $\mathbf{P}^{(KF)}(i|i)$, which equals the covariance matrix of the estimation error [5]:

$$p(\mathbf{h}^{(0)}(i)|\mathbf{r}(0), \ldots, \mathbf{r}(i)) = \mathcal{N}(\mathbf{h}^{(0)}(i); \mathbf{h}^{(KF)}(i|i), \mathbf{P}^{(KF)}(i|i)), \quad (8)$$

where the initial value $\mathbf{P}^{(KF)}(0|-1)$ is given by eq. (7).

For later use in Wiener filtering and estimator combination, the variables are transformed to the frequency domain:

$$\tilde{\mathbf{h}}^{(KF)}(i|i) = \mathbf{W}_{M\times\hat{L}_h} \mathbf{h}^{(KF)}(i|i) \quad (9)$$

$$\tilde{\mathbf{P}}^{(KF)}(i|i) = \mathbf{W}_{M\times\hat{L}_h} \mathbf{P}^{(KF)}(i|i)(\mathbf{W}_{M\times\hat{L}_h})^H. \quad (10)$$

Note that we adopted a block fading model with constant CIR for all B symbols of a block. We could have obtained individual estimates for each symbol interval $k = iB + n$ within the i-th burst by prediction, which we, however, did not do to save computations. Thus we set $\tilde{\mathbf{h}}^{(KF)}(k) = \tilde{\mathbf{h}}^{(KF)}(i|i)$ and $\tilde{\mathbf{P}}^{(KF)}(k) = \tilde{\mathbf{P}}^{(KF)}(i|i)$ for $k = iB, \ldots, (i+1)B - 1$.

3.2 Wiener Filter

The estimation of the CFR consists of two steps: first ML estimates are computed for individual or all subcarriers, second an estimate for the CFR on all subcarriers is obtained by Wiener filtering.

The received signal at subcarrier m, $m = 0, \ldots, M-1$, and symbol interval k, is given by

$$\tilde{r}_m(k) = \tilde{h}_m^{(0)}(k)\tilde{a}_m(k) + \tilde{n}_m(k), \quad (11)$$

where $\tilde{r}$, $\tilde{h}^{(0)}$ and $\tilde{n}$ denote the discrete Fourier transforms of r, $h^{(0)}$ and n, respectively.

First we consider ML estimation on the N_p subcarriers $p(j)$, $j = 1, \ldots N_p$, containing embedded pilots. Since the symbols are known we have

$$\tilde{z}'_{p(j)}(k) = \frac{\tilde{r}_{p(j)}(k)\tilde{a}^*_{p(j)}(k)}{|\tilde{a}_{p(j)}(k)|^2}; \quad j = 1, \ldots, N_p. \quad (12)$$

A ML estimate of the CFR on the remaining data subcarriers $d(q)$, $q = 1, \ldots, M - N_p$ is obtained by using the EM algorithm [6]:

$$\tilde{z}'_{d(q)}(K) = \left(\sum_{k\in\Omega_K} \sum_{\tilde{a}\in\Omega_{\tilde{a}}} \tilde{r}_{d(q)}(k)\tilde{a}^* \frac{p(\tilde{r}_{d(q)}(k)|\tilde{z}'_{d(q)}(K-1), \tilde{a})}{p(\tilde{r}_{d(q)}(k)|\tilde{z}'_{d(q)}(K-1))} \right) \cdot \left(\sum_{k\in\Omega_K} \sum_{\tilde{a}\in\Omega_{\tilde{a}}} |\tilde{a}|^2 \frac{p(\tilde{r}_{d(q)}(k)|\tilde{z}'_{d(q)}(K-1), \tilde{a})}{p(\tilde{r}_{d(q)}(k)|\tilde{z}'_{d(q)}(K-1))} \right)^{-1}. \quad (13)$$

Here, the inner sum is over all symbols of the symbol alphabet $\Omega_{\tilde{a}}$. The outer sum is over a set of ΔB successive time indices: $\Omega_K = \{iB + (K-1)\Delta B + 1, ..., iB + K\Delta B\}$, $K = 1, \dots (B-1)/\Delta B$. Thus K successive CFR estimates are equal: $\tilde{z}'_{d(q)}(k) = \tilde{z}'_{d(q)}(K)$, $k = (K-1)\Delta B + 1, \dots, K\Delta B$. The size of the time window ΔB has to be chosen as a compromise to allow for a localized and still reliable estimate. Note that we only carry out one iteration of the EM algorithm and therefore left out the iteration counter in (13). The CFR estimate $\tilde{z}'_{d(q)}(K-1)$ is taken as initial value for the EM algorithm applied to the K-th interval, where $\tilde{z}'_{d(q)}(0) = \tilde{h}^{(KF)}_{d(q)}(k)$. Note that the densities in (13) are Gaussians:

$$p(\tilde{r}_{d(q)}(k)|\tilde{z}'_{d(q)}(K-1), \tilde{a}) = \mathcal{N}(\tilde{r}_{d(q)}(k); \tilde{z}'_{d(q)}(K-1)\tilde{a}, \sigma^2_{\tilde{v}}). \tag{14}$$

Regarding the ML estimates of the CFR on individual, eq. (12), or all subcarriers, eq. (12) and (13), as "observations", the purpose of the subsequent WF is to improve estimates by exploiting knowledge about the correlation in time and frequency direction. Instead of applying a 2-dimensional filter, we first carry out a Wiener filtering on the pilot subcarriers along the time axis and second along the frequency axis. This is known to achieve almost identical performance, at, however, greatly reduced computational cost, as will be seen below. Due to the equal CFR estimates within the K-th interval and to reduce computations we pass on Wiener filtering in time direction on the data subcarriers.

Since the WF along the time axis can be designed quite easily using estimates of the Doppler frequency and noise variance, it is not considered here in detail. Let the resulting estimates be denoted by $\tilde{\mathbf{z}}(k)$. In the following we assume that ML estimates had been obtained on all subcarriers, i.e. $\tilde{\mathbf{z}}(k)$ has dimension $(M \times 1)$. We have the following linear observation model:

$$\tilde{\mathbf{z}}(k) = \mathbf{A}\tilde{\mathbf{h}}^{(0)}(k) + \tilde{\mathbf{v}}(k), \tag{15}$$

where $\mathbf{A} = \mathbf{I}_{M\times M}$ for the case considered here and $(\tilde{\mathbf{v}})_m(k) = \tilde{n}_m(k)\tilde{a}^*_m(k)/|\tilde{a}_m(k)|^2, m = 0, ..., M-1$. Furthermore, $E[\tilde{\mathbf{v}}(k)\tilde{\mathbf{v}}^H(k)] = \mathbf{\Lambda}_{\tilde{\mathbf{v}}}(k)$ is the diagonal covariance matrix of the estimation error in time direction, which is a function of the position k within a burst. However, we used the same value, the value in the middle of the burst ($k = iB + B/2$), for all k, making $\mathbf{\Lambda}_{\tilde{\mathbf{v}}} \approx \sigma^2_{\tilde{v}} \cdot \mathbf{I}_{M\times M}$ independent of the symbol index k, in order to save computations. The other assumption made here is that the estimation error variance of the ML estimates of the CFR on the data subcarriers obtained by EM are equal to estimation error variance obtained on the pilot subcarriers, which is definitely a coarse approximation.

The ML estimates are now smoothed in frequency direction by applying the WF:

$$\tilde{\mathbf{h}}^{(WF)}(k) = \mathbf{B}(k)\tilde{\mathbf{z}}(k), \tag{16}$$

where $\mathbf{B}$ is a solution of the Wiener-Hopf equation

$$\mathbf{B}(k)\mathbf{R}_{\tilde{\mathbf{z}}}(k) = \mathbf{R}_{\tilde{\mathbf{h}}\tilde{\mathbf{z}}}(k). \tag{17}$$

Using the linear model (15) we obtain

$$\mathbf{R}_{\tilde{\mathbf{z}}}(k) = E[\tilde{\mathbf{z}}(k)\tilde{\mathbf{z}}^H(k)] = \mathbf{A}\mathbf{R}_{\tilde{\mathbf{h}}}(k)\mathbf{A}^H + \mathbf{\Lambda}_{\tilde{\mathbf{v}}} \tag{18}$$

$$\mathbf{R}_{\tilde{\mathbf{h}}\tilde{\mathbf{z}}}(k) = E[\tilde{\mathbf{h}}^{(0)}(k)\tilde{\mathbf{z}}^H(k)] = \mathbf{R}_{\tilde{\mathbf{h}}}(k)\mathbf{A}^H, \tag{19}$$

where

$$\mathbf{R}_{\tilde{\mathbf{h}}}(k) = E[\tilde{\mathbf{h}}^{(0)}(k)(\tilde{\mathbf{h}}^{(0)})^H(k)] \tag{20}$$

is the correlation matrix of the unknown CFR, which has also been used as initial value of the error covariance matrix of the KF:

$$\mathbf{R}_{\tilde{\mathbf{h}}}(k) = \tilde{\mathbf{P}}^{(KF)}(0|-1). \tag{21}$$

3.3 Combination of estimators

We now have two estimates for the CFR: $\tilde{\mathbf{h}}^{(KF)}(k)$ and $\tilde{\mathbf{h}}^{(WF)}(k)$. They can be optimally combined to an estimate $\tilde{\mathbf{h}}^{(C)}$, using their respective estimation error covariances [5]:

$$(\tilde{\mathbf{P}}^{(C)})^{-1}\tilde{\mathbf{h}}^{(C)} = (\tilde{\mathbf{P}}^{(WF)})^{-1}\tilde{\mathbf{h}}^{(WF)} + (\tilde{\mathbf{P}}^{(KF)})^{-1}\tilde{\mathbf{h}}^{(KF)}, \tag{22}$$

where

$$(\tilde{\mathbf{P}}^{(WF)})^{-1} = \mathbf{R}_{\tilde{\mathbf{h}}}^{-1} + \mathbf{A}^H\mathbf{\Lambda}_{\tilde{\mathbf{v}}}^{-1}\mathbf{A} \tag{23}$$

$$(\tilde{\mathbf{P}}^{(C)})^{-1} = (\tilde{\mathbf{P}}^{(WF)})^{-1} + (\tilde{\mathbf{P}}^{(KF)})^{-1} - \mathbf{R}_{\tilde{\mathbf{h}}}^{-1} \tag{24}$$

are the covariance matrices of the estimation error of the WF and the combined estimator, respectively. Using (15) - (21) we obtain the surprisingly simple result

$$(\tilde{\mathbf{P}}^{(C)})^{-1}\tilde{\mathbf{h}}^{(C)} = \mathbf{A}^H\mathbf{\Lambda}_{\tilde{\mathbf{v}}}^{-1}\tilde{\mathbf{z}} + (\tilde{\mathbf{P}}^{(KF)})^{-1}\tilde{\mathbf{h}}^{(KF)}, \tag{25}$$

where

$$(\tilde{\mathbf{P}}^{(C)})^{-1} = \mathbf{A}^H\mathbf{\Lambda}_{\tilde{\mathbf{v}}}^{-1}\mathbf{A} + (\tilde{\mathbf{P}}^{(KF)})^{-1}. \tag{26}$$

A closer look reveals an alternative interpretation. The result is identical to the Maximum a posteriori (MAP) estimation of the mean vector $\tilde{\mathbf{h}}^{(0)}$ of correlated jointly Gaussian random variables, given a Gaussian prior according to eqs. (8) - (10) and the ML estimate $\tilde{\mathbf{z}}$ (eq. (15)) on the data [7], [8].

4. Results

4.1 Known Channel Model

In this section we present experimental results using an IEEE 802.11a-like system. A burst consists of $B = 102$ symbols, of which the first two are the known preamble and the remaining form the payload. Channels no. 7, 21, 43 and 57 of the 64 available subcarriers are reserved for pilots. The channel bandwidth in the 5 Ghz Band is chosen to 20 MHz, where the data rate is 24 MBit/s and 96 uncoded data bits form a symbol (coding rate 1/2). The used modulation is 16-QAM. The Rayleigh channel with six independent propagation paths has a power loss and delay profile of $[-1, -3, -5, -7, -9, -12]$ dB and $[0, 100, 200, 300, 400, 500]$ ns. with a Jakes' Doppler spectrum on each propagation path. This channel model represents a typical urban type of scenario. First, we assume all channel model parameters, including the terminal velocity of $v = 30.8$ km/h, to be perfectly known.

Fig. 1 shows the mean square estimation error (MSEE) of the CFR estimation and the bit error rate (BER) of the decoder, respectively. Here, the observation vector $\tilde{\mathbf{z}}(k)$ is either the estimate of the CFR solely on the pilot subcarriers after Wiener filtering along the time axis ("(ML-Pilot-WF)+KF") or on all subcarriers by using the EM algorithm ("(ML-Pilot-Data-WF)+KF"). The combined estimators are compared with a KF operating on the preamble and a WF, which estimates the CFR on all subcarriers by interpolation of the pilot subcarriers according to eqs. (16) - (21), denoted as "ML-Pilot-WF". This is the classical way of smoothing CFR estimates by exploiting the knowledge of the maximum delay $\hat{L}_h$ [2]. A lower bound for the BER is obtained by assuming the CFR to be perfectly known. As can be seen from these

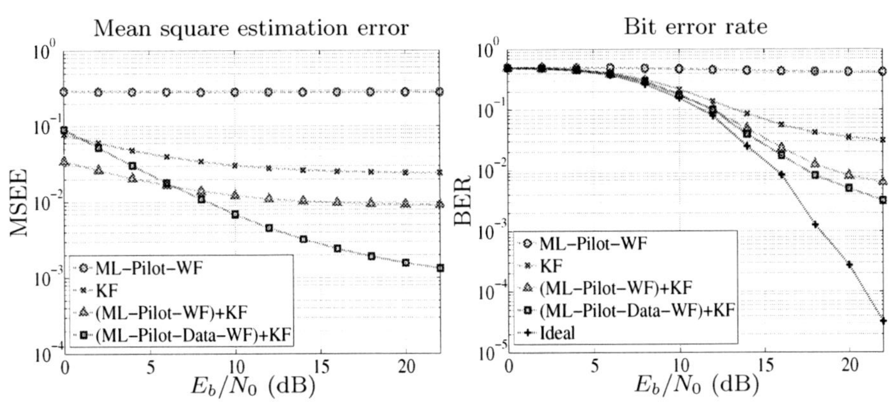

Figure 1. MSEE and BER of combined and individual estimators.

results, estimating the CFR on the pilot subcarriers alone performs very poorly. Also the KF, operating only on the preamble of a long burst, is not very effective. However, the combination of the two individually poor estimates is very powerful. This is probably due to the fact that the Frobenius norm of the covariance matrix $\tilde{\mathbf{P}}^{(KF)}(k)$ of the a posteriori density $p(\tilde{\mathbf{h}}^{(0)}(i)|\mathbf{r}(0),\ldots,\mathbf{r}(i))$, as provided by the KF, is typically much smaller than the norm of the covariance matrix $\tilde{\mathbf{P}}^{(KF)}(0|-1)$ of the corresponding a priori density $p(\tilde{\mathbf{h}}^{(0)}(i))$. The more informative covariance matrix leads to more effective estimation. The effect of the EM algorithm is also quite noticeable.

4.2 Robustness

For the experiments reported so far we assumed that the Doppler frequency, the multipath profile (number and delay of paths) and the variance of the additive noise is known. In this section we investigate the robustness of proposed channel estimators to incomplete knowledge of these parameters. Figs. 2 a) - c) show the MSEE of the proposed combined and the individual estimators:

a) As a function of the E_b/N_0 assumed by the receiver, while the true channel-sided E_b/N_0 is 14 dB.

b) As a function of terminal velocity (and thus max. Doppler frequency) assumed by the terminal, while the true terminal velocity is 30.8 km/h at $E_b/N_0 = 14$ dB.

c) For wrong assumptions concerning the multipath profile: The maximum delay (500 ns) is assumed to be known, but the profile is assumed to consist of 11 paths with relative delays of 50 ns.

For each setup, the system parameters not mentioned, are assumed to be known to the extent described in section 4.1. Figs. 2 a) - c) show that the combined estimators consistently outperform the individual estimators, when the parameters are not too far off the correct values. For setup a) the combined estimator becomes ineffective, when the assumed E_b/N_0 is smaller than about 8 dB. A false assumed velocity seems to have almost no effect at all estimation techniques.

4.3 Computational Complexity

The EM algorithm calculates the a posteriori densities for all combinations of received signal and signal in the constellation. However, the estimator combination (22) requires the inversion of $\tilde{\mathbf{P}}^{(WF)}$ and $\tilde{\mathbf{P}}^{(KF)}$ primarily. Since we have set $\tilde{\mathbf{P}}^{(KF)}$ constant within a burst and $\mathbf{\Lambda}_{\tilde{\mathbf{v}}}$ con-

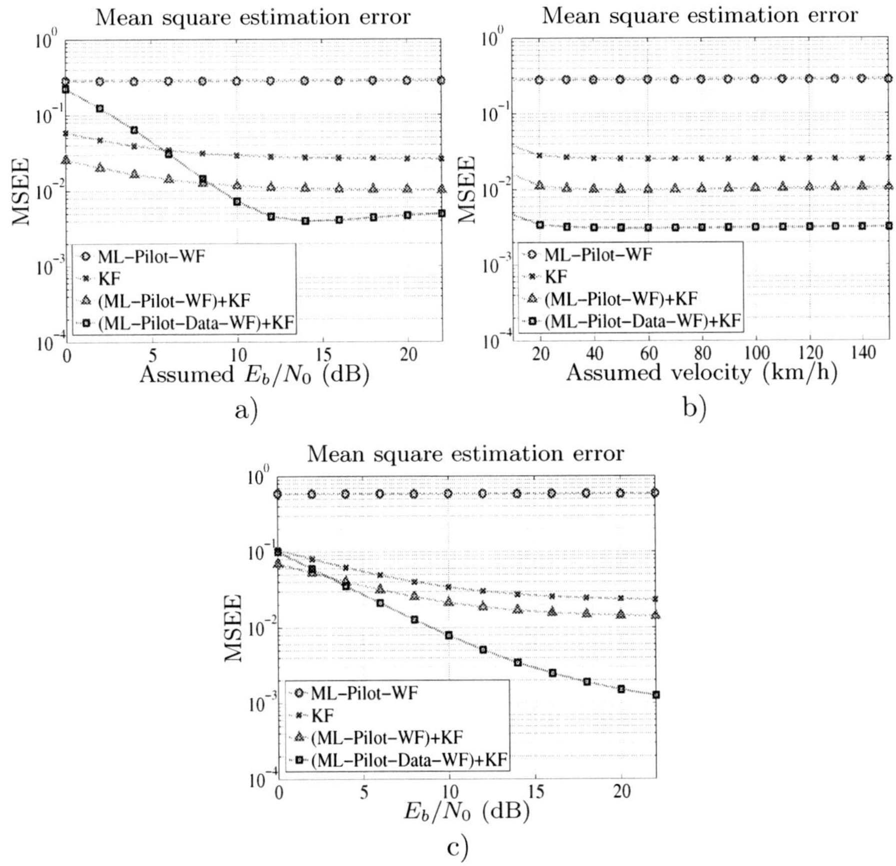

Figure 2. MSEE at different assumed parameters.

stant for the whole transmission, see remarks at the end of section 3.1 and in section 3.2, the estimator combination requires only one matrix inversion per burst, see eq. (26), resulting in only a small computational overhead due to estimator combination.

5. Conclusions and Outlook

In this paper, a combined channel estimation algorithm based on Kalman filtering in the time and Wiener filtering in the frequency domain is proposed. While the individual estimators perform rather poorly, the combined estimation turned out to be very effective. Further, robustness issues und computational complexity are briefly discussed. In future research we try to investigate the effects of phase noise and non-perfect frequency synchronization. Further, a better estimate of the estimation error variance of the ML estimate of the CFR obtained by the EM algorithm should be developed.

References

[1] Z. Wang and G. Giannakis, "Wireless multicarrier communications", *IEEE Signal Processing Magazine*, vol. 17, no.3, May 2000, pp. 29-48.

[2] A. R. S. Bahai, B. R. Saltzberg and M. Ergen, *Multi-Carrier Digital Communications: Theory and Applications of OFDM*, Springer, 2004.

[3] T. Roman, M. Enescu and V. Koivunen, "Time-domain method for tracking dispersive channels in OFDM systems", in *Proc. IEEE VTC Spring*, Jeju, Korea, 2003.

[4] K. Han, S. Lee, J. Lim and K. Sung, "Channel Estimation for OFDM with Fast Fading Channels by Modified Kalman Filter", *IEEE Trans. on Cons. Electr.*, vol. 50, no. 2. May 2004, pp. 443-449.

[5] T. Kailath, A. Sayed and B. Hassibi, *Linear Estimation*, Prentice Hall, 2000.

[6] X. Ma, H. Kobayashi and S. Schwartz, "EM-based channel estimation algorithms for OFDM", *EURASIP Journal on Applied Signal Processing*, 2004:10, pp. 1460-1477.

[7] R. Duda, P. Hart, D. Stork, *Pattern Classification*, Wiley, 2001.

[8] M. Lasry and R. Stern, "A posteriori estimation of correlated jointly gaussian mean vectors", *IEEE Trans. Pattern Analysis Machine Intell.*, vol. PAMI-6, no. 4. July 1984, pp. 530-535.

DESIGN AND PERFORMANCE ANALYSIS OF LOW-COMPLEXITY PILOT-AIDED OFDM CHANNEL ESTIMATORS

Eugene Golovins and Neco Ventura
University of Cape Town, Rondebosch, 7700, South Africa
{glvjev001,neco}@crg.ee.uct.ac.za

Abstract: In this article, we present an explicit comparative analysis of the two low-complexity channel estimation techniques, which can be used in the pilot-assisted orthogonal frequency division multiplexing (OFDM) systems, – linear minimum mean squared error (LMMSE) and least squares (LS) criteria based. Numerical results are preceded by derivation of the optimum design form to yield both robustness and minimum possible computational complexity for the given class of algorithms. The work also investigates the effect of quantity and power of pilot subcarriers on estimation accuracy and BER performance.

Key words: channel estimation; least squares; pilot symbol assisted modulation.

1. INTRODUCTION

OFDM technology has become increasingly popular in recent years because of its efficiency in reducing the severe effects of frequency-selective fading. Increased transmission rates in OFDM are achieved due to the use of spectrally efficient quadrature amplitude modulation (QAM). However coherent demodulation of the QAM signals requires knowledge or accurate estimation of the channel frequency response (CFR), in order to minimise the probability of detection error [1].

In most current application scenarios of wireless OFDM systems (e.g., fixed wireless access [2]), the propagation channel exhibits strong frequency correlation (within one OFDM symbol) and time correlation (across several symbols) properties. Priority should be addressed to the accurate frequency-domain estimation performed on the interval of one OFDM symbol [3][4][5] as it allows to achieve better performance for a given complexity than the

S. Plass et al. (eds.), Multi-Carrier Spread Spectrum 2007, 327–336.

approaches exploiting long-term time correlation of the channel that require long-drawn training periods. Channel estimates can be obtained using pilot-symbol assisted modulation (PSAM) where training data is transmitted on certain subcarriers within the OFDM band [1], allowing for CFR measurements with subsequent interpolation onto other subcarriers.

Performance of several frequency-domain interpolators (e.g., linear, trigonometric function based, etc.) is evaluated analytically in [6] for channels with different frequency selectivity characteristics. [7] proposes to use polynomial regression models to approximate CFR in both the time and the frequency domain. Frequency-domain LMMSE CFR estimator has been introduced in [3] with low-rank approximation to reduce complexity. An alternative to frequency-domain processing is the so-called DFT-methods in which CFR is transformed into the channel impulse response (CIR) and estimation benefits from the suggested finite CIR length peculiar to multipath channels by filtering out noise in CIR tail part. Several maximum likelihood schemes [4][5] have been proposed respectively and found to outperform the frequency-domain algorithms [5] while being of lesser complexity [8].

This article describes the design and performance analysis of constrained LS (CLS) and LMMSE (CLMMSE) channel estimators and has the following organisation of the remaining content. In Section 2 a general OFDM system model is presented. Section 3 includes derivation of the estimation algorithms. The complexity of these algorithms is analysed in Section 4. Section 5 contains performance evaluation results obtained by means of both theoretical calculations and simulations.

2. OFDM SYSTEM MODEL

In this work, we consider a single-input-single-output discrete-time baseband OFDM model with ideal timing and frequency synchronisation. In the transmitter, a serial stream of data bits is divided into N parallel binary streams, each of which passes through a linear modulation scheme. The OFDM symbol is formed as the result of IDFT, applied to N parallel (data and training) complex-valued modulation subsymbols X_n, $n = 0,\ldots,N-1$. The resultant waveform is converted to a serial sequence of samples. Before transmission each OFDM symbol is prepended with a cyclic prefix (CP), which is a copy of the last portion of the OFDM symbol.

In the considered scenario the channel is assumed to be slowly time-varying, i.e. the CFR is approximately constant during one OFDM symbol, so there is no loss of orthogonality between subcarriers and therefore no intercarrier interference (ICI). In [4] it is asserted that a time-varying channel can be well approximated by the time-invariant model during time interval T if $T \le 0.01/f_D$, where $f_D = f_c v / c$ is the maximum Doppler frequency, f_c is the RF carrier frequency, v is the speed of relative movement between the transmitter and the receiver, and c is the speed of light. This criterion is usually satisfied for all fixed or slow-moving high-rate wireless OFDM systems, operating in the band 2-11 GHz (e.g., [2]), as the duration of the OFDM symbol is much shorter than the coherence time of the radio channel.

Assuming the channel to be time-invariant on the interval of a single OFDM symbol, CP length N_{cp} can be selected to be large enough to accommodate a finite CIR h_m, $m = 0,...,L-1$, where the maximum sample-normalised excess delay $L-1 \le N_{cp}$. Thus, the intersymbol interference (ISI) between consecutive OFDM symbols will be eliminated. During transmission of the ith OFDM symbol CIR filter tap-gains, $h_m(i)$, can be modelled as L i.i.d. zero-mean complex Gaussian variables, with Rayleigh distribution of magnitudes and uniform distribution of phases, whereas CIR variation between symbols, i.e. $h_m(i)$ to $h_m(i+1)$, is governed by a lowpass-filtered stochastic process. Such a channel representation (often termed quasistationary) can be considered as a wide-sense stationary uncorrelated scattering (WSSUS) approximation for small Doppler spreads.

At the receiver side, after removing cyclic prefix and applying DFT to the ith OFDM symbol we get a vector of the received subsymbols:

$$\mathbf{Y}(i) = \left[Y_0(i) \quad \cdots \quad Y_{N-1}(i)\right]^T = \mathbf{X}_{[D]}(i)\mathbf{H}(i) + \boldsymbol{\Xi}(i), \tag{1}$$

where $\mathbf{X}_{[D]}(i) = \text{diag}[\mathbf{X}(i)]$ denotes a diagonal matrix with the data and pilot subsymbols $X_n(i),\ n = 0,...,N-1$; $\mathbf{H}(i) = \left[H_0(i) \quad \cdots \quad H_{N-1}(i)\right]^T$ is the CFR vector; and $\boldsymbol{\Xi}(i) = \left[\Xi_0(i) \quad \cdots \quad \Xi_{N-1}(i)\right]^T$ are the DFT-transformed white Gaussian noise (WGN) variables.

3. CHANNEL ESTIMATION

We denote the received subsymbols at the pilot positions as

$$\mathbf{Y}^{\mathrm{P}} = \mathbf{X}^{\mathrm{P}}_{[D]}\mathbf{H}^{\mathrm{P}} + \boldsymbol{\Xi}^{\mathrm{P}} = \mathbf{X}^{\mathrm{P}}_{[D]}\mathbf{C}\mathbf{H} + \boldsymbol{\Xi}^{\mathrm{P}} = \left[Y_{p_0} \quad \cdots \quad Y_{p_{P-1}}\right]^T, \tag{2}$$

where $\mathbf{X}^{\mathbf{p}}_{[D]} = \mathrm{diag}(\mathbf{X}^{\mathbf{p}})$ contains reference values of P pilot subsymbols, and $\mathbf{C}$ is the $P \times N$ -size selection matrix with the elements $C_{m,n} = \begin{cases} 1, & \text{if } n = p_m \\ 0, & \text{otherwise} \end{cases}$ that is needed to extract samples of the CFR $\mathbf{H} = [H_0 \quad \cdots \quad H_{N-1}]^T$ corresponding to the pilot subcarriers indexed $p_m, 0 \le m \le P-1$. (Hereafter OFDM symbol index i is omitted for clarity, as the processing is done on a symbol-by-symbol basis, without considering correlation with the neighbouring symbols.)

Channel estimation algorithms treat $\mathbf{Y}^{\mathbf{p}}$ as an input to yield the CFR estimate $\hat{\mathbf{H}}$ by means of a linear transformation $\hat{\mathbf{H}} = \mathbf{M}\mathbf{Y}^{\mathbf{p}}$.

3.1 CLS estimator

The LS problem of obtaining an estimate of the CIR $\mathbf{h} = [h_0 \quad \cdots \quad h_{L-1}]^T$ based on PSAM measurements is formulated as a minimisation of the quadratic difference between the received pilot subsymbols $\mathbf{Y}^{\mathbf{p}}$ and the reference pilot values $\mathbf{X}^{\mathbf{p}} = \left[X_{p_0} \quad \cdots \quad X_{p_{P-1}}\right]^T$ being affected by the assumed CFR model $\mathbf{H}$ that can be written as:

$$\begin{aligned} J(\mathbf{h}) &= \left(\mathbf{Y}^{\mathbf{p}} - \mathbf{X}^{\mathbf{p}}_{[D]}\mathbf{H}^{\mathbf{p}}\right)^H \left(\mathbf{Y}^{\mathbf{p}} - \mathbf{X}^{\mathbf{p}}_{[D]}\mathbf{H}^{\mathbf{p}}\right) \\ &= \left(\mathbf{Y}^{\mathbf{p}} - \mathbf{X}^{\mathbf{p}}_{[D]}\mathbf{CFBh}\right)^H \left(\mathbf{Y}^{\mathbf{p}} - \mathbf{X}^{\mathbf{p}}_{[D]}\mathbf{CFBh}\right), \end{aligned} \tag{3}$$

where $\mathbf{F}$ is the $N \times N$ -size Fourier matrix with elements $F_{m,n} = \exp(-\mathrm{j}2\pi mn/N)$. As the assumed CIR model is constrained by only L components, vector $\mathbf{h}$ of the size $L \times 1$ has to be zero-padded up to the size $N \times 1$ before DFT of $\mathbf{h}$ can be taken to yield the CFR $\mathbf{H}$. For that reason the $N \times L$ -size padding matrix $\mathbf{B} = \left|\mathbf{I}_{L\times L} \quad \mathbf{0}_{L\times(N-L)}\right|^T$ is used.

Finding the complex $\mathbf{h}$-respective gradient of $J(\mathbf{h})$ and setting it equal to a zero vector produces the constrained LS CIR solution, which yields the CFR estimate after the DFT conversion:

$$\begin{aligned} \hat{\mathbf{H}}^{\mathbf{CLS}} &= \mathbf{FB}\hat{\mathbf{h}}^{\mathbf{CLS}} = \mathbf{FB}\left(\mathbf{B}^H\mathbf{F}^H\mathbf{C}^H\mathbf{X}^{\mathbf{p}H}_{[D]}\mathbf{X}^{\mathbf{p}}_{[D]}\mathbf{CFB}\right)^{-1}\mathbf{B}^H\mathbf{F}^H\mathbf{C}^H\mathbf{X}^{\mathbf{p}H}_{[D]}\mathbf{Y}^{\mathbf{p}} \\ &= \mathbf{FBW_{CLS}}\mathbf{B}^H\mathbf{F}^H\mathbf{C}^H\mathbf{X}^{\mathbf{p}H}_{[D]}\mathbf{Y}^{\mathbf{p}} = \mathbf{SY}^{\mathbf{p}}, \end{aligned} \tag{4}$$

where the number of pilot subcarriers P is required to be no less than the CIR length, L, for the matrix $\mathbf{W_{CLS}}$ to be full-rank and therefore invertible.

The overall procedure of the algorithm (4) is executed in the following order: zero-padded vector with the pilot elements $Y_{p_m} X^*_{p_m}, 0 \le m \le P-1$ is translated to the time domain, where it is weighted by the small ($L \times L$) matrix $\mathbf{W}_{\mathbf{CLS}}$ (4), and converted back to the frequency domain at the final processing stage.

When the pilot subcarriers are equispaced, i.e. $C_{m,n} = \begin{cases} 1, \text{ if } n = p_m = mN/P + v \\ 0, \text{ otherwise} \end{cases}$, where $N \bmod P = 0$, $0 \le m \le P-1$, and $0 \le v \le N/P - 1$, and if pilot subsymbols $X_{p_m}, m = 0, ..., P-1$ transmitted on different subcarriers have equal constant power set to $|X_p|^2$, then $\mathbf{W}^{-1}_{\mathbf{CLS}} = \mathbf{B}^H \mathbf{F}^H \mathbf{C}^H \mathbf{X}^{\mathbf{p}H}_{[D]} \mathbf{X}^{\mathbf{p}}_{[D]} \mathbf{CFB} = P \mathbf{X}^{\mathbf{p}H}_{[D]} \mathbf{X}^{\mathbf{p}}_{[D]}$ and $\mathbf{X}^{\mathbf{p}}_{[D]} = X_p \mathbf{I}_{P \times P}$. This results in the expression for the weighting matrix $\mathbf{S}$ in (4) simplifying to

$$\mathbf{S}_{\text{EP-ES}} = \frac{1}{P X_p} \mathbf{FBB}^H \mathbf{F}^H \mathbf{C}^H \tag{5}$$

3.2 LMMSE estimator

PSAM-assisted LMMSE estimator is constructed in the form of $\hat{\mathbf{H}}^{\mathbf{MMSE}} = \mathbf{QY}^{\mathbf{p}}$, where the $N \times P$-size weighting matrix $\mathbf{Q}$ is selected in order to minimise MSE between the CFR estimate $\hat{\mathbf{H}}^{\mathbf{MMSE}}$ and the assumed CFR model $\mathbf{H}$:

$$\begin{aligned} M(\mathbf{Q}) &= \frac{1}{N} \mathrm{E}\left[\left(\hat{\mathbf{H}}^{\mathbf{MMSE}} - \mathbf{H} \right)^H \left(\hat{\mathbf{H}}^{\mathbf{MMSE}} - \mathbf{H} \right) \right] = \frac{1}{N} \mathrm{trace}\left[\mathbf{R}_{\mathbf{HH}} - \mathbf{QX}^{\mathbf{p}}_{[D]} \mathbf{CR}_{\mathbf{HH}} \right. \\ &\left. - \mathbf{R}_{\mathbf{HH}} \mathbf{C}^H \mathbf{X}^{\mathbf{p}H}_{[D]} \mathbf{Q}^H + \mathbf{Q} \left(\mathbf{X}^{\mathbf{p}}_{[D]} \mathbf{CR}_{\mathbf{HH}} \mathbf{C}^H \mathbf{X}^{\mathbf{p}H}_{[D]} + \sigma^2_{\text{wgn}} \mathbf{I} \right) \mathbf{Q}^H \right] \end{aligned} \tag{6}$$

where $\mathbf{R}_{\mathbf{HH}} = \mathrm{E}\left[\mathbf{HH}^H \right]$ is the CFR correlation matrix, and $\mathrm{E}\left[\mathbf{\Xi}^{\mathbf{p}} \mathbf{\Xi}^{\mathbf{p}H} \right] = \sigma^2_{\text{wgn}} \mathbf{I}$ is the WGN correlation matrix, with σ^2_{wgn} being equal to the noise variance.

Finding the complex $\mathbf{Q}$-respective gradient of $M(\mathbf{Q})$ and setting it equal to a zero matrix produces the weighting matrix $\mathbf{Q}$ that is optimal in the MMSE sense, yielding

$$\hat{\mathbf{H}}^{\mathbf{MMSE}} = \mathbf{QY}^{\mathbf{p}} = \mathbf{R}_{\mathbf{HH}} \mathbf{C}^H \left[\mathbf{CR}_{\mathbf{HH}} \mathbf{C}^H + \sigma^2_{\text{wgn}} \left(\mathbf{X}^{\mathbf{p}H}_{[D]} \mathbf{X}^{\mathbf{p}}_{[D]} \right)^{-1} \right]^{-1} \mathbf{X}^{\mathbf{p}\,-1}_{[D]} \mathbf{Y}^{\mathbf{p}}, \tag{7}$$

where the number of pilot subcarriers P is required to be no less than the rank of $\mathbf{R}_{\mathbf{HH}}$.

In the case of the equipowered pilot subsymbols the signal-to-noise power ratio (SNR) observed at the pilot subcarriers has a simple expression of $SNR_p = |X_p|^2 / \sigma_{\rm wgn}^2$ and $\sigma_{\rm wgn}^2 \left(\mathbf{X}_{[D]}^{\mathbf{p}H} \mathbf{X}_{[D]}^{\mathbf{p}}\right)^{-1} = \sigma_{\rm wgn}^2 |X_p|^{-2} \mathbf{I} = SNR_p^{-1}\mathbf{I}$. For OFDM systems where all subcarriers are active, i.e. are used either for data transmission (subject to the same modulation type over all data subcarriers) or training, one can show that SNR_p is related to the SNR observed at the receiver input according to the following equation:

$$SNR_p = \frac{N}{P + (N - P)\gamma^{-1}} SNR, \tag{8}$$

where $\gamma = |X_p|^2 / \mathrm{E}\left[|X_d|^2\right]$ represents the pilot-to-data power ratio.

The LMMSE estimator (7) uses a priori knowledge of $\sigma_{\rm wgn}^2$ (or *SNR*) and $\mathbf{R}_{\mathbf{HH}}$, and is optimal when these statistics of the channel are known. As will become clear from the further discussion, *SNR* value can be predefined: higher target SNRs are preferable to obtain more accurate estimates. Also the robust estimator design necessitates account for the worst correlation of the multipath channel, namely when the channel power-delay profile (PDP) is uniform [3][9]. Taking into account this assumption and constraint of the CIR length $L \le N_{\rm cp}+1$, $\mathbf{R}_{\mathbf{HH}}$ can be written in the following eigen-decomposed form that is true for sample-spaced multipath channels:

$$\mathbf{R}_{\mathbf{HH}} = \mathrm{E}\left[\ \mathbf{FBh}(\mathbf{FBh})^H\ \right] = \mathbf{FBR}_{\mathbf{hh}}\mathbf{B}^H\mathbf{F}^H, \tag{9}$$

where $\mathbf{R}_{\mathbf{hh}} = L^{-1}\mathbf{I}_{L\times L}$ represents a diagonal matrix with entries equal to the uniform PDP with unity-normalised energy. The diagonal form results from the assumed uncorrelated scattering property of the CIR components, i.e. $\mathrm{E}[h_k h_l] = 0$ for any $k \ne l$, where $0 \le k, l \le L-1$. For the non-sample spaced channels in which CIR energy is spread over all its bins after band-limiting (sampling at the receiver input), and therefore correlation between CIR samples is present, such a design matrix is also suitable provided that CIR energy is effectively concentrated in the first L samples and leakage to the remaining ones is negligible [9].

A straightforward product with the weighting matrix $\mathbf{Q}$ (7) involves NP complex multiplications that represent a considerable computational load if P is large. One way to decrease it is to apply singular value decomposition (SVD) of $\mathbf{Q}$ as proposed in [3]. We propose an alternative approach as to make use of the transform-domain processing similar to CLS implementation. Applying the matrix inversion lemma to $\mathbf{Q}$ (7) and the eigenvalue decomposition of $\mathbf{R}_{\mathbf{HH}}$ (9), we derive a closed-form expression of the constrained LMMSE estimator:

$$\hat{\mathbf{H}}^{\mathrm{MMSE}} = \mathbf{Q}\mathbf{Y}^{\mathbf{p}} = SNR_p \mathbf{F}\mathbf{B}\mathbf{R}_{\mathbf{hh}}\left[\ \mathbf{I} - SNR_p \mathbf{A}\left(\mathbf{R}_{\mathbf{hh}}^{-1} + SNR_p \mathbf{A}\right)^{-1}\ \right]\mathbf{B}^H \mathbf{F}^H \mathbf{C}^H \mathbf{X}_{[D]}^{\mathbf{p}\ -1} =$$
$$= \mathbf{F}\mathbf{B}\mathbf{W}_{\mathbf{CLMMSE}}\mathbf{B}^H \mathbf{F}^H \mathbf{C}^H \mathbf{X}_{[D]}^{\mathbf{p}\ -1}, \qquad (10)$$

where $\mathbf{A} = \mathbf{B}^H \mathbf{F}^H \mathbf{C}^H \mathbf{C}\mathbf{F}\mathbf{B}$.

When the robust design is made ($\mathbf{R}_{\mathbf{hh}} = L^{-1}\mathbf{I}$) and the pilot subcarriers are equipowered ($\mathbf{X}_{[D]}^{\mathbf{p}} = X_p \mathbf{I}$) and equispaced ($\mathbf{A} = P\mathbf{I}$), $\mathbf{Q}$ in (10) simplifies to

$$\mathbf{Q}_{\text{EP-ES}} = \frac{1}{X_p\left(P + SNR_p^{-1} L\right)}\mathbf{F}\mathbf{B}\mathbf{B}^H \mathbf{F}^H \mathbf{C}^H . \qquad (11)$$

Comparing (5) and (11) one can see that the CLS estimator represents a special case of the CLMMSE, for which $SNR_p \to \infty$. This result justifies the selection of a higher target SNR for the robust CLMMSE estimator design.

4. ALGORITHM COMPLEXITY

Approximate computational complexity of the estimators (4) and (10) based on the radix-4 FFT/IFFT pair is summarised in Table 1 and illustrated in Fig. 1 for a range of CP lengths compared to SVD implementation [3].

One can see that the FFT-based estimators in most cases have much lower (up to an order of magnitude) complexity than the SVD-based implementation. It should also be noted that for a big number of subcarriers in the OFDM spectrum with equispaced pilots, the amount of multiplications required by the FFT-based algorithms is almost independent of N_{cp}.

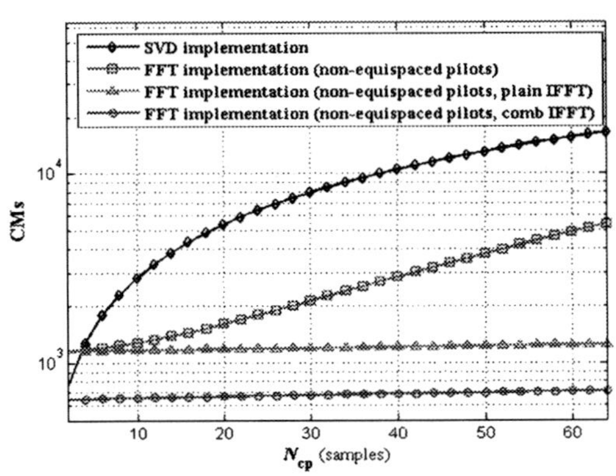

Figure 1. Dependence of the computational complexity on CP length ($N = 256$ and $P = N_{cp} + 1 = \min$ for the FFT-based algorithms).

Table 1. Computational complexity of the channel estimators

Estimator type	Complex multiplications (*CMs*) per symbol
SVD-based estimator	$N(N_{\text{cp}}+1)$
FFT-based estimator for non-equispaced pilots	$3N/2(\log_4 N-1)+(N_{\text{cp}}+1)(N_{\text{cp}}+1/2)+P$
FFT-based estimator for equispaced pilots (plain IFFT)	$3N/2(\log_4 N-1)+(N_{\text{cp}}+1)/2+P$
FFT-based estimator for equispaced pilots (comb-optimised IFFT)	$N/2(3/2\log_4 N-1)+\log_2(N_{\text{cp}}+1)+P$

5. PERFORMANCE ANALYSIS

5.1 MSE performance

For performance assessment and comparison of the CLS and CLMMSE channel estimation algorithms we use the previously defined MSE criterion:

$$\begin{aligned}\text{mse}(\mathbf{R}_{\mathbf{HH}}, SNR_p) &= \frac{1}{N}\text{E}\left[(\hat{\mathbf{H}}-\mathbf{H})^H(\hat{\mathbf{H}}-\mathbf{H})\right] = \frac{1}{N}\text{E}\left[(\mathbf{M}\mathbf{Y}^{\mathbf{P}}-\mathbf{H})^H(\mathbf{M}\mathbf{Y}^{\mathbf{P}}-\mathbf{H})\right] = \\ &= \frac{1}{N}\text{trace}\left[\mathbf{M}\mathbf{X}^{\mathbf{P}}_{[D]}(\mathbf{C}\mathbf{R}_{\mathbf{HH}}\mathbf{C}^H + SNR_p^{-1}\mathbf{I})\mathbf{X}^{\mathbf{P}H}_{[D]}\mathbf{M}^H + \mathbf{R}_{\mathbf{HH}} - 2\mathbf{M}\mathbf{X}^{\mathbf{P}}_{[D]}\mathbf{C}\mathbf{R}_{\mathbf{HH}}\right], \quad (12)\end{aligned}$$

where $\mathbf{M}=\mathbf{S}$ for the CLS estimator, and $\mathbf{M}=\mathbf{Q}=\widetilde{\mathbf{R}}_{\mathbf{HH}}\mathbf{C}^H\left[\mathbf{C}\widetilde{\mathbf{R}}_{\mathbf{HH}}\mathbf{C}^H+\widetilde{SNR}_p^{-1}\mathbf{I}\right]^{-1}\mathbf{X}^{\mathbf{P}\,-1}_{[D]}$ for the CLMMSE scheme, where $\widetilde{\mathbf{R}}_{\mathbf{HH}}$ denotes the design correlation matrix as opposed to the true CFR correlation matrix $\mathbf{R}_{\mathbf{HH}}$, and $\widetilde{SNR}_p$ is referred to as the design SNR_p setting.

5.2 Numerical results and conclusions

The simulation scenario consists of an uncoded PSAM-driven OFDM system with 256 subcarriers, Gray-mapped 16QAM modulation of the data subcarriers and equal spacing of pilots. 16 and 32 pilot subcarriers are considered with power of the pilot subsymbols being set equal to minimum and maximum in the 16QAM constellation that corresponds to the estimator design forms (5) and (11). CP length is 15 samples. A simple one-tap equaliser is used for CFR compensation at each data subcarrier before detection. The 16-tap bandlimited channel ($L=16$) is chosen with an exponentially decaying PDP and the sample-normalised rms delay spread equal to 2. Both the classical WSSUS model with $f_{\text{D}}=0.01/T$ and its quasistationary approximation (Section 2) have been simulated over more than 3600 OFDM symbols. The target energy-per-bit-to-noise-ratio (E_{b}/N_0)

for the robust CLMMSE estimator has been set to 30 dB. Numerical results are shown in Fig. 2 (there is a tight MSE match between theory (12) and simulations). Note that E_b / N_0 metric used here is linked to SNR of the PSAM-driven OFDM transmissions by equation [8]

$$\frac{E_{\mathrm{b}}}{N_0} = SNR\frac{N + N_{\mathrm{cp}}}{k(N - P)}, \tag{13}$$

where k is number of bits carried by one modulation symbol (4 for 16QAM).

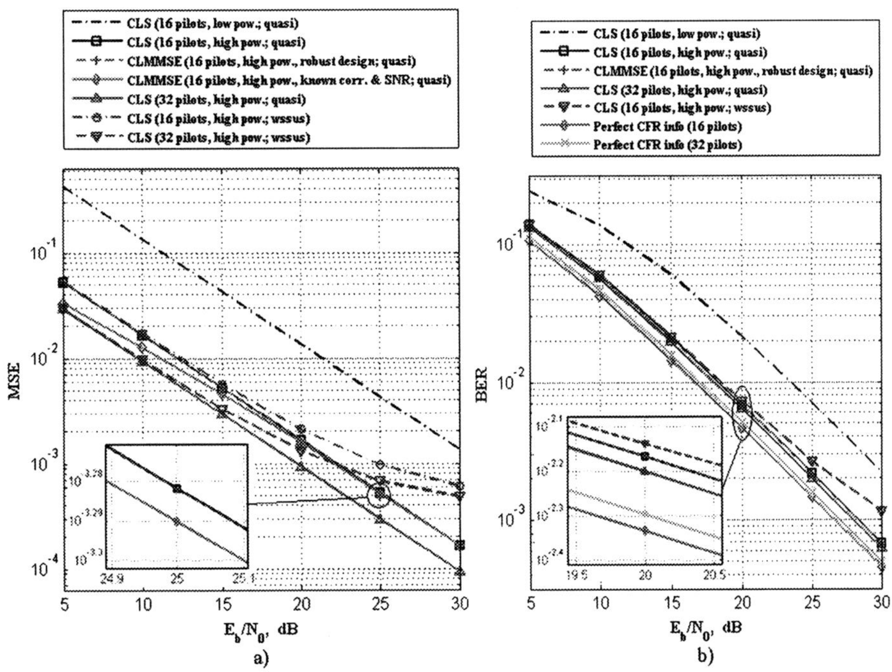

Figure 2. Performance of the pilot-aided estimators: a) MSE; b) BER.

It can be seen that CLMMSE estimator (10) with the robust design parameters exhibits the same performance as CLS (4). This is explained by the quite high design SNR setting that makes the CLMMSE estimator almost identical to CLS (compare (5) and (11)). Knowledge of the channel correlation and SNR generally allows improving performance over the lower SNR range, especially when strong correlation is observed between CIR samples (e.g., in the non-sample-spaced channels). However, both of these statistics should be accurately estimated and involved in the channel estimator that complicates its design, in particular requiring a matrix inversion in (10). Our analysis has shown that the exact knowledge of neither correlation, nor SNR alone can give substantial performance gain. Thus, CLS is an obvious choice for a robust low-complexity receiver.

Increasing the number of pilot subcarriers to improve the CFR estimation has been found very inefficient since apart from greater transmission redundancy and E_b/N_0 loss (two different, 0.4dB-spaced desired performance curves in Fig. 2b – for 16 and 32 pilots) it does not yield satisfactory BER performance gain (< 0.4dB). Furthermore, in the case of noticeable time variation of the CIR ($f_D T \geq 0.01$ in the standard WSSUS model), the difference between estimators driven by 16 and 32 pilots almost disappears at higher SNRs due to the error floor caused by ICI. Thus, the number of pilots must be a minimum – equal to the CIR length. On the contrary, most performance gain is achieved by boosting power of the pilot subsymbols (e.g., up to 5dB in BER when using high-powered instead of low-powered pilots). But when SNR at the pilot positions is high, SNR at the data subcarriers becomes low worsening detection reliability. Therefore the pilot-to-data power ratio trade-off must be searched for in the system with a given number of pilot subcarriers.

REFERENCES

[1] M. Engels. *Wireless OFDM Systems: How to Make Them Work?* IMEC, Belgium, 2002.

[2] IEEE Std 802.16-2004. Part 16: Air interface for fixed broadband wireless access systems. Oct. 2004.

[3] O. Edfords, M. Sandell, J.-J. van de Beek, S. K. Wilson, and P. O. Börjesson. OFDM channel estimation by singular value decomposition. *IEEE Transactions on Communications*, 46:931–939, Jul. 1998.

[4] P. Chen, and H. Kobayashi. Maximum Likelihood channel estimation and signal detection for OFDM systems. In *Proceedings IEEE International Conference on Communications (ICC)*, pp. 1640-1645, Apr. 2002.

[5] L. Deneire, P. Vandenameele, L. van der Perre, B. Gyselinckx, and M. Engels. A low complexity ML channel estimator for OFDM communications. *IEEE Transactions on Communications*, 51:135-140, Feb. 2003.

[6] Y. Mostofi, and D. C. Cox. Average error rate analysis for pilot-aided OFDM receivers with frequency-domain interpolation. In *Proceedings IEEE Wireless Communication and Networking Conference (WCNC)*, pp. 1421-1425, Mar. 2004

[7] M.-X. Chang, and Y. T. Su. Model-based channel estimation for OFDM signals in Rayleigh fading. *IEEE Transactions on Communications*, 50:540-545, Apr. 2002.

[8] E. Golovins, and N. Ventura. Comparative Analysis of Low Complexity Channel Estimation Techniques for the Pilot-assisted Wireless OFDM Systems. In *Proceedings Southern African Telecommunication Networks and Applications Conference (SATNAC)*, Sep. 2006.

[9] Y. (G.) Li, L. J. Cimini Jr., and N.R. Sollenberger. Robust channel estimation for OFDM systems with rapid dispersive fading channels. *IEEE Transactions on Communications*, 46:902-915, Jul. 1998.

A STUDY ON CHANNEL ESTIMATION FOR OFDM SYSTEMS USING EM ALGORITHM BASED ON MULTI-PATH DOPPLER CHANNEL MODEL

Akihiro Waku*, Masahiro Fujii**, Makoto Itami* and Kohji Itoh*

**Department of Applied Electronics, Tokyo University of Science,*
2641 Yamazaki, Noda-Shi, Chiba, Japan

awaku@itlb.te.noda.tus.ac.jp,{itami,itoh}@te.noda.tus.ac.jp

***Department of Information Science, Utsunomiya University,*
7-1-2 Yoto, Utsunomiya-Shi, Tochigi, Japan

fujii@is.utsunomiya-u.ac.jp

Abstract In OFDM mobile communication with high-speed vehicle, due to the effect of large multi-path Doppler shifts varying independently with respect to the paths, the time allowed to the conventional methods of estimation of the channel characteristics is severely limited without using separate estimates of the Doppler shifts. As a result, the quality of performance of the data demodulation remains low using the estimates insufficiently smoothed of the noise.

This report proposes a novel method using Expectation-Maximization (EM) algorithm for simultaneously estimating the Doppler shifts and the complex amplitudes of the path-wise channel responses. The results of computer simulation show capability of proposed method much superior to the conventional one.

1. Introduction

OFDM is a digital modulation scheme that achieves very efficient frequency utilization and realizes high speed data transmission with limited band width. The OFDM signal consists of many carriers that are orthogonally arranged in the frequency domain, and data symbols modulated by QPSK, QAM, etc. are transmitted on each carrier. OFDM has an ability to avoid Inter-Symbol-Interference (ISI) due to multi-path transmission, owing to the guard interval inserted between the adjacent

S. Plass et al. (eds.), Multi-Carrier Spread Spectrum 2007, 337–346.

symbols, and provides higher transmission rate than single-carrier signaling with the same bandwidth.

In OFDM mobile communication with high-speed vehicle, due to the effect of large multi-path Doppler shifts varying differently with respect to the paths, the symbol-to-symbol phase difference of the channel frequency characteristics cannot be compensated by way of the simple estimation of frequency offset. As a result, the quality of performance of the data demodulation remains low using symbol wise estimates insufficiently smoothed of the noise effect.

In order to overcome the limitation we aim at realizing a joint Maximum Likelihood(ML) estimation of multi-path parameters, i.e. Doppler shift, relative delay time and complex amplitude of each path, as well as the number of paths. And we propose in this paper to make use of Expectation-Maximization(EM) algorithm for that purpose, dealing with the complex amplitudes as hidden parameters with known statistical characteristics.

Starting from initial EM estimates of Doppler shifts of the selected paths with relative delay times corresponding to the sampling points within the delay profile, we succeeded in jointly improving the estimates of multi-path Doppler shifts and relative delay times using EM iteration algorithm. The results of computer simulation show capability of the proposed method much superior to the conventional symbol-wise estimation or simple averaging of the symbol-wise estimates.

2. Multi-Path Doppler Model

We assume the equivalent low-pass channel impulse response model which consists of discrete Q paths (see Figure 1) is given by

$$h(\tau;t) = \sum_{q=0}^{Q-1} h_q e^{j2\pi f_{D,q}t}\delta(\tau-\tau_q), \tag{1}$$

where h_q, $f_{D,q}$, τ_q are complex amplitude, Doppler frequency and relative delay time at the q-th path, respectively, and the frequency response is given by

$$H(f;t) = \sum_{q=0}^{Q-1} h_q e^{j2\pi(f_{D,q}t-f\tau_q)}. \tag{2}$$

We assume OFDM system which employs FFT operation with L points, i.e. L carriers.

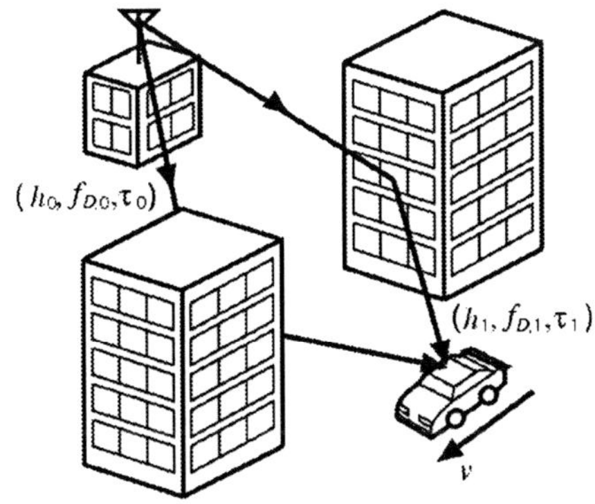

Figure 1. Multi-Path Doppler Model (in case of $Q = 2$)

The output $Y_{k,l}$ of FFT operation at the receiver for the k-th pilot symbol period and the l-th frequency carrier is expressed as

$$Y_{k,l} = H_{k,l} + N_{k,l} \tag{3}$$

and we define

$$\boldsymbol{Y} = \begin{bmatrix} Y_{0,0} & \cdots & Y_{0,L-1} \\ \vdots & \ddots & \vdots \\ Y_{K-1,0} & \cdots & Y_{K-1,L-1} \end{bmatrix}, \tag{4}$$

where K is the number of consecutive symbol periods observed for the joint estimation of the channel parameters assumed constant in those periods, and $N_{k,l}$ is the additive white Gaussian noise. Pilot symbol modulation is assumed demodulated at the receiver with time and reference carrier synchronization. Neglecting the effect of carrier phase variations due to Doppler shifts in the symbol interval whose duration is T_s, $H_{k,l}$ is approximated as

$$\begin{aligned} H_{k,l} = H(lf_0; kT_s) &= \sum_{q=0}^{Q-1} h_q e^{j2\pi(kf_{D,q}T_s - lf_0\tau_q)} \\ &= \sum_{q=0}^{Q-1} h_q e^{j2\pi(\frac{k}{K}\alpha_q - \frac{l}{L}\beta_q)} \end{aligned} \tag{5}$$

where f_0 is the adjacent carrier frequency separation, α_q is normalized Doppler frequency, β_q is normalized relative delay time, respectively, defined by

$$\alpha_q = Kf_{D,q}T_s \tag{6}$$

$$\beta_q = Lf_0\tau_q. \tag{7}$$

3. Proposed Method of Channel Estimation

For the purpose of optimally reducing the effect of noise, we employ Maximum A Posteriori(MAP) criterion which maximizes a posteriori probability based on the received signal over an interval as long as the path structure is assumed unchanged. In order to estimate the set of parameters

$$\boldsymbol{\phi} = [\underline{\phi}_0, \cdots, \underline{\phi}_{Q-1}]^T, \quad \underline{\phi}_q = [h_q, \alpha_q, \beta_q] \tag{8}$$

based on the observation $\boldsymbol{Y}$, a posteriori probability,

$$p(\boldsymbol{\phi}|\boldsymbol{Y}) = \frac{p(\boldsymbol{\phi})p(\boldsymbol{Y}|\boldsymbol{\phi})}{p(\boldsymbol{Y})} \tag{9}$$

should be maximized with respect to $\boldsymbol{\phi}$. The estimation based on this criterion needs prior information about $\boldsymbol{\phi}$. Moreover, it is computationally almost intractable jointly to estimate the set of $\boldsymbol{\phi}$ based on the received signal matrix $\boldsymbol{Y}$ using the MAP criterion. Therefore we better divide the problem into 2 stages. First we obtain ML estimates of

$$\boldsymbol{\psi} = [\underline{\psi}_0, \cdots, \underline{\psi}_{Q-1}]^T, \quad \underline{\psi}_q = [\alpha_q, \beta_q], \tag{10}$$

by EM algorithm, dealing with $\underline{h} = [h_0, \cdots, h_{Q-1}]^T$ as hidden random variables. And secondly, by regarding the estimated parameters as correct, we estimate $\underline{h}$.

3.1 Application of EM Algorithm

First, we discuss to obtain ML estimates of $\boldsymbol{\psi}$ as defined as follows

$$\hat{\boldsymbol{\psi}} = \arg\max_{\boldsymbol{\psi}} p(\boldsymbol{Y}|\boldsymbol{\psi}), \tag{11}$$

In general the number of paths Q may be a target parameter of estimation. Now EM algorithm is an estimate innovation algorithm in which conditional probability density function of $\underline{h}$ defined by $g^i(\underline{h}) = p(\underline{h}|\boldsymbol{\psi}^i, \boldsymbol{Y})$ is introduced conditioned by $\boldsymbol{\psi}^i$ and $\boldsymbol{Y}$ where $\boldsymbol{\psi}^i$ is the i-th estimate of $\boldsymbol{\psi}$. And

$$\mathcal{Q}(\boldsymbol{\psi}; \boldsymbol{\psi}^i) = \int g^i(\underline{h}) \log \frac{p(\underline{h}, \boldsymbol{Y}|\boldsymbol{\psi})}{g^i(\underline{h})} d\underline{h} \tag{12}$$

is defined. Using Jensen's inequality, we obtain,

$$\mathcal{Q}(\boldsymbol{\psi}; \boldsymbol{\psi}^i) \leq \log \int g^i(\underline{h}) \frac{p(\underline{h}, \boldsymbol{Y}|\boldsymbol{\psi})}{g^i(\underline{h})} d\underline{h} = \log p(\boldsymbol{Y}|\boldsymbol{\psi}) \tag{13}$$

in which we find that $\mathcal{Q}(\boldsymbol{\psi};\boldsymbol{\psi}^i)$ gives a lower bound to $\log p(\boldsymbol{Y}|\boldsymbol{\psi})$ and the equality is shown to hold when $\boldsymbol{\psi} = \boldsymbol{\psi}^i$, which means $\mathcal{Q}(\boldsymbol{\psi};\boldsymbol{\psi}^i)$, as the function of $\boldsymbol{\psi}$, touches from below the function $\log p(\boldsymbol{Y}|\boldsymbol{\psi})$.

Therefore, starting from that value of $\boldsymbol{\psi}$, there should be a vectorial direction of changing $\boldsymbol{\psi}$ to increase $\mathcal{Q}(\boldsymbol{\psi};\boldsymbol{\psi}^i)$, provided $\log p(\boldsymbol{Y}|\boldsymbol{\psi})$ can be increased. Now $\mathcal{Q}(\boldsymbol{\psi};\boldsymbol{\psi}^i)$ can be decomposed as

$$\mathcal{Q}(\boldsymbol{\psi};\boldsymbol{\psi}^i) = \mathcal{Q}'(\boldsymbol{\psi};\boldsymbol{\psi}^i) + \mathcal{H}, \tag{14}$$

where

$$\mathcal{Q}'(\boldsymbol{\psi};\boldsymbol{\psi}^i) = \int g^i(\underline{h}) \log p(\underline{h}, \boldsymbol{Y}|\boldsymbol{\psi}) d\underline{h}. \tag{15}$$

As $\mathcal{H} = -\int g^i(\underline{h}) \log g^i(\underline{h}) d\underline{h}$ does't depend on $\boldsymbol{\psi}$, we obtain the following iterative innovation of the value of $\boldsymbol{\psi}$.

$$\boldsymbol{\psi}^{i+1} = \arg\max_{\boldsymbol{\psi}} \mathcal{Q}'(\boldsymbol{\psi};\boldsymbol{\psi}^i) \tag{16}$$

The value obtained by the algorithm, however, is not guaranteed of global optimality. Some means is necessary in general to select scattered starting values of $\boldsymbol{\psi}$ and compare the resulted local maximum values of $\log p(\boldsymbol{Y}|\boldsymbol{\psi})$. As $\mathcal{Q}'(\boldsymbol{\psi};\boldsymbol{\psi}^i)$ is $\boldsymbol{\psi}^i$ conditional mean of $\log p(\underline{h}, \boldsymbol{Y}|\boldsymbol{\psi})$ with respect to $\underline{h}$, equation (15) is called "E-Step", and equation (16) of the maximum value search about $\boldsymbol{\psi}$ is called "M-Step". In a simple term, the proposed method with EM which consists of the following two steps.

E-Step Calculate $\mathcal{Q}'(\boldsymbol{\psi};\boldsymbol{\psi}^i)$ according to equation (15) using $g^i(\underline{h})$

M-Step Renew $\boldsymbol{\psi}^i$ as equation (16) designates.

3.2 Calculation of E-Step

We now show the calculation process of channel estimation using EM algorithm. Assuming path-independent Rayleigh fading, elements of vector $\underline{h}$ are modeled as mutually independent zero-mean complex-valued Gaussian random variables where $\mathtt{E}\{\underline{h}\} = \underline{0}, \mathtt{E}\{\underline{h}\underline{h}^H\} = \mathrm{diag}[\sigma_0^2, \cdots, \sigma_{Q-1}^2]$ hold. And the additive Gaussian noise defined in equation (3) satisfies $\mathtt{E}\{N_{k,l}\} = 0$, $\mathtt{E}\{N_{k,l}N^*_{k',l'}\} = \sigma_n^2 \delta_{k,k'}\delta_{l,l'}$. Using these statistics, the probability density function $g^i(\underline{h})$ is calculated as

$$\begin{aligned} g^i(\underline{h}) &= \frac{p(\boldsymbol{Y}|\underline{h}, \boldsymbol{\psi}^i)p(\underline{h})}{p(\boldsymbol{Y}|\boldsymbol{\psi}^i)} \\ &= \frac{C}{p(\boldsymbol{Y}|\boldsymbol{\psi}^i)} \exp\{-(\underline{h} - \overline{\underline{h}}(\boldsymbol{\psi}^i))^H \boldsymbol{v}(\boldsymbol{\psi}^i)(\underline{h} - \overline{\underline{h}}(\boldsymbol{\psi}^i)) \\ &\quad + \overline{\underline{h}}(\boldsymbol{\psi}^i)^H \boldsymbol{v}(\boldsymbol{\psi}^i)\overline{\underline{h}}(\boldsymbol{\psi}^i)\}, \end{aligned} \tag{17}$$

where

$$
\begin{aligned}
u_{q,k,l}(\underline{\psi}_q^i) &= e^{j2\pi(\frac{k}{K}\alpha_q^i - \frac{l}{L}\beta_q^i)} & (18)\\
\underline{u}_{k,l}(\boldsymbol{\psi}^i) &= [u_{0,k,l}(\underline{\psi}_0^i), \cdots, u_{Q-1,k,l}(\underline{\psi}_{Q-1}^i)]^T & (19)\\
z_q(\underline{\psi}_q^i) &= \sum_{k=0}^{K-1}\sum_{l=0}^{L-1} u_{q,k,l}^*(\underline{\psi}_q^i) Y_{k,l} & (20)\\
\underline{z}(\boldsymbol{\psi}^i) &= [z_0(\underline{\psi}_0^i), \cdots, z_{Q-1}(\underline{\psi}_{Q-1}^i)]^T & (21)\\
v_{q,q'}(\underline{\psi}_q^i, \underline{\psi}_{q'}^i) &= \frac{1}{\sigma_n^2}\sum_{k=0}^{K-1}\sum_{l=0}^{L-1} u_{q,k,l}^*(\underline{\psi}_q^i) u_{q',k,l}(\underline{\psi}_{q'}^i) + \frac{\delta_{q,q'}}{\sigma_{h,q}^2} & (22)
\end{aligned}
$$

$$
\boldsymbol{v}(\boldsymbol{\psi}^i) = \begin{bmatrix} v_{0,0}(\underline{\psi}_0^i, \underline{\psi}_0^i) & \cdots & v_{0,Q-1}(\underline{\psi}_0^i, \underline{\psi}_{Q-1}^i) \\ \vdots & \ddots & \vdots \\ v_{Q-1,0}(\underline{\psi}_{Q-1}^i, \underline{\psi}_0^i) & \cdots & v_{Q-1,Q-1}(\underline{\psi}_{Q-1}^i, \underline{\psi}_{Q-1}^i) \end{bmatrix} \tag{23}
$$

$$
\overline{\underline{h}}(\boldsymbol{\psi}^i) = \frac{1}{\sigma_n^2}\boldsymbol{v}^{-1}(\boldsymbol{\psi}^i)\underline{z}(\boldsymbol{\psi}^i). \tag{24}
$$

The factor C depends neither on $\boldsymbol{\psi}^i$ nor on $\underline{h}$. According to equation (17), $g^i(\underline{h})$ is seen to be a complex Gaussian probability density function with average $\overline{\underline{h}}(\boldsymbol{\psi}^i)$, covariance matrix $\boldsymbol{v}^{-1}(\boldsymbol{\psi}^i)$. Referring to equation (17), $\log p(\underline{h}, \boldsymbol{Y}|\boldsymbol{\psi})$ is represented as

$$
\begin{aligned}
\log p(\underline{h}, \boldsymbol{Y}|\boldsymbol{\psi}) &= \log p(\boldsymbol{Y}|\underline{h}, \boldsymbol{\psi})p(\underline{h})\\
&= \log C - (\underline{h} - \overline{\underline{h}}(\boldsymbol{\psi}^i))^H \boldsymbol{v}(\boldsymbol{\psi})(\underline{h} - \overline{\underline{h}}(\boldsymbol{\psi}^i))\\
&\quad + \overline{\underline{h}}(\boldsymbol{\psi}^i)^H \boldsymbol{v}(\boldsymbol{\psi})\overline{\underline{h}}(\boldsymbol{\psi}^i).
\end{aligned} \tag{25}
$$

Substituting equation (17) and (25) into equation (15), and retaining the terms depending on $\boldsymbol{\psi}$ and $\boldsymbol{\psi}^i$, we obtain, in place of $Q'(\boldsymbol{\psi}; \boldsymbol{\psi}^i)$,

$$
\begin{aligned}
Q''(\boldsymbol{\psi}; \boldsymbol{\psi}^i) &= \frac{2}{\sigma_n^2}\Re\{\overline{\underline{h}}^H(\boldsymbol{\psi}^i)\underline{z}(\boldsymbol{\psi})\} - \mathrm{tr}\,\{\boldsymbol{v}^{-1}(\boldsymbol{\psi}^i)\boldsymbol{v}(\boldsymbol{\psi})\}\\
&\quad - \overline{\underline{h}}^H(\boldsymbol{\psi}^i)\boldsymbol{v}(\boldsymbol{\psi})\overline{\underline{h}}(\boldsymbol{\psi}^i)\\
&= \frac{2}{\sigma_n^4}\Re\{\underline{z}^H(\boldsymbol{\psi}^i)\{\boldsymbol{v}^{-1}(\boldsymbol{\psi}^i)\}^H\underline{z}(\boldsymbol{\psi})\} - \mathrm{tr}\,\{\boldsymbol{v}^{-1}(\boldsymbol{\psi}^i)\boldsymbol{v}(\boldsymbol{\psi})\}\\
&\quad - \frac{1}{\sigma_n^4}\underline{z}^H(\boldsymbol{\psi}^i)\{\boldsymbol{v}^{-1}(\boldsymbol{\psi}^i)\}^H\boldsymbol{v}(\boldsymbol{\psi})\boldsymbol{v}^{-1}(\boldsymbol{\psi}^i)\underline{z}(\boldsymbol{\psi}^i).
\end{aligned} \tag{26}
$$

Especially, when $\boldsymbol{\psi} = \boldsymbol{\psi}^i$, the result reduces to

$$
Q'''(\boldsymbol{\psi}^i; \boldsymbol{\psi}^i) = \underline{z}^H(\boldsymbol{\psi}^i)\{\boldsymbol{v}^{-1}(\boldsymbol{\psi}^i)\}^H\underline{z}(\boldsymbol{\psi}^i). \tag{27}
$$

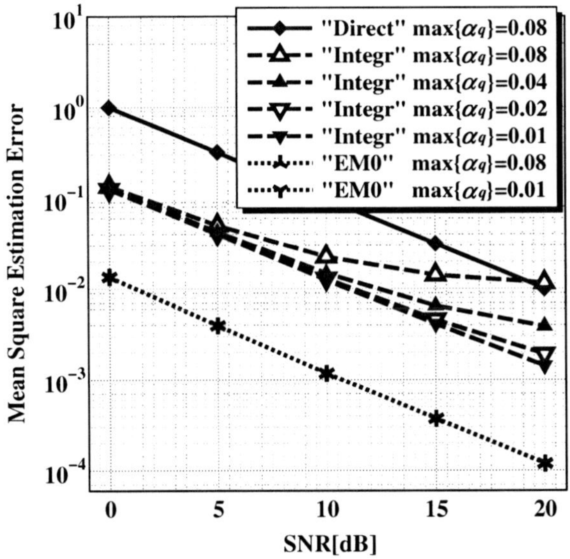

Figure 2. SNR vs MSEE in variable estimator

3.3 Estimation of the Complex Amplitude

Suppose terminating the iteration at I-th step after which no substantial improvement can be expected, and using the values $\boldsymbol{\psi}^I$ in the equation (24) regarding they are correct, we obtain the semi-optimal estimate $\underline{\hat{h}}$ of $\underline{h}$ as

$$\underline{\hat{h}} = \frac{1}{\sigma_n^2}\boldsymbol{v}^{-1}(\boldsymbol{\psi}^I)\underline{z}(\boldsymbol{\psi}^I), \quad \underline{\hat{h}} = [\hat{h}_0, \cdots, \hat{h}_{Q-1}]^T, \tag{28}$$

and the semi-optimal estimate $\hat{H}_{k,l}$ of $H_{k,l}$ is given by

$$\hat{H}_{k,l} = \sum_{q=0}^{Q-1} \hat{h}_q e^{j2\pi(\frac{k}{K}\alpha_q^I - \frac{l}{L}\beta_q^I)}. \tag{29}$$

4. Numerical Results

We consider two paths ($Q = 2$) channel models, shown in equation (1), in which the normalized Doppler frequencies $\{\alpha_q\}$ defined in equation (6) are mutually statistically independent random variables each of which is uniformly distributed within the interval $(-\max\{\alpha_q\}, \max\{\alpha_q\})$ and each path has the specified normalized relative delay time β_q. For the purpose of evaluation of performances, we assume 2 values of $\max\{\alpha_q\}$. As for the relative path delay, we assume $\beta_0 = 0.0$ and $\beta_1 = 5.0$ in

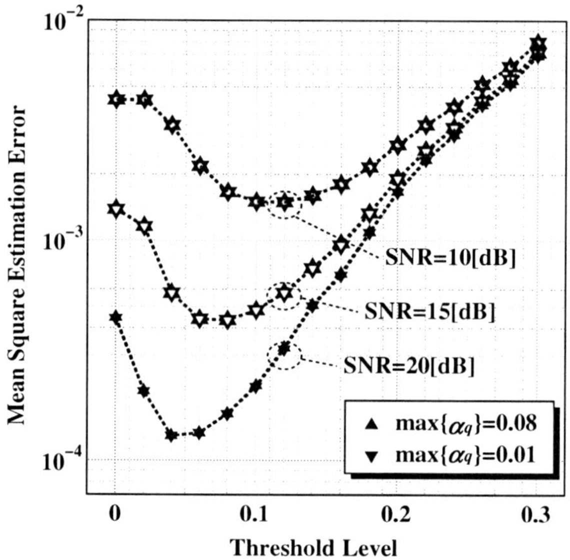

Figure 3. Threshold Level vs MSEE in "EM1"

common. The number of OFDM symbols used for estimation $K = 8$ and the number of carriers $L = 32$.

The EM algorithm starts with the initial EM estimates of the Nyquist samples of the channel impulse response.

Figure 2 compares 2 conventional methods with an EM method named "EM0" whose initial path estimates is obtained by selecting the largest 2 Nyquist samples of the initial EM estimates. Among the 2 conventional methods the "Direct" method employs $Y_{k,l}$ in equation (3) as estimates of $H_{k,l}$. In "Integr" method the estimate of $H_{k,l}$ is obtained by averaging $Y_{k,l}$ $k = 0, 1, 2, \cdots$ for K symbol intervals. The performance of "EM0" is the best, because it makes use of the estimates of normalized Doppler frequencies α_q^i in order to cancel symbol-to-symbol phase rotation to obtain better quality estimates of the path delays as well as the complex-amplitudes.

Figure 3 shows the performance of EM estimator named "EM1" which employs as initial path estimates such Nyquist samples as surpassing the preset threshold.

Figure 4 and Figure 5 show that "EM1" with the best threshold performs just as good as "EM0" inspite of its ignorance of the number of paths, and the higher the Doppler frequency becomes, the greater the number of iterations is required for high performances.

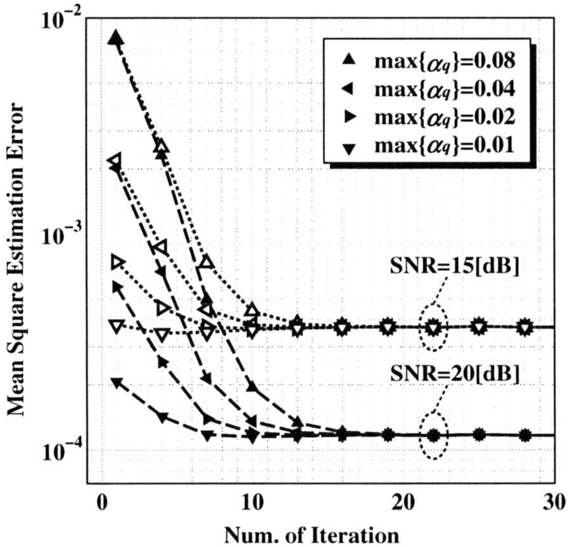

Figure 4. Num. of Iteration vs MSEE in "EM0"

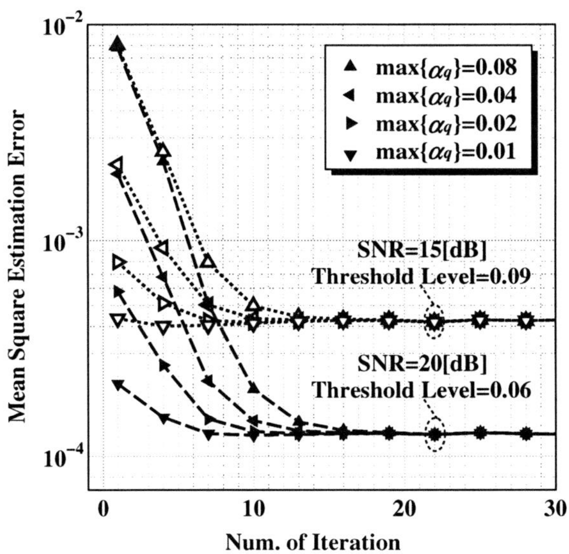

Figure 5. Num. of Iteration vs MSEE in "EM1"

5. Conclusions

This report proposed a novel method using EM algorithm of simultaneously estimating the Doppler shifts and the complex amplitudes of the path-wise channel responses. As mentioned earlier, the proposed method with EM algorithm is not guaranteed to converge globally. As for all of the results of the computer simulation, however, the convergence turned to be global. And the numerical results showed capability of proposed method much superior to the conventional one. It is, yet, to be investigated in which situation the convergence can be local and to which extent the problem can be solved by introducing, e.g. appropriate ways of scattering initial assumed estimates, especially of path delay times.

References

[1] Masahiro Fujii, Makoto Itami and Kohji Itoh: " Joint Estimation of Frequency Offset and Channel Frequency Response using EM algorithm for OFDM systems", IEICE Transactions on Fundamentals, E89-A, no. 11, pp. 3123-3130, (2006)

[2] Mitsuru Nakamura, Masahiro Fujii, Makoto Itami, Kohji Itoh "A Study on an MMSE ICI Canceller for OFDM under Doppler–spread Channel", The 14th IEEE International Symposium on Personal, Indoor and Mobile Radio Communications 2003(PIMRC2003), pp236–240, Sep. 2003.

[3] Ye (Geoffrey) Li, Leonard J. Cimini, Jr, Nelson R. Sollenberger, "Robust Channel Estimation for OFDM Systems with Rapid Dispersive Fading Channels," IEEE Trans. on Commun., vol.46, NO.7 pp. 902–915, Jul., 1998.

[4] P. Robertson and S. Kaiser, "The Effects of Doppler Spreads in OFDM(A) Mobile Radia Systems," IEEE 50th Veh. Tech. Conf. Fall, pp. 329–333, Sep. 1999.

[5] Ove Edfors, "Analysis of DFT-Based Channel Estimators for OFDM," Wireless Personal Communications., pp. 55–70, Dec., 2000.

[6] Baoguo Yang, "Analysis of Low-Complexity Windowed DFT-Based MMSE Channel Estimator for OFDM Systems," IEEE Trans.on Commun., vol.49, pp. 1977–1987, Nov., 2001.

[7] Kazuhiko Fukawa, "OFDM Channel Estimation with RLS Algorithm for Different Pilot Schemes in Mobile Radio Transmission," IEICE Trans.on Commun., vol.E86-B,No.1 pp. 266–274, Jan., 2003.

[8] R.W.Chang and R.A.Gabby, "A Theoretical Study of Performance of an Orthogonal Multiplexing Data Transmission Scheme", IEEE Trans.Comm., COM-16, pp.529-5401968

[9] S.B.Weinstein and P.W.Ebert, "Data Transmission byFrequency-Division Multiplexing using the Discrete Fourier Transform", IEEE Trans.Comm., COM-19, pp. 628-6341971

[10] B.Hirosaki, "An Analysis of Automatic Equalizer for Orthogonally Multiplexed QAM System", IEEE Trans.Comm., COM-28, pp.73-831980

[11] Won Gi Jeon,Kyung Hi Chang,and Yong Soo Cho, "An Equalization Technique for Orthogonal Frequency-Division Multiplexing Systems in Time-Variant Multipath Channels," IEEE Trans.Comm,Vol.47,NO.1,pp.27-pp.32,Jan.1999

[12] F. Dellaert, "The Expectation Maximization Algorithm," Tech. report GIT-GVU-02-20, College of Computing GVU Center, Georgia Institute of Technology, Feb. 2002.

CHANNEL ESTIMATION FOR BLOCK-IFDMA

Anja Sohl, Tobias Frank, and Anja Klein
Darmstadt University of Technology
Communications Engineering Lab
Merckstr. 25, 64283 Darmstadt, Germany
{a.sohl,t.frank,a.klein}@nt.tu-darmstadt.de

Abstract In this paper, pilot multiplexing and channel estimation is investigated for two different signal models of DFT precoded OFDMA with block-interleaved subcarrier allocation (B-IFDMA). The influence of the B-IFDMA signal model on the bit error rate performance with and without perfect channel knowledge is presented. Moreover, the Peak-to-Average Power Ratio of the transmit signal, as well as the pilot symbol overhead required for channel estimation is given for the two signal models.

1. Introduction

For the uplink of B3G/4G mobile radio systems, Discrete Fourier Transform (DFT) precoded Orthogonal Frequency Division Multiple Access (OFDMA) is under consideration as candidate because it combines the advantageous properties of OFDMA, cf. e.g. [10], with a low Peak-to-Average Power Ratio (PAPR) [6]. For DFT precoded OFDMA, there exist different possibilities of how to allocate subcarriers to a user under consideration.
Blockwise subcarrier allocation leads to the Localized FDMA (LFDMA) scheme [9, 6], which, on the one hand, provides good robustness to carrier frequency offsets and due to the possibility of interpolation between different subcarriers, low pilot symbol overhead for channel estimation (CE) in frequency domain (FD). On the other hand, only low frequency diversity can be achieved [7].
Interleaved subcarrier allocation leads to the Interleaved FDMA (IFDMA) scheme [8, 4]. It provides high frequency diversity due to the spreading of the subcarriers over the total available bandwidth [2], and, compared to other DFT precoded OFDMA schemes, IFDMA exhibits the lowest PAPR. However, IFDMA is sensitive to carrier frequency offsets [3]. Moreover, in terms of CE, IFDMA requires a higher pilot

S. Plass et al. (eds.), Multi-Carrier Spread Spectrum 2007, 347–356.

symbol overhead than LFDMA since, in general, for IFDMA interpolation between different subcarriers in FD is not possible [7].
A third possibility of subcarrier allocation, where the data of a specific user is transmitted on blocks of subcarriers that are equidistantly distributed over the available bandwidth, is currently under investigation and denoted as Block-IFDMA (B-IFDMA) [12]. In contrast to IFDMA, where a block is built by a single subcarrier, for B-IFDMA, each block consists of K_f adjacent subcarriers. Due to the blockwise allocation, B-IFDMA is assumed to exhibit higher robustness against carrier frequency offsets than IFDMA and, at the same time, maintain the advantage of high frequency diversity. Another advantageous aspect compared to IFDMA is less pilot symbol overhead required for CE because B-IFDMA supports interpolation in FD within each block of subcarriers.
In this paper, two different variants for B-IFDMA are introduced, that are denoted as Added Signal B-IFDMA and Joint DFT B-IFDMA in the following. The pilot insertion for CE as well as the CE algorithm are presented for each B-IFDMA variant for the case that interpolation is applied in FD and for the case that no interpolation is applied. The influence of the B-IFDMA variant on the performance results with perfect channel estimation (PCE) and with realistic channel estimation (RCE), and on the PAPR of the transmit signal is investigated. Moreover, the pilot symbol overhead required for CE is discussed for both B-IFDMA variants.
The paper is organized as follows. In Section 2, system models for Added Signal B-IFDMA and Joint DFT B-IFDMA are derived from the general system model of DFT precoded OFDMA. In Section 3, pilot insertion and CE are described for each B-IFDMA variant. In Section 4, the performance results with PCE and RCE, as well as the PAPR results and the pilot symbol overhead for Added Signal and Joint DFT B-IFDMA are presented. Section 5 concludes the work.

2. System Model

2.1 DFT Precoded OFDMA

In this Section, a general description of the DFT-precoded OFDMA system model is given. In the following, all signals are represented by their discrete time equivalents in the complex baseband. Further on, $(\cdot)^T$ denotes the transpose and $(\cdot)^H$ the Hermitian of a vector or a matrix. Assuming a system with K users, let $\mathbf{d}^{(k)} = (d_0^{(k)}, \cdots, d_{Q-1}^{(k)})^T$ denote a block of Q data symbols $d_q^{(k)}, q = 0, \cdots, Q-1$, at symbol rate $1/T_s$ transmitted by a user with index k, $k = 0, \cdots, K-1$. The data symbols

$d_q^{(k)}$ can be taken from the alphabet of a modulation scheme like Phase Shift Keying (PSK), that is applied to coded or uncoded bits. Let $\mathbf{F}_N$ and $\mathbf{F}_N^H$ denote the matrix representation of an N-point DFT and an N-point Inverse DFT (IDFT), respectively, where $N = K \cdot Q$ is the number of available subcarriers in the system. The assignment of Q subcarriers to a specific user can be described by a Q-point DFT precoding matrix $\mathbf{F}_Q$, an $N \times Q$ mapping matrix $\mathbf{M}^{(k)}$ and an N-point IDFT matrix $\mathbf{F}_N^H$ [2]. Thus, a DFT precoded OFDMA signal block at sample rate $1/T_c = K/T_S$ is given by

$$\mathbf{x}^{(k)} = \mathbf{F}_N^H \cdot \mathbf{M}^{(k)} \cdot \mathbf{F}_Q \cdot \mathbf{d}^{(k)}. \tag{1}$$

The insertion of a Cyclic Prefix, as well as the transmission over a channel and subsequent demodulation for an uplink scenario is given in [2] and will not be described in this work.

For B-IFDMA, two different variants can be derived as special cases of the general DFT precoded OFDMA system model. The first one, treated in Section 2.2, is based on an assignment of multiple IFDMA signals to one user and thus, named Added Signal B-IFDMA. The second one, treated in Section 2.3, is based on one joint DFT for all subcarriers assigned to a specific user and thus, named Joint DFT B-IFDMA. In the following, K_f denotes the number of subcarriers per block and T denotes the number of blocks assigned to a specific user k, with $Q = K_f \cdot T$. For $K_f = 1$, the B-IFDMA scheme merges into IFDMA.

2.2 Added Signal B-IFDMA

The Added Signal B-IFDMA signal can be obtained if K_f IFDMA signals that are mutually shifted by one subcarrier are superimposed and assigned to user k. The signal model for IFDMA is not described explicitly in this work as it has been introduced in detail in [2]. Let $\bar{\mathbf{d}}^{(m,k)}$ denote the m-th T elementary subblock of $\mathbf{d}^{(k)}$ with elements $d_t^{(m,k)} = d_{mT+t}^{(k)}$; $m = 0, ..., K_f - 1$ and $t = 0, ..., T - 1$. The $N \times T$ mapping matrix $\mathbf{M}_{B_\Sigma}^{(m,k)}$ of the m-th IFDMA signal assigned to user k is given by its elements $M_{B_\Sigma}^{(m,k)}(n,t)$ in the n-th row, with $n = 0, ..., N-1$, and t-th column

$$M_{B_\Sigma}^{(m,k)}(n,t) = \begin{cases} 1 & \text{for } n = t \cdot \frac{N}{T} + m + k \cdot K_f \\ 0 & \text{else} \end{cases} . \tag{2}$$

Thus, for user k the Added Signal B-IFDMA signal is obtained by

$$\mathbf{x}_{B_\Sigma}^{(k)} = \sum_{m=0}^{K_f-1} \mathbf{F}_N^H \cdot \mathbf{M}_{B_\Sigma}^{(m,k)} \cdot \mathbf{F}_T \cdot \bar{\mathbf{d}}^{(m,k)}, \tag{3}$$

where $\mathbf{F}_T$ denotes a $T \times T$ DFT matrix.

2.3 Joint DFT B-IFDMA

For Joint DFT B-IFDMA, the mapping matrix has to characterize a block interleaved subcarrier allocation. The Joint DFT B-IFDMA mapping matrix $\mathbf{M}_{B_J}^{(k)}$ is given by its elements $M_{B_J}^{(k)}(n,q)$ in the n-th row and q-th column

$$M_{B_J}^{(k)}(n,q) = \begin{cases} 1 & n = t \cdot \frac{N}{T} + r + k \cdot K_f \\ 0 & \text{else} \end{cases}, \tag{4}$$

with $n = 0, \cdots, N-1$, and $q = r + t \cdot K_f$, where $r = 0, ..., K_f - 1$ and $t = 0, ..., T-1$. Therefore, the Joint DFT B-IFDMA transmit signal $\mathbf{x}_{B_J}^{(k)}$ of user k becomes

$$\mathbf{x}_{B_J}^{(k)} = \mathbf{F}_N^H \cdot \mathbf{M}_{B_J}^{(k)} \cdot \mathbf{F}_Q \cdot \mathbf{d}^{(k)}. \tag{5}$$

3. Pilot Insertion and Channel Estimation

In this Section, pilot assisted CE is introduced for both variants of B-IFDMA. As a first solution, a whole B-IFDMA symbol is used for pilot transmission, which is termed symbolwise pilot insertion (PI) in the following. In order to reduce pilot symbol overhead, it is beneficial to interpolate between subcarriers within each block of K_f adjacent subcarriers. Therefore, as a second solution, a certain subset of the Q subcarriers assigned to user k is used for pilot transmission. This solution is termed subcarrierwise PI in the following.

3.1 Signal Generation

In this Section, the two possibilities of PI are described for Added Signal and Joint DFT B-IFDMA.

Symbolwise Pilot Insertion. For symbolwise PI, the unmodulated pilot sequence (PS) $\tilde{\mathbf{p}}^{(k)} = (\tilde{p}_0^{(k)}, \cdots, \tilde{p}_{Q-1}^{(k)})^T$ with Q complex elements, e.g. taken from a Constant Amplitude Zero Autocorrelation (CAZAC) sequence [1], is modulated according to the particular signal model for B-IFDMA.

- *Added Signal B-IFDMA*
  For Added Signal B-IFDMA, the unmodulated PS $\tilde{\mathbf{p}}^{(k)}$ is divided into K_f T-elementary subblocks $\bar{\mathbf{p}}^{(m,k)}$ with $m = 0, ..., K_f - 1$. The modulated PS $\mathbf{p}_{B_\Sigma}^{(k)}$ is built in analogy to the B-IFDMA signal in (3) and, thus, is given by

$$\mathbf{p}_{B_\Sigma}^{(k)} = \sum_{m=0}^{K_f-1} \mathbf{F}_N^H \cdot \mathbf{M}_{B_\Sigma}^{(m,k)} \cdot \mathbf{F}_T \cdot \bar{\mathbf{p}}^{(m,k)}. \tag{6}$$

- *Joint DFT B-IFDMA*
  For Joint DFT B-IFDMA, the modulated PS $\mathbf{p}_{B_J}^{(k)}$ of user k for symbolwise PI is given by

$$\mathbf{p}_{B_J}^{(k)} = \mathbf{F}_N^H \cdot \mathbf{M}_{B_J}^{(k)} \cdot \mathbf{F}_Q \cdot \tilde{\mathbf{p}}^{(k)}. \qquad (7)$$

The modulated PS $\mathbf{p}_{B_{\Sigma/J}}^{(k)} = (p_0^{(k)}, \cdots, p_{N-1}^{(k)})^T$ is transmitted instead of one block $\mathbf{x}_{B_{\Sigma/J}}^{(k)}$ of the B-IFDMA signal in time domain (TD), i.e. $\mathbf{x}_{B_{\Sigma/J}}^{(k)} = \mathbf{p}_{B_{\Sigma/J}}^{(k)}$. The FD representation of the modulated PS $\mathbf{p}_{B_{\Sigma/J}}^{(k)}$ of user k is given by its N-point DFT

$$\mathbf{P}_{B_{\Sigma/J}}^{(k)} = \mathbf{F}_N \cdot \mathbf{p}_{B_{\Sigma/J}}^{(k)} = (P_0^{(k)}, \cdots, P_{N-1}^{(k)})^T. \qquad (8)$$

The non-zero elements of $\mathbf{P}_{B_{\Sigma/J}}^{(k)}$ are combined in a vector $\mathbf{P}_{sy}^{(k)} = (P_{sy,0}^{(k)}, ..., P_{sy,Q-1}^{(k)})$ with elements $P_{sy,\hat{n}=(t\cdot K_f+r)}^{(k)} = P_{(t\cdot(K\cdot K_f)+r+k\cdot K_f)}^{(k)}$ for $r = 0, \cdots, K_f - 1$, $t = 0, ..., T-1$ and $\hat{n} = 0, ..., Q-1$.

Subcarrierwise Pilot Insertion. For subcarrierwise PI, N_p of the Q subcarriers allocated to a certain user are used for pilot transmission. Therefore, a PS $\tilde{\mathbf{p}}^{(k)} = (\tilde{p}_0^{(k)}, \cdots, \tilde{p}_{N_p-1}^{(k)})$ with N_p complex elements, e.g. taken from a CAZAC-sequence, is DFT precoded and mapped on the N_p subcarriers. N_p is dependent on the depth IP of interpolation in FD, e.g., for $IP = 2$ every 2nd subcarrier that is allocated to a specific user has to be used for pilot transmission. For simplicity, it is assumed that IP is always chosen such that the division $\frac{K_f}{IP}$ is integer. $N_p = T \cdot \frac{K_f}{IP}$ gives the number of subcarriers used for pilot transmission with T and K_f as defined in Section 2.

- *Added Signal B-IFDMA*
  For Added Signal B-IFDMA, the unmodulated PS $\tilde{\mathbf{p}}^{(k)}$ is divided into $\frac{K_f}{IP}$ subblocks $\bar{\mathbf{p}}^{(i,k)}$ with index $i = 0, ..., \frac{K_f}{IP} - 1$, each consisting of T elements. The transmitted PS $\mathbf{p}_{B_\Sigma}^{(k)}$ is given by

$$\mathbf{p}_{B_\Sigma}^{(k)} = \sum_{i=0}^{\frac{K_f}{IP}-1} \mathbf{F}_N^H \cdot \mathbf{M}_{p\Sigma}^{(i,k)} \cdot \mathbf{F}_T \cdot \bar{\mathbf{p}}^{(i,k)}. \qquad (9)$$

  The elements of the mapping matrix $\mathbf{M}_{p\Sigma}^{(i,k)}$ are given by

$$M_{p\Sigma}^{(i,k)}(n,t) = \begin{cases} 1 & \text{for } n = t \cdot \frac{N}{T} + k \cdot K_f + i \cdot IP \\ 0 & \text{else} \end{cases}, \qquad (10)$$

with $n = 0, ..., N-1$, $t = 0, ..., T-1$. Within this block of pilot transmission, $Q - N_p$ subcarriers remain unused and thus, a DS $\tilde{\mathbf{d}}^{(k)} = (\tilde{d}_0^{(k)}, ..., \tilde{d}_{Q-N_p-1}^{(k)})$ can be mapped on the $Q - N_p$ subcarriers. The DS $\tilde{\mathbf{d}}^{(k)}$ is divided into $K_f - \frac{K_f}{IP}$ subblocks $\bar{\mathbf{d}}^{(m,k)}$ with index $m = 0, ..., K_f - \frac{K_f}{IP} - 1$, each consisting of T elements. The transmitted DS $\mathbf{d}_{B_\Sigma}^{(k)}$ is given by

$$\mathbf{d}_{B_\Sigma}^{(k)} = \sum_{m=0}^{K_f - \frac{K_f}{IP} - 1} \mathbf{F}_N^H \cdot \mathbf{M}_{d\Sigma}^{(m,k)} \cdot \mathbf{F}_T \cdot \bar{\mathbf{d}}^{(m,k)}, \tag{11}$$

with $m = b + r \cdot (IP - 1)$; $b = 0, ..., IP - 2$; $r = 0, ..., \frac{K_f}{IP} - 1$ and the elements of the mapping matrix $\mathbf{M}_{d\Sigma}^{(m,k)}$ given by

$$M_{d\Sigma}^{(m,k)}(n,t) = \begin{cases} 1 & \text{for } n = b + r \cdot IP + t \cdot \frac{N}{T} + k \cdot K_f + 1 \\ 0 & \text{else} \end{cases} \tag{12}$$

for $n = 0, ..., N-1$ and $t = 0, ..., T-1$.

- *Joint DFT B-IFDMA*
  For Joint DFT B-IFDMA, the PS $\mathbf{p}_{B_J}^{(k)}$ of user k for subcarrierwise PI is given by

$$\mathbf{p}_{B_J}^{(k)} = \mathbf{F}_N^H \cdot \mathbf{M}_{pJ}^{(k)} \cdot \mathbf{F}_{N_p} \cdot \tilde{\mathbf{p}}^{(k)}, \tag{13}$$

with the elements $M_{pJ}^{(k)}(n, q_p)$ of the mapping matrix $\mathbf{M}_{pJ}^{(k)}$ given by

$$M_{pJ}^{(k)}(n, q_p) = \begin{cases} 1 & \text{for } n = t \cdot \frac{N}{T} + r \cdot IP + k \cdot K_f \\ 0 & \text{else} \end{cases}, \tag{14}$$

for $q_p = r + t \cdot \frac{K_f}{IP}$ and $t = 0, ..., T-1$; $r = 0, ..., \frac{K_f}{IP} - 1$; $n = 0, ..., N-1$. Within the block of pilot transmission $Q - N_p$ subcarriers remain unused. The DS $\tilde{\mathbf{d}}^{(k)}$ is mapped on the $Q - N_p$ subcarriers and the transmitted DS $\mathbf{d}_{B_J}^{(k)}$ is given by

$$\mathbf{d}_{B_J}^{(k)} = \mathbf{F}_N^H \cdot \mathbf{M}_{dJ}^{(k)} \cdot \mathbf{F}_{Q-N_p} \cdot \tilde{\mathbf{d}}^{(k)}. \tag{15}$$

The elements $M_{dJ}^{(k)}(n, q_d)$ of the mapping matrix $\mathbf{M}_{dJ}^{(k)}$ are given by

$$M_{dJ}^{(k)}(n, q_d) = \begin{cases} 1 & \text{for } n = t \cdot \frac{N}{T} + r \cdot IP + b + k \cdot K_f + 1 \\ 0 & \text{else} \end{cases}, \tag{16}$$

with $q_d = b + r \cdot (IP-1) + t \cdot \frac{K_f}{IP}$ for $t = 0, ..., T-1$; $r = 0, ..., \frac{K_f}{IP} - 1$ and $b = 0, ..., IP-2$.

The non-zero elements of $\mathbf{p}_{B_{\Sigma/J}}^{(k)}$ in FD are combined in a vector $\mathbf{P}_{sc}^{(k)} = (P_{sc,0}^{(k)}, \cdots, P_{sc,N_p-1}^{(k)})$ with elements $P_{sc,\hat{n}=(t \cdot K_f/IP+r)}^{(k)} = P_{(t \cdot \frac{N}{T}+r \cdot IP+k \cdot K_f)}^{(k)}$ for $r = 0, \cdots, K_f - 1$, $t = 0, ..., T-1$ and $\hat{n} = 0, ..., N_p - 1$. For both variants of B-IFDMA, the sequence $\mathbf{x}_{B_{\Sigma/J}}^{(k)}$ transmitted in TD for CE with subcarrierwise PI is given by the sum of PS and DS and thus, $\mathbf{x}_{B_{\Sigma/J}}^{(k)} = \mathbf{d}_{B_{\Sigma/J}}^{(k)} + \mathbf{p}_{B_{\Sigma/J}}^{(k)}$.

3.2 CE Algorithm

In this Section, a CE algorithm is introduced for symbol- and subcarrierwise PI. In the following, only one user will be considered and the index k will be omitted for simplicity. Let $\mathbf{h} = (h_0, \cdots, h_{N-1})^T$ denote the vector representation of a channel with N coefficients h_i, $i = 0, \cdots, N-1$, at sample rate $1/T_c$. The values H_n denote the complex coefficients of the Channel Transfer Function (CTF) $\mathbf{H} = \mathbf{F}_N \cdot \mathbf{h} = (H_0, \cdots, H_{N-1})^T$ and $\breve{\mathbf{N}} = \mathbf{F}_N \cdot \breve{\mathbf{n}} = (\breve{N}_0, \cdots, \breve{N}_{N-1})^T$ denotes the Additive White Gaussian Noise (AWGN) on each subcarrier in FD. The channel is assumed to be time invariant for the transmission of the PS. The values V_n in FD with $n = 0, \cdots, N-1$, received on each subcarrier after transmission over the channel $\mathbf{h}$ can be described by one complex channel coefficient H_n due to flat fading on each subcarrier in FD and are given by

$$V_n = H_n \cdot P_n + \breve{N}_n. \tag{17}$$

At the non-zero samples $P_{sy/sc,\hat{n}}$ of the PS in FD, the channel transfer coefficients $H_{\hat{n}}$ can be estimated by

$$\hat{H}_{\hat{n}} = \frac{V_{\hat{n}}}{P_{\hat{n}}} = H_{\hat{n}} + \frac{\breve{N}_{\hat{n}}}{P_{\hat{n}}}. \tag{18}$$

The estimated coefficients $\hat{H}_{\hat{n}}$ are used for the equalization of $N_t = \frac{T_c}{5}$ consecutive blocks in TD, with T_c the coherence time of the channel.

Symbolwise Pilot Insertion. Thus, for symbolwise PI, the vector of estimated channel transfer coefficients is given by

$$\hat{\mathbf{H}} = (\hat{H}_0, ..., \hat{H}_{Q-1}). \tag{19}$$

Subcarrierwise Pilot Insertion. For subcarrierwise PI, the vector $\hat{\mathbf{H}}$ of estimated channel transfer coefficients is given by

$$\hat{\mathbf{H}} = (\hat{H}_0, ..., \hat{H}_{N_p-1}). \tag{20}$$

As interpolation is possible in FD, a linear interpolation filter is applied to the estimated channel transfer coefficients. The coefficients $\bar{H}_q^{(t)}$ with $q = 0, ..., K_f - 1$ of the vector $\bar{\mathbf{H}}^{(t)}$ for each block $t = 0, ..., T-1$ consisting of K_f elements are given by

$$\bar{H}_q^{(t)} = \begin{cases} \hat{H}_{t \cdot \frac{K_f}{IP} + r} & ; q = r \cdot IP \\ \delta_{\tilde{r}} \cdot (q - \tilde{r} \cdot IP) + \hat{H}_{t \cdot \frac{K_f}{IP} + \tilde{r}} & ; q = \tilde{r} \cdot IP + b + 1 \\ \delta_{(\frac{K_f}{IP} - 2)} \cdot (q - K_f - 2IP) + \hat{H}_{\frac{K_f}{IP}(t+1)-2} & ; q = K_f - IP + b + 1 \end{cases}, \tag{21}$$

with $r = 0, ..., K_f/IP - 1$; $\tilde{r} = 0, ..., K_f/IP - 2$; $b = 0, ..., IP - 2$; $\delta_x = \frac{\hat{H}_{t \cdot K_f/IP + x} - \hat{H}_{t \cdot K_f/IP + x + 1}}{IP}$. The vector $\tilde{\mathbf{H}}$ of interpolated channel transfer coefficients is given by

$$\tilde{\mathbf{H}} = (\bar{\mathbf{H}}^{(0)}, ..., \bar{\mathbf{H}}^{(T-1)}). \tag{22}$$

4. Performance Analysis

In this Section, the performance results with PCE and RCE, as well as the PAPR and the pilot symbol overhead are given for Added Signal B-IFDMA and Joint DFT B-IFDMA. The results are valid for the parameters given in Table 1. In Figure 1, the uncoded bit error rate (BER) performance is presented for Added Signal B-IFDMA and Joint DFT B-IFDMA with $K_f = 4$ and $T = 32$ resulting in a net bit rate of 10 MBit/s per user. The results are given for PCE and for RCE with symbol- and subcarrierwise PI. The results show that Added Signal B-IFDMA and Joint DFT B-IFDMA exhibit the same performances. For CE with symbolwise PI, both schemes show a performance degradation of about 3 dB at $BER = 10^{-2}$ compared to PCE. For CE with subcarrierwise PI, the degradation increases up to 3.6 dB at $BER = 10^{-2}$ due to interpolation errors, but, compared to symbolwise PI, an 0.2 MBit/s higher net bit rate is provided due to lower pilot symbol overhead. In Table 2, the pilot symbol overhead Λ in dB is presented for subcarrierwise PI in dependency of the number K_t of consecutive time division multiple access (TDMA) slots for Added Signal and Joint DFT B-IFDMA. It is assumed, that every 2. subcarrier in FD and one TDMA slot within K_t

Table 1. Simulation Parameters

Carrier Frequency	3.7 GHz	Coding	no Coding
Bandwidth	40 MHz	Equalizer	MMSE FDE
No. of Subcarriers	1024	Guard Interval	3.6 μs
Modulation	QPSK	Channel	WINNER SCM [11], 50 km/h

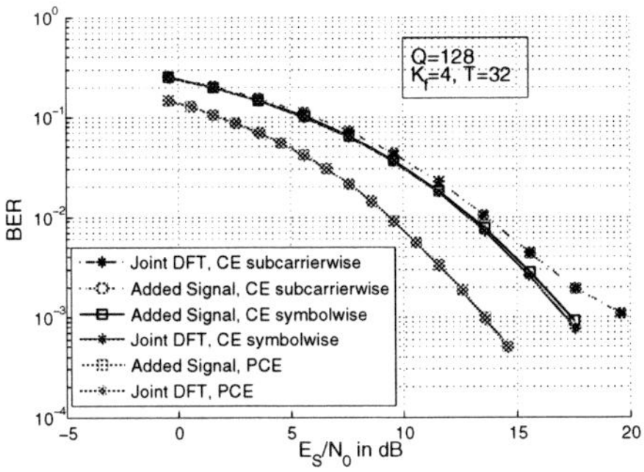

Figure 1. BER performance for added signal B-IFDMA and joint DFT B-IFDMA with PCE and RCE

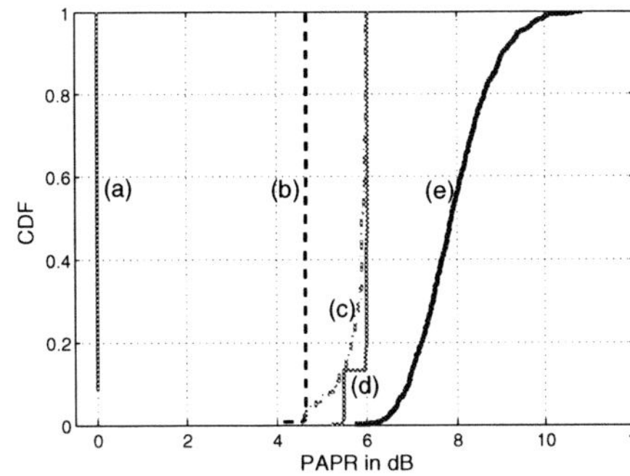

Figure 2. PAPR for (a)IFDMA, (b)Joint DFT B-IFDMA, (c)Joint DFT B-IFDMA with subcarrierwise PI, (d)Added Signal B-IFDMA with/without subcarrierwise PI, (e)OFDMA

Table 2. Pilot Symbol Overhead for $K_f = 4$

Number K_t of consecutive TDMA slots	1	3	12	24
Λ in dB for added signal B-IFDMA	3	0.8	0.18	0.09
Λ in dB for Joint DFT B-IFDMA	3	0.8	0.18	0.09

has to be used for pilot transmission. As the energy that has to be spent per data bit is increased relatively by pilot transmission, the overhead is equivalent to a Signal-to-Noise-Ratio (SNR)-degradation calculated according to the method introduced in [7]. The results in Table 2 show that Added Signal B-IFDMA and Joint DFT B-IFDMA require the same pilot symbol overhead for CE as the possibility of interpolation is identical for both. As it is always possible to interpolate within each block of K_f adjacent subcarriers, the pilot symbol overhead is independent from the data rate. The results show that the smaller the number K_t of consecutive TDMA slots, the higher the pilot symbol overhead, because the possibility of interpolation in TD decreases. Figure 2 presents the PAPR in dB for one block of Added Signal B-IFDMA with and without subcarrierwise PI, Joint DFT B-IFDMA with and without subcarrierwise PI and for IFDMA and OFDMA without PI as references. The results are given as cumulative density functions (CDF) and are valid for a net bit rate of 10 MBit/s and $K_f = 4$ for B-IFDMA. It can be seen that Joint DFT B-IFDMA exhibits a 1.4 dB lower PAPR than Added Signal B-IFDMA for 90% of all possible signals. If subcarrierwise PI is applied to the considered block, the PAPR degrades for Joint DFT B-IFDMA and both schemes show hardly a difference in PAPR for CDF values larger than 0.4.

5. Conclusion

In this work, Added Signal B-IFDMA and Joint DFT B-IFDMA have been introduced and investigated in terms of BER performance, pilot multiplexing for CE, PAPR and pilot symbol overhead. Both B-IFDMA variants provide the same BER performance for PCE as well as for the case with RCE. In terms of PAPR, Joint DFT B-IFDMA shows a clear better result than Added Signal B-IFDMA. This effect even intensifies for an increasing number K_f of subcarriers per B-IFDMA block. For subcarrierwise PI, there is hardly a difference for both schemes in terms of PAPR, but with increasing K_f, the PAPR of joint DFT B-IFDMA with subcarrierwise PI is expected to improve compared to Added Signal B-IFDMA with subcarrierwise PI. As both schemes exhibit the same subcarrier allocation, they require the same pilot symbol overhead for CE. If B-IFDMA is combined with TDMA, the pilot symbol overhead increases with decreasing number of consecutive TDMA slots.

References

[1] N. Benvenuto and G. Cherubini. *Algorithms for Communications Systems and their Applications.* John Wiley & Sons Ltd., 2002.

[2] T. Frank, A. Klein, E. Costa, and A. Kuehne. Low Complexity and Power Efficient Space-Time-Frequency Coding for OFDMA. In *Proc. of 15th Mobile & Wireless Communications Summit*, Mykonos, Greece, June 2006.

[3] T. Frank, A. Klein, E. Costa, and E. Schulz. Robustness of IFDMA as Air Interface Candidate for Future Mobile Radio Systems. In *Advances in Radio Science*, Miltenberg, Germany, Oct. 2004.

[4] T. Frank, A. Klein, E. Costa, and E. Schulz. IFDMA - A Promising Multiple Access Scheme for Future Mobile Radio Systems. In *Proc. PIMRC 2005*, Berlin, Germany, Sep. 2005.

[5] J. Lim, H. G. Myung, and D. J. Goodman. Proportional Fair Scheduling of Uplink Single-Carrier FDMA Systems. In *Proc. of PIMRC06*, Helsinki, Finland, September 2006.

[6] H. G. Myung, J. Lim, and D. J. Goodman. Peak-to-Average Power Ratio of Single Carrier FDMA Signals with Pulse Shaping. In *Proc. of PIMRC06*, Helsinki, Finland, September 2006.

[7] A. Sohl, T. Frank, and A. Klein. Channel Estimation for DFT precoded OFDMA with blockwise and interleaved subcarrier allocation. In *Proc. International OFDM Workshop 2006*, Hamburg, Germany, August 2006.

[8] U. Sorger, I. De Broeck, and M. Schnell. IFDMA - A New Spread-Spectrum Multiple-Access Scheme. In *Proc. ICC'98*, Atlanta, Georgia, USA, June 1998.

[9] E. UMTS. *TR-101 112, V3.2.0.* Sophia-Antipolis, France, April 1998.

[10] R. van Nee and R. Prasad. *OFDM for Wireless Multimedia Communications.* Artech House, 1st edition, 2000.

[11] WINNER. Final report on link level and system level channel models. Technical Report D5.4 v. 1.4, WINNER-2003-507581, November 2005.

[12] WINNERII. The winner 2 air interface: Refined multiple access concepts. Technical Report D4.6.1, WINNER II-4-027756, November 2006.

ROBUST TIME DOMAIN CHANNEL ESTIMATION FOR MIMO-OFDMA DOWNLINK SYSTEM

B. Le Saux, M. Hélard, and R. Legouable
France Telecom R&D Division, Broadband Radio Access laboratory
4 rue du Clos Courtel, 35512 Cesson-Sévigné, France
{benoit.lesaux;maryline.helard;rodolphe.legouable}@orange-ftgroup.com

Abstract This paper investigates a new time domain channel estimation for MIMO OFDMA downlink systems. Compared to classical frequency domain channel estimation, time approach allows improvements in a multi-antenna system. Nevertheless, the presence of null carriers in the spectrum leads to performance degradation because of "border effect" phenomenon. An improved time domain channel estimation for MIMO-OFDMA system is proposed for downlink transmission and compared to classical channel estimation methods. Besides, this channel estimation process can be applied in any SISO and/or MIMO multicarrier transmission.

1. Introduction

Combination of multiple-input multiple-output (MIMO) systems with orthogonal frequency division multiple access (OFDMA) has already been demonstrated to be a promising multi-user scheme for future wireless communications. Indeed, by assigning different subsets of subcarriers to users, OFDMA transmission can benefit from frequency diversity gain. In this paper, we focus our attention on channel estimation based on transmission of pilot symbols inside the frame [1]. Firstly, since downlink transmission is considered here, each user can use all pilot symbols transmitted in the frame for channel estimation in the whole bandwidth. Besides, channel estimation in multicarrier (MC) MIMO systems that is classically performed in frequency domain (FD) can also be performed in time domain (TD) [2] [3]. For FD estimation, different sets of pilot subcarriers are used for each transmit antenna while in TD estimation the same set of pilot subcarriers is shared by all transmit antennas. TD technique is known to give very good results by significantly reducing the noise on the estimated channel coefficients. Nevertheless, it was shown

S. Plass et al. (eds.), Multi-Carrier Spread Spectrum 2007, 357–366.

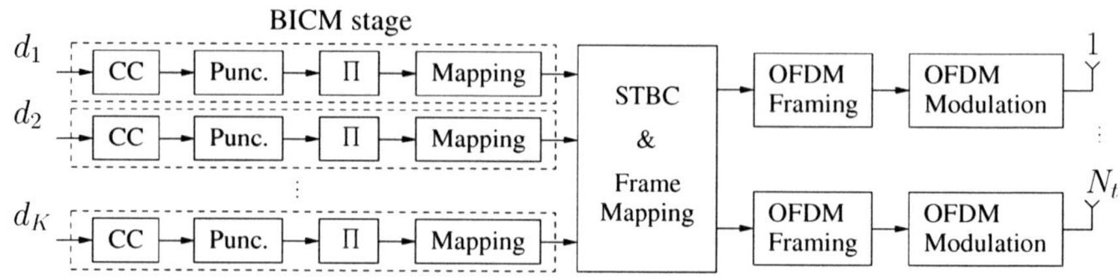

Figure 1. MIMO-OFDMA downlink transmission.

that TD estimation presents the "border effect" phenomenon that leads to a degradation of performance in the presence of null carriers at the spectrum extremities [4]. The aim of the paper is to investigate a new TD channel estimation scheme adapted to MIMO-OFDMA downlink transmission to correctly support the case of null carriers at border spectrum. Principle is to consider a truncated singular value decomposition of pilot matrix in order to avoid noise increase when channel estimation is performed in time domain.

Notation: Superscripts *, T, H and † stand for conjugation, transpose, Hermitian transpose and pseudo-inversion respectively. Operator e represents exponential function. diag$\{x\}$ stands for the diagonal matrix with the column vector x on its diagonal.

2. MIMO-OFDMA system model

2.1 Transmitter

The transmitter scheme is depicted in Figure 1 for K active users. The bit interleaved coded modulation (BICM) well adapted to transmission over flat fading Rayleigh channels is performed for each active users u, $1 \leq u \leq K$. Each user stream is then space-time encoded and mapped by considering OFDMA frame mapping. The frame mapping consists in attributing useful subcarriers to each user, either localized per block of adjacent subcarriers or distributed among useful subcarriers. The second attribution can benefit from more frequency diversity than the first one.

Then OFDM framing consists in computing N_t frames with N_t the number of transmit antennas and in inserting pilot symbols for channel estimation, null subcarriers at spectrum extremities and for DC subcarrier. Finally N_t substreams are individually OFDM modulated and simultanously transmitted from a different antenna. The OFDM symbol transmitted from the i-th transmit antenna at time index n is denoted $\mathbf{X}^i(n) \in \mathbb{C}^{N_{FFT} \times 1}$.

2.2 Channel model

The studied MIMO-OFDMA downlink system is made of N_t transmit antenna (Tx) at the base station and N_r receive antenna (Rx) for each user. Channel coefficients for each subcarrier between the i-th transmit antenna and the j-th receive antenna of user u can be expressed in the matrix form:

$$\mathrm{H}_{ji,m,u} = \sum_{l=0}^{L-1} h_{ji,l,u} e^{-j2\pi \frac{ml}{N_{FFT}}} \tag{1}$$

where $h_{ji,l,u}$ is a time varying channel tap between the base station and user u, L the number of taps and N_{FFT} the Fast Fourier Transform (FFT) size. By using OFDM modulation and demodulation, a frequency selective channel is transformed into a set of orthogonal flat Rayleigh fading sub-channels. OFDM demodulation is applied for each user u at each receive antenna j and the n-th received OFDM symbol at j-th antenna can be written as:

$$\mathbf{R}_{j,u}(n) = \sum_{i=1}^{N_t} \mathrm{diag}\left\{\sqrt{N_{FFT}}\mathbf{F}\left[\mathbf{h}_{ji,u}{}^{\mathrm{T}}, \mathbf{0}_{1\times(N_{FFT}-L)}\right]^{\mathrm{T}}\right\} \times \mathbf{X}_i(n) + \Xi_{j,u}(n) \tag{2}$$

with $\mathbf{F} \in \mathbb{C}^{N_{FFT}\times N_{FFT}}$ the unitary Discrete Fourier Transform (DFT) matrix, $\mathbf{h}_{ji,u} \in \mathbb{C}^{L\times 1}$ the vector of time varying channel tap from the i-th transmit antenna at the base station to the j-th receive antenna of user u, and $\Xi_{j,u}(n)$ the zero-mean complex Gaussian random process.

3. Channel estimation for MIMO systems

3.1 Construction of training sequences

In MIMO-OFDM system, sub-channels between each antenna link and for each useful subcarrier have to be individually estimated by the receiver to equalize received signal. Thus, construction of training sequences, *i.e.* value, position and number of pilot symbols in MIMO-OFDMA frame, has to be adapted depending on the number of Tx antenna at the base station. We distinguish two different ways to construct training sequences: frequency domain (FD) and time domain (TD). Indeed, the main difference between these two methods is where orthogonality between training sequences is obtained.

FD is based on the transmission of pilot symbols on one Tx antenna and null symbols on the other Tx antennas at the same time as depicted in Figure 2a. Orthogonality between FD training sequences is so

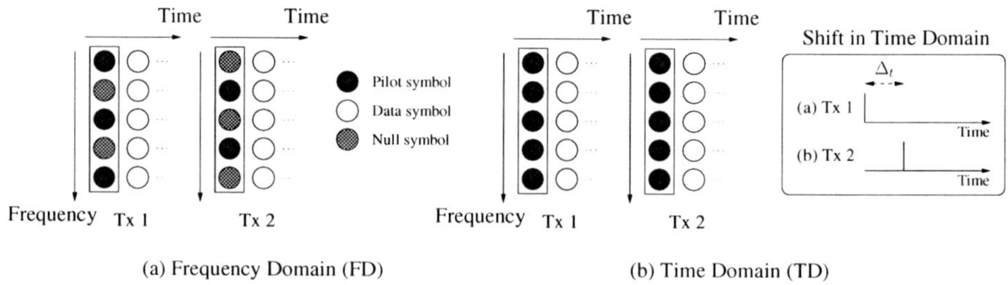

Figure 2. Frame structure for FD and TD channel estimation in MC MIMO $N_t = 2$. Full pilot OFDM symbol transmission. Impulse TD training sequences.

obtained owing to null symbols inserted in OFDM pilot symbol [5]. Contrary to FD training sequence, with TD technique, each Tx antenna uses the same subset of pilot subcarriers (see Figure 2b). Besides, orthogonality between TD training sequences is obtained in time domain owing to a time shift between training sequences. Indeed, the pilot symbol for i-th Tx antenna at p-th pilot subcarrier is obtained by:

$$c(p)^i = c(p)e^{-j2\pi\frac{p(i-1)\Delta_t}{N_p}} \tag{3}$$

with Δ_t the temporal shift such as $L \leq \Delta_t$ and $c(p)$ the training sequence transmitted in the frequency domain from Tx1 [3]. This sequence is called "seed sequence". For instance, Figure 2b describes the case where "seed sequence" is an impulse.

Finally, channel estimation between each antenna link for each useful subcarrier is then performed thanks to interpolation of pilot symbols. Optimal approach relies on LMMSE estimator which exploits correlations in time and in frequency (Wiener 2-D) [2].

3.2 Time domain channel estimation

We adapt channel estimation technique developped in [3] for MIMO-OFDM system to a MIMO-OFDMA transmission for which each user has only to estimate a subset of subcarriers. Since we consider a downlink transmission, each user can process channel estimation using all pilot symbols transmitted at the base station into the frame.

The received n-th OFDM symbol at the j-th receive antenna of user u can be expressed as follows:

$$\mathbf{R}_{j,u}(n) = \sum_{i=1}^{N_t} \text{diag}\left\{\mathbf{X}_i(n) + \mathbf{B}_i(n)\right\} \tilde{\mathbf{F}}\mathbf{h}_{j,i} + \Xi_{j,u}(n) \tag{4}$$

with $\tilde{\mathbf{F}}$ is equal to $\sqrt{N_{FFT}}$ times the first L columns of $\mathbf{F}$, $\mathbf{X}$ and $\mathbf{B} \in \mathbb{C}^{N_{FFT}\times 1}$ data and pilot vector respectively. By considering g consecutive OFDM symbols and only pilot subcarriers in a received $\mathbf{R}_{j,u}$ block OFDM symbol, eq. (4) can be written as:

$$\begin{aligned} \mathbf{R}_{j,u} &= \left[\mathbf{R}_{j,u}(0)^{\mathrm{T}}, \ldots, \mathbf{R}_{j,u}(g-1)^{\mathrm{T}}\right]^{\mathrm{T}} && (5) \\ &= \mathbf{T}\mathbf{h}_{j,u} + \mathbf{A}\mathbf{h}_{j,u} + \Xi_{j,u} && (6) \end{aligned}$$

with $\mathbf{h}_{j,u} = \left[\mathbf{h}_{j1,u}{}^{\mathrm{T}}, \ldots, \mathbf{h}_{jN_t,u}{}^{\mathrm{T}}\right]$. Matrices $\mathbf{T}$ and $\mathbf{A} \in \mathbb{C}^{(g\times N_{FFT})\times(N_t\times L)}$ are useful data matrix and pilot matrix respectively. To cancel interference terms because of data useful matrix in received block OFDM symbol, the solution commonly used consists in choosing disjoint sets of tones for training and data in each OFDM symbol.

By only considering pilot subcarriers, pilot matrix is equal to:

$$\tilde{\mathbf{A}} = \begin{bmatrix} \tilde{\mathbf{B}}^{1}_{\mathrm{diag}}(0)\tilde{\mathbf{F}}' & \ldots & \tilde{\mathbf{B}}^{N_t}_{\mathrm{diag}}(0)\tilde{\mathbf{F}}' \\ \vdots & \ldots & \vdots \\ \tilde{\mathbf{B}}^{1}_{\mathrm{diag}}(g-1)\tilde{\mathbf{F}}' & \ldots & \tilde{\mathbf{B}}^{N_t}_{\mathrm{diag}}(g-1)\tilde{\mathbf{F}}' \end{bmatrix} \tag{7}$$

where $\tilde{\mathbf{F}}' \in \mathbb{C}^{(g\times N_p)\times(N_t\times L)}$ corresponds to $\tilde{\mathbf{F}}$ with only P pilot subcarriers. Finally, the required condition to independently estimate N_t sub-channels per receive antenna is $N_t \times \Delta_t \leq N_p$ with $\Delta_t \leq L$.

The TD estimation of $\mathbf{h}_{j,u}$ required the knowledge of pilot matrix that can be computed before any transmission of data and is obtained as:

$$\hat{\mathbf{h}}_{j,u} = \tilde{\mathbf{A}}^{\dagger}\tilde{\mathbf{R}}_{j,u} = \left(\tilde{\mathbf{A}}^{\mathrm{H}}\tilde{\mathbf{A}}\right)^{-1}\tilde{\mathbf{A}}^{\mathrm{H}}\tilde{\mathbf{R}}_{j,u} \tag{8}$$

4. Improved time domain channel estimation

We consider in this part the case where the number of modulated subcarriers, *i.e.* useful subcarriers which carry data and pilot subcarriers, is different from the size of DFT process. This case is commonly implemented due to the insertion of null subcarriers at spectrum's extermities.

4.1 Influence of the subset of pilot subcarriers

In [4], Morelli evaluates the impact of only considering a subset of modulated subcarriers in the spectrum in a Single Input Single Output case (SISO). Liang extends Morelli's work in MIMO-OFDM system [6]:

$$\mathrm{MSE}^{ij}(p) = \sigma_n^2 \sum_{k_1=0}^{L}\sum_{k_2=0}^{L}\left[\tilde{\mathbf{A}}^{\mathrm{H}}\tilde{\mathbf{A}}\right]^{-1} e^{j2\pi p\frac{(k_1-k_2)}{N_{FFT}}} \tag{9}$$

When all subcarriers are available, minimizing MSE of channel estimation proposed by [3], *i.e.* $\tilde{\mathbf{A}}^{\mathrm{H}}\tilde{\mathbf{A}} = \mathcal{P}\mathbf{I}_{LN_t}$, involves $\mathrm{MSE}^{i,j}(p)$ to be constant for each pilot subcarriers. However, in case of frequency selective channel ($L > 1$) and when only a subset of subcarriers are modulated, $\tilde{\mathbf{A}}^{\mathrm{H}}\tilde{\mathbf{A}}$ is no more identity matrix and so the MSE varies with the subcarrier index. Thus, in the presence of null carriers at the spectrum's extremities, TD estimation in MC MIMO system presents the same "border effect" phenomenon than in SISO case.

Morelli proposed to design training sequence by using non-uniformly spaced pilots [4]. This principle proposed in SISO-OFDM system can be directly applied in a multi-antennas transmission but is not optimal in terms of spectral efficiency due to the increase of pilot symbols. Besides, simulation results show that problem is still present when all modulated subcarriers on pilot OFDM symbol are used as pilot subcarriers. Li proposed another construction of training sequence based on a non-uniform power distribution but leading to a too high peak average power ratio (PAPR) of the training symbol [7].

Finally, the idea developped in [6] consists in minimizing the energy in the off-diagonal of the matrix product $\tilde{\mathbf{A}}^{\mathrm{H}}\tilde{\mathbf{A}}$ in order to tend to training sequences where all subcarriers are available, *i.e.* where $\tilde{\mathbf{A}}^{\mathrm{H}}\tilde{\mathbf{A}} = \mathcal{P}\mathbf{I}_{LN_t}$. Another design called "jumping design" is so proposed and outperforms "truncate design" proposed by Larsson [8]. It consists in constructing training sequence by considering only pilot subcarriers, contrary to previous method which considers a training sequence at first and then a truncated part. Nevertheless, "border effect" is still present.

4.2 New TD channel estimation algorithm

We consider "jumping design" proposed by Liang [6] that outperforms the other proposed design with an improved TD estimation which allows to greatly reduce "border effect" phenomenon still present in [6].

By only considering pilot subcarriers, eq. (6) becomes:

$$\tilde{\mathbf{R}}_{j,u} = \tilde{\mathbf{A}}\mathbf{h}_{j,u} + \tilde{\Xi}_{j,u} \tag{10}$$

which exact solution is given by:

$$\begin{aligned} \hat{\mathbf{h}}_{j,u} &= \tilde{\mathbf{A}}^{\dagger}\left(\tilde{\mathbf{R}}_{j,u} - \Xi_{j,u}\right) &(11)\\ &= \sum_{s=1}^{r}\frac{\mathbf{u}_s^{\mathrm{H}}\tilde{\mathbf{R}}_{j,u}}{\sigma_s}\mathbf{v}_s - \sum_{s=1}^{r}\frac{\mathbf{u}_s^{\mathrm{H}}\tilde{\Xi}_{j,u}}{\sigma_s}\mathbf{v}_s &(12)\end{aligned}$$

where $\mathbf{u}_s$, $\mathbf{v}_s$ and σ_s are the left singular vectors, the right singular vectors and the singular values of $\tilde{\mathbf{A}}$ with $s = 1, \ldots r$, by assuming r rank of $\tilde{\mathbf{A}}$.

Condition number of $\tilde{\mathbf{A}}$ is defined by the ratio between the greatest and lowest singular value and determines the sensitivity of the inverse operation to inaccuracies in the data $\tilde{\mathbf{R}}^j$. In fact, high sensitivity to noise is caused by the presence of noise $\tilde{\Xi}$ in the singular subspaces $\mathbf{u}_s$ of $\tilde{\mathbf{A}}$ associated with small σ_s. Besides, if practical frame is used and for instance if $N_{FFT} = 1024$, $P = 704$, $\Delta_t = 216$, the condition number of $\tilde{\mathbf{A}}$ is approximately 1.92×10^{18}. In this case, $\tilde{\mathbf{A}}$ is considered as ill conditioned matrix. That is why, "border effect" phenomenon is observed.

Therefore, we apply a new TD estimation which principle aims at considering truncated singular value decomposition (TSVD) of $\tilde{\mathbf{A}}$. Indeed, in order to get a smaller condition number in pseudo-inverse calculation, we propose to consider only several singular values greater than a threshold Th. This TSVD allows to get rid of small singular values and so to avoid noise invreasing on $\hat{\mathbf{h}}^j$. TSVD solution is defined by:

$$\hat{\mathbf{h}}_{j,u} = \sum_{s=1}^{\mathrm{Th}} \frac{\mathbf{u}_s^{\mathrm{H}} \tilde{\mathbf{R}}_{j,u}}{\sigma_s} \mathbf{v}_s - \sum_{s=1}^{\mathrm{Th}} \frac{\mathbf{u}_s^{\mathrm{H}} \tilde{\Xi}_{j,u}}{\sigma_s} \mathbf{v}_s \tag{13}$$

The value of Th has to be chosen in order to reach a compromise between accuracy on pseudo-inverse calculation and small condition number.

5. Performance results

The different channel estimation techniques, FD estimation, classical TD estimation and proposed TD estimation are applied to a 4×2 MIMO-OFDMA system with a double-Alamouti scheme. FD estimation is applied with Wiener interpolation in frequency dimension by assuming precomputed and fixed knowledge on channel frequency correlation matrix. Classical TD estimation and proposed TD estimation correspond to channel estimation methods developped in 3.2 and in 4.2 respectively.

5.1 System Parameters

Performance are provided over frequency and time selective MIMO Spatial Channel Model Extended (SCME) typical to urban macro channel model without any spatial correlation between transmit antennas [9]. Double-Alamouti space-time coding consists in simultaneously transmitting two Alamouti codes on two blocks of two antennas [10]. Space-time matrices are the following:

$$\mathbf{X}_{\mathrm{DA}} = \begin{bmatrix} x_1 & x_2 & x_3 & x_4 \\ -x_2^* & x_1^* & -x_4^* & x_3^* \end{bmatrix}^{\mathrm{T}} \tag{14}$$

with $\mathbf{X}_{\mathrm{DA}} \in \mathbb{C}^{T \times N_t}$ double-Alamouti space-time block code.

Frame used is defined in the french RNRT OPUS project that considered physical description of a MIMO-OFDMA system [5] (see Table 1). The system parameters are issued and close to those defined in 3GPP LTE framework [11].

5.2 Simulation Results

Figure 3 shows Bit Error Rate (BER) performance and mean square error (MSE) on different subcarriers for different channel estimation algorithms. Firslty, "Border effect" phenomenon can be observed with classical TD estimation derived from (8) represented in figure by "Chest TD Classical". On the contrary, proposed TD estimation allows to reduce discontinuities with the threshold equal to Th = 0.1. Besides, proposed TD estimation allows a better MSE than FD estimation. Indeed, channel estimation of each antenna link benefits from more pilot symbols with TD estimation than with FD estimation. In fact, the more the number of transmit antennas, the more the number of required null symbols in the training sequence with FD estimation and so fewer pilot symbols are present to estimate each antenna link.

Moreover, Figure 3 shows the performance results in term of BER versus $\frac{E_b}{N_0}$. Performance are evaluated for perfect channel estimation represented in figure by "Chest Perfect", channel estimation with FD and proposed TD method with different value of Th. BER performance are in accordance with MSE measurements. Indeed, FD method leads to a degradation of 3 dB at BER = 10^{-4}. On the other hand, difference between perfect channel estimation and TD estimation is only of 0.5 dB with Th = 0.1 whereas classical TD method can not be used due

FFT size & Useful carrier	1024 & 600
Data & Pilot	85.7% & 14.3%
Temporal shift Δ_t	72
Number of users & User bandwidth	50 & 12 subcarriers
Modulation	16-QAM
Interleaver size	2880
Turbo code & Code rate	UMTS & 1/3

Table 1. System parameters

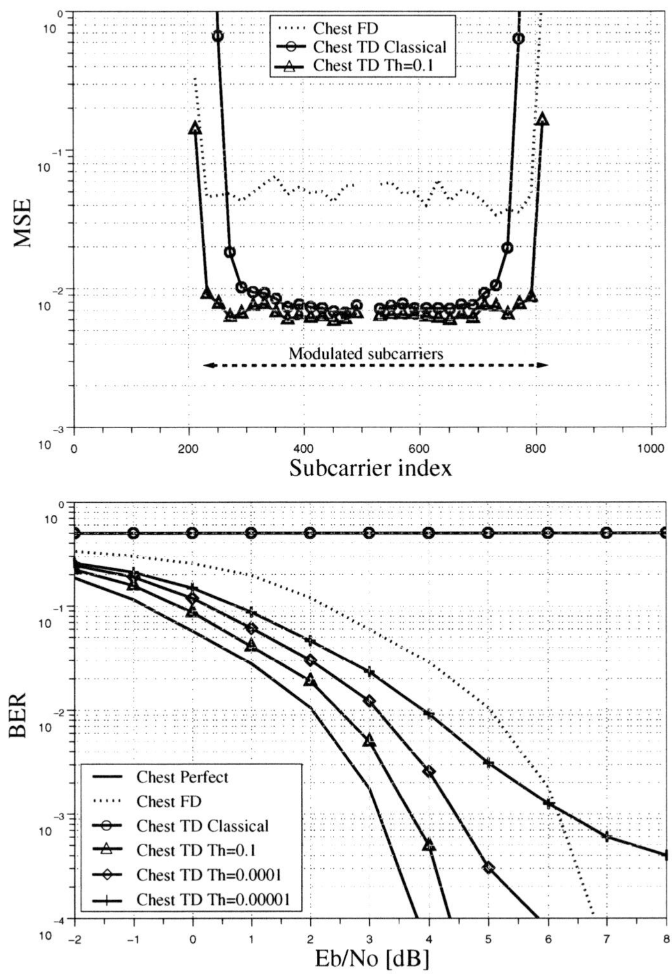

Figure 3. MSE measurement and BER performance with FD and TD channel estimation algorithms.

to "border effect" phenomenon. Performance with TD method mainly depends on the value of Th. In fact, when Th value is near float accuracy, an error floor appears but performance remains better than classical technique.

6. Conclusions

In this paper, channel estimation is investigated in time and in frequancy domain for MIMO-OFDMA transmission. Classical TD is demonstrated to be efficient only where every modulated subcarriers are dedicated to pilot symbols. Therefore it can not be used in real frame where only a subset of subcarriers is modulated due to insertion of null subcarriers at the spectrum's extremities for RF mask requirement. That is why frequency channel estimation algorithms is classically preferred but leading to significant degradation compared to perfect channel estimation. The proposed robust TD channel estimation solves this problem by applying a truncated SVD calculation on pilot matrix. Theoritical analysis and simulations results demonstrate the efficiency of the method that outperforms classical TD and FD channel estimation techniques. This channel estimation process can be applied in any SISO and/or MIMO multicarrier transmission and adapted to uplink.

References

[1] X.G. Doukopoulos and R. Legouable. Robust channel estimation via fft interpolation for multicarrier systems. submitted to IEEE Vehicular Technology Conference, May 2007.

[2] I. Tolochko and M. Faulkner. Real time lmmse channel estimation for wireless ofdm systems with transmitter diversity. volume 3, pages 1555–1559. IEEE Vehicular Technology Conference, Sept. 2002.

[3] I. Barhumi, G. Leus, and M. Moonen. Optimal training design for mimo ofdm systems in mobile wireless channels. *IEEE Transactions on Signal Processing*, 51, no. 6, Juin 2003.

[4] M. Morelli and U. Mengali. A comparison of pilot-aided channel estimation methods for ofdm systems. *IEEE Transactions on Signal Processing*, 49(12):3065–3073, Jan. 2001.

[5] W. Hachem, D. Kténas, J. Barletta, R. Legouable, N. Chapalain, L. Brunel, and A. Garot. Définition des fonctionnalités de base des couches phy/mac du démonstrateur. Technical report, RNRT OPUS, 2006.

[6] T.-J. Liang and G. Fettweis. Mimo preamble design with a subset of subcarriers in ofdm-based wlan. volume 2, pages 1032–1036. IEEE Vehicular Technology Conference, Jun. 2005.

[7] Z. Li, Y. Gai, and Y. Xu. Optimal training signals design for mimo ofdm systems with guard subcarriers. IEEE Vehicular Technology, 2006.

[8] E.G. Larsson and J. Li. Preamble design for multiple-antenna ofdm-based wlans with null subcarriers. *IEEE Signal Processing Letters*, 8:285–288, Nov. 2001.

[9] D. S. Baum, J. Hansen, G. Del Galdo, M. Milojevic, J. Salo, and P. Kyösti. An interim channel model for beyond-3g systems: extending the 3gpp spatial channel model (scm). volume 5, pages 3132–3136. IEEE Vehicular Technology Conference, May 2005.

[10] S. Baro, G. Bauch, A. Pavlic, and A. Semmler. Improving blast performance using space-time block codes and turbo-decoding. pages 1067–1071. IEEE GLOBECOM, Nov. 1998.

[11] 3GPP TSG-RAN. 3gpp tr 25.814, physical layer aspects for evolued utra (release 7). Technical report, 2006.

JOINT ITERATIVE CHANNEL ESTIMATION AND SOFT-CHIP COMBINING FOR A MIMO MC-CDMA ANTIJAM SYSTEM *

Galib Asadullah M.M. and Gordon L. Stuber
Wireless Systems Laboratory, Georgia Institute of Technology
Atlanta, GA 30332, USA
{galibmm, stuber} @ece.gatech.edu

Abstract In this paper, multiple input multiple output (MIMO) channel coefficients, jamming power, and jammer state state information (JSI) are jointly estimated to suppress the adverse effects of partial band noise jamming (PBNJ) for mlticarrier code division multiple access (MC-CDMA) systems using complex quadratic spreading sequences and cyclic delay diversity (CDD). We derive weighted least square error (LSE) and linear minimum mean square error (MMSE) channel estimators (CE) under PBNJ and investigate the constraints on cyclic delays. For the pilot symbols, hard or soft JSI is estimated to improve the channel estimates by updating the covariance matrix. On the other hand, iteratively estimated JSI is exploited to suppress jamming in the despread data symbols. Simulation results demonstrate that the proposed iterative despreading (includes JSI estimation and soft chip combining), demodulation, and decoding (IDDD) receiver with soft JSI (S-JSI) based MMSE-CE and chip-combining is very robust against jamming.

Keywords: Multicarrier code division multiple access, channel estimation, antijamming, iterative receivers.

1. Introduction

Multicarrier code division multiple access (MC-CDMA) [1] that combines code division multiple access (CDMA) and orthogonal frequency division multiplexing (OFDM) is robust against multipath fading, multiuser interference (MUI), and interference due to intentional jamming or other narrowband links co-existing in the same bandwidth [2, 3]. In [4], an uncoded constant envelope

*Prepared through collaborative participation in the Collaborative Technology Alliance for Communications & Networks sponsored by the U.S. Army Research Laboratory under Cooperative Agreement DAAD19-01-2-0011. The U.S. Government is authorized to reproduce and distribute reprints for Government purposes notwithstanding any copyright notation thereon.

S. Plass et al. (eds.), Multi-Carrier Spread Spectrum 2007, 367–376.

[5] MC-CDMA multiple input multiple output (MIMO) system with cyclic delay diversity (MC-CDMA-CDD) was introduced that achieves full spatial diversity and satisfactory coding gain.

Channel estimation is necessary for coherent demodulation, combining different receive antenna signals, and estimation of jammer state information (JSI) which indicates the presence of a jamming signal in a frequency sub-band. Conversely, knowledge of the jamming signal power and JSI can improve the quality of channel estimation. In [6], the time varying channel for a MC-CDMA system is estimated using Slepian basis expansion. A maximum likelihood (ML) channel acquisition and least mean square (LMS) channel tracking method for single transmit antenna MC-CDMA system is presented in [7]. However, neither of these papers consider transmit diversity or the presence of any jamming signal.

Chip combining with appropriate weights at the despreader can mitigate jamming in a MC-CDMA system where each symbol is spread in the frequency domain (FD). For antijam MC-CDMA systems that employ chip combining, JSI is essential for determining the chip combining weights [8]. In [5], a hard JSI estimate ($\in \{0, 1\}$) based chip combining method is used to suppress jamming signal.

This paper introduces an iterative despreading, demodulation, and decoding (IDDD) receiver with joint channel and JSI estimation for the convolutionally coded single user MC-CDMA-CDD system [4] operating under partial band noise jamming (PBNJ) and time varying multipath fading channel. Using the sufficient statistic criterion of information theory, we show that the optimum chip combining weights are inversely proportional to noise variance of the corresponding subcarriers. We iteratively estimate JSI of the subcarriers and exploit it to perform pilot assisted channel estimation or appropriate chip combining for the data symbols. Simulation results demonstrate that the proposed iterative soft JSI estimate assisted channel estimation and chip combining improves the bit error rate (BER).

The rest of the paper is organized as follows. Section 2 describes the MC-CDMA-CDD transmitter. Section 3 discusses pilot assisted channel estimation (PACE) under PBNJ. The antijamming IDDD receiver including jamming power and JSI estimation, and soft chip combining, is introduced in Section 3. Section 4 illustrates the simulation results and Section 5 concludes the paper.

2. MC-CDMA-CDD Transmitter

Fig. 1 shows the MC-CDMA-CDD transmitter with $P(\leq N_c)$ transmit antennas, where N_c is the number of subcarriers, which is also equal to the spreading factor (SF). This paper considers M-ary phase shift keying (MPSK). Pilot symbols are inserted at the interval of N_t symbols. The sets of pilot and data

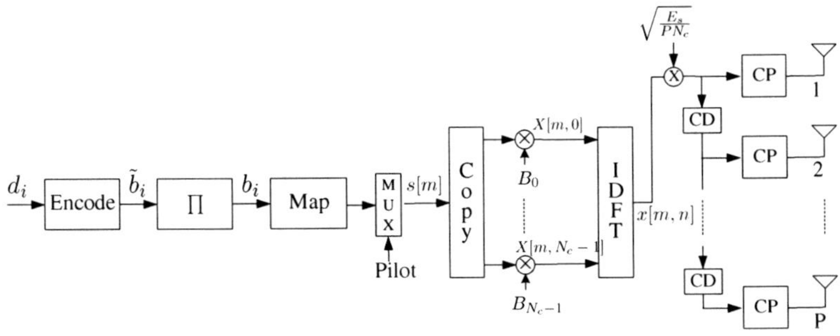

Figure 1. MC-CDMA-CDD transmitter.

symbol positions are denoted by $\mathcal{P}$ and $\mathcal{D}$, respectively. The mth mapped symbol $s[m]$ is spread by the frequency domain spreading sequence $\{B[k]\}$, $k = 0, 1, \cdots, N_c - 1$ to generate the frequency domain signal (normalized by $\sqrt{E_s}$) $X[m,k] = s[m]B[k]$, $m \in \mathcal{P} \cup \mathcal{D}$, where E_s is the average received symbol energy and $\cup$ denotes union of sets. The time and frequency domain constant-envelope spreading sequences [5] considered here are $b[n] = e^{-j\frac{\pi}{8}} e^{j\frac{\pi n^2}{N_c}}$ and $B[k] = e^{j\frac{\pi}{8}} e^{-j\frac{\pi k^2}{N_c}}$, respectively. If the corresponding time and frequency domain chips without any cyclic shifts are arranged in vector form as $\mathbf{b}^{(0)} = [b[0], b[1], \cdots, b[N_c-1]]^T$ and $\mathbf{B}^{(0)} = [B[0], B[1], \cdots, B[N_c-1]]^T$, respectively, then $\mathbf{B}^{(0)} = \mathbb{F}\,\mathbf{b}^{(0)}$, where the elements of the DFT matrix are $[\mathbb{F}]_{k,n} = e^{-j\frac{2\pi kn}{N_c}}$; $k, n = 0, 1, \cdots, N_c - 1$.

The frequency domain signal is transformed into time domain signal $x[m,n] = \frac{1}{\sqrt{N_c}} \sum_{k=0}^{N_c-1} X[m,k]\, e^{j2\pi kn/N_c} = s[m]b[n]$ by inverse DFT (IDFT). Before transmitting from the p-th antenna, the sequence $\{x[m,n]\}_{n=0}^{N_c-1}$ is cyclic delayed (CD) by d_pT_c so that the transmitted signal on the kth subcarrier is $T^{(p)}[m,k] = \sqrt{\frac{E_s}{PN_c}}\, X[m,k]\, e^{-j\frac{2\pi k d_p}{N_c}}$, where d_p is the number of unit cyclic shifts applied to the pth transmit antenna and T_c is the chip duration equal to the sampling period.

3. Antijam Receiver

The antijam receiver with channel, jamming power and JSI estimators, and chip combiner is shown in Fig. 2. We consider time varying mutipath fading channels. Considering any receive antenna $q \in \{1, \cdots, Q\}$, where Q is the total number of recieve antennas, the discrete channel frequency response corresponding to the kth subcarrier of the mth symbol and pth transmit antenna is $H^{(p,q)}[m,k] = \sum_{l=0}^{L-1} h^{(p,q)}\,[m,l]\, e^{-j2\pi kl/N_c}$, where L is the number of channel taps, and $h^{(p,q)}[m,l]$ is a zero mean complex Gaussian distributed channel

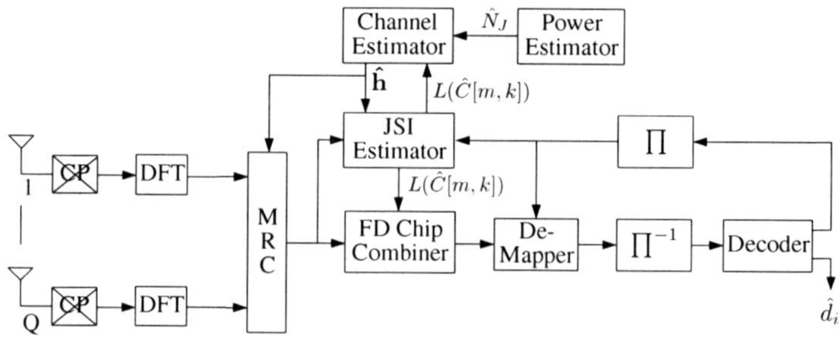

Figure 2. Iterative antijam receiver.

coefficient associated with the lth path between the pth transmit and qth receive antenna. The frequency response vector corresponding to the pth transmit and qth receive antenna pair is $\mathbf{H}^{(p,q)}[m] = \mathbb{F}\,\mathbf{h}^{(p,q)}[m]$, where the elements of the DFT matrix are $[\mathbb{F}]_{k,l} = e^{-j\frac{2\pi kl}{N_c}}$; $k = 0, 1, \cdots, N_c - 1,\ l = 0, 1, \cdots, L-1$; $\mathbf{h}^{(p,q)}[m] = [h^{(p,q)}[m,0], h^{(p,q)}[m,1], \cdots, h^{(p,q)}[m, L-1]]^T$ and $\mathbf{H}^{(p,q)}[m] = [H^{(p,q)}[m,0], H^{(p,q)}[m,1], \cdots, H^{(p,q)}[m, N_c-1]]^T$.

Pilot Assisted Channel Estimation (PACE)

After removing the cyclic prefix and applying DFT, the frequency domain signal corresponding to any pilot symbol at position $m \in \mathcal{P}$ and qth receive antenna is $Y^{(q)}[m,k] = \sum_{p=0}^{P-1}\sum_{l=0}^{L-1} h^{(p,q)}[m,l]\; T^{(p)}[m,k]\; e^{-j2\pi kl/N_c} + C[m,k]J^{(q)}[m,k] + W^{(q)}[m,k]$, where $W^{(q)}[m,k]$ are independent identically distributed (i.i.d.) additive white Gaussian noise (AWGN) samples having zero mean and variance N_0, $C[m,k] \in \{0,1\}$ is the JSI, $J^{(q)}[m,k]$ is the i.i.d. Gaussian jamming signal at the qth receive antenna having zero mean and variance $\sigma_J^2 = N_J/\eta$, N_J is the average (over whole bandwidth) power spectral density (psd), and $\eta = \frac{1}{N_c}\sum_{k=0}^{N_c-1} C[m,k]$ is the jamming fraction. The JSI associated with the mth symbol and kth subcarrier, $C[m,k] = 1$ when the kth subcarrier of the mth symbol is jammed, and $C[m,k] = 0$ when the kth subcarrier is not jammed. The received signals on different subcarriers at any receive antenna q can be arranged in a column vector as $\mathbf{Y}^{(q)}[m] = \left[Y^{(q)}[m,0], Y^{(q)}[m,1], \cdots,\ Y^{(q)}[m, N_c-1]\right]^T = \mathbf{A}[m]\,\mathbf{h}^{(\mathbf{q})}[m] + \mathbf{V}^{(\mathbf{q})}[m]$, where $\mathbf{h}^{(q)}[m] = \left[\mathbf{h}^{(1,q)^T}[m], \cdots, \mathbf{h}^{(P,q)^T}[m]\right]^T$ is the channel coefficient vector for the mth symbol, $\mathbf{h}^{(p,q)}[m] = [h^{(p,q)}[m,0], h^{(p,q)}[m,1], \cdots, h^{(p,q)}[m, L-1]]^T$ is the channel coefficient vector associated with the pth transmit and qth receive antenna pair, $\mathbf{V}^{(q)}[m] = [C[m,0]J^{(q)}[m,0] + W^{(q)}[m,0], \cdots, C[m, N_c-1]J^{(q)}[m, N_c-1] + W^{(q)}[m, N_c-1]]^T$ is the AWGN plus

PBNJ signal vector, and

$$\mathbf{A}[m] = \begin{bmatrix} T^{(1)}[m,0] & \cdots & T^{(P)}[m,0] & \cdots & T^{(P)}[m,0]\, e^{-j2\pi k(L-1)/N_c} \\ \vdots & & \vdots & & \vdots \\ T^{(1)}[m,N_c-1] & \cdots & T^{(P)}[m,N_c-1] & \cdots & T^{(P)}[m,N_c-1]\, e^{-j2\pi k(L-1)/N_c} \end{bmatrix},$$

is a $N_c \times PL$ matrix that depends on the pilot (or data) symbol, cyclic shifts, and channel delays. Because the AWGN and PBNJ signals are independent random variables, the AWGN plus PBNJ signal vector $\mathbf{V}^{(\mathbf{q})}[m]$ has zero mean and covariance $\mathbf{C}_V[m] = \sigma_J^2 \,\mathrm{diag}(C[m,0], C[m,1], \cdots, C[m,N_c-1]) + N_0\, \mathbf{I}_{N_c}$, where diag($\cdot$) denotes a diagonal matrix and $\mathbf{I}_N$ is $N \times N$ identity matrix.

The vector obtained by applying l $(0 < l < N_c - 1)$ cyclic shifts to the time domain (TD) spreading chip vector, $\mathbf{b}^{(0)} = \{b[n]\}_{n=0}^{N_c-1}$ can be expressed as $\mathbf{b}^{(l)} = [b[N_c - l], \cdots, b[N_c - 1], b[0], \cdots, b[N_c - l - 1]]^T$. The corresponding frequency domain vector is $\mathbf{B}^{(l)} = [B[0], B[1]e^{-j2\pi l/N_c}, \cdots, B[N_c - 1]e^{-j2\pi(N_c-1)l/N_c}]^T$. Now, by using $T^{(p)}[m,k] = \sqrt{\frac{E_s}{PN_c}}\, s[m]B[k]\, e^{-j\frac{2\pi k d_p}{N_c}}$, $\mathbf{A}[m]$ can be expressed as $\mathbf{A}[m] = \sqrt{\frac{E_s}{PN_c}}\, s[m]\, \mathbf{B}$. The $N_c \times PL$ matrix $\mathbf{B} = \left[\mathbf{B}^{(0)_{N_c}}, \cdots, \mathbf{B}^{(L-1)_{N_c}}, \cdots\cdots, \mathbf{B}^{(d_P)_{N_c}}, \cdots, \mathbf{B}^{(d_P+L-1)_{N_c}}\right]$ does not depend on the transmitted symbol, where $(\cdot)_N$ denotes modulo-N operation. Thus, the received frequency domain signal vector is expressed as $\mathbf{Y}^{(q)}[m] = \sqrt{\frac{E_s}{PN_c}}\, s[m]\, \mathbf{B}\, \mathbf{h}^{(\mathbf{q})}[m] + \mathbf{V}^{(\mathbf{q})}[m]$. Because both the received signal vector $\mathbf{Y}^{(q)}[m]$ and the unknown channel vector $\mathbf{h}^{(\mathbf{q})}[m]$ are Gaussian distributed, the posterior density function $f(\mathbf{h}^{(\mathbf{q})}|\mathbf{Y}^{(q)})$ is also Gaussian distributed. Therefore, for each receive antenna $q \in \{1, \cdots, Q\}$ and symbol $m \in \mathcal{P}$, the LSE and MMSE channel estimators are derived [9] as

$$\hat{\mathbf{h}}_{LSE}^{(q)}[m] = \sqrt{\frac{PN_c}{E_s}} \left(\mathbf{B}^H \mathbf{C}_V^{-1}[m]\mathbf{B}\right)^{-1} \mathbf{B}^H \left(\mathbf{C}_V^{-1}[m]\, \tilde{\mathbf{Y}}^{(q)}[m]\right) \quad \text{and} \tag{1}$$

$$\hat{\mathbf{h}}_{MMSE}^{(q)}[m] = \sqrt{\frac{E_s}{PN_c}} \left(\mathbf{C}_h^{-1} + \frac{E_s}{PN_c}\, \mathbf{B}^H \mathbf{C}_V^{-1}[m]\mathbf{B}\right)^{-1} \mathbf{B}^H \left(\mathbf{C}_V^{-1}[m]\, \tilde{\mathbf{Y}}^{(q)}[m]\right),$$

respectively, where $\mathbf{C}_h = E\{\mathbf{h}^{(q)}[m]\mathbf{h}^{(q)H}[m]\}$ is the covariance of the channel coefficient vector and $\tilde{\mathbf{Y}}^{(q)}[m] = s^*[m]\, \mathbf{Y}^{(q)}[m]$. The noise covariance matrix depends on the jamming signal variance and JSI. Therefore, knowledge about the jamming signal power and JSI will enhance the quality of the channel estimates.

Lemma 1: The LSE channel estimator exists, if the following conditions on cyclic delays are satisfied

$$\begin{aligned} &0 \le d_p \le N_c - L; && \forall p \in \{1, 2, \cdots, P\} \\ &|d_p - d_{p-1}| \ge L; && \forall p \in \{2, 3, \cdots, P\} \end{aligned} \tag{2}$$

Proof: In order for the LSE-CE to exist, $\mathbf{B}$ must be full rank, i.e., $rank[\mathbf{B}] = PL$ when $PL \le N_c$. If $N_c - L < d_p < N_c$ or $|d_p - d_{p-1}| < L$, the columns of $\mathbf{B}$ are no longer unique (independent) and the LSE-CE does not exist.

Optimum Chip Combining under PBNJ

The received frequency domain signal after removing the cyclic prefix and applying DFT at each receive antenna can be expressed as $Y^{(q)}[m,k] = \sqrt{\frac{E_s}{P}}$ $H^{(q)}[m,k]\, X[m,k] + C[m,k] J^{(q)}[m,k] + W^{(q)}[m,k]$, where $C[m,k] \in \{0,1\}$ is the JSI and $H^{(q)}[m,k] = \sum_{p=1}^{P} H^{(p,q)}[m,k]\, e^{-j\frac{2\pi k d_p}{N_c}}$. After maximal ratio combining of different receive-antenna signals, the output corresponding to the kth subcarrier of the mth symbol is $Y[m,k] = \sqrt{\frac{E_s}{P}}\, s[m] \sum_{q=1}^{Q} \left|H^{(q)}[m,k]\right|^2 B[k] + J[m,k] + W[m,k]$, where $s[m]$ $(m \in \mathcal{P} \cup \mathcal{D})$ is the transmitted pilot or data symbol, and the contributions from PBNJ and AWGN, respectively, are $J[m,k] = C[m,k] \sum_{q=1}^{Q} H^{(q)}[m,k]^* J^{(q)}[m,k]$ and $W[m,k] = \sum_{q=1}^{Q} H^{(q)}[m,k]^* W^{(q)}[m,k]$. Given the channel coefficients, each of the $J[m,k]$ and $W[m,k]$ have a zero mean Gaussian distribution with variances $\sigma^2_{J_k} = \frac{N_J}{\eta} C[m,k] \sum_{q=1}^{Q} \left|H^{(q)}[m,k]\right|^2$ and $\sigma^2_{W_k} = N_0 \sum_{q=1}^{Q} \left|H^{(q)}[m,k]\right|^2$, respectively.

Theorem: Under Gaussian distributed PBNJ and AWGN, the chip combining weights that provide a sufficient statistic to the demapper are inversely proportional to the AWGN plus jamming signal variance.

Proof: Consider the conditional likelihood function $f\left(\mathbf{Y}[m] \mid \{H^{(q)}[m,k]\}; s\right)$, where the received signal vector associated with the mth data symbol is $\mathbf{Y}[m] = [Y[m,0], \cdots, Y[m,N_c-1]]^T$ and data symbol s is a random variable. The conditional log-likelihood function can be expressed as

$$\log f\left(\mathbf{Y}[m] \mid \{H^{(q)}[m,k]\}; s\right) \propto Re\left[\left(\sum_{k=0}^{N_c-1} \frac{1}{N_0 + C[m,k]\frac{N_J}{\eta}} B^*[k]\, Y[m,k]\right) s^*\right],$$

where $Re[\cdot]$ denotes the real part of a complex number. Given any sufficient statistic, the received signal vector, $\mathbf{Y}[m]$ must be independent of $s[m]$. Using Fisher's factorization theorem [9], the sufficient statistic for any data symbol is $\mathcal{T}(\mathbf{Y}[m]) = \sum_{k=0}^{N_c-1} \frac{1}{N_0 + C[m,k]\frac{N_J}{\eta}} B^*[k]\, Y[m,k]$ which must be provided to the demapper in order to evaluate appropriate likelihood functions for calculating the LLR of the coded and interleaved bits. Thus, the optimum chip combining weights must be inversely proportional to the AWGN plus jamming signal variance of the relevant subcarriers.

After optimal chip combining at the despreader, the input signal to the demapper for the mth data symbol is $\hat{s}[m] = \sum_{k=0}^{N_c-1} \rho[m,k]\, B^*[k] \cdot Y[m,k] = \sqrt{\frac{E_s}{P}} \sum_{k=0}^{N_c-1} \rho[m,k] \sum_{q=1}^{Q} \left|H^{(q)}[m,k]\right|^2 s[m] + \tilde{J} + \tilde{W}$, where

$$\rho[m,k] = \begin{cases} \gamma_0 = \frac{1}{N_0}, & C[m,k] = 0 \\ \gamma_1 = \frac{1}{N_J/\eta + N_0}, & C[m,k] = 1 \end{cases}. \tag{3}$$

Therefore, JSI, average jamming psd, and jamming fraction must be estimated and used in chip combining in order to pass sufficient statistic to the demapper. The contributions from PBNJ and AWGN at the despreader output are $\tilde{J} = \sum_{k=0}^{N_c-1} \rho[m,k]B^*[k]J[m,k]$, and $\tilde{W} = \sum_{k=0}^{N_c-1} \rho[m,k]B^*[k]W[m,k]$, respectively.

Jamming Signal Power Estimation

We apply the signal projection [10] method which facilitates blind estimation of $\hat{N}_J$ for the constant envelope MC-CDMA-CDD system. If the conditions of lemma 1 are satisfied, matrix $\mathbf{B}$ has full rank. For simplicity, let the cyclic delays be $d_p = (p-1)L$ for $p = 1, \cdots, P$. A computationally efficient expression for the PBNJ signal psd is derived by projecting the received signal vector onto the entire signal space as $N_J = \frac{1}{N_c - PL}\left(E\left[\| \mathbf{Y}^{(q)}[m] \|^2\right] - \frac{1}{N_c}\sum_{d=0}^{PL-1} E\left[\left|\mathbf{B}^{(d)^H}\mathbf{Y}^{(q)}[m]\right|^2\right]\right) - N_0$. Thus, the estimator is

$$\hat{N}_J = \frac{1}{(N_c - PL)Q|\mathcal{S}_I|}\sum_{m\in\mathcal{S}_I}\sum_{q=1}^{Q}\left(\| \mathbf{Y}^{(q)}[m] \|^2 - \frac{1}{N_c}\sum_{d=0}^{PL-1}\left|\mathbf{B}^{(d)^H}\mathbf{Y}^{(q)}[m]\right|^2\right) - N_0, \tag{4}$$

where $\mathcal{S}_I$ is the set of symbol indexes over which averaging is performed.

JSI Estimation for Channel Estimation

The JSI estimates associated with the pilot symbols are necessary for channel estimation. Let $p(C[m,k] = c)$, $c \in \{0,1\}$ be defined as the *a posteriori* probability (APP) of the JSI corresponding to the mth symbol and kth subcarrier, where $m \in \mathcal{P}$ and $k \in \{0, 1, \cdots, N_c - 1\}$. Since the jamming signal is independent of the transmitted pilot symbol and channel coefficients,

$$\begin{aligned} p(C[m,k]=c) &\triangleq P\left(C[m,k]=c \mid Y[m,k], s[m], \{H^{(q)}[m,k]\}_{q=1}^{Q}\right) \\ &\propto P\left(Y[m,k] \mid s[m], C[m,k]=c, \{H^{(q)}[m,k]\}\right) p(C[m,k]=c), \end{aligned} \tag{5}$$

where $s[m]$, $m \in \mathcal{P}$ is the known pilot symbol and $p(C[m,k] = c)$ is the *a priori* probability that $C[m,k] = c$, $c \in \{0,1\}$. The LLR of $C[m,k]$ for the ith iteration ($i = 1, 2, \cdots$) is calculated as $L(\hat{C}^{(i)}[m,k]) = \log\left[\frac{\hat{\eta}^{(i)}[m]\,\hat{\beta}_1^{(i)}[m]}{(1-\hat{\eta}^{(i)}[m])\,\beta_0}\right]$ $+ (\beta_0 - \hat{\beta}_1^{(i)}[m]) \cdot \frac{\left|\frac{Y[m,k]}{\sqrt{E_s}} - \sqrt{\frac{1}{P}}\sum_{q=1}^{Q}|\hat{H}^{(q)(i)}[m,k]|^2\, s[m]B[k]\right|^2}{\sum_{q=1}^{Q}|\hat{H}^{(q)(i)}[m,k]|^2}$, where iteratively estimated jamming fraction is $\hat{\eta}^{(i)}[m] = \frac{1}{N_c}\sum_{k=0}^{N_C-1}\hat{C}^{(i-1)}[m,k]$ with $\hat{\eta}^{(1)}[m] = 1$ (assuming all subcarriers are equally likely to be jammed), $\hat{H}^{(q)(i)}[m,k] = \sum_{p=1}^{P}\hat{H}^{(p,q)(i)}[m,k]\,e^{-j\frac{2\pi k d_p}{N_c}}$, $\beta_0 = \frac{E_s}{N_0}$, $\hat{\beta}_1^{(i)}[m] = \frac{E_s}{N_0 + \hat{\sigma}_J^{2(i)}[m]}$, and $\hat{\sigma}_J^{2(i)}[m]$ $= \hat{N}_J/\hat{\eta}^{(i)}[m]$.

During the initial ($i = 1$) pilot aided channel estimation ($m \in \mathcal{P}$), the JSI estimates of the corresponding symbols are not available. Hence, all subcarriers are assumed equally likely for jamming, and the noise covariance matrix is approximated as $\hat{\mathbf{C}}_V^{(1)}[m] = (\hat{N}_J + N_0)\,\mathbf{I}_{N_c}$. Next, the initial channel estimates ($\hat{H}^{(p,q)(1)}[m,k]$) and the average PBNJ power ($\hat{N}_J$) are used for first JSI estimation for the pilot symbols ($m \in \mathcal{P}$). Once the JSI estimates are available, they are used for updating the noise covariance matrix and subsequently for refining the channel estimates using (1). Given the H-JSI and S-JSI estimates, the covariance matrix at the ith ($i \geq 2$) PACE iteration is updated as

$$\left[\hat{\mathbf{C}}_V^{(i)}[m]\right]_{kk} = \begin{cases} N_0 & ; \quad \hat{C}^{(i)}[m,k] = 0 \\ \hat{\sigma}_J^{2(i)}[m] + N_0 & ; \quad \hat{C}^{(i)}[m,k] = 1 \end{cases} \quad \text{and} \tag{6}$$

$$\left[\hat{\mathbf{C}}_V^{(i)}[m]\right]_{kk} = N_0\, p(\hat{C}^{(i)}[m,k] = 0) + \left(\hat{\sigma}_J^{2(i)}[m] + N_0\right) p(\hat{C}^{(i)}[m,k] = 1)\ , \tag{7}$$

respectively, which yields the channel estimates $\hat{\mathrm{h}}^{(q)(i)}[m],\ m \in \mathcal{P}$ for the ith iteration. Wiener filtering is used to interpolate the channel corresponding to the data symbols.

Soft Chip Combining

The JSI associated with the data symbols ($m \in \mathcal{D}$) is estimated iteratively by exploiting the LLR of the coded bits from the convolutional decoder and used during chip combining as described in [11]. While conventional hard JSI ($C[m,k]$) only reveals the presence of a jamming signal in a particular subchannel, the soft JSI, $L(\hat{C}^{(i)}[m,k])$, bears the same information with additional reliability information, which depends on the dominant jamming signal. Given the LLR of the JSI associated with each subcarrier at the ith IDDD iteration, we compute the most probable value of $\rho[m,k]$ as $\hat{\rho}^{(i)}[m,k] = \frac{1}{N_0}\frac{1}{1+e^{L(\hat{C}^{(i)}[m,k])}} + \frac{1}{\hat{\sigma}_J^{2(i)}[m]+N_0}\frac{1}{1+e^{-L(\hat{C}^{(i)}[m,k])}}$. The S-JSI based combining weights are then used for iterative chip combining that yields the despreader output $\hat{s}^{(i)}[m] = \sum_{k=0}^{N_c-1} \hat{\rho}^{(i)}[m,k]\, B^*[k]\, Y[m,k],\ m \in \mathcal{D}$.

4. Simulation Results and Discussions

In all simulations, code blocks of 2048 bits, $N_t = 9$, code rate $R_c = 1/2$, QPSK modulation, two transmit and two receive antennas are used. The time varying multipath Rayleigh fading channel has delay profile $0, 1, 2$ $[T_c]$ and power profile $0, -3, -6$ [dB], respectively, and $L = G = 3$. Assuming jamming is dominant, the signal to AWGN power ratio is set at $E_s/N_0 = 15$ dB. The carrier frequency, bandwidth, and number of sucarriers are $f_c = 5$ GHz, $B_f = 2$ MHz, and $N_c = 32$, respectively. Eight Wiener filter taps are used for channel interpolation. Figs. 3(a) and (b) show the BER over different signal to jamming power ratio (SJR=E_b/N_J) under full band noise jamming (FBNJ) for different antijamming receiver schemes with LSE and MMSE channel estimators, respectively. For each curve, six IDD or IDDD iterations are performed with six channel estimation iterations including the initial one. For the IDD

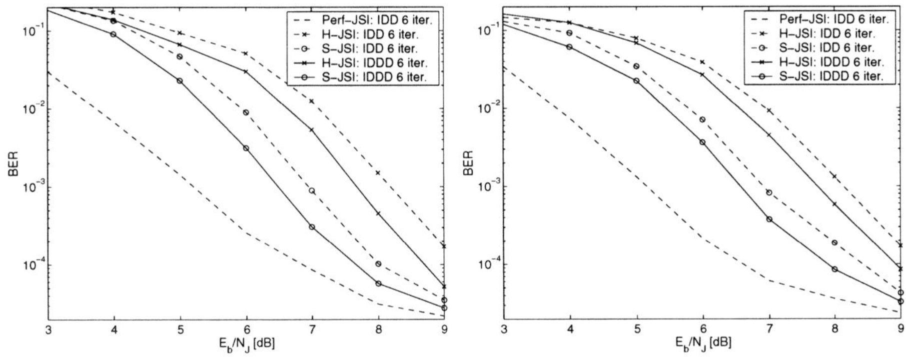

(a) LSE-CE (b) MMSE-CE

Figure 3. BER with (a) LSE and (b) MMSE channel estimators for different receiver schemes under FBNJ ($\eta = 1.0$, 6 iterations, P = 2, Q = 2, L = 3, v = 90 km/h). The S-JSI-IDDD receiver outperforms the other (H-/S-JSI-IDD, H-JSI-IDDD) receiver schemes.

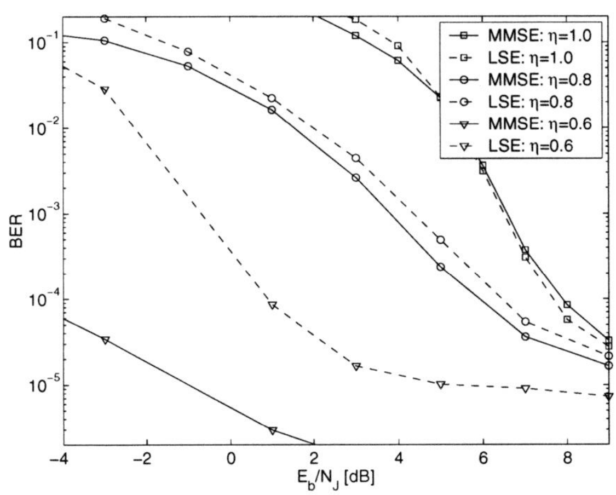

Figure 4. BER of S-JSI-IDDD receiver with S-JSI based LSE and MMSE channel estimation for different values of the jamming fraction (6 iterations, P = 2, Q = 2, v = 90 km/h).

receivers, chip combining is performed just once with either perfect or the initial JSI estimates (H-JSI or S-JSI) before beginning iterative demodulation and decoding. The effective SJR is defined as $\frac{E_b}{N_J} = \frac{E_s}{log_2(M)\,R_c N_J}\frac{N_c+G}{N_c}\frac{N_t}{N_t-1}$. The S-JSI based CE and soft chip combining yields smaller BER than conventional H-JSI based CE and chip combining, and the effect is evident both in IDD and IDDD receivers. Moreover, the IDDD schemes outperform the IDD schemes for the same type of JSI. The BER of the S-JSI based IDDD receiver with LSE and MMSE channel estimators and for different jamming fractions is plotted in Fig. 4.

5. Conclusions

We have derived iterative LSE and MMSE channel estimators and proposed a soft chip combining technique for the constant envelope MC-CDMA cyclic delay diversity system operating under PBNJ. The signal-space projection based technique is applied to our system for jamming power estimation which works without training sequence or data symbol decisions. Our proposed IDDD receiver with S-JSI based MMSE-CE and chip combining is robust against PBNJ with various jamming fractions.

Disclaimer

The views and conclusions contained in this document are those of the authors and should not be interpreted as representing the official policies, either expressed or implied, of the Army Research Laboratory or the U. S. Government.

References

[1] S. Hara and R. Prasad, "Overview of multicarrier CDMA," *IEEE Commun. Soc. Mag.*, vol. 35, pp. 126-133, Dec. 1997.

[2] G. K. Kaleh, "Frequency-Diversity Spread Spectrum Communication System to Counter Bandlimited Gaussian Interference," *IEEE Trans. Commun.*, vol. 44, pp. 886-893, July 1996.

[3] S. Zhou, G. B. Giannakis and A. Swami, "Digital Multi-Carrier Spread Spectrum Versus Direct Sequence Spread Spectrum for Resistance to Jamming and Multipath," *IEEE Trans. Commun.*, vol. 50, pp. 643-655, Apr. 2002.

[4] Galib A. M.M. and G. L. Stüber, "Performance Analysis of Constant Envelope Multicarrier CDMA MIMO Systems with Cyclic Delay Diversity," in *Proc. 10th International OFDM-Workshop*, 2005.

[5] J. Tan and G. L. Stuber, "Multicarrier Spread Spectrum System With Constant Envelope: Antijamming, Jamming Estimation, Multiuser Access," *IEEE Trans. Wirel. Commun.*, vol. 4, pp. 1527-1538, July, 2005.

[6] T. Zemen, C. F. Mecklenvrauker, J. Wehinger and R. R. Muller, "Iterative Joint Time-Variant Channel Estimation and Multi-User Detection for MC-CDMA," *IEEE Trans. Wirel. Commun.*, vol. 5, pp. 1469-1478, June, 2006.

[7] L. Sanguinetti and M. Morelli, "Channel Acquisition and Tracking for MC-CDMA Uplink Transmissions," *IEEE Trans. Veh. Technol.*, vol. 55, pp. 956-967, May 2006.

[8] M. K. Simon, J. K. Omura, R. A. Scholtz, and B. K. Levitt, *Spread Spectrum Communications Handbook*, Rev. ed., McGraw Hill Inc., 1994.

[9] S. M. Kay, *Fundamentals of Statistical Signal Processing, Volume 1: Estimation Theory*, Prentice Hall, 1993.

[10] M. Turkboylari and G. L. Stuber, "An efficient algorithm for estimating the signal-to-interference ratio in TDMA cellular systems," *IEEE Trans. Commun.*, vol. 46, pp. 728-731, June 1998.

[11] Galib A. M.M. and Gordon L. Stuber, "Soft-Chip Combining MIMO Multicarrier CDMA Antijam System," *IEEE MILCOM*, Oct. 2006.

PILOT DESIGN FOR MIMO-OFDM WITH BEAMFORMING

Michele Carta
University of Pisa, Dept. of Information Engineering
Via G. Caruso, 56122 Pisa, Italy.
michele.carta@iet.unipi.it

Ivan Cosovic, Gunther Auer
DoCoMo Euro-Labs
Landsberger Strasse 312, 80687 Munich, Germany.
{cosovic,auer}@docomolab-euro.com

Abstract In this contribution [1] we address the issue of pilot design for channel estimation in multiple input multiple output orthogonal frequency division multiplexing (MIMO-OFDM) systems applying beamforming at the transmitter.
Specifically we consider the case when a finite set of fixed beamforming vectors is available and the selection of the optimal weight for each transmit antenna is performed exploiting long-term channel information.
Capacity maximization is considered as the optimization parameter in order to achieve a proper trade-off between estimation performances and available resources for data transmission. Our analysis shows the impact of different values of pilot boost and pilot overhead on the channel estimation mean square error (MSE) and on the capacity of the system. Simulation results are given for an outdoor mobile wireless channel, different number of antennas and channel estimator filter order for a single and a multi-user transmission scenario.

Keywords: MIMO, OFDM, channel estimation, beamforming, pilot symbol design

1. Introduction

In the last decade multicarrier OFDM modulation has gained considerable attention as an efficient technique for high-data-rate transmissions over wireless frequency-selective channels. It has been adopted in several wireless stan-

[1]This work has partly been performed in the framework of the IST project IST-2003-507581 WINNER (World Wireless Initiative New Radio), which is partly funded by the European Union. This work has been performed during Michele Carta's research visit to DoCoMo Euro-Labs.

S. Plass et al. (eds.), Multi-Carrier Spread Spectrum 2007, 377–386.

dards such as digital audio and video broadcasting (DAB, DVB), 802.11a and 802.16a. Furthermore, in combination with multiple antennas at both transmitter and receiver, it is considered as a strong candidate technology for future wireless communication systems, e.g., within IST WINNER (World Wireless Initiative New Radio) project [1].

To perform coherent data detection at the OFDM receiver, estimates of the channel information are needed. The most common approach to obtain these estimates at the receiver is via pilot aided channel estimation (PACE). Pilot symbols multiplexed in the OFDM frame utilize both power and spectrum and thus a pilot based channel estimator should be designed in order to achieve a proper trade-off between estimation performances and available resources for data transmission. To this end, a suitable optimization parameter is the maximization of the channel capacity. In [2], a lower capacity bound for multiple-antennas systems in presence of block-fading channels is given. In [3], it is shown that equidistant pilot symbols maximize capacity. These results suggest that channel estimation via two-dimensional (2D) interpolation in time and frequency direction of equidistantly placed pilot symbols [4] is a good choice for capacity maximizing pilot design in MIMO-OFDM. In [5], channel estimation and consequent capacity maximization are analyzed in a MIMO-OFDM system with spatial multiplexing and no channel state information at the transmitter.

In this paper we focus on channel estimation issue when a single-stream beamforming technique is performed at the transmitter. Such a transmission strategy is preferred when dealing with propagation environment that present low-rank spatial correlation matrix. This implies that a favorable direction of transmission can be exploited in order to experience a *beamforming gain* in the signal-to-noise ratio (SNR) at the receiver. The long-term spatial correlation matrix of the channel is assumed known in order to select the optimal beamforming vector among a fixed set. An equivalent received signal model is developed in order to quantify the SNR loss due to pilot overhead and pilot boost. On the other hand, we show how effective is the beamforming technique improving channel estimator performances. Achievable capacity bound is then expressed as a function of pilot grid design parameters, number of antennas and mean square error (MSE) of the channel estimator. Results are discussed for an outdoor wireless propagation scenario considering a single- and a multi-user transmission.

The remainder of the paper is structured as follows: in Section 2 the considered MIMO-OFDM system is described; in Section 3 the MSE of a 2D pilot aided linear estimator is analyzed and spatial correlation properties of the channel are discussed. Section 4 offers a capacity analysis of the system. Finally, simulation results are presented in Section 5 and conclusions are drawn in Section 6.

2. System Model

A MIMO-OFDM system with N_{T} transmit and N_{R} receive antennas is considered. Denote with N_{c} number of used subcarriers and with L number of OFDM symbols per frame. OFDM modulation is performed by N_{IDFT}–point ($N_{\mathrm{IDFT}} \geq N_{\mathrm{c}}$) inverse DFT (IDFT), followed by insertion of a cyclic prefix of N_{CP} samples. We assume that a beamforming vector is used for transmission of a single-stream of data symbols. Its selection among a predefined static set $\mathbf{V} \in \{\mathbf{v}_1, \mathbf{v}_2, \ldots, \mathbf{v}_{N_{\mathrm{bf}}}\}$ of N_{bf} complex-valued beamforming vectors is either done exploiting long-state channel information, assumed available at the transmitter or driven by the receiver through a feedback process. Under the assumption of perfect orthogonality in time and frequency, the received sample taken at the ℓ-th OFDM symbol block on the n-th subcarrier at the m_{R}-th receive antenna is given by

$$\begin{aligned} Y_{n,\ell}^{(m_{\mathrm{R}})} &= \sqrt{E_{\mathrm{d}}}\, H_{\mathrm{eff}\ n,\ell}^{(m_{\mathrm{R}})}\, x_{n,\ell} + z_{n,\ell}^{(m_{\mathrm{R}})}\ ; \\ & 0 \leq m_{\mathrm{R}} < N_{\mathrm{R}}. \end{aligned} \tag{1}$$

where

$$H_{\mathrm{eff}\ n,\ell}^{(m_{\mathrm{R}})} = \sum_{m_{\mathrm{T}}=1}^{N_{\mathrm{T}}} H_{n,\ell}^{(m_{\mathrm{R}},m_{\mathrm{T}})}\, v_{\mathrm{tx}}^{(m_{\mathrm{T}})} \tag{2}$$

is a complex scalar factor representing the effective channel experienced at the m_{R}-th receive antenna. In (2), $H_{n,\ell}^{(m_{\mathrm{R}},m_{\mathrm{T}})}$ and $v_{\mathrm{tx}}^{(m_{\mathrm{T}})}$ denote the channel transfer function (CTF) between m_{T}-th transmit antenna and m_{R}-th receive antenna and the m_{T}-th transmit antenna weight of the $N_{\mathrm{T}} \times 1$ selected beamforming vector $\mathbf{v}_{\mathrm{tx}} = \left(v_{\mathrm{tx}}^{(1)}, \ldots, v_{\mathrm{tx}}^{(N_{\mathrm{T}})}\right)^T$. In (1), $x_{n,\ell}$ and $z_{n,\ell}^{(m_{\mathrm{R}})}$ account for the unitary-energy transmitted symbol and the additive white Gaussian noise (AWGN) sample at the m_{R}-th receive antenna with zero mean and variance N_0, respectively. E_{d} is the energy per transmitted data symbol assuming $||\mathbf{v}_{\mathrm{tx}}||^2 = 1$. The CTF $H_{n,\ell}^{(m_{\mathrm{R}},m_{\mathrm{T}})}$ is the Fourier transform of the channel impulse response (CIR) sampled at the frequency instant $f = n/T$ and the time instant $t = \ell T_{\mathrm{sym}}$, where $T_{\mathrm{sym}} = (N_{\mathrm{c}} + N_{\mathrm{CP}})T_{\mathrm{spl}}$ and $T = N_{\mathrm{c}} T_{\mathrm{spl}}$ represent the OFDM symbol duration with and without the cyclic prefix of N_{CP} samples, and T_{spl} is the sample duration. From (1), it is worth noting that in order to perform coherent detection of the data symbol $x_{n,\ell}$, channel state information required does not include the whole $N_{\mathrm{R}} \times N_{\mathrm{T}}$ channel gains matrix but only the effective channel vector of dimension $N_{\mathrm{R}} \times 1$ defined in (2). Therefore, we assume that N_{p} pilots equidistantly spaced within OFDM frame are devoted to PACE. The pilot overhead is defined as

$$\Omega_{\mathrm{p}} = N_{\mathrm{p}}/(N_{\mathrm{c}} L) \tag{3}$$

Denoting with $N_{\rm d}$ number of data symbols within a frame, we have $N_{\rm c}L = N_{\rm d}+N_{\rm p}$ and the total transmit power over all $N_{\rm T}$ antennas equals

$$P_{\rm tot} = E_{\rm d}N_{\rm d} + E_{\rm p}N_{\rm p} \tag{4}$$

Channel estimates accuracy can be enhanced by a pilot boost $S_{\rm p}$. With an energy per transmitted pilot set to $E_{\rm p} = S_{\rm p}E_{\rm d}$, the SNR at the input of the channel estimation unit increases $S_{\rm p}$ times. On the other hand, if the overall transmit power in (4) has to be kept constant, the portion dedicated to the payload information is reduced. The power dedicated to pilot symbols is determined by the pilot overhead $\Omega_{\rm p}$ and the pilot boost $S_{\rm p}$. The pilot insertion loss relative to a reference system assuming the same frame size, same overall transmission power and no overhead due to pilots is given by

$$E_{\rm d}/E_0 = 1/(1+\Omega_{\rm p}(S_{\rm p}-1)) \tag{5}$$

where E_0 is the energy per symbol of the pilot-less system.

3. Estimation Error Analysis

We consider a linear channel estimator relying on the sequence of the observed pilot symbols. The estimate of the effective channel is given by $\widehat{H}^{(m_{\rm R})}_{{\rm eff}\ n,\ell} = \mathbf{w}^T\tilde{\mathbf{y}}^{(m_{\rm R})}$ where the $M_{\rm t}M_{\rm f} \times 1$ vector $\tilde{\mathbf{y}}^{(m_{\rm R})}$ contains $M_{\rm t}$ and $M_{\rm f}$ received pilots in frequency and time, respectively, at the $m_{\rm R}$-th receive antenna and the $M_{\rm t}M_{\rm f} \times 1$ vector $\mathbf{w}$ represents an arbitrary linear estimator, e.g., 2D FIR filter.

An equivalent system model can be drawn from (1) for MSE analysis

$$\begin{aligned} Y^{(m_{\rm R})}_{n,\ell} &= \sqrt{E_{\rm d}}\,\widehat{H}^{(m_{\rm R})}_{{\rm eff}\ n,\ell}\,x_{n,\ell} + \eta^{(m_{\rm R})}_{n,\ell}; \\ & 0 \le n < N_c,\ 0 \le \ell < L,\ 0 \le m_{\rm R} < N_{\rm R}. \end{aligned} \tag{6}$$

wherein symbol $x_{n,\ell}$ represents a data or a pilot, depending on the indices n and ℓ. In (6), $\eta^{(m_{\rm R})}_{n,\ell} = \sqrt{E_{\rm d}}\,\varepsilon^{(m_{\rm R})}_{n,\ell}\,x_{n,\ell} + z^{(m_{\rm R})}_{n,\ell}$ and $\varepsilon^{(m_{\rm R})}_{n,\ell} = H^{(m_{\rm R})}_{{\rm eff}\ n,\ell} - \widehat{H}^{(m_{\rm R})}_{{\rm eff}\ n,\ell}$. In the following we assume that the estimation error $\varepsilon^{(m_{\rm R})}_{n,\ell}$ is a Gaussian random variable with zero mean and variance equal to the MSE. Furthermore, since receive antennas are uncorrelated, this variance is independent of $m_{\rm R}$ and we denote it with $\sigma^2_\varepsilon[n,\ell] = E\left[|\varepsilon_{n,\ell}|^2\right]$, dropping index $m_{\rm R}$. Finally, the variance of the effective noise term, $\eta^{(m_{\rm R})}_{n,\ell}$, equals $\sigma^2_\eta[n,\ell] = N_0 + E_{\rm d}\sigma^2_\varepsilon[n,\ell]$ and does not depend on the index $m_{\rm R}$ either. The SNR loss due to channel estimation is dependent on two factors: the estimation error $\varepsilon_{n,\ell}$ and the overhead in terms of transmit power dedicated to pilot symbols in (5) and it is given by

$$\Delta\gamma = \frac{\gamma_0}{\gamma} = \left(1 + \Omega_{\rm p}(S_{\rm p}-1) + \gamma_0\,\sigma^2_\varepsilon[n,\ell]\right). \tag{7}$$

In (7),

$$\gamma = \frac{E_{\mathrm{d}}\sigma^2_{\hat{H}_{\mathrm{eff}}}[n,\ell]}{N_0 + E_{\mathrm{d}} \cdot \sigma^2_{\varepsilon}[n,\ell]} \tag{8}$$

represents the SNR for the signal model in (6) and $\gamma_0 = E_0\sigma^2_{H_{\mathrm{eff}}}[n,\ell]/N_0$ denotes the SNR for a reference system assuming perfect channel knowledge and no overhead due to pilots. It is worth observing that $G_{\mathrm{bf}} = \sigma^2_{H_{\mathrm{eff}}}[n,\ell]$ can be defined as the beamforming gain which accounts for the improvement of the received SNR at each receive antenna with respect to a system without beamforming at the transmitter. It depends on the CTF and beamforming vector statistics and, considering zero mean valued channel gains $H^{(m_{\mathrm{R}},m_{\mathrm{T}})}_{n,\ell}$, $G_{\mathrm{bf}} = E\big[\big|H_{\mathrm{eff}\ n,\ell}\big|^2\big]$ holds true. Furthermore in (7), we assumed that $\sigma^2_{\hat{H}_{\mathrm{eff}}}[n,\ell] \approx \sigma^2_{H_{\mathrm{eff}}}[n,\ell]$.

Parametrization of the MSE. The MSE of an arbitrary 2D pilot aided scheme is given by

$$\begin{aligned}\sigma^2_{\varepsilon}[n,\ell] &= E\big[\varepsilon_{n,\ell}\varepsilon^*_{n,\ell}\big] = E\big[\big|H_{\mathrm{eff}\ n,\ell} - \widehat{H}_{\mathrm{eff}\ n,\ell}\big|^2\big] \\ &= E\big[\big|H_{\mathrm{eff}\ n,\ell}\big|^2\big] - 2\Re\big\{\mathbf{w}^H \mathbf{r}_{\tilde{\mathbf{y}}H_{\mathrm{eff}}}[n,\ell]\big\} + \mathbf{w}^H \mathbf{R}_{\tilde{\mathbf{y}}\tilde{\mathbf{y}}}\mathbf{w}\end{aligned} \tag{9}$$

with $\widehat{H}_{\mathrm{eff}\ n,\ell} = \mathbf{w}^H\tilde{\mathbf{y}}$. The 2D correlation functions $\mathbf{R}_{\tilde{\mathbf{y}}\tilde{\mathbf{y}}} = E\{\tilde{\mathbf{y}}\tilde{\mathbf{y}}^H\}$ and $\mathbf{r}_{\tilde{\mathbf{y}}H_{\mathrm{eff}}}[n,\ell] = E\{\tilde{\mathbf{y}}H^*_{\mathrm{eff}\ n,\ell}\}$ represent the auto-correlation matrix of the received pilots, $\tilde{\mathbf{y}}$, and the cross-correlation between $\tilde{\mathbf{y}}$ and the desired response $H_{\mathrm{eff}\ n,\ell}$, respectively [6]. The MSE in (9) is a function of n and ℓ. For the sake of a simple model, we choose to average the MSE over the entire sequence, such that $\sigma^2_{\epsilon}[n,\ell] \rightarrow \bar{\sigma}^2_{\epsilon}$. Assuming that the effective channel and the noise are uncorrelated, the auto-correlation function can be expressed as $\mathbf{R}_{\tilde{\mathbf{y}}\tilde{\mathbf{y}}} = \mathbf{R}_{\tilde{\mathbf{h}}_{\mathrm{eff}}\tilde{\mathbf{h}}_{\mathrm{eff}}} + \frac{N_0}{S_{\mathrm{p}}E_{\mathrm{d}}}\mathbf{I}$, where $\mathbf{R}_{\tilde{\mathbf{h}}_{\mathrm{eff}}\tilde{\mathbf{h}}_{\mathrm{eff}}}$ is the auto-correlation matrix of the $M_{\mathrm{f}}M_{\mathrm{t}}\times 1$ vector $\tilde{\mathbf{h}}_{\mathrm{eff}}$ that represents the effective channel at pilot positions excluding the AWGN term, and $\mathbf{I}$ denotes the identity matrix, all of dimension $M_{\mathrm{f}}M_{\mathrm{t}} \times M_{\mathrm{f}}M_{\mathrm{t}}$. With the pilot insertion loss of (5), the SNR at pilot positions becomes $\gamma_{\mathrm{p}} = G_{\mathrm{bf}}\,S_{\mathrm{p}}\,E_{\mathrm{d}}/N_0 = \gamma_0\,S_{\mathrm{p}}/(1+\Omega_{\mathrm{p}}(S_{\mathrm{p}}-1))$.
The MSE expression can be splitted into a noise error component $\bar{\sigma}^2_n$ and an interpolation or lag error component $\bar{\sigma}^2_i$. Dividing (6) by G_{bf} for a more convenient notation and inserting $\mathbf{R}_{\tilde{\mathbf{y}}\tilde{\mathbf{y}}}$, the noise part is inversely proportional to γ_{p} and is given by

$$\bar{\sigma}^2_n = \frac{\mathbf{w}^H\mathbf{w}}{\gamma_{\mathrm{p}}} = \frac{1}{G_{\mathrm{n}}\,\gamma_0} \cdot \frac{1+\Omega_{\mathrm{p}}(S_{\mathrm{p}}-1)}{S_{\mathrm{p}}} \tag{10}$$

where $G_{\mathrm{n}} = 1/(\mathbf{w}^H\mathbf{w})$ defines the estimator gain of $\bar{\sigma}_n^2$. According to (9) the variance of the interpolation error is

$$\bar{\sigma}_i^2 = 1 - \frac{2}{G_{\mathrm{bf}}}\Re\{\mathbf{w}^H\mathbf{r}_{\tilde{\mathbf{h}}_{\mathrm{eff}}H_{\mathrm{eff}}}\} + \frac{1}{G_{\mathrm{bf}}}\mathbf{w}^H\mathbf{R}_{\tilde{\mathbf{h}}_{\mathrm{eff}}\tilde{\mathbf{h}}_{\mathrm{eff}}}\mathbf{w}. \tag{11}$$

Note that G_{n} and $\bar{\sigma}_i^2$ are independent from the SNR if $\mathbf{w}$ is SNR-independent.

Long-Term Channel Information. The beamforming gain introduced in (8), assuming fixed beamforming vectors, reads

$$G_{\mathrm{bf}} = \mathbf{v}_{\mathrm{tx}}^T\mathbf{R}_{\mathrm{tx}}\mathbf{v}_{\mathrm{tx}}^* \tag{12}$$

where $\mathbf{R}_{\mathrm{tx}}$ is the $N_{\mathrm{T}} \times N_{\mathrm{T}}$ long-term transmit spatial correlation matrix of the channel. The beamforming vector for transmission is then selected as

$$\mathbf{v}_{\mathrm{tx}} = \arg\max_{\mathbf{v}_i \in \mathbf{V}}\{\mathbf{v}_i^T\mathbf{R}_{\mathrm{tx}}\mathbf{v}_i^*\} \tag{13}$$

In a single spatial stream transmission with long-term adaptive beamforming, the optimal beamforming vector $\mathbf{v}_{\mathrm{opt}}$ is the eigenvector corresponding to the maximum eigenvalue (i.e, the maximum G_{bf}) of the matrix $\mathbf{R}_{\mathrm{tx}}$. Thus $\mathbf{v}_{\mathrm{tx}}$ minimizes $||\mathbf{v}_i - \mathbf{v}_{\mathrm{opt}}||^2$. The higher the number of available beams, the lower the decrease in the achievable G_{bf} with respect to the dominant eigenvalue transmission.

4. Capacity Analysis

The ergodic channel capacity that includes channel estimation and pilot insertion losses in the case when channel is not known at the transmitter can be lower bounded by [2]

$$C = (1 - \Omega_{\mathrm{p}})\mathrm{E}\left[\log_2\det\left(\mathbf{I}+\mathbf{H}_{\mathrm{eff}\,n,\ell}\mathbf{H}_{\mathrm{eff}\,n,\ell}^H\frac{\gamma_0}{\Delta\gamma}\right)\right] \tag{14}$$

where the m_{R}-th element of the column vector $\mathbf{H}_{\mathrm{eff}\,n,\ell}$ is $H_{\mathrm{eff}\,n,\ell}^{(m_{\mathrm{R}})}$ and where $\mathbf{I}$ is the $N_{\mathrm{R}} \times N_{\mathrm{R}}$ identity matrix. Furthermore, the expectation is taken over the frequency and time, i.e., over indices n and ℓ, and $\Delta\gamma$ is given by (7). Exploiting (7), (10) and (11) we obtain

$$C=(1-\Omega_{\mathrm{p}})\mathrm{E}\left[\log_2\det\left(\mathbf{I}+\mathbf{H}_{\mathrm{eff},n,\ell}\mathbf{H}_{\mathrm{eff},n,\ell}^H\frac{\gamma_0}{1+\Omega_{\mathrm{p}}(S_{\mathrm{p}}-1)+\gamma_0\,\sigma_{\varepsilon}^2[n,\ell]}\right)\right]. \tag{15}$$

The capacity loss due to the pilot aided channel estimation is characterized by two factors: the corresponding SNR loss and the spectral efficiency loss due to pilot insertion. The optimum pilot boost which maximize capacity is then obtained by differentiating $\Delta\gamma$ from (7) or C from (15) with respect to $S_{\rm p}$ and setting the result to zero. This leads to

$$S_{\rm p,opt} = \sqrt{\frac{1 - \Omega_{\rm p}}{\Omega_{\rm p} G_{\rm n} G_{\rm bf}}} \tag{16}$$

It can be concluded that the optimum pilot boost $S_{\rm p,opt}$ increases if less pilots are used and/or less transmit antennas are used. It decreases if an estimator with large gain, $G_{\rm n}$, is used and/or $G_{\rm bf}$ increases. Capacity for the optimally chosen pilot boost can be easily derived. On the other hand, both $G_{\rm n}$ and $\bar{\sigma}_i^2$ depend on $\Omega_{\rm p}$, consequently closed-form determination of $\Omega_{\rm p}$ that maximizes (15) is a task of formidable complexity which is not pursued here. Simulation results are given to this aim and to show how slight can be the loss in capacity when choosing a proper set of fixed beamforming vectors compared to the dominant eigenvalue transmission case.

5. Results

Simulations have been run for a 4×4 and an 8×8 MIMO-OFDM system with $N_c = 1024$ subcarriers, $L = 12$ OFDM symbols per frame and a guard interval (GI) duration of $T_{GI} = 64{\cdot}T_{spl}$. The signal bandwidth is set to 20MHz, which corresponds to a sampling duration of $T_{spl} = 50$ns and the modulation format is 4-QAM. Transmitter is an 120 degrees aperture multi fixed beams array with equidistant radiating elements with $\lambda/2$ spacing (λ is the wavelength of the signal). The number of beamforming vectors is $N_{\rm bf} = 8$ and $N_{\rm bf} = 16$, for the 4×4 and the 8×8 system, respectively, corresponding to a same number of equidistant available angles of transmission. Antenna weights of each beamforming vector are generated according to [7].

Concerning the channel setup, two multipath propagation scenarios specified within the IST-WINNER project have been considered: the macrocell suburban propagation scenario (referred to as C1) and the macro urban propagation scenario (referred to as C2). Gains are modeled as zero-mean Gaussian processes independent from each other, with power spectra obeying to the Jakes model [9] with normalized maximum Doppler frequency of $f_D T_{sym} = 0.03$, which corresponds to a velocity of 250km/h at 5GHz carrier frequency. The related power-delay profile can be found in [8]. Spatial correlation has been introduced through $\mathbf{R}_{\rm tx}$, which depends on the power-delay profile, angle of departures of the channel and on the geometry of the transmitter [10].

Channel estimation task is carried out after all pilots of one frame have been received. A cascaded channel estimator consisting of two one dimensional (1D) estimators called 2×1D PACE is considered [4], entailing slightly worse performances than optimal 2D estimator, while being significantly less com-

plex. Uniformly distributed power delay profile and Doppler power spectrum non zero within the range $[0, \tau_\mathrm{w}]$ and $[-f_{\mathrm{D,w}}, f_{\mathrm{D,w}}]$ are considered to obtain FIR coefficients in time and frequency direction. In addition, the average SNR at the filter input, γ_w, is required, which should be equal to largest expected average SNR. We set $\tau_\mathrm{w} = T_\mathrm{CP}$, $f_{\mathrm{D,w}} T_\mathrm{sym} = 0.04$ and $\gamma_\mathrm{w} = 30$ dB.
Results are provided for a single-user (SU) transmission where all the pilots in a frame can be exploited for channel estimation and a multi-user (MU) scenario where only a set of 8 contiguous subcarriers are dedicated to each user, thus less pilots in frequency dimension are available. Pilot spacings in frequency and time are referred to as D_f and D_t, respectively, and SNR $= E_0/N_0$.

MSE performances. In Figures 1 and 2, we show the MSE performance of the cited estimator versus SNR for a SU and a MU scenario, respectively. Pilot grid is designed according to τ_w and $f_{\mathrm{D,w}}$ as shown in [5], setting $D_\mathrm{f} = 4$, $D_\mathrm{t} = 4$, $M_\mathrm{f} = 8$ and $M_\mathrm{t} = 3$ for the SU case whilst $M_\mathrm{f} = 2$ for the MU case. Results are slightly affected by the propagation scenario whilst doubling the number of transmit antennas results in increasing G_bf from 5dB to 7.7dB for C1 and from 5dB to 6dB for C2. As a consequence, $\bar{\sigma}_n^2$ decreases whereas the floor exhibited by all the curves is due to $\bar{\sigma}_i^2$, which is SNR-independent. The SU system outperforms the MU case as M_f is 4 times greater, entailing a higher filter order in frequency dimension.

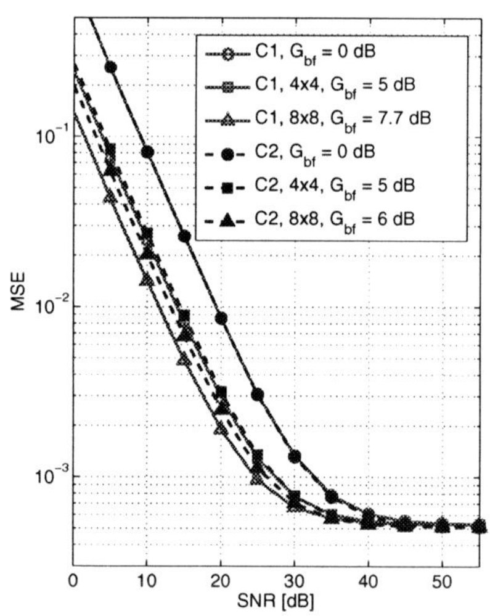

Figure 1. MSE vs SNR for a SU scenario.

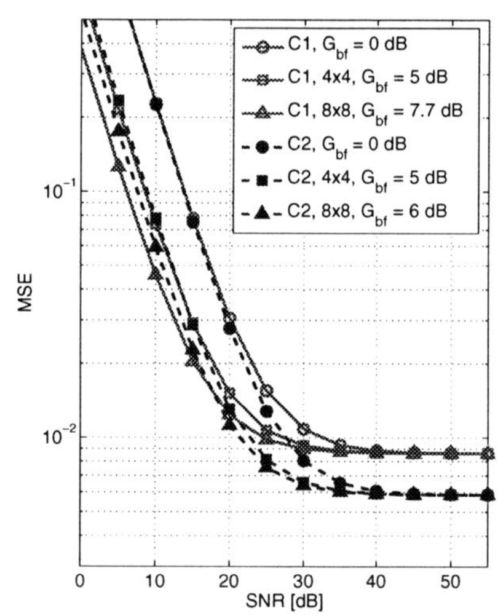

Figure 2. MSE vs SNR for a MU scenario.

Capacity analysis. Figure 3shows the capacity of the MIMO systems for the channel C1 when adopting either a set of static vectors (denoted SBF) or long-term adaptive beamforming (denoted ABF), when $\Omega_\mathrm{p} = 0$ and $S_\mathrm{p} = 1$.

Capacity loss is extremely slight and the gap is even smaller when considering an 8×8 MIMO system since a higher number of fixed beams allows to reduce $||\mathbf{v}_{\mathrm{tx}} - \mathbf{v}_{\mathrm{opt}}||^2$, approximating better the dominant eigenvector of $\mathbf{R}_{\mathrm{tx}}$. Similar results have been obtained considering channel C2.

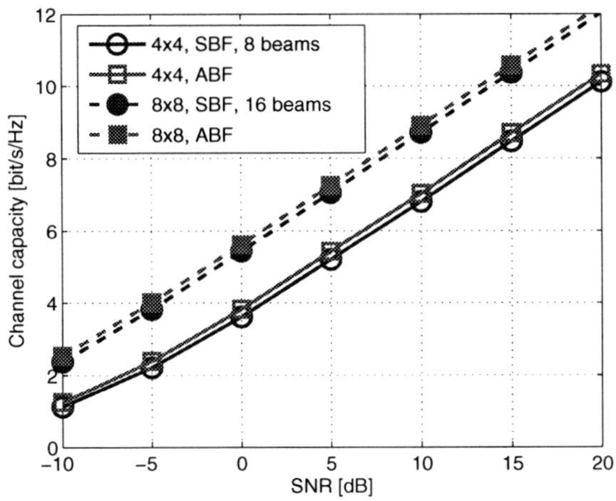

Figure 3. Capacity vs SNR for 4×4 and 8×8 MIMO-OFDM systems.

In Figure 4 the channel capacity versus pilot boost S_{p} for an 8×8 MIMO system considering different pilot grids is plotted at SNR $=$ 10 dB for the channel C1. Capacity is maximized when dealing with lower pilot overhead and adopting S_{p} which satisfies (16). Adopting low values of S_{p}, the SU scenario capacity outperforms the MU capacity, due to a lower $\bar{\sigma}_n^2$. On the other hand, for high S_{p}, capacity decreases since MSE $\simeq \bar{\sigma}_i^2$, i.e. the minimum MSE is attained but less power is devoted to data symbols.

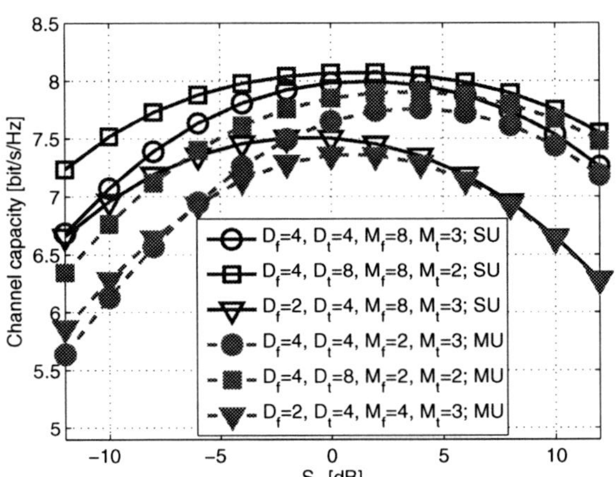

Figure 4. Capacity vs S_{p} for an 8×8 MIMO-OFDM system with different pilot grids.

6. Conclusions

A theoretical framework for pilot grid design was developed for MIMO-OFDM systems when a single-stream fixed beamforming is performed at the transmitter. A simple model has been proposed to account for SNR loss when PACE is performed. We focused on the channel estimator performances as a function of the beamforming gain and the employed pilot grid. Then, capacity maximization has been discussed, with pilot overhead and pilot boost as the design parameters and the optimal choice for the former has been derived analytically. Simulation results have been provided for an outdoor wireless propagation channel, in a single- and multi-user scenario, assessing the effectiveness of the beamforming technique for improving MSE performances and capacity of the system.

References

[1] IST-2003-507581 WINNER, D2.10 Final report on identified RI key technologies, system concept, and their assessment, Dec. 2005.

[2] B Hassibi and B M Hochwald,"How much Training is Needed in Multiple-Antenna Wireless Links?," *IEEE Trans. Information Theory*, vol. 49, pp. 951-963, Apr. 2003.

[3] S Adireddy and L Tong, "Optimal Placement of Known Symbols for Slowly Varying Frequency-Selective Channels," *IEEE Trans. Wireless. Comm.*, vol. 4, pp.1292-1296, Jul. 2005.

[4] P. Hoeher, S. Kaiser, and P. Robertson, "Two-Dimensional Pilot-Symbol-Aided Channel Estimation by Wiener Filtering," in *Proc. IEEE Intern. Conf. on Acoustics, Speech, and Signal Processing (ICASSP '97)*, vol. 3, pp. 1845-1848, 1997.

[5] I. Cosovic and G. Auer, "Capacity Analysis of MIMO-OFDM Including Pilot Overhead and Channel Estimation Errors," to appear in *Proc. IEEE International Conference on Communications (ICC'07), Glasgow, Scotland*, Jun. 2007.

[6] P. Hoeher, S. Kaiser, and P. Robertson, "Pilot-Symbol-Aided Channel Estimation in Time and Frequency," in *Proc. Communication Theory Mini-Conference within IEEE Global Telecommun. Conf. (Globecom '97), Phoenix, USA*, pp. 90-96, 1997.

[7] IST-4-027756 WINNER II D3.4.1 v1.0 The WINNER II Air Interface: Refined Spatial-Temporal Processing Solutions, Nov. 2006.

[8] IST-2003-507581 WINNER, D5.4 Final Report on Link Level and System Level Channel Models, Sep. 2005.

[9] W. C. Jakes, *Microwave Mobile Communications*, Wiley, NY, 1974.

[10] 3GPP TSG RAN WG 1, TSGR1#15 R1-00-1067, Channel model for Tx diversity simulations using correlated antennas, Aug. 2000.

CHANNEL ESTIMATION BY EXPLOITING SUBLAYER INFORMATION IN OFDM SYSTEMS

Maik Bevermeier, Tobias Ebel and Reinhold Haeb-Umbach
Department of Communications Engineering
University of Paderborn, Germany
{bevermeier,ebel,haeb}@nt.uni-paderborn.de

Abstract In this paper we investigate the possibilities to generate a postamble in block-oriented OFDM transmission without compromising the effective user data rate. The idea is to complete known bits at the end of each physical layer frame to a full OFDM symbol to be used for channel estimation. A Wiener Filter is then employed to interpolate between the channel estimates obtained on the pre- and postamble. We present experimental results for an IEEE 802.11a-like system. A comparison with purely preamble-based channel estimation shows a reduced mean square channel estimation error and bit error rate at low computational complexity and only minor reduction of the net user data rate. Results of robustness tests are also discussed.

Keywords: Orthogonal frequency division multiplexing, Wiener filtering.

1. Introduction

In recent years Orthogonal Frequency Division Multiplexing (OFDM) became more and more important in wired and wireless communication. Turning a frequency selective channel into multiple non-frequency selective narrowband channels leads to lower costs and higher data rates in comparison to classical single-carrier transmission standards [1].

In this article we examine a pilot-aided channel estimation technique based on a known preamble in front of the payload and known data, which finalize the physical layer frame (PHY). Channel estimation using the known symbols of the preamble are in widespread use. They allow to estimate the channel transfer function (CTF) for each subcarrier and assume the CTF to be constant for the duration of the payload. Their performance, however, is quite limited in the presence of high terminal velocities or large payload lengths, since the radio channel may change

S. Plass et al. (eds.), Multi-Carrier Spread Spectrum 2007, 387–396.

significantly during the payload. To compensate for this effect semi-blind algorithms have been proposed, where the initial channel estimates obtained on the preamble are further refined on the payload [2].

In this paper we investigate the possibility of using known sublayer bits at the end of the payload to generate a complete postamble OFDM symbol. To obtain a completely known final OFDM symbol, however, some user data bits have to be replaced by known bits. The amount and identity of the bits to be introduced depend on the coding scheme and other system parameters. We will show that the reduction of the net user data rate is usually considerably smaller compared to the case if a complete postamble symbol were introduced after the payload.

Once a complete postamble OFDM symbol is obtained, instantaneous channel estimation is carried out on pre- and postamble and the channel estimates on the payload are obtained by Wiener interpolation between these two estimates. The experimental results show that the lower bit error rate obtained due to improved channel estimation seems to justify the slight reduction in net user data rate.

The article is organized as follows: In the next section we describe the OFDM system model. Section 3 outlines the general structure of the physical layer used in a typical 802.11a-like system and the coding process. In Section 4, we present our proposal of using the known bits to finalize the data frame, in conjunction with the required modifications of the layer structure. Before discussing the simulation results, including robustness and complexity issues, we describe the design of the Wiener (Interpolation) Filter (WF). The paper is finished by a conclusion.

2. System Model

We consider a coherent block-oriented OFDM transmission with M sub-carriers. The $(M \times 1)$ symbol vector in frequency domain can be written as

$$\tilde{\mathbf{a}}(k) = (\tilde{a}_0(k), \ldots, \tilde{a}_{M-1}(k))^T, \tag{1}$$

where $k = iB + n$ is the symbol counter, i counts the packets and n is the symbol index within a block of B symbols. The OFDM modulated symbol vector is obtained as

$$\mathbf{x}(k) = (\mathbf{W}_{M\times M})^H \tilde{\mathbf{a}}(k), \tag{2}$$

where $\mathbf{W}_{M\times M}$ denotes the $(M \times M)$ DFT matrix, its (i,l)-entry being $(\mathbf{W})_{i,l} = \frac{1}{\sqrt{M}} \exp(-j2\pi il/M)$. $(\cdot)^H$ denotes Hermitian transpose.

Prior to transmission, a cyclic prefix is prepended and removed after transmission. The received signal at subcarrier m, $m = 0, ..., M-1$, and symbol interval k, is given by

$$\tilde{r}_m(k) = \tilde{h}_m(k)\tilde{a}_m(k) + \tilde{n}_m(k), \tag{3}$$

where $\tilde{r}$, $\tilde{h}$ and $\tilde{n}$ denote the discrete Fourier transforms of the received signal, the channel impulse response and the additive noise. We assume that the channel frequency response (CFR) $\tilde{h}_m(k)$ is a complex-valued Gaussian random variable with a Jakes' power density spectrum. Furthermore, the additive white noise $\tilde{n}_m(k)$ characterizes a complex Gaussian random variable with variance $\sigma_m^2(k)$. The different noise variances for each symbol and subcarrier result from the coding process. For example, the preamble of a WLAN system has constant energy for all subcarriers and is not coded. By contrast the payload is passed through the coding process, which results in different signal-to-noise power ratios of the subcarriers. This includes the postamble symbol to be generated from the trailing payload bits.

Throughout this paper we assume perfect timing and frequency synchronization, and absence of phase noise.

3. Physical Layer

3.1 PHY Frame Format

The structure of the physical layer (PPDU: **P**hysical Layer Convergence Procedure **P**rotocol **D**ata **U**nit) is depicted in Fig. 1 [3], [4]. The PPDU symbols can be grouped in three parts: the preamble containing known uncoded symbols, the signal field transmitted with the most robust coding and modulation scheme, and the data section, which can employ different coding and modulation formats depending on the configuration.

Figure 1. PPDU structure of an IEEE 802.11a-like system.

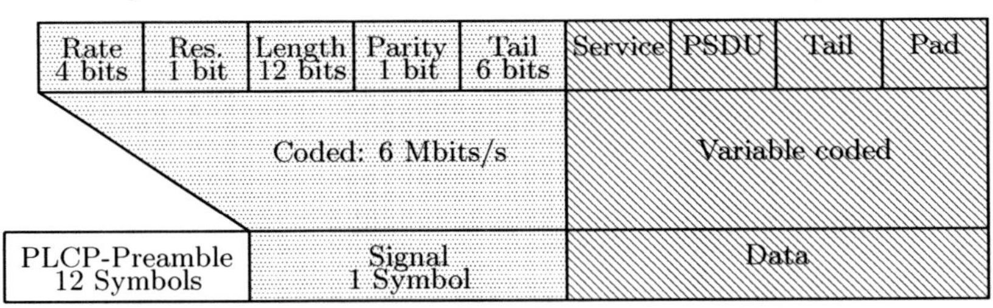

The preamble serves synchronization and channel estimation purposes. The signal field contains information about data rate, code rate, modulation format and the length of the following data section. Preamble and signal field are referred to as header and their number of symbols is denoted by B_{Head}. The data section, also called payload in the following, starts with the service field, which contains information about the scrambling sequence, see section 3.2, followed by the user data (PSDU: **P**hysical **S**ublayer Service **D**ata **U**nit). The data field is of variable length and is completed by N_{Tail} Tail bits of zero value for resetting the convolutional decoder at the receiver side. These are followed by further zero bits, called Pad bits, filling the frame to get an integer B of OFDM symbols, which is given by

$$B = \left\lceil \frac{N_{\mathrm{Service}} + N_{\mathrm{PSDU}} + N_{\mathrm{Tail}}}{N_{\mathrm{DBPS}}} \right\rceil + B_{\mathrm{Head}}. \tag{4}$$

Here, N_{Service} is the number of bits, used for the service field, which is needed for synchronization of the descrambler in the receiver and N_{PSDU} is the PSDU length. N_{DBPS} is the number of uncoded data bits per OFDM symbol. Then the number of Pad bits N_{Pad} can be calculated by

$$N_{\mathrm{Pad}} = (B - B_{\mathrm{Head}}) \cdot N_{\mathrm{DBPS}} - (N_{\mathrm{Service}} + N_{\mathrm{PSDU}} + N_{\mathrm{Tail}}). \tag{5}$$

3.2 Coding

In this section we give a short overview of the coding process, since its knowledge is required to understand the construction of the postamble described later in section 4. The encoding consists of the following four operations:

- Scrambler: a $N_{\mathrm{Scr}} = 127$ bit sequence is used for scrambling the data field. The generator polynomial is $g(x) = x^{(-7)} + x^{(-4)} + x^{(0)}$, where $[x^{(-1)}, ..., x^{(-7)}]$ are the fields of a shift register. $x^{(0)}$ is the unscrambled data bit and $x'^{(0)}$ the scrambled bit with

  $$x'^{(0)} = x^{(-4)} \oplus x^{(-7)}, \tag{6}$$

  where $\oplus$ denotes the modulo-2 addition. For initialization of the scrambler a pseudo-random sequence is used.

- Convolutional Coder: two generator polynomials are used for the coder, where $N_{\mathrm{CL}} = 7$ is the constraint length.

- Puncturer: a predefined pattern allows the selection of different code rates of 1/2, 2/3 and 3/4.

- Interleaver: the interleaver works OFDM symbol-wise, i.e. there is no interleaving across symbols.

The decoding process at the receiver is done in reverse order.

Note that the afore-mentioned signal field is always transmitted with the lowest available data rate (here 6 Mbits/s) to guarantee a reliable decoding at the receiver. This means that the modulation employed is BPSK and the code rate is 1/2.

4. Generation of Postamble

The main idea of this paper is to generate a complete OFDM symbol at the end of the data field for an improved channel estimation. Completing known sublayer frame bits like Tail and Pad bits to a full OFDM symbol results in only a small loss of efficiency compared to an extra symbol generated for channel estimation purposes only.

Interestingly, for all data rates there exist configurations, where Tail and Pad bits alone form already a complete OFDM symbol, see Tab. 1.

Table 1. Configurations, where Tail and Pad bits form a complete OFDM symbol.

Data rate (Mbits/s)	N_{DBPS}	*PSDU length (bytes)*	*Number of OFDM symbols in the Data field*
6	24	1,4,7,...,4093	2,3,4,...,1366
9	36	7,16,25,...,4093	3,5,7,...,911
12	48	4,10,16,...,4090	2,3,4,...,683
18	72	7,16,25,...,4093	2,3,4,...,456
24	96	10,22,34,...,4090	2,3,4,...,342
36	144	16,34,52,...,4084	2,3,4,...,228
48	192	22,46,70,...,4078	2,3,4,...,171
54	216	25,52,79,...,4075	2,3,4,...,152

Due to the memory of the convolutional coder we need N_{CL} additional known bits in front of the Tail bits to have full knowledge about the last OFDM symbol for the configurations given in Tab. 1.

For an arbitrary configuration it holds that $N_{\mathrm{Tail}} + N_{\mathrm{Pad}} \leq N_{\mathrm{DBPS}}$, with the equal sign for the configurations of Tab. 1. In order to obtain a full OFDM symbol to be utilized for channel estimation, we propose to replace the last N_{Extra} user data bits by zeros, where $N_{\mathrm{Extra}} = N_{\mathrm{DBPS}} + N_{\mathrm{CL}} - (N_{\mathrm{Tail}} + N_{\mathrm{Pad}})$, see Fig. 2.

Next we consider the determination of the postamble reference symbol from the bits. This is an issue, since the mapping of the bits to an OFDM

Figure 2. Proposed modification of the payload.

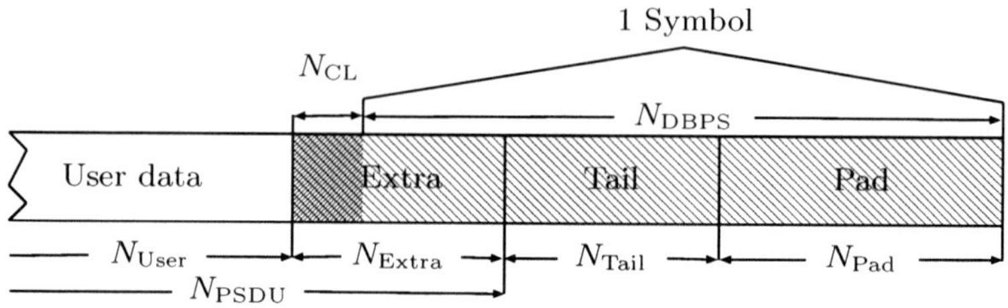

symbol depends on the selected modulation scheme and the random initialization sequence of the scrambler, which needs to be determined.

This can be accomplished as follows: First, the signal field is equalized and detected using the channel estimate obtained on the preamble, denoted by $\tilde{h}'_m(k_{\mathrm{Pre}})$. The temporal vicinity of the signal field to the preamble ensures proper equalization. Since the received signal field contains the information about the modulation scheme, the first symbol of the payload, which contains the service field, can now be detected, and the corresponding bits are obtained after deinterleaving, depuncturing and convolutional decoding.

The first 7 bits of the service field at the input to the scrambler in the transmitter are always set to zero. Using this knowledge and applying the descrambling operation according to eq. (6), $x^{(-7)} = x^{(-4)} \oplus x'^{(0)}$, where $x'^{(0)}$ contains the incoming data bits after demodulation, it is seen that after 7 steps the register fields $[x^{(-1)}, ..., x^{(-7)}]$ contain the scrambling initialization sequence we were looking for.

With this knowledge the transmitted trailing symbol of the payload can be computed simply by re-encoding the corresponding bits. This known symbol now serves as postamble on which a channel estimate $\tilde{h}'_m(k_{\mathrm{Post}})$ is calculated.

In section 6.4 we compute the loss in user data rate due to this postamble creation.

5. Interpolation

For the design of the WF we assume that the time-variance of the CFR is described by a Jakes' power density spectrum. The correlation between two symbols of distance $\Delta k T_S$ is then given by

$$\varphi(\Delta k) = J_0(2\pi f_d \Delta k T_S), \tag{7}$$

where J_0 denotes the modified Bessel function of first kind and 0-th order, f_d the Doppler frequency and T_S the duration of an OFDM symbol [5], [6].

With the signal model given in eq. (3), Maximum-Likelihood (ML) estimates of the CFR on the pre- and postamble can be obtained by

$$\tilde{h}'_m(k) = \frac{\tilde{r}_m(k)\tilde{a}^*_m(k)}{|\tilde{a}_m(k)|^2}. \tag{8}$$

where $k \in [k_{\mathrm{Pre}}, k_{\mathrm{Post}}]$. Now these ML estimates are considered as "observations" for the WF design according to the following observation model:

$$\tilde{h}'_m(k) = \tilde{h}_m(k) + \tilde{w}_m(k), \tag{9}$$

where $\tilde{w}_m(k)$ is a white Gaussian noise process.

The WF along the time direction is given by

$$\tilde{h}^{(WF)}_m(k) = \mathbf{b}_m(k)\tilde{\mathbf{h}}'_m, \tag{10}$$

where $\tilde{\mathbf{h}}'_m = [\tilde{h}'_m(k_{\mathrm{Pre}})\ \tilde{h}'_m(k_{\mathrm{Post}})]^T$. Here, $(\cdot)^T$ denotes transpose. $\mathbf{b}_m(k)$ is the solution of the Wiener-Hopf equation

$$\mathbf{b}_m(k)\mathbf{R}_m(k) = \mathbf{p}(k) \tag{11}$$

with the autocorrelation matrix

$$\mathbf{R}_m(k) = \begin{bmatrix} \varphi(0) + \sigma^2_m(k_{\mathrm{Pre}}) & \varphi(k_{\mathrm{Post}} - k_{\mathrm{Pre}}) \\ \varphi(k_{\mathrm{Pre}} - k_{\mathrm{Post}}) & \varphi(0) + \sigma^2_m(k_{\mathrm{Post}}) \end{bmatrix} \tag{12}$$

and the cross-correlation vector $\mathbf{p}(k) = [\varphi(k - k_{\mathrm{Pre}})\ \varphi(k - k_{\mathrm{Post}})]$, which is independent of the subcarrier index m. For simplicity and to save computations, we assume equal noise variances for both, pre- and postamble, i.e. $\sigma^2 = \sigma^2_m(k_{\mathrm{Pre}}) = \sigma^2_m(k_{\mathrm{Post}})\ \forall m$.

6. Results

6.1 Known Channel Model

In this section we present experimental results using the physical layer structure of an 802.11a-like OFDM system. The burst consists of 139 symbols including 12 preamble symbols for channel estimation and synchronisation, 1 symbol for the signal field and the data field comprises 126 symbols. 48 of the $M = 64$ available subcarriers carry user data, and the modulation employed is 16-QAM at a coding rate of 1/2. The bandwidth in the 5 GHz Band is chosen to be 20 MHz. The terminal velocity, which determines the cutoff frequency f_d of the Jakes' spectrum, is 30 km/h, which we assume to be perfectly known in the receiver.

The Rayleigh fading channel with six independent propagation paths has a power loss profile of $[-6.5, -7.5, -8.7, -9.8, -10.9, -12.0]$ dB and delays of $[0, 50, 100, 150, 200, 250]$ ns, which simulates a typical outdoor scenario.

Figs. 3 a) and b) show the mean square estimation error (MSEE) of the estimated CFR and the bit error rate (BER) of the decoder.

Figure 3. Comparison of different channel estimation techniques.

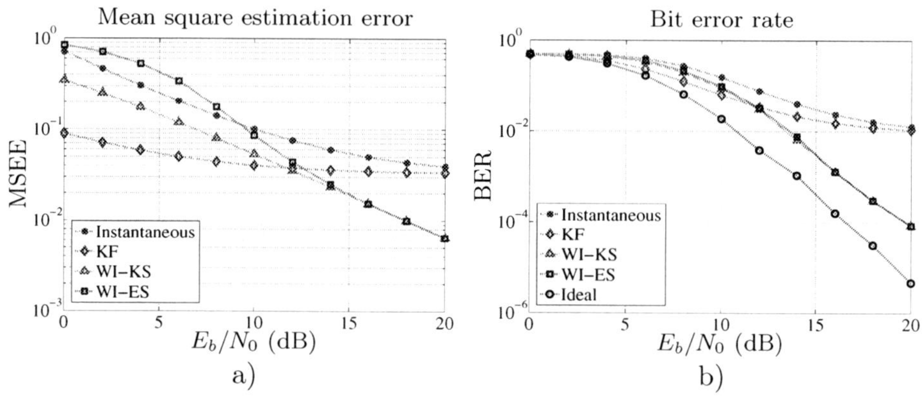

The Wiener Interpolation Filter, described by eqs. (7) - (12), with perfectly known scrambler initialization sequence (WI-KS) and with detecting the sequence on the data (WI-ES) as presented in section 4, are compared with the instantaneous ML estimate $\tilde{h}'_m(k_{\mathrm{Pre}})$ obtained on the preamble (Instantaneous) and a time domain Kalman Filter (KF), which also solely operates on the preamble [7]. A lower bound for the BER is obtained by assuming the CFR to be perfectly known (Ideal). It can be seen that the WF outperforms the KF for E_b/N_0 greater than 11 dB. Different tests have shown that this threshold decreases with the terminal velocity. For higher E_b/N_0 the KF and the ML estimator are not very effective, because their curves run in saturation. By contrast, WI-KS and WI-ES approximately follow the ideal achievable BER. For values lower than 11 dB there exists a significant difference between WI-KS and WI-ES, resulting from the poor ML estimates on the preamble. Therefore, the detection of the scrambler initialization sequence is error-prone, which can cause incorrect construction of the postamble and consequently a poor performance of the Wiener Interpolation Filter.

6.2 Robustness

In the previous section we assumed the Doppler frequency (or terminal velocity) and noise variance to be known. In the following experiments we investigate the sensitivity of the Wiener interpolation towards false assumptions on these parameters. Figs. 4 a) and b) show the MSEE of the different estimators:

a) As a function of the E_b/N_0 assumed by the receiver, while the true channel-sided E_b/N_0 is 14 dB.

b) As a function of terminal velocity (and thus max. Doppler frequency), while the true terminal velocity is 30 km/h at $E_b/N_0 = 14$ dB.

Figure 4. Sensitivity of channel estimation towards wrong assumptions on parameter values.

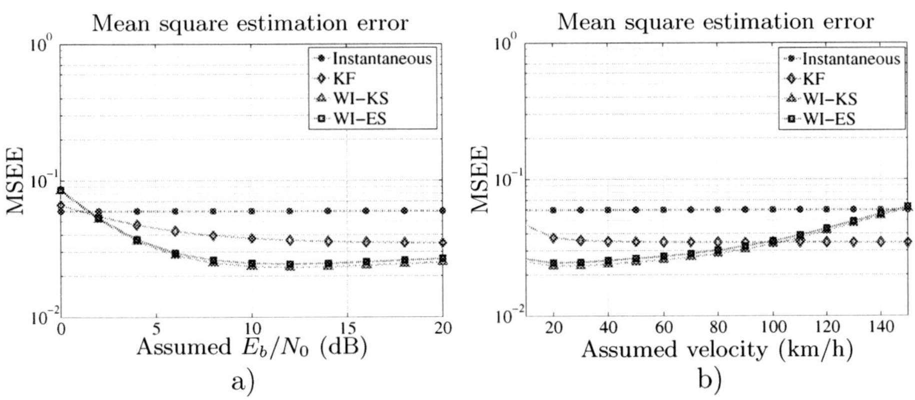

The results indicate that the WF is not very sensitive to wrong assumptions on noise variance and terminal velocity.

6.3 Computational Complexity

The postamble creation requires that the last part of the data field has to be re-encoded. However, this step needs to be done only once per session, unless the modulation format changes. For the computation of the Wiener Interpolation Filter coefficients, the matrix given in eq. (12) has to be inverted. If we assume a constant Doppler frequency and payload length, this matrix is constant however and needs only be inverted once.

6.4 Efficiency

The ratio $v = N_{\text{Extra}}/N_{\text{PSDU}}$ of user data bits to be replaced by known bits to the PSDU length is a measure of loss of efficiency.

In the most favorable cases given in Tab. 1 only N_{CL} extra bits are needed. For a data rate of 24 Mbits/s, when the PSDU consists of 1498 bytes, the efficiency loss is $v = 0.07\%$. In the worst case N_{DBPS} bits are needed, resulting in 0.79% efficiency loss. In view of the potential bit error rate performance improvement seen in Fig. 3 b), notably for large E_b/N_0, our proposal seems to be justified.

7. Conclusions and Outlook

We described a method to construct a postamble symbol in OFDM block-oriented transmission from the Tail and Pad bits of the payload and the replacement of a few user data bits by zero bits. Channel estimation is then carried out by a WF, which interpolates between the ML estimates of the pre- and postamble. Compared to channel estimation techniques based on smoothing across successive preambles, the proposed scheme does not extend processing beyond a single frame and therefore reduces processing delay and complexity. The experimental results showed significant bit error rate improvements for large E_b/N_0 compared to channel estimation schemes utilizing only the preamble.

In future research we try to improve the performance by inserting non-zero bits to increase the energy of the generated postamble and taking into account KF based estimates instead of instantaneous estimates. The effects of non-perfect synchronization and phase noise have to be investigated, too.

References

[1] Z. Wang and G. Giannakis, *Wireless multicarrier communications,* IEEE Signal Processing Magazine, vol. 17, no. 3, pp. 29-48, 2000.

[2] A. R. S. Bahai, B. R. Saltzberg and M. Ergen, *Multi-Carrier Digital Communications: Theory and Applications of OFDM,* Springer, 2004.

[3] IEEE, *Part11: Wireless LAN Medium Access Control (MAC) and Physical Layer (PHY) specifications - High-speed Physical Layer in the 5 GHz Band,* IEEE Std. 802.11a-1999.

[4] B. O'Hara and A. Petrick, *IEEE 802.11 Handbook - A Designer's Companion,* Standards Information Network IEEE Press, 2001.

[5] K.-D. Kammeyer, *Nachrichtenuebertragung,* Teubner, Ed. 3, 2004.

[6] K.-D. Kammeyer. *Berichtigungen und Ergaenzungen zum Buch Nachrichtenuebertragung,*
http://www.ant.uni-bremen.de/pub/books/nue/correction/index.html.

[7] T. Roman, M. Enescu and V. Koivunen, *Time domain method for tracking dispersive channels in OFDM systems,* Proc. IEEE VTC Spring 2003, Jeju, Korea.

JOINT DATA DETECTION AND CHANNEL ESTIMATION FOR UPLINK MC-CDMA SYSTEMS OVER FREQUENCY SELECTIVE CHANNELS

Erdal Panayirci[1*], Hakan Doğan[2], Hakan A. Çirpan[2] and Bernard H. Fleury[3]

[1] *Kadir Has University, Department of Electronics Enginering, Cibali 34083, Istanbul, Turkey*
epanayirci@khas.edu.tr

[2] *Istanbul University, Department of Electrical and Electronics Engineering, Avcilar 34850, Istanbul, Turkey*
hdogan@istanbul.edu.tr,hcirpan@istanbul.edu.tr

[3] *Aalborg University, Section Navigation and Communications, Department of Electronic Systems, Fredrik Bajers Vej 7A3 DK-9000 Aalborg, Denmark and Forschungszentrum Telekommunikation Wien (ftw.) Donau-City-Strasse 1 A-1220 Vienna, Austria*
bfl@kom.auc.dk

Abstract This paper is concerned with joint channel estimation and data detection for uplink multicarrier code-division multiple-access (MC-CDMA) systems in the presence of frequency fading channel. The detection and estimation, implemented at the receiver, are based on a version of the expectation maximization (EM) algorithm which is very suitable for the the multicarrier signal formats. Application of the EM-based algorithm to the problem of iterative data detection and channel estimation leads to a receiver structure that also incorporates a partial interference cancelation. Computer simulations show that the proposed algorithm has excellent BER end estimation performance.

Keywords: Joint data detection and channel estimation, MC-CDMA Systems, EM algorithm, Superimposed signals

*This research has been conducted within the NEWCOM Network of Excellence in Wireless Communications funded through the EC 6th Framework Programme and the Research Fund of Istanbul University under Projects T-856/02062006. This work was also supported in part by the Turkish Scientific and Technical Research Institute(TUBITAK) under Grant 104E166.

S. Plass et al. (eds.), Multi-Carrier Spread Spectrum 2007, 397–406.

1. Introduction

MC-CDMA transmission through the frequency-selective fading channels causes the signal-to-noise ratio (SNR) degradation and the occurrence of the multiple-access interference (MAI). To eliminate or reduce the resulting performance degradations, equalization and detection of the received signal can be performed at the receiver based on the complete channel information. In conventional MC-CDMA systems, MAI mitigation is accomplished at the receiver using single-user or multi-user detection schemes [1]. However, multiuser detection scheme is known to increase the bandwidth efficiency of the system drastically, although its detection complexity grows exponentially with the number of users and the number of multipaths, which makes it infeasible to implement. Several suboptimal detection techniques have proposed in literature such as linear multiuser detection[2] and iterative cancelation of MAI, either in a successive or parallel way in the received signal prior to data detection [3]. Therefore data detection and channel estimation should be performed perfectly for a good initialization of the interference cancelation detector.

In this paper we consider an efficient iterative algorithm based on the Expectation-Maximization (EM) technique for multi-user data detection and channel estimation, jointly for uplink MC-CDMA systems in the presence of frequency selective fading channels. As will be seen shortly, a partial parallel interference cancelation (PIC) is incorporated into the resulting detection algorithm [4]. The work is an extension of [5] in which joint data detection and channel estimation of downlink DS-CDMA systems were considered based on an EM algorithm in the presence of flat Rayleigh channels. The channel estimation becomes more challenging for uplink systems since each channel between every user and the base station must be estimated rather that estimating a single channel in case of a down-link transmission.

Notation: Vectors (matrices) are denoted by boldface lower (upper) case letters; all vectors are column vectors; $(.)^*$, $(.)^T$ and $(.)^H$ denote the conjugate, transpose and conjugate transpose, respectively; $\|.\|$ denotes the Frobenius norm; $\mathbf{I}_L$ denotes the $L \times L$ identity matrix; $diag\{.\}$ denotes a diagonal matrix; and finally, $tr\{.\}$ denotes the trace of a matrix.

2. Signal Model

We consider a baseband MC-CDMA uplink system with P sub-carriers and K mobile users which are simultaneously active. For the kth user, each transmit symbol is modulated in the frequency domain by means of a $P \times 1$ specific spreading sequence $\mathbf{c}_k$. After transforming by a P-point IDFT and parallel-to-serial (P/S) conversion, a cyclic prefix (CP) is inserted of length equal to at least the channel memory (L). In this work, to simplify the notation, it is assumed that the spreading factor equals to the number of sub-carriers and all users have the same spreading factor. Finally, the signal is transmitted through a multipath channel with impulse response

$$h_k(t) = \sum_{k=1}^{L} g_{k,l}\delta(t - \tau_{k,l}) \quad (1)$$

where L is the number of paths in the kth user's channel; g_{kl} and τ_{kl} are, respectively, the complex fading processes and the delay of the kth user's lth path. It is assumed that only the second-order statistics of the fading processes are known to the receiver. Also, fading can vary from symbol to symbol but remains constant over a symbol interval.

At the receiver, the received signal is first serial-to-parallel (S/P) converted, then CP is removed and DFT is then applied to the received discrete time signal to obtain the received vector expressed as

$$\mathbf{y}(m) = \sum_{k=1}^{K} b_k(m)\mathbf{C}_k\mathbf{F}\mathbf{h}_k + \mathbf{w}(m) \quad m = 1, 2, ..., M \quad (2)$$

where $b_k(m)$ denotes data sent by the user k within the mth symbol; $\mathbf{C}_k = \text{diag}(\mathbf{c}_k)$ with $\mathbf{c}_k = [c_{k1}, c_{k1}, ..., c_{kP}]^T$ where each chip, c_{ik}, takes values in the set $\{-\frac{1}{\sqrt{P}}, \frac{1}{\sqrt{P}}\}$ denoting the kth user's spreading code ; $\mathbf{F} \in \mathbb{C}^{P\times L}$ denotes the DFT matrix with the (k, l)th element given by $\frac{1}{\sqrt{P}}e^{-j2\pi kl/P}$; and $\mathbf{w}(m)$ is the $P \times 1$ zero-mean, i.i.d. Gaussian vector that models the additive noise in the P tones, with variance $\sigma^2/2$ per dimension.

Suppose M symbols are transmitted. We stack $\mathbf{y}(m)$ as $\mathbf{y}$=[$\mathbf{y}^T(1), ...$ $, \mathbf{y}^T(M)]^T$. Then the received signal model can be written as

$$\mathbf{y} = \begin{bmatrix} b_1(1)\mathbf{C}_1\mathbf{F} & \cdots & b_K(1)\mathbf{C}_K\mathbf{F} \\ \vdots & \ddots & \vdots \\ b_1(M)\mathbf{C}_1\mathbf{F} & \cdots & b_K(M)\mathbf{C}_K\mathbf{F} \end{bmatrix} \begin{bmatrix} \mathbf{h}_1 \\ \vdots \\ \mathbf{h}_K \end{bmatrix} + \begin{bmatrix} \mathbf{w}(1) \\ \vdots \\ \mathbf{w}(M) \end{bmatrix} \quad (3)$$

and can be rewritten in more succinct form

$$\mathbf{y} = \mathbf{A}\mathbf{h} + \mathbf{w} \quad (4)$$

By using the assumption; $\mathbf{h}_k \sim N(0, \mathbf{\Sigma}_{\mathbf{h}_k})$ where $\mathbf{\Sigma}_{\mathbf{h}_k} = E[\mathbf{h}_k\mathbf{h}_k^\dagger]$, we have $\mathbf{h} \sim N(0, \mathbf{\Sigma}_\mathbf{h})$ with $\mathbf{\Sigma}_\mathbf{h} = \text{diag}[\mathbf{\Sigma}_1, \cdots, \mathbf{\Sigma}_K]$.

3. Joint Data Detection and Channel Estimation

The problem of interest is to derive an iterative algorithm based on the EM technique for data detection and channel estimation jointly employing the signal model given by (2). Since the EM method has been studied and applied to a number of problems in communications over the years, the details of the algorithm will not be presented in this paper. The reader is suggested to read [6] for a general exposition to EM algorithm and [8] for its application to the estimation problem related to the work herein. The suitable approach for applying the EM algorithm for the problem at hand is to decompose the received signal in (2) into the sum [7]

$$\mathbf{y}(m) = \sum_{k=1}^{K} \mathbf{x}_k(m), \quad m = 1, 2 \cdots, M \tag{5}$$

where

$$\mathbf{x}_k(m) \quad = \quad b_k(m)\mathbf{C}_k\mathbf{F}\mathbf{h}_k + \mathbf{w}_k(m). \tag{6}$$

$\mathbf{x}_k(m)$ represents the received signal component transmitted by the kth user through the channel with impulse response $\mathbf{h}_k$. The Gaussian noise vector, $\mathbf{w}_k(m)$ ýn (6) represents the portion of $\mathbf{w}(m)$ in the decomposition defined by $\sum_{k=1}^{K} \mathbf{w}_k(m) = \mathbf{w}(m)$, whose variance is $N_0\beta_k$. The coefficients β_k determine the part of the noise power of $\mathbf{w}(m)$ assigned to $\mathbf{x}_k(m)$, satisfying $\sum_{k=1}^{K} \beta_k = 1, 0 \leq \beta_k \leq 1$.

The problem now is to estimate the transmitted symbols $\mathbf{b} = \{b_k(m)\}_{k=1, m=1}^{K,M}$ and the complex channel responses $\mathbf{h}_k$ for each user based on observed data $\mathbf{y}$. In the EM algorithm, we view the observed data $\mathbf{y}$ as the "incomplete" data, and define the "complete" data as $\chi = \{(\mathbf{x}_1, \mathbf{h}_1), (\mathbf{x}_1, \mathbf{h}_1), ...,$ $(\mathbf{x}_K, \mathbf{h}_K)\}$ where $\mathbf{x}_k = [\mathbf{x}_k(1), ..., \mathbf{x}_k(M)]^T$ for $k = 1, 2, \cdots, K$.

The EM algorithm: Given the complete data set as $\chi = \{(\mathbf{x}_1, \mathbf{h}_1), ...$ $(\mathbf{x}_2, \mathbf{h}_2), ..., (\mathbf{x}_K, \mathbf{h}_K)\}$, the loglikelihood function of the parameter vector to be estimated $\mathbf{b}$ can be expressed as

$$\log p(\chi|\mathbf{b}) = \sum_{k=1}^{K} \log p(\mathbf{x}_k, \mathbf{h}_k|\mathbf{b}_k) \tag{7}$$

where

$$\log p(\mathbf{x}_k, \mathbf{h}_k|\mathbf{b}_k) = \log p(\mathbf{x}_k|\mathbf{b}_k, \mathbf{h}_k) + \log p(\mathbf{h}_k|\mathbf{b}_k) \tag{8}$$

and, $\mathbf{b}_k = [b_k(1), b_k(2), \cdots, b_k(M)]$. We neglect the $\log p(\mathbf{h}_k|\mathbf{b}_k)$ term in (7) since the data sequence $\mathbf{b}_k$ and $\mathbf{h}_k$ are independent of each other.

The first step to implement the EM algorithm, called the *Expectation Step (E-Step)*, is the computation of the average log-likelihood function. The conditional expectation is taken over χ given the observation $\mathbf{y}$ and that b equals its estimate calculated at ith iteration.

$$Q\left(\mathbf{b}|\mathbf{b}^{(i)}\right) = E\left\{\log p(\chi|\mathbf{b}|\mathbf{y}, \mathbf{b}^{(i)}\right\} \tag{9}$$

Taking into account the special form of $\log p(\boldsymbol{\chi}|\mathbf{b})$ in (7), Eq. (8) can be decomposed as

$$Q\left(\mathbf{b}|\mathbf{b}^{(i)}\right) = \sum_{k=1}^{K} Q_k(\mathbf{b}_k|\mathbf{b}^{(i)}) \tag{10}$$

where

$$\begin{aligned} Q_k(\mathbf{b}_k|\mathbf{b}^{(i)}) &= E\left\{\log p(\mathbf{x}_k, \mathbf{h}_k|\mathbf{b}_k)|\mathbf{y}, \mathbf{b}^{(i)}\right\} & (11)\\ &= E\left\{\log p(\mathbf{x}_k|\mathbf{b}_k, \mathbf{h}_k)|\mathbf{y}, \mathbf{b}^{(i)}\right\}. & (12)\end{aligned}$$

Note that (11) follows from (8).

Neglecting the terms independent of $\mathbf{b}_k$, From (6), $\log p(\mathbf{x}_k|b_k, \mathbf{h}_k)$ can be calculated as

$$\log p(\mathbf{x}_k|\mathbf{b}_k, \mathbf{h}_k) \sim \sum_{m=1}^{M} \Re\{b_k(m)\mathbf{h}_k^{\dagger}\mathbf{F}^{\dagger}\mathbf{C}_k^T\mathbf{x}_k(m)\}. \tag{13}$$

Inserting (13) in (12), we have for $Q_k(\mathbf{b}_k|\mathbf{b}^{(i)})$

$$Q_k(\mathbf{b}_k|\mathbf{b}^{(i)}) = \sum_{m=1}^{M} \Re\{b_k(m)(\mathbf{h}_k^{\dagger}\mathbf{F}^{\dagger}\mathbf{C}_k^T\mathbf{x}_k(m))^{[i]}\} \tag{14}$$

where

$$(\mathbf{h}_k^{\dagger}\mathbf{F}^{\dagger}\mathbf{C}_k^T\mathbf{x}_k(m))^{[i]} \triangleq E\left\{\mathbf{h}_k^{\dagger}\mathbf{F}^{\dagger}\mathbf{C}_k^T\mathbf{x}_k(m)|\mathbf{y}, \mathbf{b}^{(i)}\right\} \tag{15}$$

$(\mathbf{h}_k^{\dagger}\mathbf{F}^{\dagger}\mathbf{C}_k^T\mathbf{x}_k(m))^{[i]}$ can be calculated by applying the conditional expectation rule as

$$\begin{aligned} (\mathbf{h}_k^{\dagger}\mathbf{F}^{\dagger}\mathbf{C}_k^T\mathbf{x}_k(m))^{[i]} &= E\{\mathbf{h}_k^{\dagger}E(\mathbf{F}^{\dagger}\mathbf{C}_k^T\mathbf{x}_k(m)|\mathbf{y}, \mathbf{b}^{(i)}, \mathbf{h})|\mathbf{y}, \mathbf{b}^{(i)}\}\\ &= E\{\mathbf{h}_k^{\dagger}\mathbf{F}^{\dagger}\mathbf{C}_k^T E(\mathbf{x}_k(m)|\mathbf{y}, \mathbf{b}^{(i)}, \mathbf{h})|\mathbf{y}, \mathbf{b}^{(i)}.\} & (16)\end{aligned}$$

The conditional distribution of $\mathbf{x}_k(m)$ given $\mathbf{y}$, $\mathbf{h}$ and $\mathbf{b} = \mathbf{b}^{(i)}$ is Gaussian with mean

$$\begin{aligned} E(\mathbf{x}_k(m)|\mathbf{y}, \mathbf{b}^{(i)}, \mathbf{h}) &= (b_k(m))^{[i]}\mathbf{C}_k\mathbf{F}\mathbf{h}_k & (17)\\ &+ \beta_k\left(\mathbf{y}(m) - \sum_{j=1}^{K}(b_j(m))^{[i]}\mathbf{C}_j\mathbf{F}\mathbf{h}_j\right)\end{aligned}$$

where $(b_k(m))^{[i]} \triangleq E(b_k(m)|\mathbf{y}, \mathbf{b}^{(i)}, \mathbf{h})$. Inserting (17) in (15) and using the properties $\mathbf{F}^{\dagger}\mathbf{F} = \mathbf{I}$ and $\mathbf{C}_k^T\mathbf{C}_k = \frac{1}{P}\mathbf{I}$. we can rewrite (15) as

$$(\mathbf{h}_k^{\dagger}\mathbf{F}^{\dagger}\mathbf{C}_k^T\mathbf{x}_k(m))^{[i]} = \frac{1}{P}(b_k(m))^{[i]}E\{\mathbf{h}_k^{\dagger}\mathbf{h}_k|\mathbf{y},\mathbf{b}^{(i)}\} + \beta_k E\{\mathbf{h}_k^{\dagger}|\mathbf{y},\mathbf{b}^{(i)}\}\mathbf{F}^{\dagger}\mathbf{C}_k^T\mathbf{y}(m)$$

$$-\beta_k \sum_{j=1, j\neq k}^{K} (b_j(m))^{[i]} E\{\mathbf{h}_k^\dagger \mathbf{F}^\dagger \mathbf{C}_k^T \mathbf{C}_j \mathbf{F} \mathbf{h}_j | \mathbf{y}, \mathbf{b}^{(i)}\}. \quad (18)$$

On the other hand, since $\mathbf{w} \sim N(0, \mathbf{\Sigma}^2 \mathbf{I})$ and the prior pdf of $\mathbf{h}$ is chosen as $\mathbf{h} \sim N(0, \mathbf{\Sigma_h})$, we can write the conditional pdf's of $\mathbf{h}$ given $\mathbf{y}$ and $\mathbf{b}^{(i)}$ as

$$\begin{aligned} p(\mathbf{h}|\mathbf{y}, \mathbf{b}^{(i)}) &\sim p(\mathbf{y}|\mathbf{h}, \mathbf{b}^{(i)}) p(\mathbf{h}) \\ &\sim \exp\left[-\frac{1}{\sigma^2}(\mathbf{y} - \mathbf{A}\mathbf{h})^\dagger (\mathbf{y} - \mathbf{A}\mathbf{h}) - \mathbf{h}^\dagger \mathbf{\Sigma_h}^{-1} \mathbf{h}\right]. \end{aligned} \quad (19)$$

After some algebra it can be shown that

$$p(\mathbf{h}|\mathbf{y}, \mathbf{b}^{(i)}) \sim N(\boldsymbol{\mu}_\mathbf{h}^{(i)}, \mathbf{\Sigma}_\mathbf{h}^{(i)}) \quad (20)$$

where

$$\begin{aligned} \boldsymbol{\mu}_\mathbf{h}^{(i)} &= \frac{1}{\sigma^2} \mathbf{\Sigma}_\mathbf{h}^{(i)} \mathbf{A}^{(i)\dagger} \mathbf{y} \\ \mathbf{\Sigma}_\mathbf{h}^{(i)} &= \left[\mathbf{\Sigma_h}^{-1} + \frac{1}{\sigma^2} (\mathbf{A}^{(i)})^\dagger \mathbf{A}^{(i)}\right]^{-1} \end{aligned} \quad (21)$$

and the matrix $\mathbf{A}$ is defined in (3)and (4).

Now let us compute the terms on the right hand side of Eq. (17). We calculate $E\{\mathbf{h}_k^\dagger \mathbf{h}_k | \mathbf{y}, \mathbf{b}^{(i)}\}$ as

$$(|\mathbf{h}_k\|^2)^{[i]} \triangleq E\{\mathbf{h}_k^\dagger \mathbf{h}_k | \mathbf{y}, \mathbf{b}^{(i)}\} = \mathrm{tr}\left[\mathbf{\Sigma}_\mathbf{h}^{(i)}[k,k] + \boldsymbol{\mu}_h^{(i)}[k] \boldsymbol{\mu}_h^{(i)}[k]^\dagger\right] \quad (22)$$

where $\mathbf{\Sigma_h}[i,j]$ denotes the (i,j)th element of the matrix $\mathbf{\Sigma_h}$ Second expectation in (18) can be computed as

$$(\mathbf{h}_k)^{[i]} \triangleq E\{\mathbf{h}_k | \mathbf{y}, \mathbf{b}^{(i)}\} = \boldsymbol{\mu}_h^{(i)}[k]. \quad (23)$$

Finally, to calculate the last expectation $E\{\mathbf{h}_k^\dagger \mathbf{F}^\dagger \mathbf{C}_k^T \mathbf{C}_j \mathbf{F} \mathbf{h}_j | \mathbf{y}, \mathbf{b}^{(i)}\}$ in (18), define $\mathbf{\Psi}_j \triangleq \mathbf{C}_j \mathbf{F}$, $\mathrm{s}_j \triangleq \mathbf{\Psi}_j \mathbf{h}_j$. It then follows that

$$\mathbf{\Sigma}_s^{(i)} \triangleq E[\mathrm{s}\mathrm{s}^\dagger | \mathbf{y}, \mathbf{b}^{(i)}] = E[\mathbf{\Psi} \mathbf{h} \mathbf{h}^\dagger \mathbf{\Psi}^\dagger | \mathbf{y}, \mathbf{b}^{(i)}] = \mathbf{\Psi} \mathbf{\Sigma}_\mathbf{h}^{(i)} \mathbf{\Psi}^\dagger. \quad (24)$$

Therefore,

$$E\{\mathbf{h}_k^\dagger \mathbf{F}^\dagger \mathbf{C}_k^T \mathbf{C}_j \mathbf{F} \mathbf{h}_j | \mathbf{y}, \mathbf{b}^{(i)}\} = E[\mathrm{s}^\dagger \mathrm{s} | \mathbf{y}, \mathbf{b}^{(i)}] = \mathrm{tr}\left[\mathbf{\Sigma}_s^{(i)}[k,j] + \boldsymbol{\mu}_s^{(i)}[k] \boldsymbol{\mu}_s^{(i)}[j]^\dagger\right] \quad (25)$$

where, $\boldsymbol{\mu}_s^{(i)} \triangleq \boldsymbol{\Psi}\boldsymbol{\mu}_h^{(i)}$

Maximization-Step (M-Step): The second step to implement the EM algorithm is the *M-Step* where the parameter $\mathbf{b}$ is updated at the $(i+1)$th iteration according to

$$\mathbf{b}^{(i+1)} = \arg\max_{\mathbf{b}} Q(\mathbf{b}|\mathbf{b}_i) = \sum_{k=1}^{K} Q_k(\mathbf{b}_k|\mathbf{b}^{(i)}). \quad (26)$$

M-Step can be performed by maximizing the terms $Q_k(\mathbf{b}_k|\mathbf{b}^{(i)})$ individually, as follows

$$\mathbf{b}_k^{(i+1)} = \arg\max_{\mathbf{b}_k} Q_k(\mathbf{b}_k|\mathbf{b}_i) \quad (27)$$

Moreover, when no coding is used, it follows from (27) that each component of $\mathbf{b}_k^{(i+1)}$ can be separately obtained by maximizing the corresponding summation in the right-hand expression, as follows

$$b_k^{(i+1)}(l) = \mathrm{sgn}\left[\Re\{(\mathbf{h}_k^{\dagger}\mathbf{F}^{\dagger}\mathbf{C}_k^T\mathbf{x}_k(m))^{[i]}\}\right]. \quad (28)$$

We finally have

$$\begin{aligned} b_k^{i+1}(l) &= \mathrm{sgn}\left[\Re\left\{\frac{1}{P}b_k^{[i]}(m)\|\boldsymbol{\mu}_h^{(i)}[k]\|^2(1-\beta_k)\right.\right. \\ &+ \left.\left.\beta_k(\mathbf{h}_k^{(i)\dagger}\boldsymbol{\Psi}_k^T\left[\mathbf{y}(m) - \sum_{j=1,j\neq k}^{K} b_j^{[i]}(m)\boldsymbol{\Psi}_j\boldsymbol{\mu}_h^{(i)}[j]\right]\right\}\right]. \quad (29) \end{aligned}$$

As a conclusion, Equation (29) can be interpreted as joint channel estimation and data detection with partial interference cancelation. At each incretion step during data detection, the interference reduced signal is fed into a single user receiver consisting of a conventional coherent detector. As a result, a K-user optimization problem have been decomposed into K independent optimization problems which can e solved in a computationally feasible way. Finally we remark that this paper is an extension of the work presented by Fleury to the problem of joint channel estimation and data detection for the uplink multi-carrier CDMA systems operating in the presence of the frequency selective channels.

4. Simulations

In this section, performance of uplink MC-CDMA system based on proposed receiver is simulated for frequency selective channels. In simulations, we assume that all users are received the same power. The orthogonal Walsh sequences selected as spread code and processing gain equals to the number of subcarriers Nc = 16. The number of users is selected as U=8 and each user send frame ,that is composed of T preamble bits, and F data bits, over mobile fading channel. Wireless channels between mobiles antennas and receiver antenna are assumed to be complex gaussian channel of length L and it has distribution $N(0, Ch)$. For all simulations weight coefficients in (28) are chosen to be equal, i.e., $B_k = 1/U$

At the receiver, initial MMSE channel estimate is obtained by using T preamble bits while channel covariance matrix C_h is assumed to be know. Initial MMSE estimate of F data bits is computed from the observation of $\mathbf{y}$ while assuming we have estimated channel coefficients. We refer to this method for obtaining $\mathbf{h}$ and $\mathbf{b}$ as MMSE separate detection and estimation (MMSE-SDE). In the simulations, if the output of the (MMSE-SDE) is applied to paralel interference cancellation (PIC) receiver ,that we will compare with EM-JDE, is referred as Combined MMSE-PIC receiver. Moreover, we also simulated MMSE and Combined MMSE-PIC detectors for the perfect channel state information cases that are referred as CSI-MMSE and CSI-Combined MMSE-PIC respectively.

Fig. 1 compares the BER performance of the proposed EM-JDE approach with MMSE-SDE, Combined-MMSE-PIC, CSI-MMSE and CSI-Combined MMSE-PIC. For fair comparison we simulated Combined-MMSE-PIC, CSI-Combined MMSE-PIC and EM-JDE receivers for three of iterations. It is observed that the proposed EM-JDE outperforms the MMSE-SDE,Combined-MMSE-PIC as well as CSI-MMSE and approaches the CSI-Combined MMSE-PIC for higher Eb/No values.

To investigate initialization effect on the EM-JDE and Combined-MMSE-PIC, we also demonstrate BER performance of both algorithms for different preamble bits in Fig. 2. Lower preamble bits provide poor initial estimates, resulting in a more BER performance degradation in Combined-MMSE-PIC as compared to EM-JDE systems. On the other for higher preamble bits it was observed that performance difference between Combined-MMSE-PIC and EM-JDE decreases because of considerably good initial estimates of channel coefficients.

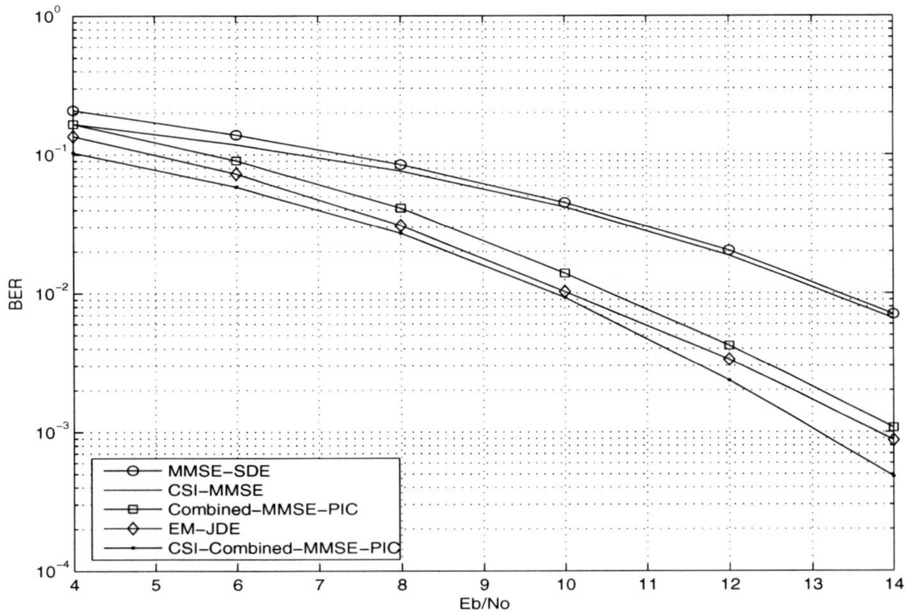

Figure 1. BER performance in frequency selective channels (T=4,F=40,L=8,U=8)

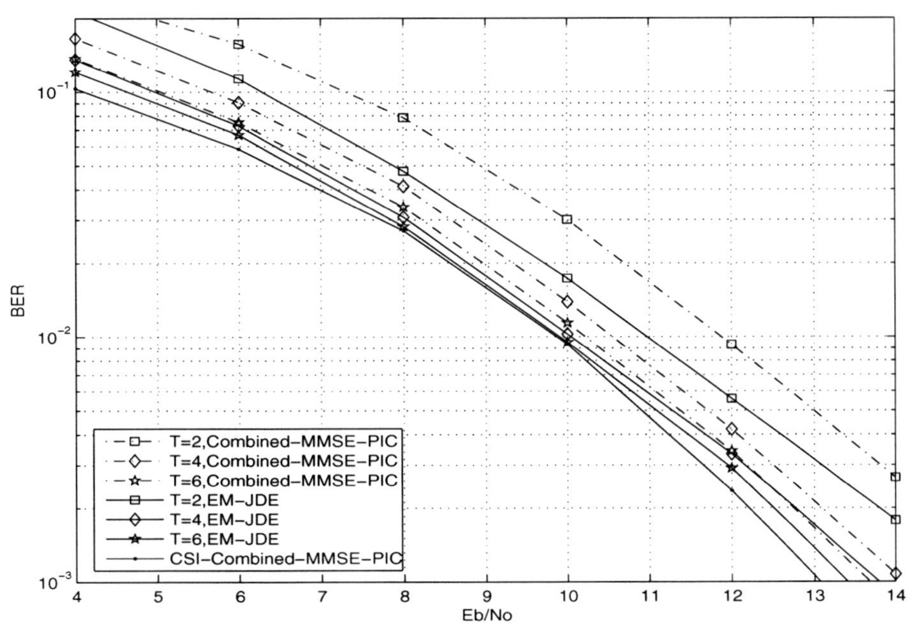

Figure 2. Behavior of the BER performance as a function of used preamble bits (F=40,L=8,U=8)

References

[1] S. Hara and R. Prasad, "Overview of Multicarrier CDMA," *IEEE Communication Magazine*, pp. 126Ű133, December 1997.

[2] Z. Xie. R.T. Short and C.K.Rushfort1i: "Family of suboptimum detectors for coherent multiuser communications." *IEEE Journal on Selected Areas in Communications*, vol. 8, no. 4, pp. 683-690, May 1990.

[3] V. Kuhn, "Combined MMSE-PIC in coded OFDM-CDMA systems," *IEEE GLOBECOM "01*, pp. 231-235, Nov. 2001.

[4] S. Iraji, T. Sipilĺa and J. Lilleberg, "Channel Estimation and Signal Detection for MC-CDMA in Multipath Fading Channels," *in Proc. IEEE Int. Symp. on Personal, Indoor and Mobile Radio Commun. (PIMRC)*, September 1993

[5] A. Kocian and B. H. Fleury, "EM-based joint data detection and channel estimation of DS-CDMA signals," *IEEE Trans. Commun.*, vol. 51, no. 10, pp. 1709Ű-1720, Oct. 2003.

[6] G.K. Kaleh and R. Valet, "Joint parameter estimation and symbol detection for linear or nonlinear unknown channels," *IEEE Trans. Commun.* vol. 42, No. 7, pp. 2406-2413, July 1994.

[7] M. Feder and E. Weinstein, "Parameter Estimation of superimposed signals using the EM algorithm,"*IEEE Tran. on Acoustic, Speech and Signal Processing*, Vol. 36, pp. 477-489, April 1988.

[8] H. Dogan, H. A. Cirpan and Erdal Panayirci, "Iterative Channel Estimation and Decoding of Turbo Coded SFBC-OFDM Systems," *IEEE Trans. Wireless Commun.*, accepted for publication in Oct. 2006.

ON THE PERFORMANCE OF MC-CDMA SYSTEMS WITH PARTIAL EQUALIZATION IN THE PRESENCE OF CHANNEL ESTIMATION ERRORS

Flavio Zabini
WiLab, Univ. of Bologna, Italy
fzabini@deis.unibo.it

Barbara M. Masini
WiLab, Univ. of Bologna, Italy
bmasini@deis.unibo.it

Andrea Conti
WiLab, ENDIF, Univ. of Ferrara
aconti@ieee.org

Abstract The paper investigates the performance of a multi-carrier code division multiple access (MC-CDMA) system adopting a partial equalization (PE) technique at the receiver. It has been recently shown that, in spite of the fact that the PE technique has the same complexity of the well known maximal ratio combining, orthogonality restoring combining and equal gain combining techniques, the PE significantly improves the performance in terms of bit error probability (BEP). Previous works considered PE with ideal channel state information (CSI) and evaluate the PE parameter which minimizes the BEP; here we relax this assumption considering the presence of CSI errors. In particular, we analytically derive the mean BEP for the downlink of a MC-CDMA system and the optimal PE parameter, when imperfect CSI is assumed. Numerical results show that the optimum choice of the PE with ideal CSI provides significant performance improvement also in the presence of CSI errors.

S. Plass et al. (eds.), Multi-Carrier Spread Spectrum 2007, 407–416.

1. Introduction

Multi-carrier transmission is proved to be an efficient technique in counteracting frequency selective fading in high data rate mobile communication (see, e.g., [1], [3], [4], [5], [6]). In this work we consider the architecture of multi-carrier code division multiple access (MC-CDMA) presented in [1] and [7], where the spreading is done in the frequency-domain and Walsh-Hadamard (W-H) codes are used with the spreading factor equal to the number of subcarriers[1]. In particular, we investigate the down-link detection in frequency selective fading channels using the partial equalization (PE) technique presented in [9]. Although the system is assumed to be synchronous and different users undergo the same channel, at the receiver side the orthogonality between spreading sequences is corrupted by the fading. In this situation the choice of equalization becomes a crucial point impacting the trade-off between the residual amount of multiple access interference and the increasing in thermal noise. Within the family of linear combining techniques (i.e., low complex solutions to be implemented in the mobile terminal) equal gain combining (EGC), maximum ratio combining (MRC), orthogonal restoring combining (ORC) and threshold ORC (TORC) are usually considered: when the system is noise-limited, the best choice is given by MRC represents the best choice; otherwise the system is interference-limited, ORC can be exploited to cancel the interference with the counter effect of emphasizing the impact of noise. For this reason, with TORC a threshold is introduced to cancel the contributions of those sub-channels highly corrupted by the noise [12]. In [8], the optimum solution in minimum mean square error (MMSE) sense is presented, but its evaluation is cumbersome in no-full-load cases and also requires the knowledge of the signal-to-noise ratio (SNR) and the number of active users. The PE technique consists in choosing the complex weight referring to the l-th subcarrier as given by:

$$G_l = \frac{H_l^*}{|H_l|^{1+\beta}}, \qquad \beta \in [-1, 1] \tag{1}$$

where H_l is the l-th complex channel coefficient and β is a partial equalization parameter to be optimized. Note that when $\beta = 0, -1, 1$, (1) coincides with EGC, MRC and ORC channel gains, respectively. In [9], it is shown that the PE technique allow to obtain bit error probability (BEP) close to the optimum MMSE, but with the same complexity of EGC, MRC, ORC or TORC. In addition, optimum values of β minimizing the BEP con be found out. The above mentioned techniques require all the knowledge of the channel state information (CSI). In [9] the CSI was assumed perfectly estimated, whereas in this work we extend the analytical framework by considering also the impact of channel estimation error. Also in this case we find out the optimum β, that is the value of β minimizing the BEP, and $\overline{\gamma}$, M, N_u as well as the statistic of CSI error. We make the following common assumption (see [1] and [7]):

- the system remains always synchronous (different users and sub-carriers have the same delay that can be perfectly compensated);
- flat Rayleigh fading, uncorrelated among sub-channels, thus all the complex channel coefficients are independent identically distributed (i.i.d.) zero-mean Gaussian random variables (r.v.'s). This is the case of sufficiently spaced sub-carriers (more then a coherence bandwidth) or in the presence of frequency interleaving (see, e.g., [2]);
- the number of sub-carriers is sufficiently high (e.g., in practical systems such as DVB-T, 2k or 8k subcarrier are considered)to justify the use of the central limit theorem (CLT) and the law of large numbers (LLN);
- the channel estimation error is assumed to be complex Gaussian distributed with zero-mean and normalized variance $\varepsilon \doteq \sigma_E^2/\sigma_H^2$, where σ_H^2 will be defined in the following.

Note that the framework allows a completely analytical evaluation of the performance with careful investigation of dependencies between system parameters, the comparison with previous results appeared in the literature, and a lower-bound of the performance in realistic scenarios. We notice that the independence between channel coefficients referred to different sub-carriers implies that also estimation errors for different sub-carriers are independent.

2. System Models

Following the previously mentioned assumptions and the procedure adopted in [9], the received signal for BPSK modulation can be written as:

$$r(t) = \sqrt{\frac{2E_b}{M}} \sum_{k=0}^{N_u-1} \sum_{i=-\infty}^{+\infty} \sum_{m=0}^{M-1} \alpha_m c_m^{(k)} a^{(k)}[i] g_d(t - iT_b) \cos(2\pi f_m t + \vartheta_m) + n(t), \tag{2}$$

where E_b is the transmitted energy per bit, k, i and m are user, data and subcarrier indexes, respectively, α_m is the random amplitude coefficient related to the mth sub-channel (that we suppose identically Rayleigh distributed with $\mathbb{E}\left\{\alpha_m^2\right\} = 2\sigma_H^2$, being σ_H the parameter of the distribution, the same for all m) and ϑ_m is the random phase uniformly distributed within $[0, 2\pi[$, $c_m^{(k)}$ is the mth chip of orthogonal spreading sequences for user k(taking value ± 1), $a^{(k)}[i]$ is the data transmitted during the ith time-symbol, $g_d(t)$ is a rectangular pulse waveform, with duration $[0, T + T_d]$ and unitary energy, T_b the bit-time, and f_m is the subcarrier-frequency. In particular, $T_b = T + T_g$ is the total OFDM symbol duration, with a time-guard T_g (inserted between consecutive multi-carrier symbols to eliminate the residual inter symbol interference (ISI) due to the channel delay spread T_d).

2.1 Channel estimation errors model

We assume that the estimated complex channel coefficients are:

$$\hat{H_m} = H_m + E_m, \tag{3}$$

being $H_m = X_m + jY_m = \alpha_m e^{j\vartheta_m}$ the m-th complex channel coefficients, and E_m the channel estimation error considered complex Gaussian distributed with real and imaginary parts given by $X_{Em} = \mathcal{N}(0, \sigma_e^2)$ and $Y_{Em} = \mathcal{N}(0, \sigma_e^2)$. It is straightforward to derive $\hat{\alpha_m} \doteq |\hat{H_m}|$ and $\hat{\vartheta_m} \doteq \angle \hat{H_m}$.

2.2 Decision Variable

The decision variable is given by:

$$\begin{aligned} z^{(n)} &= \overbrace{\sqrt{\frac{E_b \delta_d}{M}} \sum_{l=0}^{M-1} \Theta_l(\beta)\, a^{(n)}}^{r(n)} + \\ &+ \overbrace{\sqrt{\frac{E_b \delta_d}{M}} \sum_{l=0}^{M-1} \sum_{k=0, k\neq n}^{N_u - 1} \Theta_l(\beta) c_l^{(n)} c_l^{(k)} a^{(k)}}^{I^{(n)}} + \overbrace{\sum_{l=0}^{M-1} \hat{\alpha}_l{}^{-\beta} n_l}^{N^{(n)}}, \end{aligned} \tag{4}$$

where:

$$\Theta_l(\beta) \doteq \alpha_l \hat{\alpha}_l{}^{-\beta} \cos(\vartheta_l - \hat{\vartheta}_l). \tag{5}$$

By making use of the CLT, the properties of the orthogonal codes, and exploiting the independence of the data, we derive the useful term $r^{(n)}$, the interference term $I^{(n)}$ and the noise term $N^{(n)}$ as independent Gaussian r.v.'s with distributions:

$$r^{(n)} \sim \mathcal{N}\left(\sqrt{E_b \delta_d M} \mathbb{E}\left\{\Theta_l(\beta)\right\}, E_b \delta_d \zeta_\beta\right), \tag{6}$$

$$I^{(n)} \sim \mathcal{N}\left(0, E_b\delta_d(N_u-1)\zeta_\beta\right), \tag{7}$$

$$N^{(n)} \sim \mathcal{N}\left(0, M\frac{N_0}{2}\mathbb{E}\left\{\hat{\alpha}_l{}^{-2\beta}\right\}\right), \tag{8}$$

where

$$\zeta_\beta \doteq \mathbb{E}\left\{\Theta_l^2(\beta)\right\} - \left(\mathbb{E}\left\{\Theta_l(\beta)\right\}^2. \tag{9}$$

Therefore, the decision variable (3) leads to the following expression of the BEP conditioned to $r^{(n)}$ as:

$$P_b|_r = \frac{1}{2}\text{erfc}\left(\frac{\sqrt{\frac{E_b\delta_d}{M}}\sum_{l=0}^{M-1}\Theta_l(\beta)}{\sqrt{2E_b\delta_d(N_u-1)\zeta_\beta + M\mathbb{E}\left\{\hat{\alpha}_l{}^{-2\beta}\right\}N_0}}\right). \tag{10}$$

2.3 Bit Error Probability

By involving the LLN[2] we can write the following expression for unconditioned BEP, thus averaged over fast fading:

$$P_b \simeq \frac{1}{2}\text{erfc}\left(\sqrt{\frac{E_b\delta_d(\mathbb{E}\left\{\Theta_l(\beta)\right\})^2}{2E_b\delta_d\frac{N_u-1}{M}\zeta_\beta + \mathbb{E}\left\{\hat{\alpha}_l{}^{-2\beta}\right\}N_0}}\right), \tag{11}$$

where:

$$\mathbb{E}\left\{\hat{\alpha}_l{}^{-2\beta}\right\} = (2\sigma_H^2)^{-\beta}\Gamma(1-\beta)\left[1+\frac{\sigma_E^2}{\sigma_H^2}\right]^{-\beta} \tag{12}$$

$$\mathbb{E}\left\{\Theta_l(\beta)\right\} = (2\sigma_H^2)^{\frac{1-\beta}{2}}\Gamma\left(\frac{3-\beta}{2}\right)\Pi\left(\frac{\sigma_E^2}{\sigma_H^2},\beta\right) \tag{13}$$

$$\zeta_\beta(\alpha) = (2\sigma_H^2)^{1-\beta}\Gamma(2-\beta)\Sigma\left(\frac{\sigma_E^2}{\sigma_H^2},\beta\right) - (2\sigma_H^2)^{1-\beta}\Gamma^2\left(\frac{3-\beta}{2}\right)\Pi^2\left(\frac{\sigma_E^2}{\sigma_H^2},\beta\right) \tag{14}$$

The $\Gamma(x)$ is the Eulero-Gamma function and

$$\Pi\left(\frac{\sigma_E^2}{\sigma_H^2},\beta\right) \doteq \left(1+\frac{\sigma_E^2}{\sigma_H^2}\right)^{\frac{1-\beta}{2}} - \frac{\frac{\sigma_E^2}{\sigma_H^2}}{\left(1-\frac{\sigma_E^2}{\sigma_H^2}\right)\left(1+\frac{\sigma_E^2}{\sigma_H^2}\right)^2}\left[\frac{1+\frac{\sigma_E^2}{\sigma_H^2}}{1+\frac{3}{4}\frac{\sigma_E^2}{\sigma_H^2}}\right]^{\frac{3-\beta}{2}} \tag{15}$$

and

$$\Sigma\left(\frac{\sigma_E^2}{\sigma_H^2},\beta\right) \doteq \left(1+\frac{\sigma_E^2}{\sigma_H^2}\right)^{1-\beta} - \frac{2\frac{\sigma_E^2}{\sigma_H^2}}{\left(1-\frac{\sigma_E^2}{\sigma_H^2}\right)\left(1+\frac{\sigma_E^2}{\sigma_H^2}\right)^2}\left[\frac{1+\frac{\sigma_E^2}{\sigma_H^2}}{1+\frac{3}{4}\frac{\sigma_E^2}{\sigma_H^2}}\right]^{2-\beta} + $$
$$+ \sqrt{\frac{\pi}{8}\frac{\frac{\sigma_E^2}{\sigma_H^2}}{1+\frac{\sigma_E^2}{\sigma_H^2}}}\frac{\frac{\sigma_E^2}{\sigma_H^2}}{\left(1-\frac{\sigma_E^2}{\sigma_H^2}\right)\left(1+\frac{\sigma_E^2}{\sigma_H^2}\right)^2}\left[\frac{1+\frac{\sigma_E^2}{\sigma_H^2}}{1+\frac{7}{8}\frac{\sigma_E^2}{\sigma_H^2}}\right]^{\frac{3}{2}-\beta} \times$$
$$\times S\left(\frac{\frac{1}{8}\frac{\sigma_E^2}{\sigma_H^2}}{1+\frac{7}{8}\frac{\sigma_E^2}{\sigma_H^2}},\beta\right)\frac{\Gamma\left(\frac{3}{2}-\beta\right)}{\Gamma\left(2-\beta\right)} \tag{16}$$

being

$$S(z,\beta) \doteq 3\,{}_2F_1\left[\frac{3-2\beta}{4},\frac{5-2\beta}{4},1,z^2\right] + 4z\left(\frac{3}{2}-\beta\right) \times$$
$$\times\ {}_2F_1\left[\frac{5-2\beta}{4},\frac{7-2\beta}{4},1,z^2\right] +$$
$$+ \frac{1}{2}z\left(\frac{3}{2}-\beta\right)\,{}_2F_1\left[\frac{5-2\beta}{4},\frac{7-2\beta}{4},2,z^2\right] +$$
$$+ 2z^2\left(\frac{3}{2}-\beta\right)\left(\frac{5}{2}-\beta\right)\,{}_2F_1\left[\frac{7-2\beta}{4},\frac{9-2\beta}{4},2,z^2\right] \tag{17}$$

and ${}_2F_1(a,b,c,x)$ the Hypergeometric function. By substituting (12), (13) and (14) in (11) we have:

$$P_b = \frac{1}{2}\mathrm{erfc}\left\{\left[\frac{1}{\overline{\gamma}}\frac{\Gamma(1-\beta)\left(1+\frac{\sigma_E^2}{\sigma_H^2}\right)^{-\beta}}{\Gamma^2\left(\frac{3-\beta}{2}\right)\Pi^2\left(\frac{\sigma_E^2}{\sigma_H^2},\beta\right)} + 2\frac{N_u-1}{M}\times \right.\right. \tag{18}$$
$$\left.\left. \times \frac{\Gamma(2-\beta)\Sigma\left(\frac{\sigma_E^2}{\sigma_H^2},\beta\right) - \Gamma^2\left(\frac{3-\beta}{2}\right)\Pi^2\left(\frac{\sigma_E^2}{\sigma_H^2},\beta\right)}{\Gamma^2\left(\frac{3-\beta}{2}\right)\Pi^2\left(\frac{\sigma_E^2}{\sigma_H^2},\beta\right)}\right]^{-\frac{1}{2}}\right\},$$

where:

$$\overline{\gamma} \doteq \frac{1}{L_P}\frac{E_b}{N_0}\frac{1}{1+\frac{T_d}{T}} = \frac{2\sigma_H^2 E_b \delta_d}{N_0}. \tag{19}$$

As a benchmark, it can be noticed that for $\sigma_E^2/\sigma_H^2 = 0$, the (18) reduces to the expression in the absence of channel estimation error obtained in [9], where comparison with simulation results were also provided.

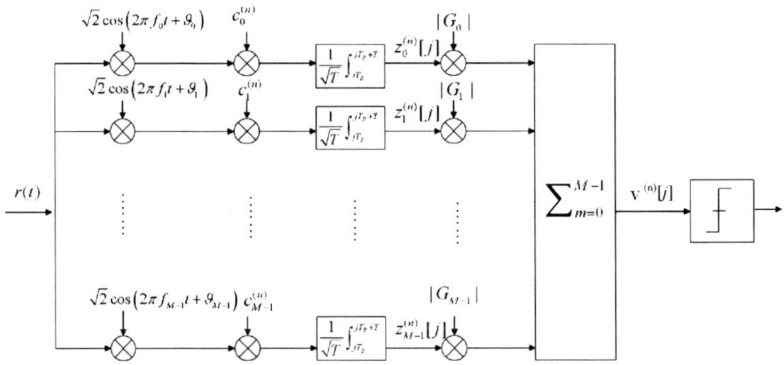

Figure 1. The receiver's block scheme

2.4 Numerical Results

This section reports numerical results obtained with the above described framework. In Fig. 2 the mean BEP, averaged over fast fading, versus the PE parameter, β, ranging from -1 to 1, is shown. A mean SNR $\overline{\gamma} = 10$dB is assumed for half-loaded system ($M = 1024$ and $N_u = 512$), when the error parameter ε is equal to 0% (ideal CSI), 0.25%, 1%, 4%, 9%, 16% and 25%. It can be noticed that β strongly affects the performance, which is deteriorated by the increasing of ε. Moreover, note that the value of β minimizing the mean BEP is close to the optimal value in the case of ideal CSI.

In Fig.3, the mean BEP is shown as a function of the mean SNR varying the error on the CSI and for $M = 1024$, $N_u = 512$; two cases are considered for the PE parameter: $\beta = 0.5$ fixed and the optimum β obtained through numerical minimization of the mean BEP. Note that, in these particular conditions, the adoption of a fixed $\beta = 0.5$ gives a performance close to the optimum for all values of the mean SNR and the CSI error.

The mean BEP as a function of the number of users for $M = 1024$ sub-carriers ($log_2 N_u = 10$ represents the full-load case) is plotted in Fig. 4 for $\overline{\gamma} = 8$dB and $\varepsilon = 0\%$ and 16%. Here, EGC ($\beta = 0$) is compared with the case of optimum PE parameter, which improves the performance. This figure allows the evaluation of the maximum system load for a given mean BEP: as an example result, in the case of ideal CSI for a target mean BEP of 10^{-3}, the number of active users can be up to about 16 for EGC and 40 with the optimum β. In the presence of CSI errors with $\varepsilon = 16\%$ the EGC does not reach the target BEP whereas this can be done with optimum β for a maximum number of active users of about 16.

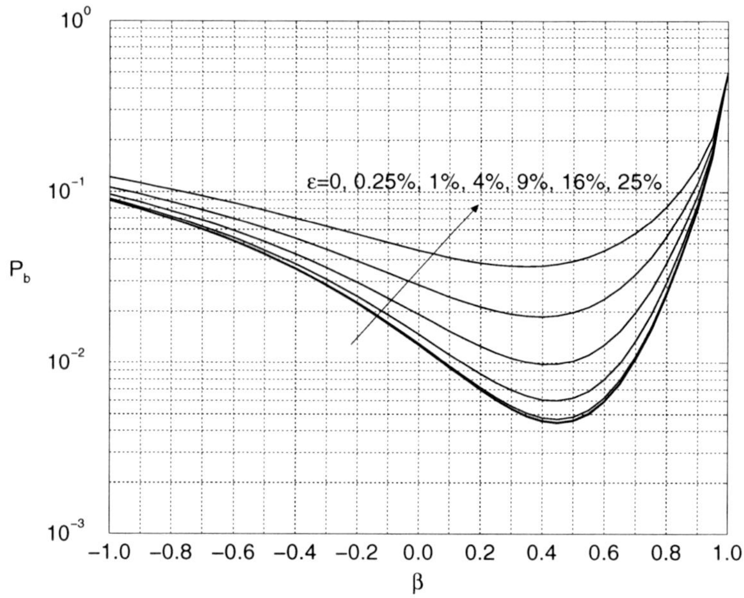

Figure 2. BEP vs β for $\overline{\gamma} = 10dB$, $M = 1024$, $N_u = 512$ and various ε.

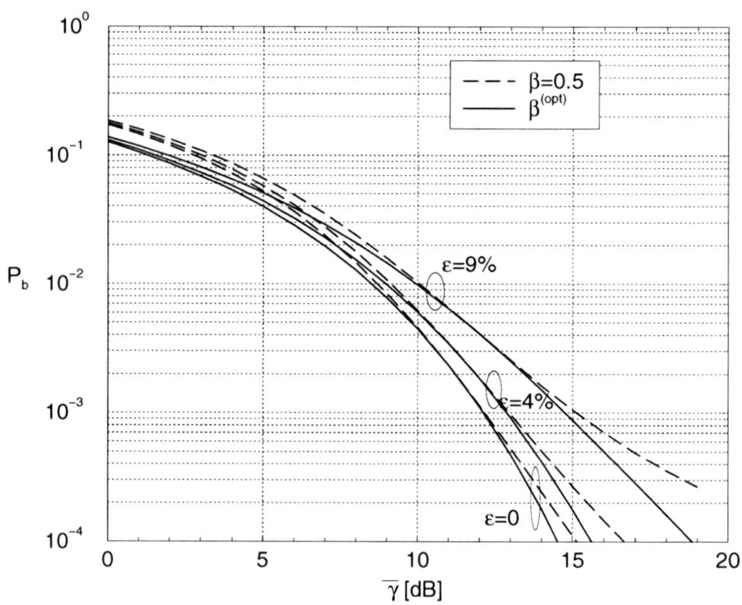

Figure 3. BEP vs $\overline{\gamma}$ for different values of β and ε with $M = 1024$ and $N_u = 512$.

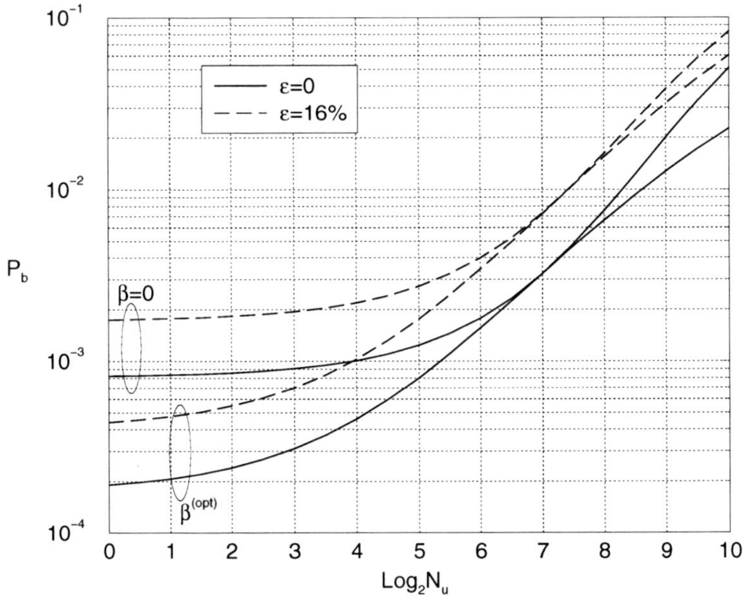

Figure 4. BEP vs N_u for $\overline{\gamma} = 8dB$, and two values of ε. The cases with EGC ($\beta = 0$) and $\beta^{(opt)}$ are considered

3. Conclusions

In this work the performance of a MC-CDMA system has been analitically investigated when adopting a partial equalization (PE) technique. The framework extend what previously derived under the assumption of ideal CSI at the receiver; in particular, the presence of CSI errors has been considered. The benefit of PE technique for various error standard deviations is given and its relation with the system load and the SNR is derived. In spite of the fact that the PE technique has the same complexity of the well known maximal ratio combining, orthogonality restoring combining and equal gain combining techniques, the PE significantly improves the performance in terms of BEP. In particular, we analytically derived the mean BEP for the downlink of a MC-CDMA system and numerically evaluated the impact of the optimal PE parameter, when imperfect CSI is assumed. Numerical results show that the optimum choice of the PE with ideal CSI provides significant performance improvement also in the presence of CSI errors. As a performance figure, the framework enables the evaluation of the maximum system load for a given mean BEP, SNR and PE parameter. Moreover it opens tha way to the adaptation of the PE parameter tracking slow processes evolutions.

Notes

1. Note that this structure of MC-CDMA is also known as OFDM-CDMA and has to not be confused with MC-DS-CDMA, presented in [10] and [11], where the spreading is done in the time-domain.

2. We approximate $\sum_{l=0}^{M-1} \Theta_l(\beta)$ with $M\mathbb{E}\{\Theta_l(\beta)\}$

References

[1] N. Yee, J.-P. Linnartz and G. Fettweis, Multi-Carrier-CDMA in indoor wireless networks, in *Conference Proceedings PIMRC '93*, Yokohama, Sept, 1993. p 109-113

[2] S. Kaiser, K. Fazel, "A Flexible Spread-Spectrum Multi-Carrier Multiple-Access System for Multi-Media Applications Personal", *Indoor and Mobile Radio Communications, 1997.* 'Waves of the Year 2000'. PIMRC '97., The 8th IEEE International Symposium on Volume 1, 1-4 Sept. 1997 Page(s): 100 - 104 vol.1 Digital Object Identifier 10.1109/PIMRC.1997.624371

[3] K. Fazel, "Performance of CDMA/OFDM for mobile communication system", in *2nd International Conference on Universal Personal Communications, 1993. 'Personal Communications: Gateway to the 21st Century'. Conference Record, 12-15 Oct. 1993 Pages:975 - 979 vol.2*

[4] K. Fazel, S. Kaiser "Multi-Carrier and Spread Spectrum Systems", Wiley, ISBN: 0-470-84899-5.

[5] S. Hara and R. Prasad, "An Overview of multicarrier CDMA", in *IEEE 4th International Symposium on Spread Spectrum Techniques and Applications Proceedings, 22-25 Sept. 1996, Pages:107 - 114 vol.1*

[6] S. Kaiser, "On the performance of different detection techniques for OFDM-CDMA in fading channels", in *Globecom '95 pp.2059-2063 Nov. 1995*

[7] N. Yee and J.-P. Linnartz, "BER of multi-carrier CDMA in an indoor Rician fading channel"; in *Conference Record of The Twenty-Seventh Asilomar Conference on Signals, Systems and Computers, 1993., 1-3 Nov. 1993 Pages:426 - 430 vol.1*

[8] N. Yee and J.-P. Linnartz, "Wiener filtering for Multi-Carrier CDMA", in *IEEE / ICCC conference on Personal Indoor Mobile Radio Communications (PIMRC)and Wireless Computer Networks (WCN), The Hague, September 19-23, 1994, Vol. 4, pp. 1344-1347*

[9] A. Conti, B. M. Masini, F. Zabini and O. Andrisano, "On the Down-link Performance of Multi-Carrier CDMA Systems with Partial Equalization", in *IEEE Transactions on Wireless Communications, Gen. 2007.*

[10] S. Kondo and L.B. Milstein, On the use of multicarrier direct sequence spread spectrum systems, in *Military Communications Conference, 1993. MILCOM '93. Conference record. 'Communications on the Move'., IEEE,11-14 Oct. 1993 Pages:52 - 56 vol.1*

[11] S. Kondo and L.B. Milstein, "Performance of multicarrier DS CDMA systems", in *IEEE Transactions on Communications, Volume: 44, Issue: 2, Feb. 1996 Pages: 238 - 246*

[12] N. Yee and J.-P. Linnartz, Controlled equalization of multi-carrier CDMA in an indoor Rician fading channel in *Vehicular Technology Conference, 1994 IEEE 44th ,8-10 June 1994 Pages:1665 - 1669 vol.3*

[13] M. K. Simon, "Probability Distributions Involving Gaussian Random Variables", Kluwer Academic Publishers, ISBN:1-4020-7058-6.

CROSS-COUPLED RAO-BLACKWELLIZED PARTICLE AND KALMAN FILTERS FOR THE JOINT SYMBOL-CHANNEL ESTIMATION IN MC-DS-CDMA SYSTEMS

Julie Grolleau, Audrey Giremus, Eric Grivel
Equipe Signal et Image, UMR CNRS 5218 IMS - Dpt LAPS, Universite Bordeaux 1
351 cours de la liberation, 33405 Talence cedex, France
julie.grolleau@etu.u-bordeaux1.fr,{audrey.giremus, egrivel}@u-bordeaux1.fr

Abstract This paper deals with the joint symbol-channel estimation for quasi-synchronous Multi-carrier Direct-Sequence Code Division Multiple Access (MC-DS-CDMA) systems over Rayleigh fading channels. To solve this non-linear problem, Rao-Blackwellized particle filters have proved efficient. In this framework, our contribution is twofold. 1) Instead of using an autoregressive (AR) model which does not match the bandlimitation of the theoretical power spectrum density (PSD) of a Rayleigh channel, we suggest modeling the channel by a low-pass filtered version of the so-called stochastic sinusoidal process. It consists of sinusoids in quadrature with random magnitudes modeled as AR processes. By suitably choosing the AR parameters, this combination has the advantage of providing a stationnary process whose PSD is bandlimited and has two peaks at the maximum Doppler frequency for any AR order. 2) The estimation of the model parameters is included in the joint symbol-channel estimation process. For this purpose, a deterministic Rao-Blackewellized Particle Filter which jointly estimates the symbols and the channels is cross-coupled with a Kalman Filter which yields the AR parameters.

1. Introduction

In mobile communications, high speeds of terminals and scatterers cause Doppler effects that can seriously affect the reception performance. Thus, joint channel/symbols estimation is a major challenge for reliable wireless transmissions. Among the previous works related to the joint symbol detection and channel estimation for CDMA systems, three families of approaches have emerged:

• **The optimal receiver minimizing the symbol error probability.** It is based on linear suppressed-carrier modulations over several nonselective Rayleigh fading channels [5]. However, its main drawback is the compu-

S. Plass et al. (eds.), Multi-Carrier Spread Spectrum 2007, 417–426.

tational complexity that grows exponentially as a function of the number of transmitted symbols. Therefore, the following suboptimal algorithms have been considered.

• **The two-stage receivers.** The channel estimation stage consists in using an adaptive filter such as the Recursive Least Square (RLS) or the Least Mean Suare (LMS). The channel estimate is then incorporated in the sequence detection stage such as the Viterbi Detector (VD), a PerSurvivor Processing Detector [7] or a Decision feedback [6]. Nevertheless, in the latter case, deep fades lead to bursts of opposite decisions [12]. Furthemore, it should be noted that the RLS and LMS algorithms are not well suited for fast fading. Therefore, both adaptive estimators are outperformed by model based methods such as Kalman filtering.

• **Particle filtering based receivers.** Recently, the opportunity of using non linear filtering approaches to simultaneously estimate the symbols and the channel has been investigated. More particulary, a great deal of interest has been paid to Rao-Blackwellized particle filtering techniques [9]-[11]. They aim at selecting the most likely transmitted symbols according to a particle filter scheme and estimating the channels conditionnally to the particles by means of Kalman filtering. In that way, the variance of both estimates is known to be significantly decreased. Up to now, in [9]-[11], the authors have considered AR Moving Average (ARMA) or a first-order AR process, whose parameters are assumed to be known or preliminary estimated, so as to model the channel.

In this paper, we focus our attention on a quasi-synchronous Multi-Carrier Direct Sequence Code Division Access (MC-DS-CDMA) system over Rayleigh fading channel. In the new receiver based on Rao-Blackwellized particle filtering techniques, our contribution is twofold:

Firstly, we investigate the relevance of a new model for Rayleigh channels which better matches the theoretical spectral properties of Rayleigh channels. In [1], Jakes has shown that the real and imaginary parts of a Rayleigh channel are decorrelated and have the same power spectral density (PSD). This latter is bandlimited, U-shaped and exhibits twin peaks at $\pm f_{d_n}$ where f_{d_n} denotes the normalized maximum Doppler frequency of the channel. AutoRegressive (AR) or ARMA models are usually chosen since they are linear and simple [12], but this approach may be questionable. Indeed, as the PSD of the real and imaginary parts is bandlimited, the channel process is deterministic according to the Kolmogoroff-Szëgo formula[1] [13]. For this reason, in several works [12] and [4], modeling a Rayleigh channel by a finite order AR process leads to various problems. On the one hand, when using a second order AR model whose parameters are estimated by solving the Yule-Walker equations, the PSD twin peaks are localized at $\pm\frac{f_{d_n}}{\sqrt{2}}$ instead of $\pm f_{d_n}$ [12]. On the other hand, when increasing the AR order, the channel autocorrelation matrix involved in the

Yule-Walker equations becomes ill-conditionned [4]. To overcome this problem, the authors suggest "slightly" modifying the properties of the channel by considering the sum of the theoretical fading process and a zero-mean white process whose variance ϵ is very small (e.g., $\epsilon = 10^{-7}$ for $f_{d_n} = 0.01$). Nevertheless, whatever the model order may be, the twin PSD peaks never reach $\pm f_{d_n}$. Therefore, we propose to model a Rayleigh channel by a low-pass filtered version of a sum of two sinusoids in quadrature at the maximum Doppler frequency and whose amplitudes are AR processes. By suitably choosing the AR parameters, this combination has the advantage of providing a stationnary process whose PSD is bandlimited and has two peaks at the maximum Doppler frequency for any AR order.

Secondly, we propose to include a way to estimate the AR parameters in the proposed joint symbol-channel estimator. As particle filters tend to degenerate when coping with static parameters, we cross-couple the deterministic Rao-Blackwellized filter with a Kalman filter to yield the parameter estimation.

The remainder of the paper is organized as follows: in section 2, we introduce the state-space representation of the system. In section 3, the receiver structure is presented. In section 4, the relevance of the proposed channel model and the joint estimator is illustrated by numerical results.

2. System model

The number of carriers M of the MC-DS-CDMA system is assigned to 3 or 4 so that the signal along each carrier is propagated through a nonselective channel and the carriers are sufficiently spaced with respect to the channels' coherence bandwidth. The maximum Doppler frequencies of the channels are assumed to be known and the user signatures orthogonal.

In the following, let us focus our attention on the first-user symbol detection. The proposed receiver operates as follows: a decorrelating filter is first used to eliminate the Multiuser Access Interferences (MAI). It does not depend on the fading channels. Thus, the filter output for the first user is independent of other users' symbol.

Then, the output signal is demodulated to yield M observations of the first-user symbols transmitted over M independent channels. Hence, the observation along the mth carrier can be expressed as follows:

$$y^m(n) = h^m(n)d(n) + v^m(n) \text{ for } m = 1, ..., M \qquad (1)$$

where $d(n)$ is the BPSK symbol transmitted at time n by the first user and $v^m(n)$ a white complex Gaussian noise with variance $2\sigma_v^2$. Moreover, the channel over the mth carrier, denoted $h^m(n)$, is modeled as follows:

$$h^m(n) = g^m(n) * (a^m(n)\text{cos}(2\pi f_{dn}^m n) + b^m(n)\text{sin}(2\pi f_{dn}^m n)) \qquad (2)$$

where $a^m(n)$ and $b^m(n)$ are two independent complex p^{th} order AR processes. As the real and imaginary parts of the channel are decorrelated and have the same PSD, the sequences $a^m(n)$ and $b^m(n)$ have the same real AR parameters, denoted $\{c_i^m\}_{i=1,...,p}$. In addition, $g^m(n)$ is the Lth order finite impulse response[2] of the low-pass filter whose cut-off frequency is the normalized maximum Doppler frequency $f_{d_n}^m$ along the mth carrier.

Let us now introduce the so-called $4LM \times 1$ process vector $\underline{x}(n)$, the $pM \times 1$ parameter vector $\underline{c}$ and the $2M \times 1$ observation vector $\underline{y}(n)$ defined as follows:

$$\underline{x}(n) = [\underline{a}_R^1(n)\underline{a}_I^1(n)\underline{b}_R^1(n)\underline{b}_I^1(n) \cdots \underline{a}_R^M(n)\underline{a}_I^M(n)\underline{b}_R^M(n)\underline{b}_I^M(n)]^T \tag{3}$$

$$\underline{c} = [c_1^1 \;\cdots\; c_p^1 \;\cdots\; c_1^M \;\cdots\; c_p^M]^T \tag{4}$$

$$\underline{y}(n) = [y_R^1(n)y_I^1(n) \cdots y_R^M(n)y_I^M(n)]^T \tag{5}$$

where the subscripts $._R$ and $._I$ stand for the real and the imaginary part of the complex value and $\underline{a}_R^m(n)$ contains the L last values of the AR process $a_R^m(n)$.

Given (2), the real and the imaginary parts of $a^m(k)$ and $b^m(k)$ for $k = n - L + 1, ...n$ have to be known in order to estimate $h^m(n)$ over the mth carrier at time n. So, the issue at hand is to jointly estimate at each time n, the process vector $\underline{x}(n)$, the parameter vector $\underline{c}$ and the first-user symbol $d(n)$ by using the observation vector $\underline{y}(n)$.

The state space representation of the system (1) and (2) is hence based on the state vector $X(n) = \left[\underline{x}(n)^T \; d(n) \; \underline{c}\right]^T$. In a Bayesian framework, its estimation is obtained from the a posteriori distribution $p\left(X(n)|\underline{y}(1), ..., \underline{y}(n)\right)$. When dealing with linear Gaussian state space models, the Kalman filter makes it possible to recursively compute this target distribution. However, the joint symbol-channel-parameter estimation is "strongly" non-linear and hence does not admit a closed-form solution. Therefore, we propose to approximate the a posteriori probability density funtion by means of particle filtering techniques.

A classical particle filter sequentially approximates $p\left(X(n)|\underline{y}(1), ..., \underline{y}(n)\right)$ by a set of weighted random samples called particles. Given the high dimension of the state vector $X(n)$, a great amount of particles is necessary to obtain a good approximation of this target distribution. However, as the real and imaginary parts of the AR processes $a^m(n)$ and $b^m(n)$ for $m = 1, ..., M$ are Gaussian random processes, the state space representation is linear Gaussian conditionnally to the symbols and the AR parameters. Consequently, the process vector $\underline{x}(n)$ can be marginalized out and its estimation can be carried out by using Kalman filtering. Thus, only the remainder of the state vector is estimated by means of a particle filter. In that case, each particle represents a possible value of the pair $\left[d(n) \; \underline{c}^T\right]$ and is associated with a Kalman filter which provides the conditional estimation of $\underline{x}(n)$. Then, combining these conditional estimates according to the particle weights yields the a posteriori estimation of the process vector. This strategy has the advantage of analytically solving a part of the estimation problem (by means of Kalman filtering) so that the number of particles can be significantly reduced without impairing the algorithm accuracy.

Nevertheless, the nonlinear state variables include static parameters stacked in $\underline{c}^T$ and Particle filters classically experiences degenerancy issues when tracking time-invariant parameters. One solution would consist in introducing an artificial dynamic on the parameters, at the expense of the estimation accuracy. To overcome this limitation, we propose to cross-couple a deterministic Rao-Blackwellized particle filter to jointly estimate $\underline{x}(n)$ and $d(n)$ with a Kalman filter to yield the estimation of $\underline{c}$, as illustrated in figure 3.

Joint channel/symbol estimation. As the symbols $d(n)$ take their values in a finite set, i.e $\{-1, 1\}$, they can be estimated by means of a deterministic particle filter [8]. Unlike classical particle filtering methods, it consists in exhaustively exploring the possible values of $d(n)$ to approximate the a posteriori distribution $p\left(d_{0:n}|y_{0:n}\right)$ as follows:

$$p\left(d_{0:n}|y_{0:n}\right) \approx \sum_{i=1}^{N} \sum_{t\in\{-1,1\}} \tilde{\omega}^{(i,t)}(n)\delta\left(d_{0:n} - d_{0:n}^{(i,t)}\right) \tag{6}$$

where N is the number of particles at time $n-1$, $d_{0:n} = (d(0), \cdots, d(n))$ and $y_{0:n} = \left(\underline{y}(0), \cdots, \underline{y}(n)\right)$. At time n, the particle offsprings are expressed as $d_{0:n}^{(i,t)} = \left(d_{0:n-1}^{(i)}, t\right)$ where $d_{0:n-1}^{(i)}$ stands for the ith particle at time $n-1$. Finally, $\tilde{\omega}^{(i,t)}(n)$ is the normalized importance weight at time n associated with particle (i, t) and $\delta\left(d_{0:n} - d_{0:n}^{(i,t)}\right)$ denotes the Dirac delta centered in $d_{0:n} = d_{0:n}^{(i,t)}$.

As for channel estimation, the posterior probability density $p\left(\underline{x}(n)|y_{0:n}\right)$ can be expressed as follows:

$$p\left(\underline{x}(n)|y_{0:n}\right) = \int_{d_{0:n}} p\left(\underline{x}(n), d_{0:n}|y_{0:n}\right) \mathrm{d}\left(d_{0:n}\right) \tag{7}$$

Applying the Bayes' rule yields:

$$p\left(\underline{x}(n)|y_{0:n}\right) = \int_{d_{0:n}} p\left(\underline{x}(n)|d_{0:n}, y_{0:n}\right) p\left(d_{0:n}|y_{0:n}\right) \mathrm{d}\left(d_{0:n}\right) \tag{8}$$

Thus, incorporing (6) in (8) leads to:

$$p\left(\underline{x}(n)|y_{0:n}\right) \approx \sum_{i=1}^{N} \sum_{t\in\{-1,1\}} \tilde{\omega}^{(i,t)}(n) p\left(\underline{x}(n)|d_{0:n}^{(i,t)}, y_{0:n}\right) \tag{9}$$

As $\underline{x}(n)$ is represented by a linear Gaussian state space conditionnally to $d_{0:n}^{(i,t)}$, $p\left(\underline{x}(n)|d_{0:n}^{(i,t)}, y_{0:n}\right)$ is analytically tractable. The mean and variance of this Gaussian distribution, denoted $m^{(i,t)}(n|n)$ and $P^{(i,t)}(n|n)$ respectively, can be computed by recursively applying a Kalman filter associated to $d_{0:n}^{(i,t)}$. In the sequel, its Kalman gain is denoted $K^{(i,t)}(n)$ and its innovation $\underline{e}^{(i,t)}(n)$. Therefore, $p\left(\underline{x}(n)|y_{0:n}\right)$ is approximated by a mixture of $2N$ Gaussian distributions computed by a bank of $2N$ Kalman filters associated to the $2N$ particles. The importance weights are evaluated up to a normalizing constant as follows:

$$\tilde{\omega}^{(i,t)}(n) \propto p\left(\underline{y}(n)|y_{1:n-1}, d_{0:n}^{(i,t)}\right) p(d^{(i,t)}(n)) \tag{10}$$

where the symbols are assumed to be equally distributed, i.e, $p(d^{(i,t)}(n) = t) = \frac{1}{2}\ \forall t$ for a BPSK modulation and $p\left(\underline{y}(n)|y_{1:n-1}, d_{0:n}^{(i,t)}\right)$ is supplied by the Kalman filter conditional to the particle $d_{0:n}^{(i,t)}$.

At that stage, to avoid the degeneracy and the exponential growth of the number of particles, the set of particles is resampled according to the particle weights. As a result, particles with low normalized weights are discarded whereas those with high ones are multiplied. Then, only N particles among $2N$ are selected. The symbol estimate $\hat{d}_n$ is chosen as the particle associated with the highest normalized weight.

Parameter estimation. Once the posterior estimate $\hat{\underline{x}}(n/n)$ is obtained, one can update the parameters $\underline{c}$ by using a Kalman filter. Indeed, the posterior estimate $\hat{\underline{a}}_R(n/n) = \left[\hat{a}_R^1(n/n) \ \cdots \ \hat{a}_R^M(n/n)\right]$ of the real part of the AR processes[3] $(a^m(n))_{m=1,\dots,M}$ linearly depends on $\underline{c}^T$ as follows:

$$\hat{\underline{a}}_R(n/n) = -\Psi\underline{c} + \underline{e}^*(n) \tag{11}$$

where Ψ is a block diagonal of size $M \times Mp$ whose m^{th} block on its diagonal is a $1 \times p$ vector corresponding to the truncation at length p of $\underline{a}_R^m(n-1|n-1)$. Moreover, $\underline{e}^*(n) = H\underline{e}'(n)$ is a zero-mean Gaussian white noise with variance $\sigma_e(n) = \sum_{i=1}^{N}\sum_{t\in\{-1,1\}}\tilde{\omega}^{(i,t)}(n){\sigma_e}^{(i,t)}(n)$ where:

- H is a $M \times 4LM$ matrix whose $4L(m-1)+1^{th}$ element of the mth row is set to 1 and the others equal to 0.
- $\underline{e}'(n) = \sum_{i=1}^{N}\sum_{t\in\{-1,1\}}\tilde{\omega}^{(i,t)}(n)K^{(i,t)}(n)\underline{e}^{(i,t)}(n)$ is a weighted mean of the product of Kalman gains $K^{(i,t)}(n)$ and the innovation processes $\underline{e}^{(i,t)}(n)$.
- ${\sigma_e}^{(i,t)}(n) = K^{(i,t)}(n)\sigma^{(i,t)}\left(K^{(i,t)}(n)\right)^T$ where $\sigma^{(i,t)}$ is the variance of the innovation process $\underline{e}^{(i,t)}(n)$ of the Kalman filter corresponding to the $(i,t)^{th}$ selected particle.

The relation (11) is the observation equation of the Kalman filter dedicated to the AR parameters. As they are time invariant, its state equation satisfies:

$$\underline{c}(n) = \underline{c}(n-1) \tag{12}$$

Based on this state-space representation, at time n, the Kalman filter dedicated to the AR parameters provides the estimation of $\underline{c}$ which will be exploited by the bank of Kalman filters to yield the a priori estimation of the AR processes.

3. Simulation Results

In this section, we study the relevance of the proposed estimator when the channels are modeled either by AR processes or by filtered stochastic sinusoidal (SS) processes. For this purpose, a MC-DS-CDMA system with $M = 3$ carriers is considered. $N = 100$ symbols are transmitted and 10 training symbols are used to help intializing the algorithm. The channels are generated by the Jakes' simulator [1]. The normalized maximum Doppler frequency varies from 0.0005 to 0.250. Various scenarii are studied for different values of p and L.

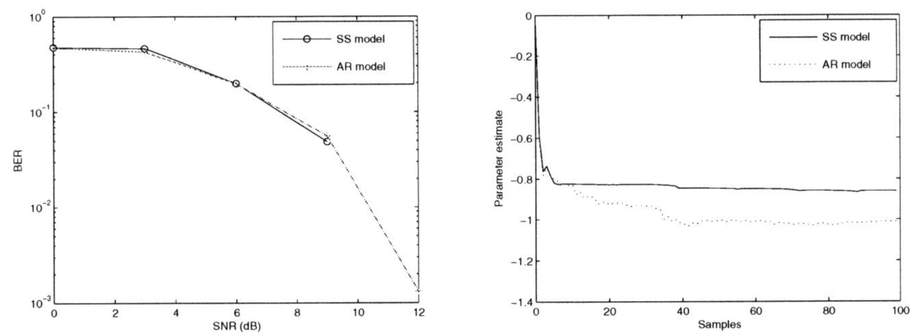

Figure 1. BER performance of the receiver (left) and consistency parameter estimation on the first carrier when SNR=15dB (right) for the two proposed 1st order models when $f_{d_n}^1 = 0.05$, $f_{d_n}^2 = 0.02$ and $f_{d_n}^3 = 0.03$

• **Comments for low Doppler frequencies** (i.e. $\left(f_{d_n}^m\right)_{m=1:3} < 0.1$). AR modeling and sinusoidal stochastic modeling for channels yield identical receiver performance in terms of BER. A 1st order model is sufficient to obtain good performance as shown in figure 1. Moreover, concerning the sinusoidal stochastic model, no filter has to be used. Indeed, when f_{d_n} is close to zero, the Gibb's phenomenon due to the truncated impulse response of a low-pass filter, affects a large band of the frequencies relatively to f_{d_n}. In that case, we suggest modeling the channel by 1st order AR process because of the lower number of state elements to be estimated than when modeling the channel by 1st order sinusoidal stochastic process.

• **Comments for high Doppler frequencies** (i.e. $\left(f_{d_n}^m\right)_{m=1:3} > 0.1$). The

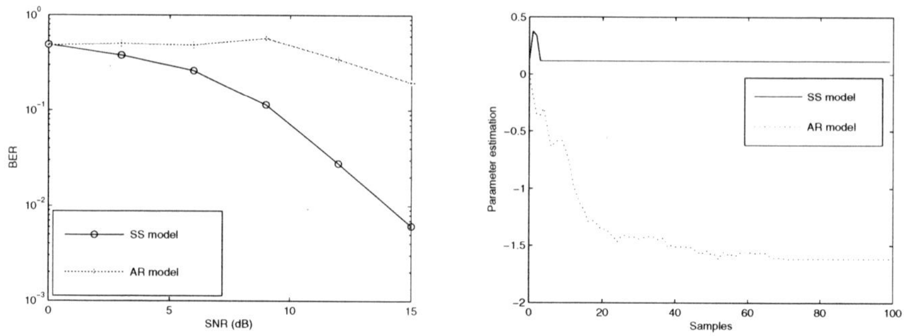

Figure 2. BER performance of the receiver (left) and consistency parameter estimation on the first carrier when SNR=15dB (right) for the two proposed 1st order models when $f^1_{d_n} = 0.15$, $f^2_{d_n} = 0.12$ and $f^3_{d_n} = 0.13$

order of the impulse response low-pass filter is set to be $L = 10$. This choice is a compromise between reducing filter roll-off and reducing the state vector length in the state-space representation of the system. Figure 2 illustrates the BER performance of the estimator when the channel is modeled by a 1st order AR process or a 1st order filtered sinusoidal stochastic process for $f^1_{d_n} = 0.15$, $f^2_{d_n} = 0.12$ and $f^3_{d_n} = 0.13$. The filtered sinusoidal stochastic modeling for channel provides better performance in terms of BER than AR modeling. In both cases, the parameter estimates converge as shown in figure 2. However, when the channel is modeled by an AR process, the system is instable as the pole corresponding to the parameter estimate is outside the unity circle in the z-plane. Increasing the model orders does not provide better results than when using 1st order model. This may be that when increasing the order, the number of parameters to be estimated is higher, the convergence of the parameter estimates require more time than when $p = 1$.

4. Conclusion

In this paper, we propose a Rao-Blackwellized particle filter cross-coupled with a Kalman filter to solve the challenging problem of joint symbols, channels and model parameter estimation for MC-DS-CDMA systems over Rayleigh nonselective fading channels. The algorithm takes advantage of the structure of the state-space representation to analytically solve a part of the estimation problem. It also provides estimates of the static parameters appearing in this non linear representation. Simulation results have shown the effectiveness of the method when modeling channels by AR processes or by filtered sinusoidal stochastic processes. Moreover, we have shown that while AR channel modeling is efficient for slow fading channels, the sinusoidal stochatic model provides better results than the AR model for fast fading channels. Moreover, the model offers the possibility to estimate the maximum Doppler frequency of the M channels. We are curently investigating a method based on particle filter to include this frequency estimation.

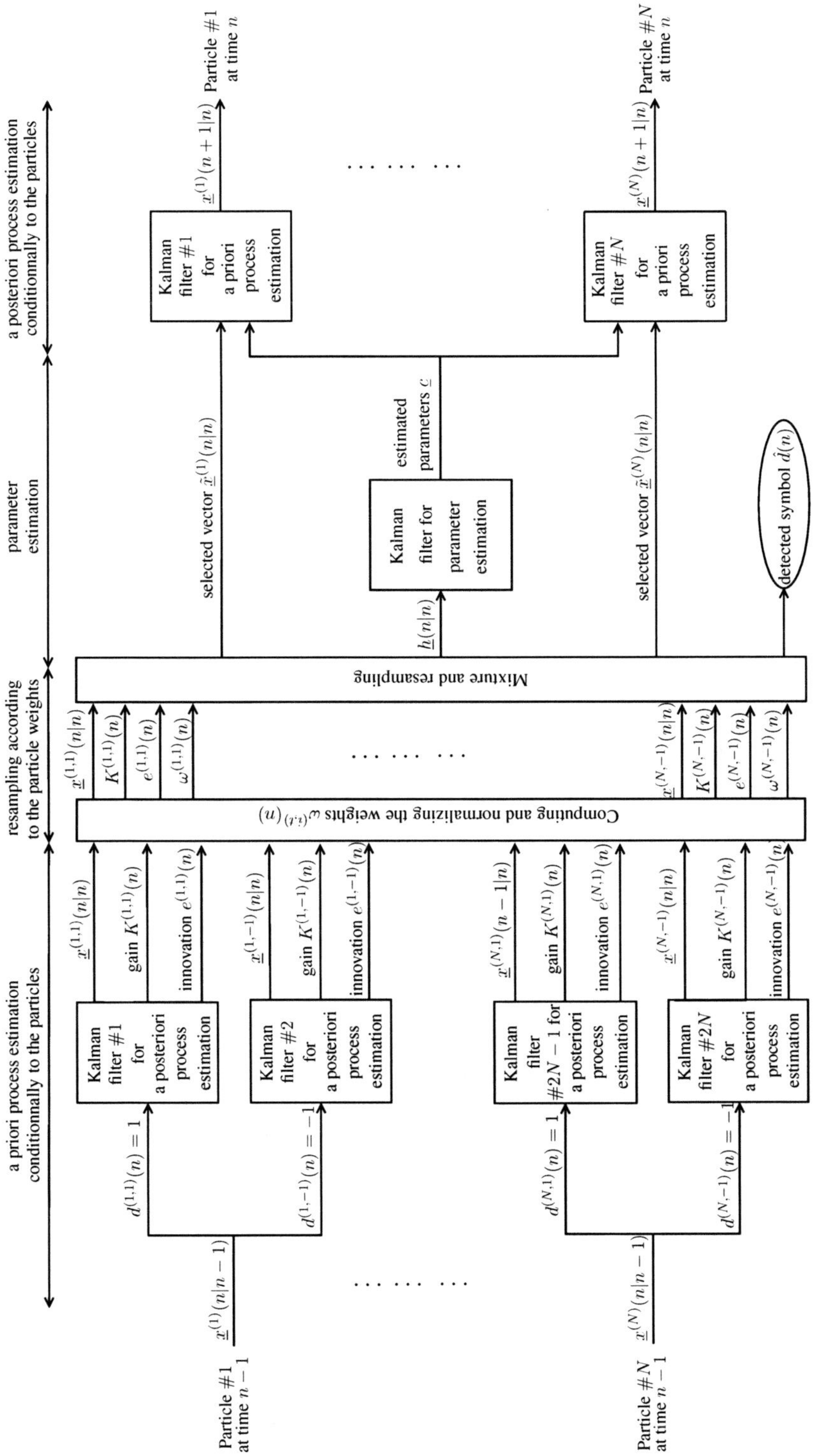

Figure 3. Joint channel/symbols/parameter estimator

Notes

1. $\lim_{p\mapsto\infty}\sigma^2 = \exp\left[\frac{1}{2\pi}\int_{-\pi}^{\pi}\ln S_h^{theo}(e^{j\omega})d\omega\right]$.
2. To make the computation tractable, this impulse response is truncated at the order $L > p$.
3. As the real or the imaginary parts of $(a^m(n))_{m=1,\ldots,M}$ or $(b^m(n))_{m=1,\ldots,M}$ have the same AR parameters, one can choose any real process to estimate $\underline{c}$.

References

[1] W. C. Jakes, "Microwave Mobile Communications", *Wiley-Interscience, New-year, 1974.*

[2] Z. Chen, "Bayesian Filtering: from Kalman Filters to Particle Filters, and Beyond", http://users.isr.ist.utl.pt/jpg/tfc0607/chen_bayesian.pdf, *Canada, 2003.*

[3] B. D. O. Anderson and J. B. Moore, "Optimal Filtering", *Prentice-Hall, New Jersey, 1979.*

[4] K. E. Baddour and N. C. Beaulieu, "Autoregressive Modeling for Fading Channel Simulation", *IEEE Trans. On Wireless Commun., vol. 4, pp. 1650-1662, July 2005.*

[5] P. Y. Kam, "Optimal Detection of Digital Data over the Nonselective Rayleigh Fading Channel with Diversity Reception", *IEEE Trans. on Commun., vol. 39, pp. 214-219, Feb. 1991.*

[6] Y. Liu and S. D. Blostein, "Identification of Frequency Nonselective Fading Channels using Decision Feedback and Adaptive Linear Prediction", *IEEE Trans. on Commun., vol. 43, pp. 1484-1492, Feb. 1995.*

[7] R. Raheli, A. Polydoros and C. Tzou, "Per-survivor Processing: a general approach to MLSE in uncertain environments", *IEEE Trans. on Commun., vol. 43, pp. 1354-364, Feb. 1995.*

[8] E. Punskaya, "Sequential Monte Carlo Methods for Digital Communications", *Ph.D. thesis, University of Cambridge, 2003.*

[9] E. Punskaya, A. Doucet and W. J. Fitzgerald, "Particle Filtering for Joint Symbol and Parameter Estimation in DS Spread Spectrum Systems", *Proceedings of ICASSP, Hong Kong, Apr. 2003.*

[10] X. Wang, R. Chen, and D. Guo, "Delayed-Pilot Sampling for Mixture Kalman Filter with Application in Fading Channels", *IEEE Trans. on Sig. Proc., vol.50, no. 2, Feb. 2002.*

[11] Y. Huang, J. Zhang, I. T. Luna, P. M. Djuric and D. P. R. Padillo, "Adaptive Blind Multiuser Detection over Flat Fast Fading Channels using Particle Filtering", *Proceeding of Globecom, 2004.*

[12] H. Wu and A. Duel-Hallen, "Multiuser Detectors with Disjoint Kalman Channel Estimators for Synchronous CDMA Mobile Radio Channels", *IEEE Trans. on Commun. vol. 48, no. 5, May 2000.*

[13] A. Papoulis, "Predictable Processes and Wold's Decomposition: a Review", *IEEE Trans. on Acous. Speech and Signal Proc., vol. ASSP-33, no. 4, Aug. 1985.*

KALMAN VS H_∞ ALGORITHMS FOR MC-DS-CDMA CHANNEL ESTIMATION WITH OR WITHOUT A PRIORI AR MODELING

Ali Jamoos, Julie Grolleau, and Eric Grivel
Equipe Signal et Image, UMR CNRS 5218 IMS - Dpt LAPS, Universite Bordeaux 1
351 cours de la liberation, 33405 Talence cedex, France
{ali.jamoos, julie.grolleau}@etu.u-bordeaux1.fr, egrivel@ims-bordeaux1.fr

Hanna Abdel-Nour
Department of Electronics Engineering, Al-Quds University
B.O.Box 20002, Jerusalem, Palestine
habdalnour@eng.alquds.edu

Abstract This paper deals with the estimation of time-varying Multi-Carrier Direct-Sequence Code Division Multiple Access (MC-DS-CDMA) fading channels using a training-aided scheme. Our approach consists in using an optimal filtering based on a linear state-space model of the fading channel system. In that case, two issues have to be investigated: 1) what kind of optimal filtering can be used? 2) how to estimate the state-space matrices?

Thus, Kalman filtering can be considered. It is optimal in the H_2 sense providing the underlying state-space model is Gaussian and accurate. However, as these assumptions may no longer be satisfied in real cases, we propose to study the relevance of H_∞ filtering. More particularly, when an explicit AR model is used for the channel, our first solution consists in estimating the fading channel and its AR parameters by means of two-cross-coupled H_∞ filters. Instead of AR model based-estimators, our second contribution is to view the channel estimation as a realization issue where the state-space matrices are estimated by using subspace methods for system identification without any *a priori* explicit model for the channel.

1. Introduction

Multi-Carrier Direct-Sequence Code Division Multiple Access (MC-DS-CDMA) is a multiplexing technique that combines the advantages of both multi-carrier modulation and DS-CDMA. In MC-DS-CDMA systems, due to user mobility, each carrier is subject to Doppler shifts resulting in time-varying fading. Thus, the estimation of the fading process over each carrier is essential to achieve optimal diversity combining and coherent symbol detection at the receiver.

S. Plass et al. (eds.), Multi-Carrier Spread Spectrum 2007, 427–436.

The time-varying fading process over each carrier is usually modeled as a zero-mean wide-sense stationary complex Gaussian process. Its real and imaginary parts are uncorrelated. In addition, according to the Jakes model, their theoretical Power Spectral Densities (PSD), denoted by Ψ_{hh}, is bandlimited, U-shaped and exhibit twin peaks at $\pm f_d$. Therefore, the corresponding discrete-time autocorrelation function $R_{hh}(n)$ is a zero-order Bessel function of the first kind:

$$R_{hh}(n) = J_0(2\pi f_d T_b |n|) \tag{1}$$

where f_d is the maximum Doppler frequency and T_b the symbol period.

In recent papers [1] [2], the fading process has been modeled as a pth order autoregressive process, denoted by AR(p) and defined as follows:

$$h(n) = -\sum_{i=1}^{p} a_i h(n-i) + u(n) \tag{2}$$

where $\{a_i\}_{i=1,\ldots,p}$ are the AR model parameters and $u(n)$ is the zero-mean complex white Gaussian driving process with variance σ_u^2.
Using a low-order AR model for the channel is debatable. On the one hand, some authors (e.g., [2] [3]) choose this model arguing for its simplicity, especially when dealing with 1st or 2nd order AR process. On the other hand, a deterministic model should be used for the channel due to the bandlimited nature of its PSD according to Kolmogoroff-Szego formula [4]. It should be noted that in between solutions have been also studied. Firstly, a sub-sampled AR Moving Average (ARMA) process followed by a multistage interpolator has been considered for channel simulation [5]. Nevertheless, only a very high down-sampling factor leads to a PSD in which no frequency band is equal to 0. Secondly, Baddour *et al.* [1] suggest using high-order AR processes (e.g. $p \geq 50$) for channel simulation. For this purpose, they "slightly" modify the properties of the channel by considering the sum of the theoretical fading process and a zero-mean white process whose variance ϵ is very small. Then, the AR parameters are estimated with the Yule-Walker (YW) equations based on the modified autocorrelation function $R_{hh}^{mod}(n) = J_0(2\pi f_d T_b |n|) + \epsilon\delta(n)$.

Taking into account the above results, we first propose to use an AR model whose order is high enough to approximate the channel. Kalman filtering can therefore be carried out for channel estimation [2]. However, as the maximum Doppler frequency f_d is usually unknown, the channel AR parameters must be estimated from the available noisy observations. Among the existing methods, Tsatsanis *et al.* [3] suggest estimating the AR parameters from the channel covariance estimates by means of a YW estimator. However, the method results in biased estimates. To avoid this drawback, the Expectation-Maximization (EM) algorithm can be used [6] and often implies a Kalman smoothing. Nevertheless, since it operates repeatedly on a batch of data, it results in large storage requirements and high computational cost. In addition, its success depends on the initial conditions. As an alternative, two recursive filters can be cross-coupled to solve the so-called dual estimation issue [7], i.e. the estimations of both the AR process and its parameters. Each time a new observation is available, the first filter uses the latest estimated AR parameters to estimate the signal, while the second filter uses the estimated signal to update the AR parameters. According to [8], this approach can be viewed as a sequential version of the EM algorithm. Recently, a variant [9] based on two interacting Kalman filters has been developed in which the variance of the innovation process in the first filter is used to define the gain of the second filter. As this solution can be seen as a recursive instrumental variable technique, consistent estimates of the AR parameters are obtained. Meanwhile, we have analyzed the relevance of this approach to estimate MC-DS-CDMA fading channels in [10].

Using Kalman filtering is of interest, but several assumptions must be fulfilled. Thus, Kalman filtering is optimal in the H_2 sense providing the underlying state-space model is accurate. Moreover, the initial state, the driving process and the measurement noise must be independent, white and Gaussian. However, these assumptions do not always hold in practice. As an alternative, H_∞ estimation techniques, initially developed in the framework of control [11], can be considered. The estimation criterion is to minimize the worst possible effects of the noise disturbances (i.e., the initial state, the driving process and the measurement noise) on the estimation error. In that sense and according to [11], H_∞ filtering is more robust against the noise disturbances and modeling approximations than Kalman filtering. Recently, Cai *et al.* [12] have proposed a channel estimation scheme for Orthogonal Frequency Division Multiplexing (OFDM) wireless systems based on two-serially-connected H_∞ filters. The first one is used for AR parameter estimation and the second one for fading process estimation. Nevertheless, the AR parameter estimates are biased since they are estimated directly from the noisy data.

In this paper, our first contribution consists in proposing a structure based on two-cross-coupled H_∞ filters for the joint estimations of the fading process and its AR parameters (see Fig.1). Then, instead of AR model based-estimators, our second contribution is to view the channel estimation as a realization issue by using subspace methods for system identification, initially developed in the field of control [13]. In that case, no *a priori* explicit model for the channel is necessary. One only assumes that the system is modeled by a linear Gaussian state-space representation. The transition matrix, the observation vector and the variances of both the driving process and the additive noise are directly estimated from the noisy observations. Once this so-called realization issue is solved, a Kalman or a H_∞ filter can be used to estimate the fading process.

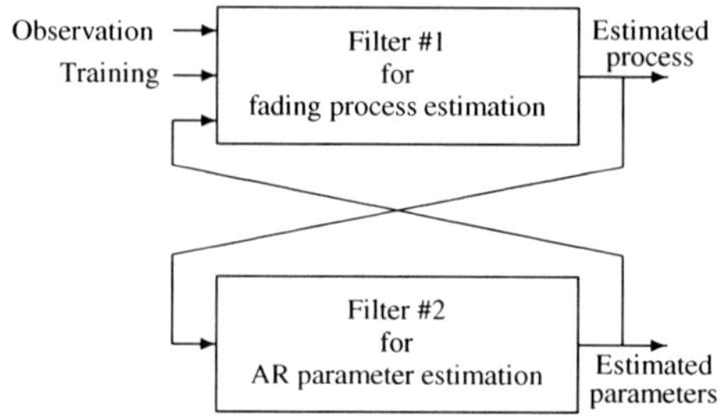

Figure 1. One of our solutions: two-cross-coupled Kalman or H_∞ filters.

2. MC-DS-CDMA System Model

A downlink MC-DS-CDMA system with BPSK modulation is considered based on M carriers and involving K active users. The transmitted signal is assumed to propagate through a time-varying frequency-selective Rayleigh fading channel. By suitably choosing the number of carriers and their spacing, each carrier can be assumed to undergo independent flat fading [14]. Therefore, the continuous-time received signal at the mth carrier is given by:

$$r_m(t) = \sum_{n=-\infty}^{+\infty} \sum_{k=1}^{K} d_k(n) c_k(t - nT_b) h_m(n) e^{j2\pi f_m t} + b_m(t) \tag{3}$$

where $d_k(n) \in \{\pm 1\}$ is the nth data bit of the kth user, $c_k(t)$ the spreading waveform, f_m the mth carrier frequency and $b_m(t)$ a zero-mean additive while complex Gaussian noise process. In addition, the fading processes $\{h_m(n)\}_{m=1,2,\cdots,M}$ are i.i.d. complex Gaussian random processes.

To retrieve the desired symbol sequence $d_1(n)$ of the first user, from the received signals $\{r_m(t)\}_{m=1,2,\ldots,M}$, the receiver structure proposed in [10] is used. It operates as follows. Firstly, the demodulated signal over the mth carrier is processed with a chip-matched filter followed by a chip rate sampler. The collected samples during one bit interval are then processed with a decorrelating detector. It yields the following observation:

$$y_m(n) = d_1(n) h_m(n) + w_m(n) \tag{4}$$

where $w_m(n)$ is a zero-mean additive white complex Gaussian noise with variance σ_w^2. Finally, Maximal Ratio Combining (MRC) provides:

$$\hat{d}_1(n) = \text{sgn}\left(\text{Re}\left(\sum_{m=1}^{M} h_m^*(n) y_m(n)\right)\right) \tag{5}$$

As the fading processes $\{h_m(n)\}_{m=1,2,\ldots,M}$ are unknown, we propose to estimate them by using a training based approach. Thus, in the time interval allocated to the transmission of the training sequence, the data modulation can be wiped out by multiplying (4) with the training symbols, as follows:

$$z(n) = d_1(n)y(n) = h(n) + v(n) \tag{6}$$

where the noise $v(n)$ is white with variance $\sigma_v^2 = \sigma_w^2$. It should be noted that, for the sake of simplicity, the carrier subscript is dropped hereafter.

As the real part $h_R(n)$ and the imaginary part $h_I(n)$ of the complex process $h(n)$ have the same statistics given in (1), only the real part $z_R(n)$ of the observations $z(n)$ will be considered to estimate the channel dynamics. For this purpose, the fading channel system is described by the following state-space model:

$$\begin{cases} \underline{x}(n) = \Phi \underline{x}(n-1) + \underline{u}(n) \\ z_R(n) = \underbrace{H\underline{x}(n)}_{h_R(n)} + v_R(n) \end{cases} \tag{7}$$

where $\underline{x}(n)$ is a real state vector of size $p \times 1$, Φ and H are real matrices of appropriate dimensions, $\underline{u}(n)$ is a $p \times 1$ zero-mean white noise vector with covariance matrix $Q_{\underline{u}}$ and $v_R(n)$ is the real part of $v(n)$ with variance $\sigma_{v_R}^2$. The relation between $\underline{x}(n)$ and $h_R(n)$ can be known or not, depending on whether an explicit model for the channel is adopted or not. In section 3, an AR model is used whereas no *a priori* model for the channel is considered in section 4.

3. Kalman vs H_∞ using AR model for the channel

Given (2) and (6), the state vector and the system matrices in the state-space representation of the system (7) can be explicitly expressed as follows:

$$\underline{x}(n) = \begin{bmatrix} h_R(n) & h_R(n-1) & \cdots & h_R(n-p+1) \end{bmatrix}^T \tag{8}$$

$$\Phi = \begin{bmatrix} -a_1 & -a_2 & \cdots & -a_p \\ 1 & 0 & \cdots & 0 \\ & \ddots & & \vdots \\ 0 & \cdots & 1 & 0 \end{bmatrix}, H = [1 \ \underbrace{0 \ \cdots \ 0}_{p-1}], \underline{u}(n) \triangleq \Gamma u_R(n) \tag{9}$$

where $\Gamma = H^T$ and $u_R(n)$ is the real part of $u(n)$ with variance $\sigma_{u_R}^2$.
On the one hand, Kalman filtering provides an estimate $\hat{\underline{x}}(n)$ of the state vector $\underline{x}(n)$ by minimizing $E[(\underline{x}(n) - \hat{\underline{x}}(n))^T(\underline{x}(n) - \hat{\underline{x}}(n))]$. On the other hand, H_∞ filtering focuses on the estimation of a specific linear combination of the state vector components:

$$c(n) = L\underline{x}(n) \tag{10}$$

where L is a $1 \times p$ linear transformation operator. Here, as $h_R(n)$ has to be estimated, $L = \begin{bmatrix} 1 & 0 & \cdots & 0 \end{bmatrix}$.

H_∞ filtering provides the estimate $\hat{h}_R(n) = L\hat{\underline{x}}(n)$, by minimizing the H_∞ norm of the transfer operator $\mathcal{T}$ that maps the noise disturbances $u_R(n)$, $v_R(n)$ and the initial state error $\underline{e}_0 = \underline{x}(0) - \hat{\underline{x}}(0)$ to the estimation error $e(n) = h_R(n) - \hat{h}_R(n) = L(\underline{x}(n) - \hat{\underline{x}}(n))$, as follows:

$$J_\infty = \sup_{u_R(n), v_R(n), \underline{x}(0)} \frac{\sum_{n=0}^{N-1} |e(n)|^2}{\underline{e}_0^T P_0^{-1} \underline{e}_0 + \sum_{n=0}^{N-1} \left(Q_u^{-1} |u_R(n)|^2 + R_v^{-1} |v_R(n)|^2 \right)} \quad (11)$$

where N is the number of available data samples. In addition, $P_0 > 0$, $Q_u > 0$ and $R_v > 0$ are weighting parameters which are tuned by the designer to achieve performance requirements.

However, as a closed-form solution to the above optimal H_∞ estimation problem does not always exist, a suboptimal strategy is usually considered:

$$J_\infty < \gamma^2 \quad (12)$$

where $\gamma > 0$ is a prescribed level of disturbance attenuation. At that stage, there exists an H_∞ channel estimator $\hat{h}_R(n)$ for a given γ if there is a stabilizing symmetric positive definite solution $P(n) > 0$ to the following Riccati-type equation:

$$P(n+1) = \Phi P(n) C^{-1}(n) \Phi^T + \Gamma Q_u \Gamma^T, \qquad P(0) = P_0 \quad (13)$$

where: $$C(n) = I_p - \gamma^{-2} L^T L P(n) + H^T R_v^{-1} H P(n)$$

The H_∞ channel estimator is given by:

$$\hat{h}_R(n) = L\hat{\underline{x}}(n) \quad (14)$$

$$\hat{\underline{x}}(n) = \Phi \hat{\underline{x}}(n-1) + K(n)\alpha(n), \quad \hat{\underline{x}}(0) = 0 \quad (15)$$

where the so-called innovation process $\alpha(n)$ and the H_∞ filter gain $K(n)$ are respectively given by:

$$\alpha(n) = z_R(n) - H\Phi\hat{\underline{x}}(n-1) \quad \text{and} \quad K(n) = P(n) C^{-1}(n) H^T R_v^{-1} \quad (16)$$

It should be noted that the matrix $P(n)$ can be seen as an upper bound of the error covariance matrix in the Kalman filter theory, i.e. $E[(\underline{x}(n) - \hat{\underline{x}}(n))(\underline{x}(n) - \hat{\underline{x}}(n))^T] \leq P(n)$. Due to (13), the H_∞ channel estimator has a computational cost slightly higher than Kalman's one. If the weighting parameters Q_u, R_v and P_0 are respectively chosen to be $\sigma_{u_R}^2$, $\sigma_{v_R}^2$ and the initial error covariance matrix of $\underline{x}(0)$, then the H_∞ estimator reduces to a Kalman one as $\gamma \longrightarrow +\infty$.

To estimate the AR parameters $\{a_i\}_{i=1,\ldots,p}$ from the real part of the estimated fading process $\hat{h}_R(n)$ (see Fig.1), equations (14) and (15) are firstly combined to express $\hat{h}_R(n)$ as a function of the AR parameters:

$$\hat{h}_R(n) = L\Phi\hat{\underline{x}}(n-1) + LK(n)\alpha(n) = -\hat{\underline{x}}^T(n-1)\underline{\theta}(n) + \nu(n) \quad (17)$$

where $\underline{\theta}(n) = [\ a_1 \quad a_2 \quad \cdots \quad a_p\ ]^T$ and $\nu(n) = LK(n)\alpha(n)$.
When the channel is assumed stationary, the AR parameters are time-invariant:

$$\underline{\theta}(n) = \underline{\theta}(n-1) \tag{18}$$

Equations (17) and (18) hence define a state-space representation for the estimation of the AR parameters. By denoting the estimation error due to the estimation of the AR parameters as $e_{\underline{\theta}} = \hat{\underline{x}}^T(n-1)(\underline{\theta}(n) - \hat{\underline{\theta}}(n))$, a second H_∞ filter can be used to recursively estimate $\underline{\theta}(n)$ such as to achieve certain level of disturbance attenuation $\gamma_{\underline{\theta}} > 0$. In this filter, two weighting parameters, denoted by R_ν and $P_{\underline{\theta}_0}$, must be also tuned.
By analogy with Kalman filter theory, Q_u can be tuned as follows [10]:

$$Q_u(n) = \lambda Q_u(n-1) + (1-\lambda)DM(n)D^T \tag{19}$$

where $D = (\Gamma^T\Gamma)\Gamma^T$, $M(n) = P(n) - \Phi P(n-1)\Phi^T + K(n)|\alpha(n)|^2 K^T(n)$, and λ is the forgetting factor.
In addition, R_v is assigned to $\sigma^2_{v_R}$ while R_ν is tuned as follows:

$$R_\nu = LK(n)(H^T P(n) H + R_v)K^T(n)L^T \tag{20}$$

Moreover, P_0 and $P_{\underline{\theta}_0}$ are assigned to the identity matrix (i.e., $P_0 = P_{\underline{\theta}_0} = I_p$).

4. Estimation of the state-space matrices without a priori AR modeling

In the previous section, a canonical state-space representation of the system was considered. In this section, we review the subspace methods [13] that make it possible to identify the state-space representation of a stochastic process directly from the noisy observations without any *a priori* explicit model for the channel. The core of the subspace methods for identification is to estimate, from noisy observations, the extended ($s > p$) observability matrix of a state-space representation (7), defined as follows:

$$\Gamma_s = \left[H^T \quad (H\Phi)^T \quad \cdots \quad \left(H\Phi^{s-1}\right)^T \right]^T \tag{21}$$

For this purpose, the $2s \times (N - 2s + 1)$ Hankel matrix $Z_{0/2s-1}$ constructed from the N observations $z_R(k)$ is introduced:

$$Z_{0/2s-1} = \begin{bmatrix} Z_{0/s-1} \\ Z_{s/s} \\ Z_{s+1/2s-1} \end{bmatrix} = \begin{bmatrix} z_R(0) & z_R(1) & \cdots & z_R(N-2s) \\ z_R(1) & z_R(2) & \cdots & z_R(N-2s+1) \\ \vdots & \vdots & \ddots & \vdots \\ z_R(s) & z_R(s+1) & \cdots & z_R(N-s) \\ z_R(s+1) & z_R(s+2) & \cdots & z_R(N-s+1) \\ \vdots & \vdots & \ddots & \vdots \\ z_R(2s-1) & z_R(2s) & \cdots & z_R(N-1) \end{bmatrix}$$

Van Overschee *et al.* [13] proved that the extended observability matrix Γ_s can be estimated from the two following orthogonal projections:

$$O_s = Z_{s/2s-1}/Z_{0/s-1} \quad \text{and} \quad O_{s-1} = Z_{s+1/2s-1}/Z_{0/s} \tag{22}$$

When $N' \triangleq N - 2s + 1 \to \infty$, they satisfy:

$$O_s = \Gamma_s X_s \quad \text{and} \quad O_{s-1} = \Gamma_{s-1} X_{s+1} \tag{23}$$

where X_s and X_{s+1} respectively denote the state sequence generated by a bank of non-steady Kalman filters working in parallel on each of the columns of the block Hankel matrix $Z_{0/s-1}$ and $Z_{0/s}$. Moreover, one has:

$$\text{Span}(X_s^T) = \text{Span}(O_s^T) \quad \text{and} \quad \text{Span}(\Gamma_s) = \text{Span}(O_s) \tag{24}$$

To obtain Γ_s and X_s, the authors in [13] propose to use the Singular Value Decomposition (SVD) of the weighted orthogonal projection $W_1 O_s W_2$:

$$W_1 O_s W_2 = \begin{bmatrix} U_1 & U_2 \end{bmatrix} \begin{bmatrix} S_1 & 0 \\ 0 & 0 \end{bmatrix} \begin{bmatrix} V_1^T \\ V_2^T \end{bmatrix} \tag{25}$$

where S_1 is a diagonal matrix that contains the p non-zero singular values of $W_1 O_s W_2$. $W_1 \in \mathbf{R}^{s \times s}$ and $W_2 \in \mathbf{R}^{N' \times N'}$ are two weighting matrices such that W_1 is full rank and $\text{rank}(Z_{0/s-1}) = \text{rank}(Z_{0/s-1} W_2)$.

The choice of W_1 and W_2 determines the state-space basis in which the representation will be identified. Different choices have been investigated, leading to various published methods. Then, Γ_s can be obtained as follows:

$$\Gamma_s = W_1^{-1} U_1 S_1^{1/2} \tag{26}$$

At that stage, H and Φ can be easily retrieved. In addition, Van Overschee *et al.* proposed a way to estimate the covariance matrix $Q_{\underline{u}}$ and the variance $\sigma_{v_R}^2$ from the least square residuals, $\rho_{w(k)}$ and $\rho_{v(k)}$, defined as follows:

$$\begin{pmatrix} \rho_{w(k)} \\ \rho_{v(k)} \end{pmatrix} = \begin{bmatrix} X_{s+1} \\ Z_{s/s} \end{bmatrix} - \begin{bmatrix} \Phi \\ H \end{bmatrix} \hat{X}_s \text{ and } \frac{1}{N'} \begin{pmatrix} \rho_{w(k)} \\ \rho_{v(k)} \end{pmatrix} \begin{pmatrix} \rho_{w(k)} \\ \rho_{v(k)} \end{pmatrix}^T = \begin{bmatrix} Q & S \\ S & \sigma_v^2 \end{bmatrix} \tag{27}$$

Let us now focus on the N4SID method [13] where $W_1 = I_{s \times s}$ and $W_2 = I_{N' \times N'}$ for channel estimation. In the framework of channel estimation, the number of available observations corresponds to the number of pilot bits. Therefore, N is finite and has to be as low as possible. Van Overschee *et al.* studied the behavior of the subspace identification methods when N is finite. They suggest using the RQ decomposition $\frac{1}{\sqrt{N'}} Z_{0/2s-1}$ to compute the orthogonal projections O_s and O_{s-1}.

$$\frac{1}{\sqrt{N'}} \begin{bmatrix} Z_{0/s-1} \\ Z_{s/s} \\ Z_{s+1/2s-1} \end{bmatrix} = \begin{bmatrix} R_{11} & 0 & 0 \\ R_{21} & R_{22} & 0 \\ R_{31} & R_{32} & R_{33} \end{bmatrix} \begin{bmatrix} Q_1^T \\ Q_2^T \\ Q_3^T \end{bmatrix} \tag{28}$$

Thus, O_s and O_{s-1} can be written as follows:

$$\hat{O}_s = \begin{bmatrix} R_{21} \\ R_{31} \end{bmatrix} Q_1^T \text{ and } \hat{O}_{s-1} = \begin{bmatrix} R_{31} & R_{32} \end{bmatrix} \begin{bmatrix} Q_1^T \\ Q_2^T \end{bmatrix} \tag{29}$$

In addition, as Q_1 is an orthogonal matrix, the SVD of O_s can be calculated by means of the SVD of $\begin{bmatrix} R_{21} & R_{31} \end{bmatrix}^T$. One has:

$$\begin{bmatrix} R_{21} \\ R_{31} \end{bmatrix} = \begin{bmatrix} \hat{U}_1 & \hat{U}_2 \end{bmatrix} \begin{bmatrix} \hat{S}_1 & 0 \\ 0 & \hat{S}_2 \end{bmatrix} \begin{bmatrix} \hat{V}_1^T \\ \hat{V}_2^T \end{bmatrix} \tag{30}$$

where $\hat{S}_1$ corresponds to the p dominant singular values. The estimate of Γ_s corresponds to:

$$\hat{\Gamma}_s = \hat{U}_1 \hat{S}_1^{1/2} \tag{31}$$

As the N4SID method is consistent [13], Γ_{s-1} and X_{s+1} can be estimated from $\hat{\Gamma}_s$. Therefore, using (27) makes it possible to get the estimation of H, Φ, $Q_{\underline{u}}$ and $\sigma_{v_R}^2$.

5. Simulation Results and Conclusions

A comparative study on the estimation of MC-DS-CDMA fading channels is carried out between the proposed approaches (i.e., two-cross-coupled H_∞ filters, subspace identification + Kalman, subspace identification + H_∞), the two-cross-coupled Kalman filters [10] and the standard RLS and LMS channel estimators [15]. A downlink MC-DS-CDMA system with $M = 2$ carriers and $K = 10$ active users is considered. The variance of the additive noise is assumed to be known for every method except the subspace identification based approaches, which estimate it. According to Fig. 2, the proposed approaches yield much lower BER than the standard LMS and RLS estimators.

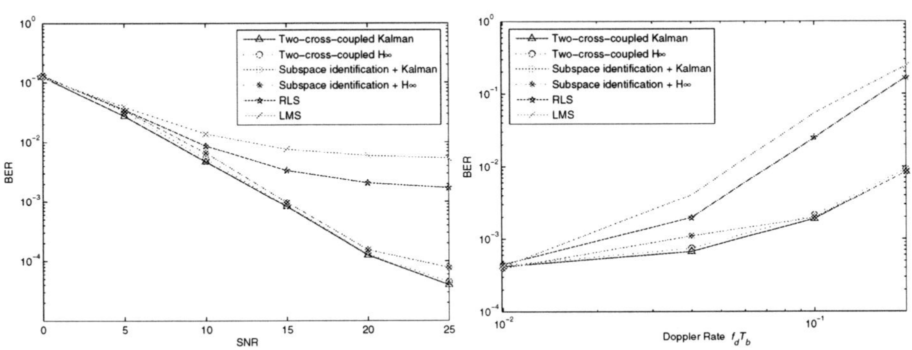

(a) BER versus SNR. $f_dT_b = 0.05$.

(b) BER versus Doppler rate. SNR=15 dB.

Figure 2. BER versus either SNR or Doppler rate f_dT_b for the various channel estimators.

In addition, the H_∞ filter based methods provide approximately the same BER as the Kalman based ones while requiring slightly higher computational cost. Nevertheless, the H_∞ filtering has the advantage of relaxing the restrictive Gaussian assumptions imposed by Kalman filtering.
On the one hand, the cross-coupled estimators provide slightly better results than the subspace identification based ones, whenever using Kalman or H_∞ filters. This may be due to the fact that the cross-coupled estimators use the exact value of the additive noise variance. On the other hand, the subspace identification based methods have the advantages of estimating the additive noise variance and requiring no *a priori* explicit model for the channel.

References

[1] K. E. Baddour and N. C. Beaulieu, "Autoregressive modeling for fading channel simulation," *IEEE Trans. On Wireless Communications*, vol. 4, pp. 1650-1662, July 2005.

[2] C. Komninakis, C. Fragouli, A. H. Sayed, and R. D. Wesel, "Multi-input multi-output fading channel tracking and equalization using Kalman estimation," *IEEE Trans. Signal Process.*, pp. 1065-1076, May 2002.

[3] M. Tsatsanis, G. B. Giannakis, and G. Zhou, "Estimation and equalization of fading channels with random coefficients," *Signal Processing*, vol. 53, pp. 211-229, Sept. 1996.

[4] A. Papoulis and S. U. Pillai, *Probability, Random Variables and Stochastic Processes*, McGraw-Hill, New York, 2002.

[5] D. Schafhuber, G. Matz, and F. Hlawatsch, "Simulation of wideband mobile radio channels using subsampled ARMA models and multistage interpolation," in *Proceedings of IEEE-SSP*, Singapore, Aug. 6-8, 2001.

[6] M. Deriche, "AR parameter estimation from noisy data using the EM algorithm," *Proc. of the IEEE-ICASSP*, Adelaide, April 1994, pp. 69-72.

[7] B. D. O. Anderson and J. B. Moore, *Optimal Filtering*, Prentice Hall, NJ, 1979.

[8] S. Gannot and M. Moonen, "On the application of the unscented Kalman filter to speech processing," *Proceedings of IWAENC*, Kyoto, Japan, September 2003.

[9] D. Labarre, E. Grivel, Y. Berthoumieu, E. Todini and M. Najim, "Consistent estimation of autoregressive parameters from noisy observations based on two interacting Kalman filters," *Signal Processing*, vol. 86, pp. 2863-2876, October 2006.

[10] A. Jamoos, D. Labarre, E. Grivel, and M. Najim, "Two cross coupled Kalman filters for joint estimation of MC-DS-CDMA fading channels and their corresponding autoregressive parameters," in *Proceedings of EUSIPCO*, Antalya, Turkey, Sept. 4-8, 2005.

[11] B. Hassibi, A. H. Sayed and T. Kailath, *Indefinite Quadratic Estimation and Control: A Unified Approach to an H_2 and H_∞ Theories*, SIAM, 1999.

[12] J. Cai, X. Shen, and J. W. Mark, "Robust channel estimation for OFDM wireless communication systems - an H_∞ approach," *IEEE Trans. On Wireless Communications*, vol. 3, pp. 2060-2071, November 2004.

[13] P. Van Overschee and B. De Moor, *Subspace identification for linear systems, theory, implementation, applications*, Kluwer Academic Publisher, 1996.

[14] S. Kondo and L. Milstein, "Performance of multicarrier DS-CDMA systems," *IEEE Trans. on Communications*, vol. 44, pp. 238-246, February 1996.

[15] D. N. Kalofonos, M. Stojanovic, and J. G. Proakis, "Performance of adaptive MC- CDMA detectors in rapidly fading Rayleigh channels," *IEEE Trans. On Wireless Communications*, vol. 2, pp. 229-239, March 2003.

Lecture Notes in Electrical Engineering

Digital Noise Monitoring of Defect Origin
Telman Aliev
ISBN 978-0-387-71753-1, 2007

Multi-Carrier Spread Spectrum 2007
Simon Plass, Armin Dammann, Stefan Kaiser, and K. Fazel
ISBN 978-1-4020-6128-8, 2007

Printed in the United States
79740LV00001B/94-108

9 781402 061288